VOYAGES

THROUGH THE UNIVERSE

This recent image from the Hubble Space Telescope shows "cometary knots" in the Helix Nebula, a shell of glowing gas expelled by a dying star about 450 light years away in the constellation of Aquarius. A wind of hot gas from the star collides with the shell of denser gas the star emitted about 10,000 years earlier. The collision fragments the inner part of the shell into denser, finger-like droplets that superficially resemble comets we see in our solar system. However, the heads of the "comets" in the Helix Nebula are typically twice the size of our entire planetary system (twice the diameter of the orbit of Pluto). Each tail stretches away from the star for about 100 billion miles. (R. O'Dell, K. Handron & NASA) *(Inset)* A view of the full Helix Nebula taken with the Anglo-Australian telescope. (© 1979 Anglo-Australian Telescope Board)

VOYAGES

THROUGH THE UNIVERSE

Volume II: Stars, Galaxies, and Cosmology

Andrew Fraknoi

Chair, Astronomy Department
Foothill College

David Morrison

Chief, Space Science Division
NASA Ames Research Center

Sidney Wolff

Director
National Optical Astronomy Observatories

SAUNDERS COLLEGE PUBLISHING
Harcourt Brace College Publishers

FORT WORTH PHILADELPHIA SAN DIEGO NEW YORK ORLANDO AUSTIN

SAN ANTONIO TORONTO MONTREAL LONDON SYDNEY TOKYO

Text Typeface: New Caledonia
Composition: Progressive Information Technologies
Publisher: John Vondeling
Acquisitions Editor: Jennifer Bortel
Developmental Editor: Jennifer Bortel
Picture Development Editor: George Semple
Managing Editor: Carol Field
Senior Project Editor: Anne Gibby
Copy Editor: Judy Patton
Proofreader: Michele Gitlin
Manager of Art and Design: Carol Bleistine
Art Director: Carol Bleistine
Illustration Supervisor: Susan Kinney
Art and Design Coordinator: Kathleen Flanagan
Text and Cover Designer: Ruth A. Hoover
Text Artwork: George V. Kelvin/Science Graphics
            Rolin Graphics
Director of EDP: Tim Frelick
Production Manager: Alicia Jackson
Marketing Manager: Marjorie Waldron
Editorial Assistant: Tara Pauliny
Marketing Coordinator: Karen Milstein

Cover Credit: Hubble Space Telescope image. Gas pillars in M16, *Eagle Nebula,* a nearby star-forming region 7,000 light years away, in the constellation Serpens. (Jeff Hester and Paul Scowen, Arizona State University, and NASA)
Frontispiece: *Helix Nebula.* (© 1979 Anglo-Australian Telescope Board/David Malin)

Printed in The United States of America

Voyages Through the Universe

ISBN: 0-15-504534-2

Library of Congress Catalog Card Number: 96-68673

6 7 8 9 0 1 2 3 4 5   032   10 9 8 7 6 5 4 3 2 1

About the Authors

Andrew Fraknoi is the Chair of the Astronomy Department at Foothill College near San Francisco and an Educational Consultant for the Astronomical Society of the Pacific (where he directs Project ASTRO, a program to bring astronomers into elementary and junior high school classrooms). From 1978 to 1992 he was Executive Director of the Society, as well as Editor of *Mercury* Magazine and the *Universe in the Classroom* Newsletter. He has taught astronomy and physics at San Francisco State University, Cañada College, and the University of California Extension Division. He is author of *The Universe in the Classroom*, co-author of *Effective Astronomy Teaching and Student Reasoning Ability*, and scientific editor of *The Planets and The Universe*, two collections of science and science-fiction literature. In the past 22 years he has presented over 400 public lectures on astronomical topics. For five years he was the lead author of a nationally syndicated newspaper column on astronomy, and he appears regularly on radio and television explaining astronomical developments. He has received the Annenberg Foundation Prize of the American Astronomical Society and the Klumpke-Roberts Prize of the Astronomical Society of the Pacific for his contributions to the public understanding of astronomy. Asteroid 4859 was named Asteroid Fraknoi in 1992 in recognition of his work in astronomy education.

David Morrison received his Ph.D. from Harvard University. He was at the University of Hawaii from 1969 to 1988, where his positions included Professor of Astronomy, Chair of the Astronomy Graduate Program, Director of the Infrared Telescope Facility at Mauna Kea Observatory, and University Vice-Chancellor for Research and Graduate Education. Dr. Morrison currently heads the space science program at the NASA Ames Research Center. His primary research interests are in planetary science. Dr. Morrison is the author of more than 120 professional articles and of several books, including *The Planetary System, Cosmic Catastrophes, Exploring Planetary Worlds,* and three other astronomy texts from Saunders. He has served as President of the Astronomical Society of the Pacific, Chair of the Astronomy Section of the American Association for the Advancement of Science, and President of the Planetary Commission of the International Astronomical Union. Dr. Morrison has received the Klumpke-Roberts Prize of the Astronomical Society of the Pacific for contributions to public understanding of science, and two medals for Outstanding Leadership from NASA for his contributions to the Galileo mission and to protecting the Earth from asteroid impacts. A celestial object, Asteroid 2410 Morrison, is named for him.

Sidney C. Wolff received her Ph.D. from the University of California at Berkeley, and then joined the Institute for Astronomy at the University of Hawaii. During the seventeen years Dr. Wolff spent in Hawaii, the Institute for Astronomy developed Mauna Kea into the world's premier international observatory. Dr. Wolff became Associate Director of the Institute for Astronomy in 1976 and Acting Director in 1983. She earned international recognition for her research, particularly on stellar atmospheres—the evolution, formation, and composition of stars. In 1984, she was named Director of the Kitt Peak National Observatory, and in 1987 became Director of the National Optical Astronomy Observatories. She is the first woman to head a major observatory in the United States. As Director of NOAO, Dr. Wolff and her staff of 460 oversee facilities used annually by nearly 1000 visiting scientists. Recently, Dr. Wolff has also been acting as Director of the Gemini Project, which is an international program to build two state-of-the-art 8-m telescopes. Dr. Wolff has served as President of the Astronomical Society of the Pacific and is the second woman to be elected President of the American Astronomical Society. She is also a member of the Board of Trustees of Carleton College, a liberal arts school that excels in science education. Dr. Wolff is the author of more than 70 professional articles and a book, *The A-Type Stars: Problems and Perspectives.*

Preface for the Student

In college textbooks, there is a long tradition that the preface of the book is read by the instructor and the rest of the book by the student. Still, many students begin reading the preface (it does come first) and then wonder why it doesn't say much to them.

So, we begin our book with a preface for student readers. It's not a preface about the subject matter of astronomy, which is introduced in the Prologue, but a preface that tells you a little about the book and gives you some hints for the effective study of astronomy. (Your professor will probably have other, more specific suggestions for doing well in your class.)

Astronomy, the study of the universe beyond the confines of our planet, is one of the most exciting and rapidly changing branches of science. Even scientists from other fields often confess to having had a lifelong interest in astronomy, though they may now be doing something more practical, such as biology, chemistry, or engineering. There are fewer than 10,000 *professional* astronomers in the world; but astronomy has a large group of *amateur astronomers* who spend many an evening with a telescope under the stars observing the sky, and who occasionally make a discovery, such as a new comet or exploding star.

Many people are fascinated just to read about bizarre objects that astronomers are uncovering, such as black holes and quasars. Others are intrigued by the scientific search for planets or life in other star systems. And many people like to follow the challenges of space exploration, such as the repair of the Hubble Space Telescope by the Shuttle astronauts, or the Galileo mission to probe the giant planet Jupiter. Hearing about an astronomical event in the news media may be what first sparked your interest in taking an astronomy course.

But some of the things that make astronomy so interesting also make it a challenge for the beginning student. The universe is a big place full of objects and processes that do not necessarily have familiar counterparts here on Earth. Like a visitor to a new country, it will take you a while to feel familiar with the territory or the local customs. Astronomy, like other sciences, has its own special vocabulary, and keeping up with the pace of discovery in astronomy is a monumental challenge.

To assist students taking their first college-level course in astronomy, we have built a number of special features into this book, and we invite you to make use of them:

- All technical terms are printed in **boldface** type the first time they are used and clearly defined in the text; their definitions are listed alphabetically in Appendix 3 (the glossary), so you can refer to them at any time. The summaries at the end of each chapter also include these boldface terms as a review.

- The book begins with a historical summary of astronomy and then surveys the universe, starting at home and finishing with the properties of the entire cosmos. But don't worry if your instructor doesn't assign all the chapters or doesn't assign the chapters in order. Throughout the book "directional signs" lead you to earlier material you need to know before tackling the current section.

- We use tables to bring together numerical data for your convenience. For example, some tables summarize the important properties of each planet in the solar system (including the Earth). Students who want to see more of the data that astronomers use can investigate the appendices at the back of the book, which give the latest information on many aspects of astronomy.

- Figure captions clearly describe what phenomena or objects students are looking at. In many textbooks, captions are afterthoughts, with only a few words of description, but in this book, we have scrutinized each figure and asked what would help clarify the diagram or image.

- Each chapter ends with a summary of the essential points in the chapter, plus review questions, thought questions, and problems to help you "process" what you have learned.

- Suggestions for further reading are included for students who want or need to learn more about a particular topic. These books and articles are written at the same introductory level as this text.

- Appendix 1 is a guide to some of the more interesting astronomy sites on the World Wide Web.

Here are a few suggestions for studying astronomy that come from good teachers and good students from around the country:

- First, the best advice we can give you is to be sure to leave enough time in your schedule to study the material in this class *regularly*. It sounds obvious, but it is not very easy to catch up with a subject like astronomy by trying to do everything just before an exam. Try to put aside some part of each day, or every other day, when you can have uninterrupted time for reading and studying astronomy.

- Try to read each assignment in the book twice, once before it is discussed in class, and once afterwards. Take notes or use a highlighter to outline ideas that you may want to review later. Also, take some time to coordinate the notes from your reading with the notes you take in class. Many students start college without good note-taking habits. If you are not a good note-taker, get some help. Many colleges and universities have student learning centers that offer short courses, workbooks, or videos on developing good study habits. Take a little time and find out what your school has to offer.

- Form a small astronomy study group with people in your class; get together as often as you can and discuss the topics that may be giving group members trouble. Make up sample exam questions and make sure everyone in the group can answer them confidently. If you have always studied alone, you may at first resist this idea, but don't be too hasty to say no. Study groups are very effective ways of discussing new information, or learning a foreign language, or studying law or astronomy.

- Before each exam, do a concise outline of the main ideas discussed in class and presented in your text. Compare your outline with those of other students as a check on your own study habits.

- If you find a topic in the text or in class especially difficult or interesting, don't hesitate to make use of the resources in your library for additional study.

- *Don't be too hard on yourself!* If astronomy is new to you, many of the ideas and terms in this book will be unfamiliar. And astronomy is like any new language; it may take a while to become a good conversationalist. Practice as much as you can, but also realize that it is natural to be overwhelmed by the vastness of the universe and the variety of things that are going on in it.

We hope you enjoy reading this text as much as we enjoyed writing it. We are always glad to hear from students who have used the text and invite you to send us your reactions to the book and suggestions for how we can improve future editions. We promise you we will read and consider every serious letter we receive. You can send your comments to Andrew Fraknoi, Astronomy Department, Foothill College, 12345 El Monte Rd., Los Altos Hills, CA 94022, USA. (Please note that we will not send you the answers to the chapter problems or do your homework for you, but all other thoughts are welcome.)

Andrew Fraknoi, David Morrison, and Sidney Wolff
July 1996

Preface for the Instructor

Voyages **Through the Universe** is a new astronomy text produced with today's students in mind—it is designed for non-science majors who may even be a little intimidated by science, and who approach astronomy with more interest than experience. With features that make it appropriate for everyone from university business majors to first-time community college students, **Voyages** is written to draw in and engage all readers, while preserving the accuracy and timeliness that our colleagues expect of us.

This new text is based in part on **Realm of the Universe,** by the late George Abell, David Morrison, and Sidney Wolff, but has been completely rethought and rewritten to make it an even more useful tool for teaching and learning astronomy. We have made the language friendly and inviting and have used examples drawn from everyday experience. Vignettes from the lives of astronomers and occasional touches of humor make this a book that students will actually *enjoy* reading.

Organization and Special Features

The book is not too long for a one-semester course, yet not so brief that important topics have been omitted. It does not overwhelm the student with detail, but instead focuses on the major threads and overarching ideas that illuminate the relationships among the various branches of astronomy. We probably have a bit less jargon than most textbooks, but we have not sacrificed any of the key concepts that you would want to see students learn in a basic course.

We have worked hard to include many of the latest ideas and discoveries in astronomy, not merely for their novelty, but for their value in advancing the quest for a coherent understanding of the universe. In each case, we have tried to fit the latest research results into a wider context and to explain clearly what they mean. Among the recent topics included in the text are the discovery of planets around other stars, the results from both the Galileo probe and Comet Shoemaker-Levy 9 impacts on Jupiter, the discovery of a number of the icy members of the Kuiper belt, new candidates for black holes in our galaxy, recent measurements of the age of the universe from several different observing groups, clearer evidence for supermassive black holes at the centers of galaxies, and much more.

We portray astronomy as a human endeavor and have tried to include descriptions and images of some of the key men and women who have created our science over the years. Illustrations also include many of the latest images from the Hubble Space Telescope and other space instruments, as well as an up-to-date collection of color images from ground-based observatories around the world. Many of the planetary images are second- or third-generation corrected views, not merely the first releases rushed out for the news media. Full-color diagrams are used as teaching tools, not as cosmetic devices. Figure captions contain full explanations of what the student should be seeing and understanding.

- The book is written as a coherent story. However, because the authors know that many instructors follow an order of topics that is different from theirs, the sections are modular.

- We do not expect students to remember every concept introduced in previous chapters. Unobtrusive verbal "sign posts" are inserted throughout to help students find where a concept was defined or explained in detail or to briefly review a key idea that may have been introduced many chapters ago. We want every student to use the book as an easy navigational tool through the world of astronomy. A complete glossary is supplied in Appendix 3.

- A carefully written Prologue introduces the basic ideas and vocabulary of astronomy and makes sure all students start their study of the universe at the same point. The Epilogue summarizes key ideas about cosmic evolution and then applies them to the quest for life elsewhere.

- Appendix 1, written with David Bruning, lists a wide range of useful World Wide Web sites in astronomy that are accessible to students.

Special Sections and Boxes

The chapters in **Voyages Through the Universe** feature a number of highlighted sections designed to help non-science students appreciate the breadth of astronomy without distracting them from the main narrative.

- *Making Connections.* These special boxes show how astronomy connects to students' experiences with other fields of human endeavor and thought, from poetry to engineering, from popular culture to natural disasters.

- *Thinking Ahead.* Each chapter begins with a stimulating question about the material that follows.

- *Voyagers in Astronomy.* These profiles of noted astronomers focus not only on their work, but on their lives and human dimensions.

- *Astronomy Basics.* Fundamental science ideas and terms that other texts just assume students know are explained carefully.

- *Seeing for Yourself.* Students get familiar with the sky and everyday astronomical phenomena through observations using simple equipment.

- *Chapter Summary.* A concise overview that enumerates all important ideas and lists important new terms in boldface.

- *Review Questions, Thought Questions,* and numerical *Problems.* Questions that allow you a wide latitude for testing student understanding. Many can be used directly for discussion sections or essay exams.

RedShift CD-ROM

The award-winning *RedShift* software (Version 1, 2), published by Maris Multimedia, expands **Voyages Through the Universe** from a static presentation to a dynamic simulation of many aspects of astronomy. The dual-platform CD-ROM allows students to

- view realistic models of the planets and main satellites in the solar system

- identify over 300,000 stars, nebulae, and galaxies
- simulate astronomical events over the course of 15,000 years

- view more than 700 full-screen photographs (including a number by David Malin)

- access the *Penguin Dictionary of Astronomy* (with over 2,000 entries)

- navigate through surface maps of Earth, Moon, and Mars.

We have included some end-of-chapter exercises using *RedShift* (written by David Bruning of *Astronomy* magazine.)

RedShift is a valuable learning tool while your course is in progress. It is also an enjoyable piece of recreational software which students can use to explore the universe long after their academic experience is completed. The CD-ROM may be packaged with the text for a very low price.

Ancillaries

In addition to *RedShift,* qualified adopters of **Voyages Through the Universe** can receive

- *The Cosmos in the Classroom: A Resource Guide for Teaching Astronomy* by Andrew Fraknoi. This manual is a rich compilation of teaching ideas and resources for both novice and veteran instructors. It includes annotated listings of the best non-technical books and articles in astronomy organized by subject; listings of outstanding slides, videos, and software for each chapter, with addresses and phone numbers of suppliers; topics for discussion and for writing papers; resource guides for exploring the lives of astronomers; ideas for historical and interdisciplinary topics and on-site observation; and helpful appendices to make a teacher's job smoother.

- *The Saunders Internet Guide for Astronomy* by David Bruning, Randy Reddick, and Elliot King. This wonderful new handbook for both instructors and students is a thorough, up-to-date introduction to the internet and the World Wide Web. The first part reviews the history and current state of the internet and explains everything from simple e-mail to Multi-User Dungeons, with special attention to those applications useful in higher education. The second part features a thorough annotated listing of World Wide Web sites related to astronomy and astronomy education.

- *The Voyages Instructor's Manual/Test Bank.* The instructor's manual test bank contains answers to thought questions and problems in the textbook, as well as a host of multiple-choice test questions for use in a variety of classroom settings.

- *ExaMaster Computerized Test Bank* for Windows and Macintosh. ExaMaster features all the questions from the printed test bank in a format that allows instructors to edit them, add questions, and print assorted versions of the same test.

- The *Saunders Astronomy Transparency Collection.* This collection contains 205 overhead transparencies of conceptually based artwork. The enlarged reproductions contain figures from Saunders' astronomy textbooks, as well as supplemental illustrations that complement the text figures. A detailed guide arranged by topic accompanies the collection. These images are also available as slides.

- The *Saunders Astronomy Collection Supplement, 1994,* features 25 additional images from the Hubble Space Telescope and major observatories.

- *Voyages Through the Universe Transparency Collection.* This collection includes 25 of the most current Hubble and ground-based telescope photographs as well as striking and informative illustrations from the textbook.

- *Saunders MediaActive CD-ROM to accompany Voyages Through the Universe.* This exclusive CD-ROM contains all the diagrams and tables from the textbook and is a superb presentation tool to be used with such software packages as Powerpoint™, Persuasion™, and Saunders' LectureActive™.

- *Saunders Astronomy Videodisc: The Solar System.* This two-sided CAV disc contains 1000 still images and 45 minutes of video showing some of the best spacecraft images of the solar system. A bar code manual and LectureActive™ presentation software accompany every disc.

The Registered Adopters Program

Instructors who fill out a brief survey can become **Registered Voyagers Adopters.** There is no cost involved, and registered adopters will receive

- invitations to special events at astronomy meetings (where they can talk with the authors and other adopters)

- first access to new ancillary materials

- updates on new developments in astronomy, as well as new teaching techniques and tools

- an opportunity to have direct input into the planning of future editions

To become a registered adopter, talk with your Saunders representative, call us toll-free at 1-800-939-7377 (ask for Karen Milstein), e-mail us at voyages@saunderscollege.com, or write to: Karen Milstein, Marketing Coordinator, Saunders College Publishing, Public Ledger Building, Suite 1250, 150 S. Independence Mall West, Philadelphia, PA 19106.

Saunders College Publishing may provide complimentary instructional aids and supplements or supplement packages to those adopters qualified under our adoption policy. Please contact your sales representative for more information. If as an adopter or potential user you receive supplements you do not need, please return them to your sales representative or send them to:

Attn: Returns Department
Troy Warehouse
465 South Lincoln Drive
Troy, MO 63379

Let Us Hear from You

Unlike stars and planets, textbooks in astronomy do not exist in a vacuum. All three authors have benefited tremendously over the years from the advice of colleagues and students who teach astronomy. We want to be sure that we continue to make changes and updates in Saunders' astronomy texts that will be the most useful to you.

Therefore, we welcome comments and suggestions about the text and the ancillaries and ideas for how future materials can be made more effective. Address your cards and letters to: Andrew Fraknoi, Astronomy Dept., Foothill College, 12345 El Monte Rd., Los Altos Hills, CA 94022 or e-mail: FRAKNOI@ADMIN.FHDA.EDU.

Acknowledgments

We would like to thank the many colleagues and friends who have provided information, images, and encouragement, including: Charles Avis, Charles Bailyn, Bruce Balick, Judy Barrett, Roger Bell, Michael Bennett, Roy Bishop, Chip Clark, Richard Dreiser, George Djorgovski, Alex Filippenko, Lola Fraknoi, Michael Friedlander, Alan Friedman, Ian Gatley, Paul Geissler, Margaret Geller, Cheryl Gundy, Heidi Hammel, Bob Havlen, Todd Henry, Scott Hildreth, George Jacoby, William Keel, Geoff Marcy, Jeff McClintock, Edward McNevin III, Michael Merrill, Jacqueline Mitton, Janet Morrison, Donald Osterbrock, Michael Perryman, Carle Pieters, Edward Purcell, Axel Quetz, Jessica Richter, Dennis Schatz, Rudy Schild, Maarten Schmidt, Joe Tenn, Ray Villard, Althea Washington, Adrienne Wasserman, and Richard Wolff.

David Bruning played an essential role in the development of the text, by writing the "Using RedShift" sections, as well as the first draft of Appendix 1. George Kelvin rendered some of the fine color diagrams that grace the book. We are grateful to Bill Hartmann, John Spencer, Don Davis, and Don Dixon for permission to reproduce their astronomical paintings, and to David Malin for his assistance and his superb astronomical photographs.

We benefited very much from the suggestions of the following reviewers of preliminary drafts of the text:

Grady Blount
Texas A&M University, Corpus Christi

Michael Briley
University of Wisconsin, Oshkosh

David Buckley
East Stroudsburg University

John Burns
Mt. San Antonio College

Paul Campbell
Western Kentucky University

Eugene R. Capriotti
Michigan State University

George L. Cassiday
University of Utah

John Cunningham
Miami-Dade Community College

Grace Deming
University of Maryland

Miriam Dittman
DeKalb College

Gary J. Ferland
University of Kentucky

George Hamilton
Community College of Philadelphia

Adrian Herzog
California State University, Northridge

Ronald Kaitchuck
Ball State University

William C. Keel
University of Alabama

Steven L. Kipp
Mankato State University

Jim Lattimer
SUNY, Stonybrook

Robert Leacock
University of Florida

Terry Lemley
Heidelberg College

Bennett Link
Montana State University

Charles H. McGruder III
Western Kentucky University

Stephen A. Naftilan
Claremont Colleges

Anthony Pabon
DeAnza College

Cynthia W. Peterson
University of Connecticut

Andrew Pica
Salisbury State University

Terry Richardson
College of Charleston

Margaret Riedinger
University of Tennessee, Knoxville

Jim Rostirolla
Bellevue Community College

Michael L. Sitko
University of Cincinnati

John Stolar
West Chester University

Charles R. Tolbert
University of Virginia

Steve Velasquez
Heidelberg College

David Weinrich
Moorhead State University

David Weintraub
Vanderbilt University

Mary Lou West
Montclair State University

Dan Wilkins
University of Nebraska, Omaha

J. Wayne Wooten
Pensacola Junior College

No book project of this complexity could succeed without the diligent efforts of many people at the publisher's. We very much appreciate the assistance of John Vondeling, Vice-President and Publisher; Jennifer Bortel, Acquisitions Editor; Anne Gibby, Senior Project Editor; Carol Bleistine, Manager of Art and Design; Alicia Jackson, Production Manager; George Semple, Picture Development Editor; Margie Waldron, Director of Marketing; Tara Pauliny, Editorial Assistant; Karen Milstein, Marketing Coordinator; and other members of the staff at Saunders College Publishing, for graciously accommodating the needs of three authors with busy schedules and strong opinions.

We dedicate this book to Alex Fraknoi, who came into the universe during the same period this book was planned and born. May a future edition of the text still be helping students with their exploration of the universe by the time he gets to college.

Andrew Fraknoi, David Morrison, and Sidney Wolff
July 1996

Contents Overview

VOLUME I: THE SKY, LIGHT, AND PLANETS

Please note that Chapters 1–14 are available in a separate volume, and as a part of the complete student edition.

Prologue and Brief Tour of the Universe   *1*

1   Observing the Sky: The Birth of Astronomy   *17*

2   Orbits and Gravity   *42*

3   Earth, Moon, and Sky   *59*

4   Radiation and Spectra   *83*

5   Astronomical Instruments   *107*

6   Other Worlds: An Introduction to the Solar System   *131*

7   Earth as a Planet   *147*

8   Cratered Worlds: The Moon and Mercury   *161*

9   Earth-like Planets: Venus and Mars   *185*

10   The Giant Planets   *209*

11   Rings, Moons, and Pluto   *229*

12   Comets and Asteroids: Debris of the Solar System   *249*

13   Cosmic Samples and the Origin of the Solar System   *269*

14   The Sun: A Garden-Variety Star   *287*

VOLUME II: STARS, GALAXIES, AND COSMOLOGY

15   The Sun: A Nuclear Powerhouse   *309*

16   Analyzing Starlight   *327*

17   The Stars: A Celestial Census   *341*

18   Celestial Distances   *361*

19   Between the Stars: Gas and Dust in Space   *377*

20   The Birth of Stars and the Search for Planets   *397*

21   Stars: From Adolescence to Old Age   *417*

22   The Death of Stars   *435*

23   Black Holes and Curved Spacetime   *459*

24   The Milky Way Galaxy   *479*

25   Galaxies   *501*

26   Quasars and Active Galaxies   *519*

27   The Organization of the Universe   *537*

28   The Big Bang   *559*

Epilogue: Cosmic Evolution and Life Elsewhere   *585*

Contents

About the Authors   *v*

Preface for the Student   *vii*

Preface for the Instructor   *ix*

Please note that Chapters 1–14 are available in a separate volume, and as a part of the complete student edition.

Prologue and Brief Tour of the Universe   *1*

1 Observing the Sky: The Birth of Astronomy   *17*

1.1 The Sky Above   *18*
1.2 Ancient Astronomy   *23*
1.3 Astrology and Astronomy   *28*
1.4 The Birth of Modern Astronomy   *130*
Making Connections: *Testing Astrology   30*
Astronomy Basics: *What's Your Angle?   21*
Astronomy Basics: *How Do We Know the Earth Is Round?   35*
Seeing for Yourself: *Observing the Planets   36*

2 Orbits and Gravity   *42*

2.1 The Laws of Planetary Motion   *44*
2.2 Newton's Great Synthesis   *47*
2.3 Universal Gravity   *47*
2.4 Orbits in the Solar System   *49*
2.5 Motions of Satellites and Spacecraft   *50*
2.6 Gravitation with More Than Two Bodies   *52*
Making Connections: *Astronomy and the Poets   55*

3 Earth, Moon, and Sky   *59*

3.1 Earth and Sky   *60*
3.2 The Seasons   *61*
3.3 Keeping Time   *66*
3.4 The Calendar   *68*
3.5 Phases and Motions of the Moon   *70*
3.6 Ocean Tides and the Moon   *73*
3.7 Eclipses of the Sun and Moon   *76*
Making Connections: *Astronomy and the Days of the Week   70*

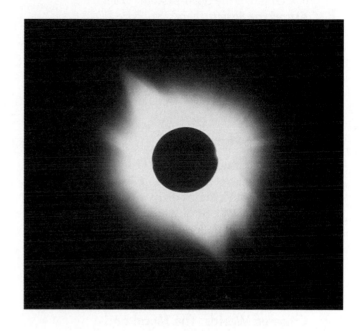

Voyagers in Astronomy: *George Darwin and the Slowing of the Earth   75*
Seeing for Yourself: *How to Observe Solar Eclipses   79*

4 Radiation and Spectra   *83*

4.1 The Nature of Light   *84*
4.2 The Electromagnetic Spectrum   *87*
4.3 Spectroscopy in Astronomy   *92*
4.4 The Structure of the Atom   *94*
4.5 Formation of Spectral Lines   *97*
4.6 The Doppler Effect   *100*
Making Connections: *The Rainbow   102*

5 Astronomical Instruments   *107*

5.1 Telescopes   *108*
5.2 Optical Detectors and Instruments   *111*
5.3 Optical and Infrared Observatories   *114*
5.4 Radio Telescopes   *118*
5.5 Observatories Outside the Earth's Atmosphere   *122*

Astronomy Basics: *How Astronomers Use Telescopes* **111**
Making Connections: *Choosing Your Own Telescope* **126**
Voyagers in Astronomy: *George Ellery Hale: Master Telescope Builder* **115**

6 Other Worlds: An Introduction to the Solar System 131

6.1 Overview of our Planetary System **132**
6.2 Composition and Structure of Planets **137**
6.3 Dating Planetary Surfaces **140**
6.4 Origin of the Solar System **141**
Astronomy Basics: *There's No Place Like Home* **138**
Making Connections: *Names in the Solar System* **136**
Voyagers in Astronomy: *Carl Sagan: Solar System Advocate* **143**

7 Earth as a Planet 147

7.1 The Global Perspective **148**
7.2 The Crust of the Earth **150**
7.3 The Earth's Atmosphere **154**
7.4 Life and Chemical Evolution **157**
7.5 Cosmic Influences on the Evolution of Earth **159**
Voyagers in Astronomy: *Alfred Wegener: Catching the Drift of Plate Tectonics* **154**

8 Cratered Worlds: The Moon and Mercury 161

8.1 General Properties of the Moon **168**
8.2 The Lunar Surface **171**
8.3 Impact Craters **173**
8.4 The Origin of the Moon **176**
8.5 Mercury **177**
Making Connections: *What a Difference a Day Makes* **178**
Seeing for Yourself: *Observing the Moon* **181**

9 Earth-like Planets: Venus and Mars 185

9.1 The Nearest Planets: An Overview **186**
9.2 The Geology of Venus **189**
9.3 The Massive Atmosphere of Venus **194**
9.4 The Geology of Mars **196**
9.5 Martian Polar Caps, Atmosphere, and Climate **200**
9.6 Divergent Planetary Evolution **203**
Voyagers in Astronomy: *Percival Lowell: Dreaming of an Inhabited Mars* **187**

10 The Giant Planets 209

10.1 Exploring the Outer Planets **210**
10.2 The Jovian Planets **213**
10.3 Atmospheres of the Jovian Planets **216**
10.4 Magnetospheres **222**
Making Connections: *Engineering and Space Science: Teaching an Old Spacecraft New Tricks* **212**
Voyagers in Astronomy: *James Van Allen: Several Planets Under His Belt* **224**

11 Rings, Moons, and Pluto 229

11.1 Ring and Satellite Systems **230**
11.2 The Galilean Satellites and Titan **231**
11.3 Triton and Pluto **236**
11.4 Planetary Rings **240**
Voyagers in Astronomy: *Clyde Tombaugh: From the Farm to Fame* **239**
Making Connections: *Astronomy and Future Tourism The Seven Wonders of the Solar System* **243**

12 Comets and Asteroids: Debris of the Solar System 249

12.1 Asteroids **250**
12.2 Asteroids Near and Far **254**
12.3 The "Long-Haired" Comets **256**
12.4 Origin and Evolution of Comets **261**
Voyagers in Astronomy: *Edmund Halley: Astronomy's Renaissance Man* **257**
Making Connections: *Comet Hunting as a Hobby* **264**

13 Cosmic Samples and the Origin of the Solar System 269

13.1 Meteors **270**
13.2 Meteorites: Stones from Heaven **272**
13.3 Formation of the Solar System **276**
13.4 Planetary Evolution **280**
Seeing for Yourself: *Showering with the Stars* **273**
Making Connections: *Some Striking Meteorites* **275**

14 The Sun: A Garden-Variety Star 287

14.1 Outer Layers of the Sun **288**
14.2 The Active Sun Introduced **294**
14.3 The Sunspot Cycle **296**
14.4 Activity Above the Photosphere **297**
14.5 Is the Sun a Variable Star? **301**
Making Connections: *Solar Flares and Their Effects on Earth* **300**
Voyagers in Astronomy: *Art Walker: Doing Astronomy in Space* **302**
Seeing for Yourself: *Observing the Sun* **305**

15 The Sun: A Nuclear Powerhouse 309

15.1 Thermal and Gravitational Energy 310
15.2 Mass, Energy, and the Special Theory of Relativity 312
15.3 The Interior of the Sun: Theory 316
15.4 The Interior of the Sun: Observations 320
Astronomy Basics: What's Watt? 310
Voyagers in Astronomy: Albert Einstein 313
Making Connections: Fusion on Earth 323

16 Analyzing Starlight 327

16.1 The Brightness of Stars 328
16.2 Colors of Stars 330
16.3 The Spectra of Stars 330
Voyagers in Astronomy: Annie Cannon: Classifier of the Stars 333

17 The Stars: A Celestial Census 341

17.1 A Stellar Census 342
17.2 Stellar Masses 344
17.3 Diameters of Stars 349
17.4 The H–R Diagram 352
Making Connections: Astronomy and Mythology: Algol the Demon Star and Perseus the Hero 350
Voyagers in Astronomy: Henry Norris Russell 353

18 Celestial Distances 361

18.1 Fundamental Units of Distance 362
18.2 Surveying the Stars 363
18.3 Variable Stars: One Key to Cosmic Distances 368
18.4 The H–R Diagram and Cosmic Distances 371
Astronomy Basics: Naming Stars 365
Making Connections: Parallax and Space Astronomy 367
Voyagers in Astronomy: John Goodricke 369

19 Between the Stars: Gas and Dust in Space 377

19.1 The Interstellar Medium 378
19.2 Interstellar Gas 379
19.3 A Model of the Interstellar Gas 383
19.4 Cosmic Dust 385
19.5 Cosmic Rays 391
Astronomy Basics: Naming the Nebulae 378
Making Connections: Cocktails in Space 382
Voyagers in Astronomy: Edward Emerson Barnard 386

20 The Birth of Stars and the Search for Planets 397

20.1 Star Formation 398
20.2 The H–R Diagram and the Study of Stellar Evolution 405
20.3 Evidence That Planets Form Around Other Stars 407
20.4 Planets Beyond the Solar System: Search and Discovery 409

21 Stars: From Adolescence to Old Age 417

21.1 Evolution from the Main Sequence to Giants 418
21.2 Star Clusters 420
21.3 Checking Out the Theory 422
21.4 Further Evolution of Stars 425
21.5 The Evolution of More-Massive Stars 427
Astronomy Basics: Stars in Your Little Finger 426
Making Connections: The Red Giant Sun and the Fate of the Earth 428

22 The Death of Stars 435

22.1 The Death of Low-Mass Stars 436
22.2 Evolution of Massive Stars: An Explosive Finish 442
22.3 Pulsars and the Discovery of Neutron Stars 449
22.4 The Evolution of Binary Star Systems 452
Voyagers in Astronomy: Subrahmanyan Chandrasekhar 438
Making Connections: Supernovae in History 444

23 Black Holes and Curved Spacetime 459

23.1 Principle of Equivalence 460
23.2 Spacetime and Gravity 463
23.3 Tests of General Relativity 465
23.4 Time in General Relativity 467
23.5 Black Holes 468
23.6 Evidence for Black Holes 472
Making Connections: Gravity and Time Machines 471

24 The Milky Way Galaxy 479

24.1 The Architecture of the Galaxy 480
24.2 Spiral Structure of the Galaxy 486
24.3 Stellar Populations in the Galaxy 487
24.4 The Mass of the Galaxy 489
24.5 The Nucleus of the Galaxy 490
24.6 The Formation of the Galaxy 494
Voyagers in Astronomy: Harlow Shapley: Mapmaker to the Stars 482
Making Connections: Light Pollution and the Milky Way 495

25 Galaxies 501

25.1 The Great Nebula Debate 502
25.2 Types of Galaxies 502
25.3 Properties of Galaxies 507
25.4 The Extragalactic Distance Scale 509
25.5 The Expanding Universe 511
Voyagers in Astronomy: Edwin Hubble: Expanding the Universe 503
Astronomy Basics: Constants of Proportionality 513

26 Quasars and Active Galaxies 519

26.1 Quasars 520
26.2 Active Galaxies 524
26.3 The Power Behind the Quasars 527
26.4 Gravitational Lenses 531
Making Connections: Quasars and the Attitudes of Astronomers 533

27 The Organization of the Universe 537

27.1 The Distribution of Galaxies in Space 538
27.2 The Evolution of Galaxies: The Observations 545
27.3 The Evolution of Galaxies: The Theories 551
27.4 A Universe of (Mostly) Dark Matter? 553
Voyagers in Astronomy: Margaret Geller: Cosmic Surveyor 544
Astronomy Basics: Why Galaxies Collide and Stars Rarely Do 547

28 The Big Bang 559

28.1 The Age of the Expanding Universe 560
28.2 The Geometry of Spacetime 563
28.3 Models of the Universe 565
28.4 The Beginning of the Universe 569
28.5 The Inflationary Universe 576
Making Connections: What Might It Be Like in the Distant Future? 568

Epilogue: Cosmic Evolution and Life Elsewhere 585

APPENDICES:

1 Astronomy on the World Wide Web A-1

2 Sources of Astronomical Information A-5

3 Glossary A-7

4 Powers-of-Ten Notation A-18

5 Units Used In Science A-20

6 Some Useful Constants for Astronomy A-21

7 Physical Data for the Planets A-22
   Orbital Data for the Planets A-22

8 Satellites of the Planets A-23

9 Upcoming (Total) Eclipses A-25

10 The Nearest Stars A-26

11 The Brightest Stars A-27

12 The Brighter Members of the Local Group A-28

13 The Chemical Elements A-29

14 The Constellations A-31

15 The Messier Catalog of Nebulae and Star Clusters A-33

Index I-1

Star Charts

Galileo Orbiter Mission

In June 1996, the Galileo Orbiter began its two-year mission to survey the four large satellites of Jupiter. Although problems with the spacecraft antenna will limit the quantity of data radioed back to Earth, it is apparent that the quality of the images and spectra is superb, promising spectacular advances beyond the photos returned by the two Voyager spacecraft during their flybys 17 years earlier. Reproduced on these pages are examples of the first Galileo images of the moons Ganymede, Europa, and Io.

See Chapter 11 for more on these intriguing worlds.

Figure 1
A very high resolution view of the older, heavily cratered terrain on Ganymede, showing a region only a few miles across. The complex, hilly terrain and indications of the presence of both light- and dark-colored materials on the surface were among the new information revealed as the spacecraft passed much closer to Ganymede than had been possible with Voyager. (NASA/JPL)

Figure 2
Two of the new better-detail Galileo photos of Ganymede are superposed on the image of the same area transmitted by Voyager in 1979. The mountainous areas of the satellite are now seen to be much more rugged than had been suggested previously, indicating a more violent geological history for this moon and suggesting the action of plate tectonics of Ganymede at some time in the distant past. (NASA/JPL)

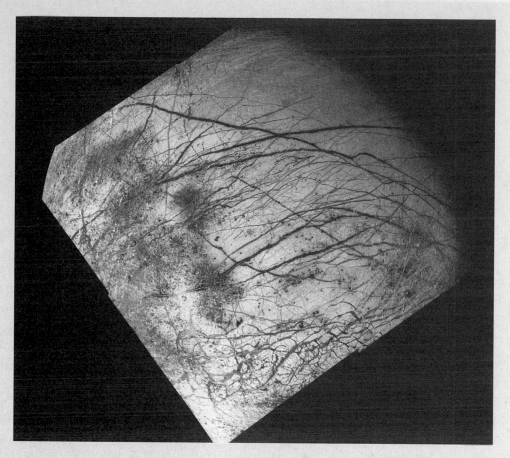

Figure 3
The first Galileo views of ice-covered Europa were better than the Voyager images but still far inferior to the two photos of Ganymede shown above. Still, the complex cracks on the surface hint of the possibility of a global ocean beneath the icy crust. If Europa really has large bodies of liquid water today, it is the only other location in the solar system beside our own planet Earth that possesses liquid water. (NASA/JPL)

Figure 4
Three early views of Io, imaged by the Galileo spacecraft from a great distance, show the multihued surface of this volcanically active satellite and demonstrate that its color has indeed changed in the 17 years since Voyager. These changes are the result of several large-scale volcanic eruptions that have taken place since the moon was last mapped. (NASA/JPL)

VOYAGES

THROUGH THE UNIVERSE

The Sun, just rising above the horizon, with the silhouette of the mirror of the McMath–Pierce Solar Telescope at Kitt Peak National Observatory. This mirror tracks the Sun across the sky and directs sunlight downward into the rest of the telescope's optics. (National Optical Astronomy Observatories)

CHAPTER 15

The Sun: A Nuclear Powerhouse

Thinking Ahead

Why does the Sun shine and keep shining? It is a prodigious generator of energy, and has been so ever since first watched by human eyes. But we believe it has been shining much longer than that. How can we get a sense of just how long it has been producing energy, and thus begin to understand the mechanism that powers our star?

> We do not argue with the critic who urges that the stars are not hot enough for this process [nuclear fusion]; we tell him to go and find a hotter place . . .
>
> Arthur Eddington in *Internal Constitution of the Stars* (1926)

How much energy does the Sun radiate into space? With modern instruments, we have measured its output with great precision and found that our star puts out about 4×10^{26} watts (W). That's a very large figure; what does it mean in human terms?

The current population of our planet is about 6 billion (6×10^9) people. Suppose for a moment that all of us simultaneously turned on a thousand 100-W light bulbs. Each person on Earth would then be lit up like a Hollywood movie theater on opening night! But all those bulbs still only total 6×10^{14} W. To use as much energy as the Sun produces, we would have to find 670 billion worlds like the Earth, all doing the same stunt. In other words, the Sun is a big, big cosmic light bulb.

In considering the Sun's energy mechanism, it is not enough to determine how much energy it uses in a second. You might think of all sorts of ways to generate a huge amount of energy if you only had to do it for a second, or even for a minute. But if you had to sustain a tremendous output of energy for a year or a billion years, then you would need a steady, reliable generator.

So to understand what powers the Sun, we first have to ask how long the Sun has been shining. This is not an easy question to answer: there is nothing "built into" the structure or appearance of the Sun that allows us to pinpoint its age directly. But as we saw in Chapter 13, we have several lines of evidence indicating that the Sun formed at roughly the same time as the planetary system—about 4.5 billion years ago.

Our task, then, is to explain not only how the Sun generates so many watts now, but also what mechanism has allowed it to do so for billions of years. When geologists and biologists began to uncover clues about the great age of the Earth in the late 19th and early 20th centuries, this appeared to be an insoluble problem.

Scientists first looked at sources already familiar to them here on Earth. This seemed a reasonable strategy, since the Sun and the Earth contain the same types of atoms, albeit in different proportions. But, as we will see, none of the energy sources known at the time could explain the Sun's longevity. It was only after scientists discovered how to tap the energy stored in the nuclei of atoms that they finally identified the source of the Sun's energy.

Figure 15.1
Wood fire. (Visuals Unlimited/Doug Sokell)

ASTRONOMY BASICS

What's Watt?

Just a word about the units we are using. A watt (W) is a unit of power—that is, of energy used or given off *per unit time.* You know from everyday experience that it's not just how much energy you expend, but how long you do it. Burning 10 Calories (Cal) in 10 min takes a very different kind of exercise than burning 10 Cal in an hour. Watts tell you the *rate* at which energy is being used; for example, a 100-W bulb uses 100 joules (j) of energy every second.

And how big is a joule? A 160-lb (73-kg) astronomy instructor running at about 10 mi/h (4.4 m/s) from a football player who just flunked the midterm has a motion energy of about 700 J.

15.1

Thermal and Gravitational Energy

Nineteenth-century scientists knew of two possible sources for the Sun's energy: heat, or thermal energy, and gravitational energy. The source of heat energy most familiar to us here on Earth is the burning (the chemical term is *oxidation*) of wood, coal, gasoline, or other fuel (Figure 15.1). However, we know exactly how much energy the burning of these materials can produce, and can thus calculate that even if the immense mass of the Sun

consisted of a burnable material like coal or wood, our star could not produce energy at its present rate for more than a few thousand years. That's not enough time for Earth to evolve even a virus, to say nothing of a life-form as complicated as an astronomy student! Besides, geologists have found fossils in rocks that are 3.5 billion years old, so the temperature of the Earth (and the heat output of the Sun) must have been suitable to sustain life as long ago as that.

In addition, results from 20th-century spectroscopy make the coal and wood suggestions completely untenable. At the temperatures found in the Sun, nothing like solid wood or coal could survive. And the dominant elements in the Sun, as we saw in the preceding chapter, are hydrogen and helium.

Conservation of Energy

Other 19th-century attempts to determine what makes the Sun shine used the law of conservation of energy. Simply stated, this law says that energy cannot be created or destroyed, but can be transformed from one type to another—say from heat energy to mechanical energy. The steam engine, which was the key to the industrial revolution, is a good example. In it, the hot steam from a boiler drives the movement of a piston, converting heat energy to motion energy.

Motion can also be transformed into heat. If you clap your hands vigorously at the end of an especially good astronomy lecture, your palms become hotter. If you rub ice on the surface of a table, the heat produced by friction melts the ice.

In the 19th century, scientists considered the possibility that the source of the Sun's heat might be the mechanical motion of meteorites falling into it. Calculations show, however, that in order to produce the total amount of energy emitted by the Sun, the mass in meteorites that

would have to fall into it every 100 years would equal the mass of the Earth. The resulting increase in the Sun's mass would, according to Kepler's third law, change the period of the Earth's orbit by 2 s per year. Such a change would be easily measurable and is not, in fact, occurring. This source of the Sun's energy was then ruled out.

Gravitational Contraction as a Source of Energy

As an alternative, the German scientist Hermann von Helmholtz and the British physicist Lord Kelvin (Figure 15.2), in about the middle of the 19th century, proposed that the Sun might produce energy by the conversion of gravitational energy to heat. They suggested that the outer layers of the Sun might be "falling" inward because of the force of gravity. In other words, they proposed that the Sun could be shrinking in size, and staying hot and bright as a result.

To imagine what would happen, picture the outer layer of the Sun starting to fall inward. This outer layer is a gas made up of individual atoms, all moving about in random directions. Temperature, which measures the amount of stored heat energy, depends on the speed of the atoms: higher speeds equal higher temperatures. If this layer starts to fall, the atoms acquire an additional velocity because of the falling motion. As the outer layer falls inward, it also contracts, moving the atoms closer together. Collisions become more likely, and some of them transfer the velocity associated with the falling motion to other atoms, increasing their velocities and hence increasing the temperature of this layer of the Sun. Other collisions excite electrons within the atoms to higher energy

orbits. When these electrons return to their normal orbits, they emit photons, which can then escape from the Sun.

Kelvin and Helmholtz calculated that a contraction of the Sun at a rate of only about 40 m per year would be enough to produce the amount of energy that it is now radiating. Over the span of human history, the decrease in the Sun's size from such a slow contraction would be undetectable. If we assume that the Sun began its life as a large, diffuse cloud of gas, then we can calculate how much energy has been radiated by the Sun during its entire lifetime as it has contracted from a very large diameter down to its present size. That amount of energy is on the order of 10^{42} J. Since the solar luminosity is 4×10^{26} W (J/s) or about 10^{34} J per year, contraction could keep the Sun shining at its present rate for roughly 100 million years.

In the 19th century, 100 million years at first seemed plenty long enough, since the Earth was then widely thought to be much younger than this. But toward the end of that century and into the 20th, geologists and physicists showed that the Earth (and hence the Sun) is actually much older. Contraction therefore cannot be the primary source of solar energy.

In the same way, every other process suggested to explain the Sun's energy failed to account for the known facts about the Sun, including its age. Scientists were thus confronted with a puzzle of enormous proportions. Either an unknown type of energy was responsible for the most important energy source known to humanity, or estimates of the span of time that the solar system (and life on Earth) had been around had to be seriously modified. Charles Darwin, whose theory of evolution required a longer time span than the theories of the Sun seemed to offer, was discouraged by these results and continued to worry about them until his death in 1882.

It was only in the 20th century that the true source of the Sun's energy was identified. The two key pieces of information required to solve the puzzle were the structure of the nucleus of the atom, and the fact that mass can be converted into energy.

Mass, Energy, and the Special Theory of Relativity

As we have seen, energy cannot be created or destroyed, but only converted from one form to another. One of the remarkable conclusions derived by Albert Einstein when he developed the theory of relativity is that mass can be considered another form of energy and can therefore be converted to energy. This remarkable equivalence is expressed in one of the most famous equations in all of science:

$$E = mc^2$$

where E is the symbol for energy, m is the symbol for mass, and c, the constant that relates the two, is the speed of light. Note that this equation is very similar in form to

$$\text{inches} = \text{feet} \times 12 \quad \text{or} \quad \text{cents} = \text{dollars} \times 100$$

That is, it is a conversion formula that allows you to calculate the conversion of one thing, mass, to another, energy. The conversion factor in our case turns out to be not 12 or 100, but another constant quantity, the speed of light squared. By the way, mass does not have to *travel* at the speed of light (or the speed of light squared, which is impossible in nature) for this conversion to occur. The factor of c^2 is just the number that must be used to relate mass and energy.

Notice that this formula does not tell you *how* to convert mass into energy, just as the formula for cents does not tell you where to exchange coins for a dollar bill. The formulas merely tell you what the equivalent values are if you succeed in making the conversion. When Einstein first derived his famous formula in 1905, no one had the faintest idea how to convert mass into energy in any practical way. Einstein himself tried to discourage speculation that the conversion of atomic mass into energy would be feasible in the near future. Today, as a result of developments in nuclear physics, we convert mass into energy in power plants, nuclear weapons, and—in high-energy physics—experiments in particle accelerators.

Because c^2, the speed of light squared, is a very large quantity, the conversion of even a small amount of mass results in a very great amount of energy. For example, the complete conversion of 1 g of matter (about 1/14 oz) would produce as much energy as the combustion of 15,000 barrels of oil.

Scientists soon realized that the conversion of mass to energy is the source of the Sun's heat and light. With Ein-stein's equation $E = mc^2$, we can calculate that the amount of energy radiated by the Sun could be produced by the complete conversion of about 4 million tons of matter to energy inside the Sun each second. Four million tons per second sounds like a lot when compared to Earthly things, but bear in mind that the Sun is a very big reservoir of matter. In fact, we will see that the Sun contains more than enough mass to continue shining at its present rate for billions of years.

But this *still* does not tell us how mass can be converted to energy. To understand how the conversion actually occurs in the Sun, we need to explore the structure of the atom a bit further.

Elementary Particles

The fundamental components of matter are called **elementary particles.** The most familiar of these are the proton, neutron, and electron—the particles that make up ordinary atoms (see Section 4.4).

Protons, neutrons, and electrons are by no means all the particles that exist. First, for each kind of particle, there is a corresponding but opposite *antiparticle*. If the particle carries a charge, its antiparticle has the opposite charge. The antielectron is the *positron,* which has the same mass as the electron but is positively charged. Likewise, the antiproton has a negative charge. The remarkable thing about such antimatter is that when a particle comes in contact with its antiparticle, the two annihilate each other, turning into pure energy.

Since our world is made exclusively of ordinary particles of matter, any antimatter cannot survive for very long. But individual antiparticles are found in cosmic rays (particles that arrive at the top of the Earth's atmosphere from space) and can be formed in particle accelerators. In fact, when we create matter from energy in our high-energy physics labs, we always get half matter and half antimatter.

Science fiction fans may be familiar with antimatter from the *Star Trek* television series and films. The Starship *Enterprise* is propelled by the *careful* combining of matter and antimatter in the ship's engine room. According to $E = mc^2$, the complete annihilation of matter and antimatter can produce a huge amount of energy; but keeping the antimatter fuel from touching the ship before it is needed must be a big problem. No wonder that Scotty, the chief engineer in the original TV show, always looked worried!

The existence of another type of particle was originally suggested in 1933 by physicist Wolfgang Pauli. Energy seemed not to be conserved in certain types of nuclear reactions, in violation of the law of conservation of energy. Rather than accept an overthrow of one of the fundamental ideas in science, Pauli suggested that a new and so far undetected particle, named the **neutrino,** was carrying away the "missing" energy. He suggested that neutrinos were particles with zero mass that, like photons, moved with the speed of light.

Albert Einstein

For a large part of his life, Albert Einstein was one of the most recognized celebrities of his day. Strangers stopped him on the street, and people all over the world wrote asking him for endorsements, advice, and assistance. In fact, when Einstein and the great film star Charlie Chaplin met in California, they found they shared similar feelings about the loss of privacy that comes with fame. Einstein's name was a household word despite the fact that most people did not understand the ideas that had made him famous.

Einstein was born in 1879 in Ulm, Germany. Legend has it that he did not do well in school (even in arithmetic), and thousands of students have since justified a bad grade by referring to this story. Alas, like many legends, this one is not true. Records indicate that although he tended to rebel against the authoritarian teaching style in vogue in Germany at that time, Einstein was a good student.

After graduating from the Federal Polytechnic Institute in Zurich, Switzerland, Einstein at first had trouble getting a job (even as a high-school teacher), but he eventually became an examiner in the Swiss Patent Office. Working in his spare time, without the benefit of a university environment but using his superb physical intuition, he wrote four papers in 1905 that would ultimately transform the way physicists looked at the world.

One of these, which earned him the Nobel Prize in 1921, set part of the foundation of *quantum mechanics*, the rich, puzzling, and remarkable theory of the subatomic

Albert Einstein in 1905. (Permission granted by the Albert Einstein Archives, The Hebrew University of Jerusalem, Israel)

realm. But his most important paper presented the *special theory of relativity*, a re-examination of space, time, and motion that added a whole new level of sophistication to our understanding of these concepts. $E = mc^2$ was a relatively minor part of this theory, added in a later paper.

In 1916 Einstein published his *general theory of relativity*, which was, among other things, a fundamentally new description of gravity (see Chapter 23). When this theory was confirmed by measurements of the "bending of starlight" during a 1919 eclipse (the *New York Times* headline read, "Lights All Askew in the Heavens"), Einstein became world-famous.

In 1933, to escape Nazi persecution, Einstein left his professorship in Berlin and settled in the United States at the newly created Institute for Advanced Studies at Princeton. He remained there until his death in 1955, writing, lecturing, and espousing a variety of intellectual and political causes. For example, he agreed to sign a letter written by Leo Szilard and other scientists in 1939, alerting President Roosevelt to the dangers of allowing Nazi Germany to develop the atomic bomb first. In 1952 Einstein was offered the second presidency of Israel. In declining the position, he said, "I know a little about nature and hardly anything about men."

When Pauli proposed this idea, neutrinos had not yet been detected because they interact very weakly with other matter. Most of them can pass completely through a star or planet without being absorbed. The Earth is more transparent to a neutrino than the thinnest and cleanest pane of glass is to a photon of light. As we will see, this "antisocial" behavior of neutrinos makes them both frustrating and very important for scientists studying the Sun and other stars.

The elusive neutrino was finally detected in 1956, and so we know that this particle really does exist. Experiments are not, however, sufficiently precise to prove that neutrinos have exactly zero mass. If neutrinos turn out to have even a tiny mass, this fact could have interesting consequences for cosmology (see Chapters 27 and 28) and for models of the Sun's interior (see Section 15.4).

Some of the properties of the proton, electron, neutron, and neutrino are summarized in Table 15.1. (Other

TABLE 15.1
Properties of Some Elementary Particles

Particle	Mass (kg)	Charge
Proton	1.67265×10^{-27}	$+1$
Neutron	1.67495×10^{-27}	0
Electron	9.11×10^{-31}	-1
Neutrino	0	0

subatomic particles have been produced by experiments with particle accelerators, but they do not play a role in the generation of solar energy.)

The Atomic Nucleus

The nucleus of an atom is not just a loose collection of elementary particles. Inside the nucleus, particles are held together by a very powerful force called the *strong nuclear force*. This is a short-range force, only able to act over distances about the size of the atomic nucleus. A quick thought experiment shows how important this force is. Take a look at your little finger and think of the atoms composing it. Among them is carbon, one of the basic elements of life. Focus your imagination on the nucleus of one of your carbon atoms. It contains six protons, which have a positive charge, and six neutrons, which are neutral. Thus the nucleus has a net charge of six positives; if only the electrical force were acting, the protons in this and every carbon atom would find each other very repulsive and fly apart.

The strong nuclear force is an attractive force, stronger than the electromagnetic force, and it keeps the particles of the nucleus tightly bound together. We saw earlier that if under the force of gravity a star "shrinks"—bringing its atoms closer together—gravitational energy is released. In the same way, if particles come together under the strong nuclear force, and unite to form an atomic nucleus, some of the nuclear energy is released. The energy given up in such a process is called the binding energy of the nucleus.

When such binding energy is released, the resulting nucleus has slightly less mass than the sum of the masses of the particles that came together to form it. In other words, the energy comes from the loss of mass. This slight deficiency in mass is always only a small fraction of the mass of one proton. But because each bit of lost mass can provide quite a bit of energy (remember $E = mc^2$), this nuclear energy release can be quite a potent mechanism.

The behavior of the nuclear force is more complicated than gravity, but during the 20th century, physicists have been able to investigate and establish the properties of atomic nuclei. It turns out that the binding energy is greatest for atoms with a mass near that of the iron nucleus (with a combined number of protons and neutrons equal to 56), and less for both the lighter and the heavier atoms. Iron, therefore, is the most stable element (since it gives up the most energy when it forms, it is the hardest nucleus to "undo").

What this means is that, in general, when light atomic nuclei come together to form a heavier one (up to iron), mass is lost and energy released. This joining together of atomic nuclei is called nuclear **fusion.**

Energy can also be produced by breaking up heavy atomic nuclei into lighter ones (down to iron); this process is called nuclear *fission* (see Figure 15.3). Nuclear fission was the process we learned to use first—in atomic bombs, and in nuclear reactors used to generate electric power—and it may thus be more familiar to you. It also sometimes occurs spontaneously, in natural radioactivity. But fission requires big, complex nuclei, whereas we know that the stars are made up predominantly of small, simple nuclei. So we must look to fusion to explain the Sun and the stars.

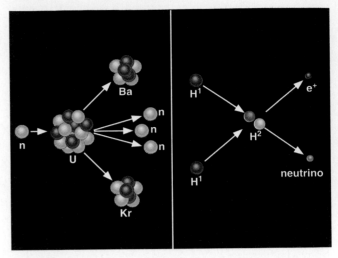

Figure 15.3

In fission, a larger nucleus breaks into two smaller components. Here a nucleus of uranium, with 92 protons and 143 neutrons, is shown undergoing fission into two smaller nuclei of barium (56 protons) and krypton (36 protons). In fusion, smaller nuclei bind together to make a larger one. We also see two nuclei of hydrogen (one proton each) fuse into a heavier hydrogen nucleus (one proton and one neutron), accompanied by the emission of a positron and a neutrino.

Nuclear Attraction Versus Electrical Repulsion

So far, this sounds like a very attractive prescription for making energy: roll some nuclei together and join them via nuclear fusion. This will cause them to lose some of their mass, which then turns into energy. However, every nucleus, even simple hydrogen, has protons in it—and protons all have positive charges. Since like charges repel via the electrical force, the closer we get two nuclei to one another, the more they repel. It's true that if we can get them within "striking distance" of the nuclear force, they will then come together with a much stronger attraction. But that striking distance is very tiny, about the size of a nucleus. How can we get nuclei close enough to participate in fusion?

The answer turns out to be heat—tremendous heat—which speeds the protons up enough to overcome the electrical forces that try to keep protons apart. Inside the Sun, as we saw, the most common element is hydrogen, whose nucleus is a single proton. Two protons can fuse only in regions where the temperature is greater than about 10 million K, and the speed of the protons averages around 1000 km/s or more. (In old-fashioned units, that's over 2 million mi/h!) Such extreme temperatures are reached only in the regions surrounding the center of the Sun, which has a temperature of 15 million K. Calculations show that nearly all of the Sun's energy is generated within about 150,000 km of its core, or within less than 10 percent of its total volume.

Even at these high temperatures, it is exceedingly difficult to force two protons to combine. On average, a proton will rebound from other protons in the Sun's crowded core for about 14 billion years, at the rate of 100 million collisions per second, before it fuses with a second proton. This is, however, only the average waiting time. Some of the enormous number of protons in the Sun's inner region are lucky and take only a few collisions to achieve a fusion reaction: they are the protons responsible for producing the energy radiated by the Sun. Since the Sun is about 4.5 billion years old, most of its protons have not yet been involved in fusion reactions.

Nuclear Reactions in the Sun's Interior

The Sun, then, taps the energy contained in the nuclei of atoms through nuclear fusion. Let's look at what happens in more detail. Deep inside the Sun, four hydrogen atoms fuse to form a helium atom. The helium atom is slightly less massive than the four hydrogen atoms that combine to form it, and that lost mass is converted to energy.

The initial steps required to form one helium nucleus from four hydrogen nuclei are shown in Figure 15.4. First, two protons combine to make a *deuterium* nucleus, which is an isotope (or version) of hydrogen that contains one proton and one neutron. In effect, one of the original protons has been converted to a neutron in the fusion reaction. Electric charge has to be conserved in nuclear reactions, and it is conserved in this one. The positive charge originally associated with one of the protons is carried away by a positron.

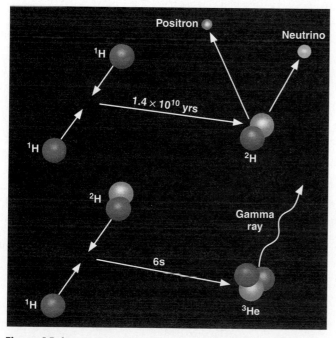

Figure 15.4
The first two steps in the process of fusing hydrogen into helium in the Sun (the p–p chain).

Since it is antimatter, this positron will instantly collide with an electron, and both will be annihilated, producing pure electromagnetic energy in the form of gamma rays. Now what happens to this gamma ray, which has been created in the center of the Sun? It finds itself in a world crammed full of fast-moving nuclei and electrons. The photon collides with particles of matter and transfers some of its energy to them. The result of this process is generally to lower the energy of the gamma-ray photon.

Such interactions happen to the gamma ray again and again and again, until—as it makes its way slowly toward the outer layers of the Sun—its energy becomes so reduced that it is no longer a gamma ray, but an x ray (recall Section 4.2). And later, as it loses more energy, it becomes an ultraviolet photon. The Sun is so full of particles (targets for the photon to hit) that it takes on the order of a million years for the average photon to emerge from the Sun's photosphere. By that time, most of the photons have given up enough energy to be ordinary light—and they are the sunlight we see coming from our star. (To be precise, each gamma-ray photon is ultimately converted into many separate lower-energy photons of sunlight.)

In addition to the positron, the fusion of two hydrogen atoms to form deuterium results in the emission of a neutrino. Because neutrinos interact so little with ordinary matter, those produced by fusion reactions near the center of the Sun travel directly to the Sun's surface and then on toward the Earth. Neutrinos move at the speed of light and they get out of the Sun only about 2 s after they are created (see Figure 15.9).

The next step in forming helium from hydrogen is to add a proton to the deuterium nucleus and create a helium nucleus that contains two protons and one neutron. In the process, some mass is again lost, and more gamma radiation is emitted. Such a nucleus is helium because an element is defined by the number of its protons; any nucleus with two protons is called helium. But this form of helium, which we call helium 3, is not the isotope we see in the Sun's atmosphere or on Earth. That helium has two neutrons and two protons and hence is called helium 4.

To produce helium 4, helium 3 combines with another just like it in the third step of fusion (illustrated in Figure 15.5). Note that two protons are left over from this step; they come out of the reaction ready to collide with other protons and thus to start step 1 all over again.

The P–P Chain

The reactions in the Sun can be described succinctly through the following nuclear formulas:

$$^1H + {}^1H \longrightarrow {}^2H + e^+ + \nu$$

$$^2H + {}^1H \longrightarrow {}^3He + \gamma$$

$$^3He + {}^3He \longrightarrow {}^4He + 2\,{}^1H$$

where the superscripts indicate the total number of neu-

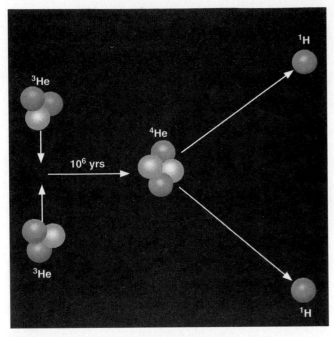

Figure 15.5
The third step in the fusion of hydrogen into helium in the Sun. Note that two of the products of the second step (see Figure 15.4) must combine before the third step becomes possible.

trons and protons in the nucleus, e^+ is the positron, ν is the neutrino, and γ indicates that gamma rays are emitted. Note that steps 1 and 2 must happen twice before step 3 can occur.

Although, as we discussed, the first step in this chain of reactions is very difficult and generally takes a long time, the other steps happen more quickly. After the deuterium nucleus is formed, it survives an average of only about 6 s before being converted to ^{3}He. About a million years after that, the ^{3}He nucleus will combine with another to form ^{4}He.

We can compute the amount of energy these reactions generate by calculating the difference in initial and final mass. The masses of hydrogen and helium atoms in the units normally used by scientists are 1.007825 u and 4.00268 u, respectively. (The unit of mass, u, is defined to be 1/12 the mass of an atom of carbon, or approximately the mass of a proton.) Here we include the mass of the entire atom, not just the nucleus, because electrons are involved as well. When hydrogen is converted to helium, two positrons are created (remember, step 1 happens twice), and these are annihilated with two free electrons, adding to the energy produced.

$$4 \times 1.007825 = 4.03130 \ u \text{ (mass of initial hydrogen}$$
$$\text{atoms)}$$
$$- \ 4.00268 \ u \text{ (mass of final helium atom)}$$
$$\overline{0.02862 \ u} \text{ (mass lost in the}$$
$$\text{transformation)}$$

The mass lost, 0.02862 u, is 0.71 percent of the mass of the initial hydrogen. Thus if 1 kg of hydrogen is converted

into helium, the mass of the helium is only 0.9929 kg, and 0.0071 kg of material is converted into energy. The velocity of light is 3×10^8 m/s, so the energy released by the conversion of just 1 kg of hydrogen to helium is

$$E = 0.0071 \times (3 \times 10^8)^2$$
$$= 6.4 \times 10^{14} \text{ J}$$

This amount, the energy released when a single kilogram of hydrogen undergoes fusion, is more than ten times the Earth's annual consumption of electricity and fossil fuels.

To produce the Sun's luminosity of 4×10^{26} J, some 600 million tons of hydrogen must be converted to helium *each second*, of which about 4 million tons turn from matter into energy. As large as these numbers are, the store of hydrogen (and thus of nuclear energy) in the Sun is still more enormous and can last a *long* time.

At the temperatures inside the Sun and in less-massive stars, most of the energy is produced by the reactions we have just described, and this set of reactions is called the **proton–proton cycle** (or sometimes, the p–p chain). It is called a cycle because the two protons produced in step 3 can fuse with other protons to initiate step 1 again. In the proton–proton cycle, protons collide directly with other protons to build into helium nuclei.

In hotter stars, another set of reactions, called the *carbon–nitrogen–oxygen (CNO) cycle*, accomplishes the same net result. In the CNO cycle, carbon and hydrogen nuclei collide to initiate a series of reactions that form nitrogen, oxygen, and ultimately helium. The nitrogen and oxygen nuclei do not survive but interact to form carbon again. Therefore, the outcome is the same as in the proton–proton cycle: four hydrogen atoms disappear, and in their place a single helium atom is created. The CNO cycle plays only a minor role in Sun but is the main source of energy at temperatures above 15×10^6 K.

Thus we have solved the puzzle that so worried scientists at the end of the 19th century. The Sun can maintain its high temperature and energy output for billions of years through the fusion of the simplest element in the universe, hydrogen. Because most of the Sun (and, as we will see, the other stars) is made of hydrogen, it is an ideal "fuel" for powering a star. As will be discussed in the coming chapters, we can define a star as a ball of gas capable of getting its core hot enough to initiate the fusion of hydrogen. There are balls of gas that lack the mass required to do this (Jupiter is a local example); like so many hopefuls in Hollywood, they will never ever be stars.

The Interior of the Sun: Theory

Fusion of protons will occur in the center of the Sun only if the temperature exceeds 10 million K. How do we know whether the Sun is actually this hot? To determine what the interior of the Sun is like, it is necessary to resort to mathematical calculations. In effect, astronomers teach a computer everything they know about the physical processes going on in the Sun's interior. The computer then calculates the temperature and pressure at every point inside the Sun and determines what nuclear reactions, if any, are taking place. The computer can also calculate how the Sun will change with time.

After all, the Sun *must* change. In its center, the Sun is slowly depleting its supply of hydrogen and creating helium instead. Will this composition change have measurable effects? Will the Sun get hotter? Cooler? Larger? Smaller? Brighter? Fainter? Ultimately, the changes in the center could be catastrophic, since eventually all the hydrogen fuel hot enough for fusion will be exhausted. Either a new source of energy must be found, or the Sun will cease to shine. We will describe the fate of the Sun in Chapters 21 and 22. For now, let's look at some of the things we must teach the computer about the Sun in order to carry out the calculations.

The Sun Is a Gas

The Sun is so hot that the material in it is gaseous throughout. Astronomers are grateful for this because a hot gas is easier to describe mathematically than are other configurations of matter. The particles that constitute a gas are in rapid motion, frequently colliding with one another. This constant bombardment is the *pressure* of the gas (Figure 15.6). More particles within a given volume of gas produce more pressure, because the combined impact of the moving particles increases with their number. The pressure is also greater when the molecules or atoms are moving faster. Since their rate of motion is determined by the temperature of the gas, higher temperatures produce higher pressure.

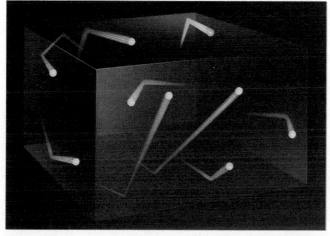

Figure 15.6
Gas pressure. The particles in a gas are in rapid motion and produce pressure through collisions with the surrounding material. Here particles are shown bombarding the sides of an imaginary container.

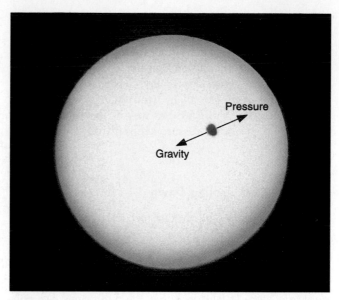

Figure 15.7
Hydrostatic equilibrium. In the interior of a star, the inward force of gravity is exactly balanced at each point by the outward force of gas pressure.

The Sun Is Stable

The Sun, like the majority of other stars, is stable: it is neither expanding nor contracting. Such a star is said to be in a condition of equilibrium. All the forces within it are balanced, so that at each point within the star the temperature, pressure, density, and so on, are maintained at constant values. We will see in Chapters 21 and 22 that even these stable stars, including the Sun, are changing as they evolve, but such evolutionary changes are so gradual that to all intents and purposes the stars are still in a state of equilibrium.

The mutual gravitational attraction between the masses of various regions within the Sun produces tremendous forces that tend to collapse the Sun toward its center. Yet we know from the history of the Earth that the Sun has been emitting approximately the same amount of energy for billions of years, and so clearly has managed to resist collapse for a very long time. The gravitational forces must therefore be counterbalanced by some other force, and that force is the pressure of the gases within the Sun (Figure 15.7). To exert enough pressure to prevent the Sun from collapsing due to the force of gravity, the gases at its center must be maintained at a temperature of 15 million K. So we can conclude that the Sun's temperature is high enough to fuse protons from the fact that it is not contracting.

If the internal pressure in a star were not great enough to balance the weight of its outer parts, the star would collapse somewhat, contracting and building up the pressure inside. If the pressure were greater than the weight of the overlying layers, the star would expand, thus decreasing the internal pressure. Expansion would stop, and equilibrium would be reached, when the pressure at every internal point again equaled the weight of the stellar layers above that point. An analogy is an inflated balloon, which will expand or contract until an equilibrium is reached between the pressure of the air inside and of that outside. This condition is called **hydrostatic equilibrium.** Stable stars are all in hydrostatic equilibrium; so are the oceans of the Earth, as well as the Earth's atmosphere. The air's own pressure keeps it from falling to the ground.

The Sun Is Not Cooling Down

Heat always flows from hotter to cooler regions. Therefore, as energy filters outward toward the surface of a star, it must be flowing from inner, hotter regions. The temperature cannot ordinarily get cooler as we go inward in a star, or energy would flow in and heat up those regions until they were at least as hot as the outer ones. We conclude that the temperature is highest at the center of a star, dropping to successively lower values toward the stellar surface. (The high temperature of the Sun's chromosphere and corona may therefore appear to be a paradox. But remember from Chapter 14 that these high temperatures are believed to be maintained by magnetic heating.)

The outward flow of energy through a star, however, robs it of its internal heat, and would result in a cooling of the interior gases if that energy were not replaced. Similarly, a hot iron begins to cool as soon as it is unplugged from its source of electric energy. Therefore a source of energy must exist within each star. In the Sun's case, we have seen that this energy source is the fusion of hydrogen to form helium.

Heat Transfer in a Star

Since the nuclear reactions that generate the Sun's energy occur deep within it, there must be ways to transport heat from the center of the Sun to its surface. There are three ways in which heat can be transported. In *conduction*, atoms or molecules pass on their energy by colliding with others nearby; this happens when the handle of a metal spoon heats up as you stir a hot cup of coffee. In *convection*, currents of warm material rise, carrying their energy with them to cooler layers; this happens when hot air from a fireplace rises in a room. In *radiation*, photons of energy move away from hot material and are absorbed by some other material to which they convey some or all of their energy; you can feel this when you put your hand close to the coils of an electric heater, allowing infrared photons to heat up your hand. Conduction and convection are both important in the interiors of planets. In stars, which are much more transparent, radiation and convection are important, while conduction can usually be ignored.

Stellar **convection** occurs as currents of hot gas flow up and down through the star (Figure 15.8). Such currents travel at moderate speeds and do not upset the con-

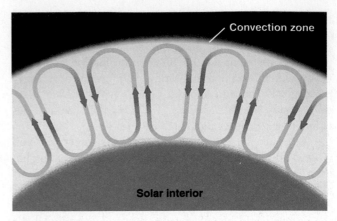

Figure 15.8
Rising convection currents carry heat from the Sun's interior to its surface, while cooler material sinks downward. Of course, nothing in a real star is as simple as diagrams in textbooks suggest.

dition of hydrostatic equilibrium. Nor do they result in a net transfer of mass either inward or outward. Nevertheless, they carry heat very efficiently outward through a star. In the Sun, convection turns out to be important in the central regions and near the surface.

Unless convection occurs, the only significant mode of energy transport through a star is by electromagnetic **radiation.** Radiation is not an efficient means of energy transport in stars, because gases in their interiors are very opaque—that is, a photon does not go far (in the Sun, typically about 0.01 m) before it is absorbed. The absorbed energy is always re-emitted, but it can be re-emitted in any direction. A photon absorbed when traveling outward in a star has almost as good a chance of being reradiated back toward the center of the star as toward its surface.

A particular quantity of energy, therefore, zigzags around in an almost random manner and takes a long time to work its way from the center of the star to the surface (Figure 15.9). As discussed earlier, in the Sun the time required is on the order of a million years. If the photons were not absorbed and re-emitted along the way, they would travel at the speed of light and could reach the surface in a little over 2 s, just as the neutrinos do.

The measure of matter's ability to absorb radiation is called its *opacity*. It should be no surprise that the Sun's gases are opaque. If they were completely transparent, we would be able to see all the way through the Sun. The processes by which atoms and ions can interrupt the flow of energy—such as by becoming ionized—were discussed in Section 4.5. In addition, individual electrons can scatter radiation helter-skelter. For a given temperature, density, and composition of a gas, all of these processes can be taken into account, and the opacity can be calculated. The computations are very complicated and thus require powerful computers, but these are now as much part of the arsenal of astronomers as are powerful telescopes.

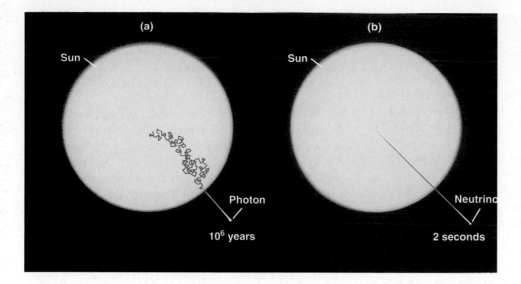

Figure 15.9
(a) Photons generated by fusion reactions in the solar interior travel only a short distance before they are absorbed. The re-emitted photons usually have lower energy and may travel in any direction. As a consequence, it takes about a million years for energy to make its way from the center of the Sun to its surface. (b) In contrast, neutrinos do not interact with matter but traverse the Sun at the speed of light, reaching the surface in only a little more than 2 s.

Model Stars

Scientists use the principles we have described to calculate what the Sun's interior is like. These physical ideas are expressed as mathematical equations that are solved to determine the values of temperature, pressure, density, and other physical quantities throughout the stellar interior. The set of solutions so obtained, based on a specific set of physical assumptions, is called a theoretical model for the interior of the Sun.

Figure 15.10 schematically illustrates the Sun's interior according to the best theoretical model. Energy is generated through fusion in the core of the Sun, which extends only about one-quarter of the way to the surface. This core contains about one-third of the total mass of the Sun, however. At the center, the temperature reaches a maximum of approximately 15 million K, and the density is nearly 150 times the density of water. The energy generated is transported toward the surface by radiation until it reaches a point about 70 percent of the distance from the center to the surface. At this point convection begins, and energy is transported the rest of the way primarily by rising columns of hot gas.

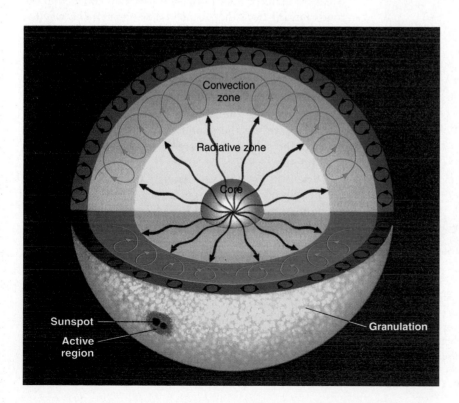

Figure 15.10
The interior structure of the Sun. Energy is generated in the core by the fusion of hydrogen to form helium. This energy is transmitted outward by radiation—that is, by the absorption and re-emission of photons. In the outermost layers, energy is transported mainly by convection.

The Interior of the Sun: Observations

Recall that when we observe the Sun's photosphere, we are not seeing very deeply into our star—certainly not into the regions where energy is generated. At first it seemed there was no way to check on the predictions of our models for the inside of the Sun. Recently, however, astronomers have devised two types of measurements that can be used to study the solar interior directly. One technique involves the analysis of tiny changes in the motion of small regions at the Sun's surface. The other relies on the measurement of the neutrinos emitted by the Sun.

Solar Pulsations

Astronomers have discovered that the Sun pulsates—that is, it alternately expands and contracts—just as your chest expands and contracts as you breathe. This pulsation is very slight, but it can be detected by measuring the *radial velocity* of the solar surface—the speed with which it moves toward or away from us. The velocity of different parts of the Sun is observed to change in a regular way, first toward the Earth, then away, then toward, and so on. Accurate measurements show that various regions on the Sun's surface with diameters ranging from 4000 to 15,000

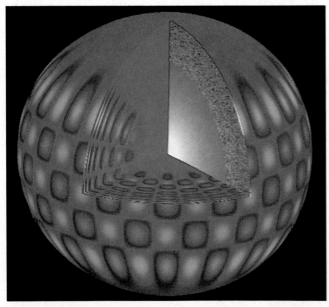

Figure 15.11
New observational techniques permit astronomers to measure small differences in velocity at the Sun's surface to infer what the deep solar interior is like. In this computer simulation, red shows surface regions that are moving away from the observer; blue marks regions moving toward the observer. Note that the velocity changes penetrate deep into the Sun's interior. (National Optical Astronomy Observatories)

km fluctuate back and forth this way. The pulsation cycle takes between 2.5 and 11 min, with the dominant periods lasting about 5 min (Figure 15.11). Rather than resembling a single person breathing, the Sun is more like a huge crowd of people, all breathing in and out at different rates.

We now know that these velocity fluctuations are produced by adding together millions of individual patterns of oscillation. Individual oscillations have velocities as small as 20 cm/s, and the combined sum of the velocities of all the patterns is only a few hundred meters per second. Since it takes only about 5 min to complete a full cycle from maximum to minimum velocity and back again, the change in the size of the Sun measured at any given point is no more than a few kilometers.

The remarkable thing is that these small velocity variations can be used to determine what the interior of the Sun is like. The motion of the Sun's surface is caused by waves that reach it from deep in the interior. Study of the amplitude and period of the velocity changes produced by this motion yields information about the temperature, density, and composition of the layers where the waves were generated. The situation is somewhat analogous to the use of earthquakes to infer the properties of the Earth's interior. For this reason, studies of solar oscillations are referred to as **solar seismology.**

It takes about an hour for waves to traverse the Sun, so they, like neutrinos, provide information about what the solar interior is like at the present time. In contrast, the emerging light that we measure now was generated about a million years ago in the core.

Solar seismology has yielded some important results. Measurements of solar pulsation show that convection extends inward from the surface 30 percent of the way toward the center; we have used this information in drawing Figure 15.10. The observed oscillations also indicate that the abundance of helium inside the Sun, except in the center where nuclear reactions have converted hydrogen to helium, is about the same as at the surface. That result is important to astronomers, because it means we are correct when we use the abundances of the elements measured in the solar atmosphere to construct models of the solar interior.

Scientists are now developing tools to make even better measurements of solar pulsations; these can be used to probe the structure of the Sun close to its center, where nuclear reactions are taking place. To do this, astronomers have set up a network of stations around the Earth to continuously monitor solar oscillations. This is needed because observations from any single station are interrupted by local sunset, but the Sun is always up somewhere in the world. The name given to this worldwide network of telescopes is the Global Oscillations Network Group (GONG) project (Figure 15.12). Observations began in 1995, and the goal is to continue the measurements for another 11 years in order to cover a complete solar cycle.

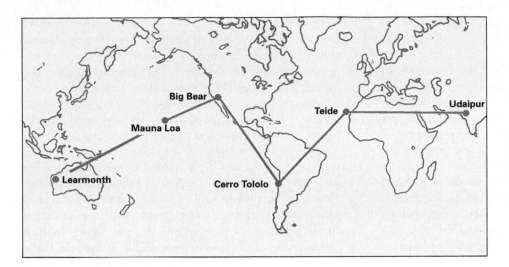

Figure 15.12
A map of the stations for the Global Oscillations Network Group (GONG) around the world. (National Optical Astronomy Observatories)

Solar Neutrinos

The second technique for obtaining information about the Sun's interior involves the detection of a few of those elusive neutrinos created during nuclear fusion. Recall from our earlier discussion that neutrinos very rarely interact with matter, and that most of the neutrinos created in the center of the Sun make their way directly out of the Sun and to the Earth at the speed of light. As far as neutrinos are concerned, the Sun is transparent. Unfortunately for those trying to "catch" some neutrinos, so is the Earth and everything on it.

About 3 percent of the total energy generated by nuclear fusion in the Sun is carried away by neutrinos. So many protons react inside the Sun's core that, scientists calculate, 35 million billion (3.5×10^{16}) solar neutrinos pass through each square meter of the Earth's surface every second. If we can devise a way to detect some of these solar neutrinos, then we can obtain information directly about what is going on in the center of the Sun.

On very, very rare occasions, a neutrino *will* interact with another atom. Several experiments have been devised to detect these interactions. The first of these uses the element chlorine, whose nucleus can be turned into a radioactive argon nucleus (an isotope called argon-37) by an interaction with a neutrino. Because the argon is radioactive, its presence can be picked up with sensitive detectors built around the tank. However, since the interaction happens so rarely, a huge amount of chlorine is needed.

Raymond Davis, Jr., and his colleagues at Brookhaven National Laboratory placed a tank containing nearly 400,000 liters of cleaning fluid (C_2Cl_4) 1.5 km beneath the Earth's surface in a gold mine at Lead, South Dakota (Figure 15.13). A mine was chosen so that the surrounding material of the Earth would keep cosmic rays from reaching the cleaning fluid and creating false signals. Calculations show that solar neutrinos should produce about one atom of argon-37 in the tank each day.

The results of this experiment, running since 1970, are that only about *one-third* as many neutrinos reach the Earth as are predicted by standard models of the solar interior. Astronomers were surprised by this result, since we thought that by now we had a pretty good understanding of both neutrinos and the Sun's interior. This demon-

Figure 15.13
The neutrino experiment operated by Raymond Davis and his colleagues one mile underground in the Homestake Gold Mine in South Dakota. The large vat contains cleaning fluid, a relatively inexpensive source of chlorine nuclei. (Brookhaven National Lab)

strates how important it is not to rest on your laurels in science, but to continue doing new experiments whenever possible.

A second experiment carried out by Japanese astronomers, who looked for the interaction of neutrinos with water, for the first time determined the direction from which the neutrinos are arriving and confirmed that they actually do come from the Sun. That's good news, but, alas, the Japanese also found fewer neutrinos than expected by models of the solar interior.

To change chlorine into argon, you need a neutrino of high energy. Calculations show that fewer than 1 percent of the neutrinos produced in the Sun have energies high enough to be detected by the chlorine, and the water experiment is likewise sensitive only to neutrinos with very high energies. The number of high-energy neutrinos that emerge from the Sun depends on the temperature of its interior. In a cooler core, the reactions that produce them occur less often. Therefore, if we lower the temperatures used in our models of the Sun and change some other details of the complex calculations that go into the models, we might conceivably be able to account for the deficiency of high-energy neutrinos.

Tinkering with the details of the calculations cannot, however, alter the predicted number of low-energy neutrinos. Nearly all of the solar energy, as well as nearly all of the solar neutrinos, are produced in the first step of the proton–proton cycle, when two protons combine to form deuterium (see Section 15.2). Since we know how much total energy is emitted by the Sun, we know very accurately how many times per second two protons combine to form deuterium. We therefore also know how many times per second the associated low-energy neutrinos are produced by this interaction. For a definitive test, scientists need experiments to look for the low-energy neutrinos associated with the proton–proton reaction.

It turns out that these low-energy neutrinos interact with the rare metallic element gallium. In the early 1990s the first results of two gallium experiments, one in Russia and the other in Italy, were reported. Our models predict that solar neutrinos, passing through a container of 60 tons of gallium, will convert about 16 atoms of normal gallium to radioactive gallium during approximately one month of operation. The results of the two gallium experiments agree: the number of low-energy neutrinos is about two-thirds the value predicted by standard solar models.

Is there something wrong with our solar models? Or, as many scientists are beginning to think, is it more likely that we are wrong about the properties of neutrinos? Physicists have shown that three types of neutrinos exist, and standard theory assumes that all three types have no mass. There is actually no laboratory evidence that the mass of a neutrino is exactly zero. If it has even a tiny mass, then it is possible for one type of neutrino to change into another type on its journey from the center of the Sun to the photosphere and on to the Earth. Fusion in the Sun produces only one type of neutrino, the so-called electron neutrino, and chlorine, water, and gallium experiments are sensitive only to this one type. If some electron neutrinos change to another type on their way from the center of the Sun to the Earth, then the experiments performed so far would have missed those neutrinos.

To test this idea, we need a new experiment that can measure all types of neutrinos. If the total number emitted by the Sun is found to be greater than the number of electron neutrinos, then we would have evidence that the mass of the neutrino is not zero. (As mentioned earlier, such a result would have important implications for other areas of astronomy as well.) One experiment of this kind, which involves heavy water (water with the hydrogen atoms replaced by deuterium), is now being built in Canada; the results should be available in the late 1990s.

Until very recently, all of the hints that neutrinos might have mass have come from measurements of solar neutrinos, over which scientists have no control. The solar measurements have inspired attempts to use laboratory accelerators to determine whether or not one type of neutrino can change into another; as we have seen, such a conversion can happen only if neutrinos have mass.

The first experiment to report success was completed at Los Alamos in 1995. This experiment was designed to produce a large quantity of one type of neutrino (technically the muon antineutrino). A detector consisting of 167 tons of baby oil was then set up 30 m away to look for a different type of neutrino (the electron antineutrino) while keeping everything else except neutrinos and antineutrinos from reaching the detector. The electron antineutrino, and no other type, produces a reaction in the baby oil that generates light. The scientists reported seeing seven flashes of light caused by electron antineutrinos. There should have been none if neutrinos have exactly zero mass. Obviously, this experiment has to be repeated, and new ones devised, before we can be sure that this early result is correct.

It will be years before we have final results of the new experiments on solar pulsations and neutrinos. Until then, we will not know which is wrong—some aspect of our solar models, or the assumption that the neutrino has no mass.

While it may seem discouraging that there are questions for which definitive answers are not yet available, science often works this way. Observations lead to the development of a model. This model then suggests a number of other measurements that can be made and predicts the outcome of those measurements. Frequently the predictions are incorrect, and the models must be modified to take into account the new measurements. And so science moves forward by successive approximations, each step providing a better and more complete description of what is actually occurring. Rather than finding the lack of final answers discouraging or frustrating, scientists find such situations exhilarating and challenging. The possibility of learning something heretofore unknown is what attracts many scientists to research.

Fusion on Earth

Wouldn't it be wonderful if we could duplicate the Sun's energy mechanism in a controlled way on Earth? (We have already duplicated it in an *uncontrolled* way in hydrogen bombs, but we hope our storehouse of these will never be used.) Fusion energy would have many advantages: it would use hydrogen (or deuterium, which is heavy hydrogen) as fuel, and there is plenty of hydrogen in the Earth's lakes and oceans. Water is much more evenly distributed around the world than is oil or uranium, meaning that a few countries would no longer hold an energy advantage over the others. And unlike fission, which leaves dangerous by-products, the nuclei that result from fusion are perfectly safe.

The problem is that, as we saw, it takes very, very high temperatures for nuclei to overcome their electrical repulsion and undergo fusion. When the first hydrogen bombs were exploded in tests in the 1950s, the "fuses" to get them hot enough were fission bombs. Interactions at such temperatures are difficult to sustain and control. Thus far, fusion experiments have generally required more energy to start them and keep them under control than is produced by the fusion itself.

Among the techniques now being tried in laboratories in the United States and around the world are keeping the hot gas under the control of powerful magnetic fields in what scientists call a "magnetic bottle," and using highly focused laser beams to compress and heat up pellets of solid deuterium. These experiments are at the forefront of our technology, and scientists estimate that it will take several decades before artificial fusion reactors become practical.

In 1989, two teams of scientists (both in Utah) made an announcement that simply astonished astronomers and physicists familiar with fusion. The two groups claimed to have achieved fusion at room temperatures, with equipment so simple it could easily be duplicated in a high-school science lab. They used an electrochemical cell in which electric current passes from one metal surface to another through a chemical solution. Both groups thought they had found evidence of deuterium fusion in their equipment—and their results, quickly dubbed "cold fusion," became a media sensation.

Alas, cold fusion did not hold up under the intense scrutiny to which all new scientific ideas and procedures are subjected. Other groups around the world could not duplicate the results. Even the original experiments failed to show many other characteristics associated with fusion. It turned out that the groups in Utah were probably seeing chemical reactions, not nuclear ones, and the initial results were discredited.

It appears that if we want to duplicate fusion on Earth we have to do what the Sun does: find a way to produce temperatures and pressures high enough to get hydrogen nuclei on intimate terms with one another. Many teams of scientists are hard at work solving the formidable technological challenges involved, although getting funding is always difficult for projects whose "payoff" is decades away. Still, perhaps by the time your children or grandchildren take an astronomy class in college, fusion will be a reality instead of a dream.

Tokamak Fusion Test Reactor (TFTR). (Princeton University, Plasma Physics Laboratory)

Summary

15.1 The Sun produces an enormous amount of energy every second. The Earth is 4.5 billion years old, so the Sun must have been shining for at least that long. Neither chemical burning nor gravitational contraction can account for the energy radiated by the Sun all this time.

15.2 Solar energy is produced by interactions of **elementary particles**—that is, protons, neutrons, electrons, and **neutrinos.** Specifically, the source of the Sun's energy is the **fusion** of hydrogen to form helium. The series of reactions required to convert hydrogen to helium is called the **proton–proton cycle.** A helium atom is about 0.71 percent less massive than the four hydrogen atoms that combine to form it, and that lost mass is converted to energy.

15.3 Even though we cannot see inside the Sun, it is possible to calculate what its interior must be like. As input for these calculations, we use what we know about the Sun. It is made entirely of hot gas. Apart from some very tiny changes, the Sun is neither expanding nor contracting (it is in **hydrostatic equilibrium**), but instead puts out energy at a constant rate. Fusion of hydrogen occurs in the center of the Sun, and the energy generated is carried to the surface by **radiation** and **convection.** A solar model describes the structure of the Sun's interior. Specifically, it describes how pressure, temperature, mass, and luminosity depend on distance from the center of the Sun.

15.4 Studies of solar oscillations (**solar seismology**) and neutrinos provide observational data about the Sun's interior. The technique of solar seismology has so far shown that the composition of the interior is much like that of the surface (except in the core, where some of the original hydrogen has been converted to helium), and that the convection zone extends 30 percent of the way from the Sun's surface to its center. Our solar models predict that we should be able to detect more neutrinos than we do. This result may indicate that the mass of the neutrino is not exactly zero.

Review Questions

1. How do we know the age of the Sun?

2. Explain how we know that the Sun's energy is not supplied either by chemical burning, as in fires here on Earth, or by gravitational contraction (shrinking).

3. What is the ultimate source of energy that makes the Sun shine?

4. How is a neutrino different from a neutron? List all the ways you can think of.

5. Describe in your own words what is meant by the statement that the Sun is in hydrostatic equilibrium.

6. Two astronomy students travel to South Dakota. One stands on the Earth's surface and enjoys some sunshine. At the same time, the other descends into the gold mine where neutrinos are detected, arriving in time to measure the creation of a new radioactive argon nucleus. Although the photon at the surface and the neutrinos in the mine arrive at the same time, they have had very different histories. Describe the differences.

7. Why do measurements of the number of neutrinos emitted by the Sun tell us about conditions deep in the solar interior?

8. Why have astronomers needed to do more than one experiment to measure neutrinos from the Sun?

Thought Questions

9. A friend who has not had the benefit of an astronomy course suggests that the Sun must be full of burning coal to shine as brightly as it does. List as many arguments as you can against this hypothesis.

10. Which of the following transformations is (are) fusion and which is (are) fission? (See Appendix 13 for a list of the elements.)

    a. helium to carbon
    b. carbon to iron
    c. uranium to lead
    d. boron to carbon
    e. oxygen to neon

11. Why is a higher temperature required to fuse hydrogen to helium by means of the CNO cycle than is required by the process that occurs in the Sun, which involves only isotopes of hydrogen and helium?

12. Explain what it means when we say that the Earth's oceans are in hydrostatic equilibrium. Now suppose you are a scuba diver. Would you expect the pressure to increase or decrease as you dive below the surface to a depth of 200 ft? Why?

13. What mechanism transfers heat away from the surface of the Moon? If the Moon is losing energy in this way, why does it not simply become colder and colder?

14. Suppose you are standing a few feet away from a bonfire on a cold fall evening. Your face begins to feel hot. What is the mechanism that transfers heat from the fire to your face? (*Hint:* Is the air between you and the fire hotter or cooler than your face?)

15. Give some everyday examples of the transport of heat by convection and by radiation.

16. Suppose the proton–proton cycle in the Sun were to slow down suddenly and generate energy at only 95 percent of its current rate. Would an observer on the Earth see an immediate decrease in the Sun's brightness? Would she immediately see a decrease in the number of neutrinos emitted by the Sun?

17. Do you think that nuclear fusion takes place in the atmospheres of stars? Why or why not?

18. Why is fission not an important energy source in the Sun?

19. Suppose a proton interacted with another proton an average of once every 100 million years rather than once every 14 billion years. In general terms, how do you think the structure of the Sun would change? (*Hint:* Think about what determines the pressure in the core of the Sun.)

20. The GONG experiment is designed to monitor the Sun's oscillations on a continuous basis. In order to save money, this experiment was designed to make use of the minimum possible number of telescopes. It turns out that if the sites are selected carefully, the Sun can be observed all but 6 percent of the time with only six observing stations. What factors have to be taken into consideration in selecting the observing sites? Can you suggest six general geographic locations that would optimize the amount of time that the Sun can be observed?

Problems

21. According to the text, a contraction of the Sun by 40 m per year would be enough to account for its current output of energy. Suppose you can measure the Sun's diameter with an accuracy of 1 percent. How long will you have to make measurements in order to determine whether or not the Sun is contracting at this rate?

22. Verify that roughly 600 million tons of hydrogen must be converted to helium in the Sun each second to explain its energy output. (*Hint:* Recall Einstein's most famous formula, and remember that for each kilogram of hydrogen, 0.0071 kg of mass is converted to energy.)

23. If an observed oscillation of the solar surface has a period of 10 min, and the average radial velocity is 1 m/s in and out, calculate the total displacement of the surface involved in this particular oscillation mode. What percent is this of the Sun's total radius?

24. Suppose you brought together a proton and an antiproton. What would happen? How much energy would result? (Take the mass of a proton to be 1.7×10^{-27} kg.)

25. Now suppose you brought together a 100-kg linebacker from the National Football League and an antiproton (ignoring the air around them). What would happen? How much energy would result?

Suggestions for Further Reading

Badash, L. "The Age of the Earth Debate" in *Scientific American,* Aug. 1989, p. 90.

Bahcall, J. "Where are the Solar Neutrinos" in *Astronomy,* Mar. 1990, p. 40.

Davies, P. "Particle Physics for Everybody" in *Sky & Telescope,* Dec. 1987, p. 582.

Fischer, D. "Closing In on the Solar Neutrino Problem" in *Sky & Telescope,* Oct. 1992, p. 378.

Goldsmith, D. *The Astronomers.* 1991, St. Martin's Press. Chapter 6 is on "Why Stars Shine."

Harvey, J. et al. "GONG: To See Inside Our Sun" in *Sky & Telescope,* Nov. 1987, p. 470.

Heppenheimer, T. *Man-made Sun: The Quest for Fusion Power.* 1984, Little Brown. Discusses our attempts to replicate the Sun's energy mechanism on Earth.

Kaler, J. *Stars.* 1992, Scientific American Library/W. H. Freeman. Good modern introduction by an astronomer on how stars work.

LoPresto, J. "Looking Inside the Sun" in *Astronomy,* Mar. 1989, p. 20. On helioseismology.

Rousseau, D. "Case Studies in Pathological Science" in *American Scientist,* Jan/Feb. 1992, p. 54. Explores the story of "cold fusion."

Sutton, C. *Spaceship Neutrino.* 1992, Cambridge U. Press. Definitive book on neutrinos.

Stars differ from one another in their luminosities and masses. This time-exposure shows stars of many different luminosities near the north celestial pole, turning with the sky as the Earth rotates under them; they appear to be circling the McMath Telescope on Kitt Peak near Tucson. (William Livingston/National Optical Astronomy Observatories)

CHAPTER 16

Analyzing Starlight

Thinking Ahead

The nearest star (other than the Sun) is so far away that the fastest spacecraft the human race has built so far would take almost 100,000 years to get there. And all the other stars are even farther away. Yet we are a curious and ambitious species, and we want to know how those stars are born, live out their lives, and die. How can we learn about objects that are so remote and beyond our physical grasp?

Twinkle, twinkle little star,
I don't wonder what you are,
For by spectroscopic ken
I know that you are
hydrogen.

Anonymous, quoted by D. Bush in *Science and English Poetry* (1950).

Scientists love categories. Our scientific understanding grows when we find the right pigeonholes in which to place the objects we are studying, whether they are butterflies, human diseases, or subatomic particles. In the same way, the stars can be classified according to their properties—how big they are, how much material they contain (their mass), and how much light they give off. As we learn more about distinguishing the stars (in this and the next chapters), we will use these categories to begin assembling the pieces of the main puzzle we are interested in solving—their life stories.

We begin our voyage to the stars by describing how astronomers use the light emitted by stars to deduce what they are like. For example, we will discover that stars come in a range of sizes and masses. There are stars so large that the orbits of many of the planets in our solar system would easily fit inside them. We will see strange dying stars called white dwarfs that have been compressed until they are a thousand times denser than water. Figuring out how all the different kinds of stars fit into the big picture has been one of the most exciting and productive areas of astronomy in the last century.

Figure 16.1

A time-exposure showing the constellation Orion as it rises over Kitt Peak National Observatory. To make the photo more interesting, the photographer began with a short exposure, giving the dots at the right. Then the camera shutter was left open for a long exposure, during which the Earth's turning motion drew out the light of each star into a long line. The colors of the various stars are caused by their different temperatures. (National Optical Astronomy Observatories)

Very simple observations are enough to show that not all stars are alike. You can make some of these observations yourself, especially if you get away from city lights or use a pair of binoculars. Take a good look at the night sky. Not all the stars are the same brightness or color. Although most are too dim to excite our color vision, a few brighter stars do show distinct colors to our eyes. Some appear to be blue-white; others are obviously red. In the winter sky, a good constellation for seeing star colors is Orion, the hunter (Figure 16.1).

In Chapter 4 we discussed that in a glowing gas, color is a good indication of temperature, with red being the coolest, and blue and violet being the hottest. Yellow, the color of our Sun, is toward the middle of the visible spectrum. Since stars are giant balls of glowing gas, we can deduce that those that look blue-white are hotter than the Sun, while red ones are cooler.

Other questions about stars can be answered only by means of careful observations with large telescopes. Since the stars are suns, it should not surprise you that the same techniques, including spectroscopy, used to study the Sun can also be used to find out what stars are like.

16.1

The Brightness of Stars

Perhaps the most important characteristic of a star is its **luminosity**—the total amount of energy that it emits per second. In Chapter 15 we saw that the Sun (pictured as a big lightbulb) puts out a tremendous amount of radiation every second. (And there are stars far more luminous than the Sun out there.) Later, we will see that if we can measure how much energy a star emits, and if we also know its mass, we can calculate how long it can continue to shine before exhausting its nuclear energy and beginning to die.

Note that luminosity is how much energy a star gives off each second, *not* how much energy reaches our eyes or telescopes on Earth. Stars are very democratic in the way they emit energy: the same amount goes in every direction in space. So only a minuscule fraction of the energy given off by a star actually reaches an observer on Earth. We call the amount of a star's energy that reaches us here its **apparent brightness.** If you look at the night sky, you see a wide range of apparent brightnesses among the stars. Most stars, in fact, are so dim that you need a telescope to detect them.

If all stars were the same luminosity—if they were *standard bulbs*—we could use the difference in their apparent brightness to tell us something we very much want to know: how far away they are. Imagine you are in a big concert hall or ballroom that is dark except for a few dozen 25-watt (W) bulbs placed in fixtures around the walls. Since they are all 25-W bulbs, their luminosity is the same. But from where you are standing in one corner, they do *not* have the same apparent brightness. Those close to you look brighter (more of their light reaches your eye), while those far away look dimmer (their light has spread out more before reaching you). In the same way, if all the stars had the same luminosity, we could immediately say that the bright-looking stars were close-by and the dim-looking ones were far away.

To pin this down more precisely, recall from Chapter 4 that we know the way in which light gets dimmer with distance. The energy we receive is inversely proportional to the square of the distance. If, for example, we have two stars of the same brightness and one is twice as far away as the other, it will look four times dimmer than the closer one. If it is three times further away, it will look nine (three squared) times dimmer, and so forth.

Alas, the stars do not have the courtesy to come in one standard luminosity. (Actually, we are pretty glad about that, because having many different types of stars makes the universe a much more interesting place.) But this means that if a star looks dim in the sky, we cannot tell whether it appears dim because it has a low luminosity but is relatively close-by, or because it has a high luminosity but is very far away.

(By the way, although stars in general did not turn out to be "standard bulbs," astronomers have not given up on the idea of finding some types of objects in the cosmos that all come with the same luminosity "built-in." We will return to our quest for standard bulbs in future chapters as our voyages take us into more distant regions of the universe.)

Because the stars are not standard bulbs, we cannot just read off their luminosities from their apparent brightnesses. We must first compensate for the dimming effects of distance on light, and to do that we must know how far away they are. Distance is among the most difficult of all astronomical measurements; we will return to how it is determined in Chapter 18, after we have learned more about stars. In this chapter, we now turn to the problem of measuring apparent brightness.

The Magnitude Scale

The branch of observational astronomy that deals with the measurement of stellar brightness and luminosity is called photometry (from the Greek *photo*, "light," and -*metry*, "to measure"). As we saw in Chapter 1, astronomical photometry began with Hipparchus. Around 150 B.C. he erected an observatory on the island of Rhodes in the Mediterranean. There he prepared a catalog of nearly 1000 stars that included not only their positions but also estimates of their apparent brightness.

Hipparchus did not have photographic plates or any instruments that could measure brightness accurately, so he simply made estimates with his eye. He sorted the stars into six brightness categories, which he called **magnitudes.** He referred to the brightest stars in his catalog as first-magnitude stars, while those so faint he could barely see them were sixth-magnitude stars.

During the 19th century, astronomers attempted to make the scale more precise by establishing exactly how much the brightness of a sixth-magnitude star differs from that of a first-magnitude star; measurements showed that we receive about 100 times more light from a star of the first magnitude than from one of the sixth. Based on this measurement, astronomers then defined an accurate magnitude system in which a difference of five magnitudes corresponds exactly to a brightness ratio of 100:1.

This means stars that differ by about one magnitude differ in brightness by a factor of about 2.5—or, to be more precise, 2.512, which is the fifth root of 100. Thus a fifth-magnitude star gives us 2.512 times as much light as one of the sixth magnitude, a fourth-magnitude star gives 2.512 times as much light as a fifth-magnitude star, and so forth.

Today, many astronomers (and astronomy students) wish we had a less old-fashioned and cumbersome system, and many researchers have stopped using magnitudes. But tradition plays a strong role in astronomy, as it does in numerous other areas of human endeavor. Magnitudes are still used on star charts and by amateur astronomers, and news stories about astronomy may well include references to magnitudes. So even though your authors are sorely tempted to stop teaching the concept of magnitudes (and maybe start a trend), we still include it since it is commonly used.

The brightest stars, those that were traditionally referred to as first-magnitude stars, actually turned out (when measured accurately) not to be identical in brightness. For example, the brightest star in the sky, Sirius, sends us about ten times as much light as the average first-magnitude star. On the modern magnitude scale, Sirius has been assigned a magnitude of −1.5. Several of the planets appear even brighter. Venus, at its brightest, is of magnitude −4.4, while the Sun has a magnitude of −26.2.

Figure 16.2 shows the range of observed magnitudes from the brightest to the faintest, along with the actual magnitudes of several well-known objects. The important fact to remember when using magnitudes is that the system goes backwards: the *larger* the magnitude, the *fainter* the object you are observing!

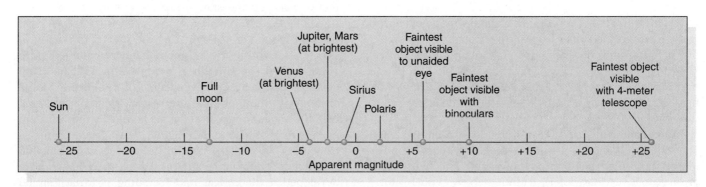

Figure 16.2
The apparent magnitudes of several well-known objects. The faintest magnitudes that can be detected by the eye, binoculars, and a large telescope are also shown.

Other Units of Brightness

Although the magnitude scale is still used for visual astronomy, it is not used at all in newer branches of the field. In radio astronomy, for example, no equivalent of the magnitude system has been defined. Rather, the radio astronomer measures the amount of energy being collected by the radio telescope, and expresses the brightness of each source in terms of the energy reaching the surface of the Earth in, for example, watts per square meter.

Similarly, most researchers in the fields of infrared, x-ray, and gamma-ray astronomy use energy units rather than magnitudes to express the results of their measurements. Nevertheless, astronomers in all fields are careful to distinguish between the total *luminosity* of the source (even when that luminosity is all in x rays) and the amount of energy that happens to reach us on Earth. After all, the luminosity is a really important characteristic that tells us a lot about the object in question, whereas the energy that reaches the Earth is an accident of cosmic geography.

In order to make the comparison among stars easy, in this text we express the luminosity of other stars in terms of the Sun's luminosity. For example, the luminosity of Sirius is 23 times that of the Sun. We use the symbol L_{Sun} to denote the Sun's luminosity; hence that of Sirius can be written as 23 L_{Sun}.

16.2

Colors of Stars

Stars are not all the same color because they are not all the same temperature. As we have seen, a hotter star has the blue colors dominating its light output, while a cool star emits most of its light energy at red wavelenths. To make such a discussion more precise, astronomers have devised quantitative methods for characterizing the color of a star and then using those colors to determine stellar temperatures.

Color Indices

In order to find the exact color of a star, astronomers normally measure the stellar brightness through filters, each of which transmits only the light from a particular narrow band of wavelengths (colors). A crude example of a filter in everyday life is a piece of red cellophane, which, when held in front of your eyes, only lets the red colors of light through.

One commonly used set of filters in astronomy measures stellar brightness at three wavelengths corresponding to ultraviolet, blue, and yellow light. The filters are given names: U (ultraviolet), B (blue), and V (visual—for yellow). These filters transmit light near the wavelengths of 360 nm, 420 nm, and 540 nm respectively. The bright-

ness measured through each filter is usually expressed in magnitudes. The difference between any two of these magnitudes—say, between the blue and the visual magnitudes (B − V)—is called a **color index.**

If we could somehow take a star, observe it, and then move it much farther away, its apparent brightness (magnitude) would change. But since its magnitude would change by the same amount for all colors, its color index (which is two magnitudes subtracted from each other) would remain the same. This idea should make sense to you. After all, you know from everyday experience that the color of an object appears the same no matter how far away it is. Color indices of stars, therefore, provide measures of their intrinsic or true colors, and do not depend on where the observer happens to be.

The way in which color is related to the *temperatures* of stars is given by Wien's law (see Chapter 4). Hot stars emit more energy in the blue and ultraviolet part of the spectrum. Cool stars are brighter at red and infrared wavelengths. By agreement among astronomers, the ultraviolet, blue, and visual magnitudes of the UBV system are adjusted to give a color index of zero to a star with a surface temperature of about 10,000 K.

The B − V color indices of stars range from −0.4 for the bluest to more than +2.0 for the reddest. The corresponding range in surface temperature is from about 50,000 K to 2000 K. Our Sun's surface temperature is about 6000 K; its dominant color is a slightly greenish yellow; and its color index is about 0.62. It looks more yellow seen from the Earth's surface because our planet's air molecules scatter some of the green light out of the beams of sunlight that reach us, leaving more yellow behind.

16.3

The Spectra of Stars

Measuring colors is only one way of analyzing starlight. Instead of filters we can use a spectrograph to spread out the light into a spectrum (see Chapters 4 and 5). By examining that spectrum, astronomers can determine what elements make up the stars—just as they did for the Sun (see Chapter 14). As early as 1823, the German physicist Joseph Fraunhofer observed that the spectra of stars have dark lines crossing a continuous band of colors (Figure 16.3). In 1864 an English astronomer, Sir William Huggins, succeeded in identifying some of the lines in stellar spectra with those of known terrestrial elements, showing that the same chemical elements found in the Sun and planets exist in the stars (Figure 16.4). Since then, astronomers have worked hard to perfect experimental techniques for obtaining and measuring spectra, and have developed a theoretical understanding of what can be learned from spectra. Today, spectroscopic analysis is one of the cornerstones of astronomical research.

Formation of Stellar Spectra

When the spectra of different stars were first observed, astronomers found that they were not all identical. Since the dark lines are produced by the chemical elements present in the stars, astronomers first thought that the spectra differ from one another because stars are not all made of the same chemical elements. This hypothesis turned out to be wrong. *The primary reason that not all stellar spectra look alike is that stars have different temperatures.* Most stars have nearly the same composition as the Sun, with only a few exceptions.

Hydrogen, for example, is by far the most abundant element in all stars (except those at advanced stages of evolution; see Chapter 22), but lines of hydrogen are not seen in the spectra of some types of stars. In the atmospheres of the hottest stars, for example, hydrogen atoms are completely ionized. Because the electron and the proton are separated, ionized hydrogen cannot produce absorption lines.

In the atmospheres of the coolest stars, hydrogen atoms have their electrons attached and can switch energy levels to produce lines. However, practically all of the hydrogen atoms are in the lowest energy state (unexcited) in these stars, and thus can absorb only those photons able to lift an electron from that first energy level to a higher level. The photons absorbed in this way produce a series of absorption lines that lie in the ultraviolet part of the spectrum and hence cannot be studied from the Earth's surface. What this means is that if you observe the spectrum of a very hot or very cool star with a typical telescope on the surface of the Earth, the most common element in that star will not show any lines.

The hydrogen lines in the visible part of the spectrum (the Balmer lines; see Chapter 4) are strongest in stars with intermediate temperatures—not too hot and not too cool. Calculations show that the optimum temperature for producing visible hydrogen lines is about 10,000 K. At this temperature, an appreciable number of hydrogen atoms are excited to the second energy level. They can then absorb additional photons, rise to still higher levels of excitation, and produce a dark absorption line. They are less conspicuous in the spectra of both hotter and cooler stars, even though hydrogen is just about equally abundant in all the stars. Similarly, every other chemical element, in each of its possible stages of ionization, has a characteristic temperature at which it is most effective in producing absorption lines in any particular part of the spectrum.

Classification of Stellar Spectra

Because a star's temperature determines which absorption lines are present in its spectrum, we can use the spectrum to measure its surface temperature. Astronomers have identified seven principal patterns of lines organized by temperature; these are called **spectral classes.** From hottest to coldest, the seven main spectral classes are designated O, B, A, F, G, K, and M.

At this point you are probably looking at these letters with amazement and asking yourself why astronomers didn't call the spectral types A, B, C, and so on. It's a long story (partly discussed in a moment) in which tradition won out over common sense. To help them remember this crazy order of letters, generations of astronomy students have used the mnemonic "Oh be a fine girl, kiss me." (Today, when many astronomy students are women, you can easily substitute "guy" for "girl.") Other mnemonics, which we hope will not be relevant for you, include "Oh brother, astronomers frequently give killer midterms" and "Oh boy, an F grade kills me!"

Each of these spectral classes is further subdivided into ten subclasses designated by numbers. A B0 star is the hottest type of B star; a B9 star is the coolest type of B star and is only slightly hotter than an A0 star. And just one more item of vocabulary: for historical reasons, astronomers call *all* elements heavier than helium "metals," even though most of these elements do not show metallic properties. (If you are getting annoyed at the strange jargon astronomers use, just bear in mind that every field of human knowledge develops its own peculiar vocabulary.

Figure 16.4
William Huggins (1824–1910) was the first to identify the lines in the spectrum of a star other than the Sun; he also took the first spectrogram, or photograph of a stellar spectrum. (Mary Lea Shane Archives of the Lick Observatory)

TABLE 16.1
Spectral Classes for Stars

Spectral Color Class	Approximate Temperature (K)	Principal Features	Examples
O Violet	>28,000	Relatively few absorption lines. Lines of doubly ionized nitrogen, triply ionized silicon, and lines of other highly ionized atoms.	10 Lacertae
B Blue	10,000–28,000	Lines of neutral helium, singly and doubly ionized silicon, singly ionized oxygen, and magnesium. Hydrogen lines more pronounced than in O-type stars.	Rigel Spica
A Blue	7500–10,000	Strong lines of hydrogen. Lines of singly ionized magnesium, silicon, iron, titanium, calcium, and others. Lines of some neutral metals show weakly.	Sirius Vega
F Blue to white	6000–7500	Hydrogen lines weaker than in A-type stars, but still conspicuous. Lines of singly ionized calcium, iron and chromium, plus lines of neutral iron and chromium are present, as are lines of other neutral metals.	Canopus Procyon
G White to yellow	5000–6000	Lines of ionized calcium are most conspicuous spectral features. Many lines of ionized and neutral metals are present. Hydrogen lines are weaker than in F-type stars. Bands of the molecule CH are strong.	Sun Capella
K Orange to red	3500–5000	Lines of neutral metals predominate. The CH bands are still present.	Arcturus Aldebaran
M Red	<3500	Strong lines of neutral metals and molecular bands of titanium oxide dominate.	Betelgeuse Antares

Try reading a legal document these days without training in the law!)

As Figure 16.5 shows, in the hottest O stars (those with temperatures over 28,000 K), only lines of ionized helium and highly ionized atoms of other elements are conspicuous. Hydrogen lines are strongest in A stars with atmospheric temperatures of about 10,000 K. Ionized metals provide the most conspicuous lines in stars with temperatures from 6000 to 7500 K (spectral type F). In the coolest M stars (below 3500 K), absorption bands of titanium oxide and other molecules are very strong. The sequence of spectral classes is summarized in Table 16.1. The spectral class assigned to the Sun is G2.

To see how spectral classification works, let's use Figure 16.5. Suppose you have a spectrum in which the hy-

drogen lines are about half as strong as those viewed in an A star. Looking at the red lines in our figure, you see that the star could be either a B star or a G star. But if the spectrum also contains helium lines, then it is a B star, whereas if it contains lines of ionized iron and other metals, it must be a G star.

Much of the pioneering work on the classification of stellar spectra was carried out at Harvard University in the first decades of the 20th century. The basis for these studies was a monumental collection of nearly a million photographic spectra of stars, obtained from many years of observations made at Harvard College Observatory in Massachusetts, as well as at its remote observing stations in South America and South Africa.

Working with this data base, Annie Cannon (see

Figure 16.5
A graph showing the strength of absorption lines of different elements as we move from hot (left) to cool (right) stars along the sequence of spectral types.

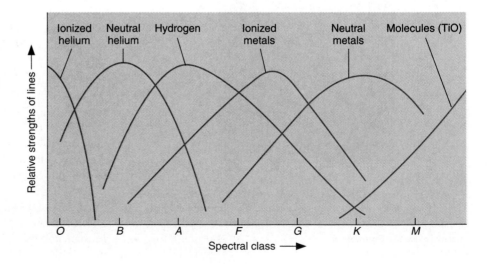

Annie Cannon: Classifier of the Stars

Annie Jump Cannon was born in Delaware in 1863. In 1880 she went to Wellesley College, one of the new breed of U.S. colleges opening up to educate young women. Only five years old at the time, Wellesley had the second student physics lab in the country and provided excellent training in basic science. After college Cannon spent a decade with her parents but was very dissatisfied, longing to do scientific work. With her mother's death in 1893, she was able to return to Wellesley as a teaching assistant, and also to take courses at Radcliffe, the women's college associated with Harvard.

This was a time when the director of the Harvard Observatory, Edward C. Pickering, needed lots of help with his ambitious program of classifying stellar spectra. Pickering quickly discovered that educated young women could be hired as assistants for one-third or one-fourth the salary paid to men, and would often put up with working conditions and repetitive tasks that men with the same education would not tolerate. (We should emphasize that astronomers were not alone in reaching such conclusions about the relatively new idea of women working outside the home: women were exploited and underestimated in many fields. This is a legacy from which our society is just beginning to emerge.)

Cannon wanted to do astronomy, and she was quickly hired by Pickering to help with the classification of spectra. After a while she became so good at it that she could visually ex-amine and determine the spectral types of several hundred stars per hour (dictating her conclusions to an assistant). She made many discoveries while investigating the Harvard photographic plates, including 300 variable stars (stars whose luminosity changes; see Chapter 18). But her main legacy is a marvelous catalog of spectral types for hundreds of thousands of stars, which serves as a foundation for much of 20th-century astronomy.

In 1911 a visiting committee of astronomers reported that "she is the one person in the world who can do this work quickly and accurately," and urged Harvard to give Cannon an official appointment in keeping with her skill and renown. Not until 1938, however, did the university appoint her an astronomer at Harvard; she was then 75 years old.

Cannon received the first honorary degree from Oxford awarded to a woman, and became the first woman to be elected an officer of the American Astronomical Society, the main professional organization of astronomers in this country. She generously donated the money from one of the major prizes she had won to found a special award for women in astronomy, now known as the Annie Jump Cannon Prize. True to form, she continued classifying stellar spectra almost to the very end of her life in 1941.

Annie Jump Cannon
(1863–1941).
(©1983 Astronomical Society of the Pacific)

"Voyagers in Astronomy" box) personally measured some 400,000 stars and assigned them to spectral classes. Because the Harvard workers did not know that temperature is the primary factor in determining the appearance of a stellar spectrum, they at first simply named the spectral classes with letters of the alphabet according to the complexity of the spectra, with type A being the simplest. It was Annie Cannon who initially glimpsed a better classification system, and focused on just a few letters from the original system. Now that we understand that spectral classes depend on the temperature of the stars, we always list the types not in order of how many lines are seen, but rather from hottest to coolest: O, B, A, F, G, K, and M.

During the 20th century, the emphasis has shifted from classification to interpretation of stellar spectra. One of the most important early contributions was the work of Cecilia Payne, then a doctoral student at Harvard. Payne's work, more than any other, proved that all stars really are composed principally of hydrogen, with the differences in their spectra resulting primarily from their different temperatures.

Effects of the Size of a Star

As we will see in Chapter 17, stars come in a wide variety of sizes. At some periods in their lives stars can expand to enormous dimensions, with correspondingly low densities and pressures in their atmospheres. Stars of such exaggerated size are called **giants.** Luckily for the astronomer who wishes to distinguish them from run-of-the-mill stars (such as our Sun), giants can be identified from a study of their spectra.

The pressure in a stellar photosphere affects its spectrum. At the very low densities that occur in the extended, tenuous photospheres of these giant stars, the pressures are also low. Ionized atoms and electrons pass close enough together to recombine much less often than they do in stars with higher pressures. Think of automobile traffic. Collisions are much more likely during rush hour, when the density of cars is high. Low-density gases, therefore, maintain a higher average degree of ionization than do high-density gases of the same temperature. Careful study of their spectra can tell which of two stars at the same temperature has higher pressure (and is thus more compressed), and which has lower pressure (and thus must be extended.)

Abundances of the Elements

Dark lines of a majority of the known chemical elements have now been identified in the spectra of the Sun and stars. Identification of the spectral lines of one of the chemical elements, in the neutral state or in one of its ionized states, immediately tells us that the star must contain that substance.

Of course, astronomy textbooks such as ours always make these things sound a bit easier than they really are. First of all, it has taken many years of careful laboratory work on Earth to determine at what precise wavelengths heated gases of each element have their spectral lines. Long books and computer listings have been compiled to show the lines that can be seen at each temperature. Second, stellar spectra often show many lines from a number of elements, and we have to be careful to sort them out correctly and beware of overlaps between lines of different elements (Figure 16.6). And third, as we saw in Chapter 4, the motion of the star changes the location of each of the lines, and therefore needs to be factored into our analysis. So, in practice, doing spectroscopy is a demanding, sometimes frustrating task that requires both training and skill. The situation has been helped considerably in recent years by automation and computers, but astronomers still need to know what they are doing when analyzing faint spectra from distant objects.

Note that the *absence* of an element's spectral lines does not necessarily mean that the element itself is absent. As we saw, the temperature and pressure in a star's atmosphere will determine what types of atoms are able to produce absorption lines. Only if the physical conditions in a star's photosphere are such that lines of an element *should* (according to calculations) be visible, can we conclude that the absence of observable spectral lines implies low abundance of the element.

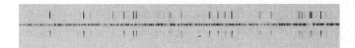

Figure 16.6
Spectrum of a variable star, showing a large number of absorption lines. The star's spectrum is the dark band crossed by white lines in the center; above and below it are some comparison lines (in this case for an iron arc near the telescope) to help astronomers measure at what wavelength each line of the star's spectrum falls. (Image courtesy of Roger Bell, University of Maryland)

Once due allowance has been made for the temperature and pressure in a star's photosphere, analyses of the amount of light subtracted from its spectrum by the absorption lines can yield information regarding the relative abundances of the various chemical elements whose lines appear. It is these kinds of detailed analyses that have shown that the relative abundances of the different chemical elements in the Sun (see Table 14.2) and in most stars are approximately the same.

Hydrogen makes up about three-quarters of the mass of most stars. Hydrogen and helium together make up from 96 to 99 percent of the mass; in some stars they amount to more than 99.9 percent. Among the 4 percent or less of "heavy elements," neon, oxygen, nitrogen, carbon, magnesium, argon, silicon, sulfur, iron, and chlorine are among the most abundant. Generally, but not invariably, the elements of lower atomic weight are more abundant than those of higher atomic weight.

Take a careful look at this list of elements. Among those contained by stars in greatest abundance are hydrogen and oxygen (which make up water); add carbon and nitrogen and you are starting to write the prescription for the chemistry of an astronomy student. We are made of elements that are common in the universe—just mixed together in a far more sophisticated form (and a much cooler environment) than is a star.

The spectrum of the Sun has been studied more thoroughly than that of any other star, and we know most about the abundances of the elements there. Appendix 13 lists how common each element is in the universe (compared to hydrogen); these estimates are based primarily on our investigation of the Sun. Some very rare elements, however, have not been detected in the Sun. Our estimate of the amounts of these elements in the universe is based on laboratory measurements of their abundance in primitive meteorites, which are considered representative of unaltered material condensed from the solar nebula (see Chapter 13).

Radial Velocity

When we measure the spectrum of a star, we determine the *wavelengths* of each of its lines. If the star is not moving with respect to the Sun, then the wavelengths corresponding to each element will be the same as those we measure in a laboratory here on Earth. But if stars are moving toward or away from us, we must consider the Doppler effect (see Section 4.6). We should see the spectral lines of moving stars shifted either toward the red end of the spectrum, if the star is moving away from us, or toward the blue (violet) end if it is moving toward us. Such motion, along an imaginary line connecting the star and the observer, is called **radial velocity** and is usually measured in kilometers per second.

William Huggins, pioneering yet again, made the first radial velocity determination of a star in 1868. He observed the Doppler shift in one of the hydrogen lines in the spectrum of Sirius and found that this star is approaching the solar system. Today, the radial velocity can be measured for any star bright enough for its spectrum to be observed. As we will see in the next chapter, radial velocity measurements of double stars are crucial in deriving stellar masses.

Proper Motion

For the sake of completeness, we should note that stars do not necessarily move directly toward or away from us. They can also move perpendicular to our line of sight; such motion is referred to as *proper motion*. If stars have some motion in this perpendicular direction, we see it as a change in the relative positions of the stars on the "dome" of the sky. However, the proper motion of a star is almost always too small to measure with much precision, even with a telescope, in a single year. For that reason we do not notice any change in the positions of the bright stars during the course of a human lifetime.

If we could live long enough, however, the changes would become obvious. For example, some 50,000 years or so from now, terrestrial observers (if humans haven't wiped themselves out by then) will find the handle of the Big Dipper unmistakably more bent than it is now (Figure 16.7).

The star with the largest proper motion is Barnard's star, our second closest neighbor, named after the American astronomer who first measured its movement. Its position in the sky changes by the width of the Moon in just two centuries (Figure 16.8). Unfortunately, Barnard's star is too faint to be seen with the unaided eye.

We measure proper motion in terms of how far the star moves in the sky in arcseconds per year. In order to convert this angular motion to a velocity, we need to know how far away the star is. If two stars are moving at the same velocity perpendicular to our line of sight, the closer one will appear to move farther across the sky in a year's time. As an analogy, imagine you are standing at the edge of a freeway. Cars will appear to whiz past you. If you watch the cars from a vantage point half a mile away, they will move much more slowly across your field of vision.

In order to know the true velocity of a star in space—that is, its total speed and the direction in which it is mov-

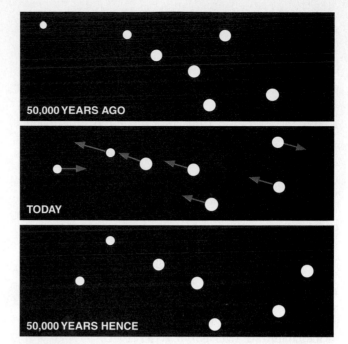

Figure 16.7
The change in the appearance of the Big Dipper due to proper motion of the stars over 100,000 years.

ing—we must know its radial velocity, proper motion, and distance.

Rotation

We can also use the Doppler effect to measure how fast a star rotates. If an object is rotating, then (unless its axis of rotation happens to be pointed exactly toward us) one of its sides is approaching us while the other is receding. This is clearly the case for the Sun or a planet; we can observe the light from either the approaching or receding edge of these nearby objects, and measure directly the Doppler shifts that arise from the rotation.

Stars, however, are so far away that they all appear as unresolved points. The best we can do is analyze the light from an entire stellar disk at once. Even so, the part of the light, including the spectral lines, that comes from the side of the star rotating towards us is shifted to shorter wavelengths, and the part from the opposite edge of the

Figure 16.8
Two photographs of Barnard's star, showing its proper motion over a period of 22 years.
(Yerkes Observatory)

Astronomy and Philanthropy

Throughout the history of astronomy, contributions from wealthy patrons of the science have made an enormous difference for building new instruments and carrying out long research projects. Edward Pickering's stellar classification project, which was to stretch over several decades, was made possible by major donations from Anna Draper, the widow of Henry Draper, a physician who was one of the most accomplished amateur astronomers of the 19th century and the first man to successfully photograph the spectrum of a star. Anna Draper gave several hundred thousand dollars (a lot more money then than today!) to Harvard Observatory; as a result, the great spectroscopic survey is still known as the Henry Draper Memorial, and many stars are still referred to by their "HD" numbers in that catalog.

In the 1870s the eccentric piano-builder and real estate magnate James Lick decided to leave some of his fortune to build the world's largest telescope. (This was actually his second plan for a memorial; fortunately he had been talked out of his first plan, which was to build the world's largest pyramid in downtown San Francisco as a monument to himself.) When, in 1887, the pier to house the telescope was finished, Lick's body was entombed in it. Atop the

James Lick
(Mary Lea Shane Archives
of the Lick Observatory)

foundation rose a 36-in. refractor, for many years the main instrument at the Lick Observatory near San Jose.

As we saw in Chapter 5, the Lick telescope remained the largest in the world until 1897, when George Ellery Hale persuaded railroad millionaire Charles Yerkes to build a 40-in. telescope near Chicago.

More recently Howard Keck, whose family made their fortune in the oil business, gave $70 million dollars from his family foundation to the California Institute of Technology to help build the world's largest telescope atop the 14,000-ft peak of Mauna Kea in Hawaii (see Chapter 5). The Keck Foundation was so pleased with what is now called the Keck telescope that they gave $74 million more to build "Keck II," another 10-m reflector on the same volcanic peak.

Now if any of you reading this book become millionaires and billionaires, and astronomy has sparked your interest, do keep an astronomical instrument or project in mind as you make out your will. But frankly, private philanthropy could not possibly support the full enterprise of scientific research in astronomy. Much of our exploration of the universe is financed by such federal agencies as the National Science Foundation and NASA. In this way, all of us, through a very small share of our tax dollars, get to be philanthropists for astronomy.

---

star is shifted to longer wavelengths. You can think of each spectral line that we observe as a sum or composite of spectral lines originating from different parts of the star's disk, all of which are moving at different speeds with respect to us. Each point has its own Doppler shift, so the absorption line we see is actually much wider in the spectrum than if the star were not rotating. If a star is rotating rapidly, all its spectral lines should be quite broad. In fact, astronomers call this effect line-broadening, and the amount of broadening can tell us the speed at which the star rotates (Figure 16.9).

Measurements of the widths of spectral lines show that many stars hotter than the Sun rotate in periods of only a day or two. The Sun, with its rotation period of about a month, rotates rather slowly. The rotation of most stars cooler than the Sun is slower still, and often cannot be measured with our present techniques.

Summary of Stellar Spectra

In 1835 the French philosopher Auguste Comte wrote a paper in which he asserted that it would never be possible by any means to study the chemical composition of stars. Yet within a few decades the development of astronomical spectroscopy, together with a rapidly maturing understanding of the ways atoms absorb and emit radiation, provided the tools to accomplish this "impossible" task.

In this chapter we have seen that spectrum analysis is an extremely powerful technique that allows the astronomer to learn all kinds of things about a star: its detailed chemical composition, as well as the temperature and pressure in its atmosphere. From the pressure, we get clues about its size. We can also measure its radial velocity and estimate its rotation. It is no wonder that astronomers spend much of their time obtaining and analyzing spectra.

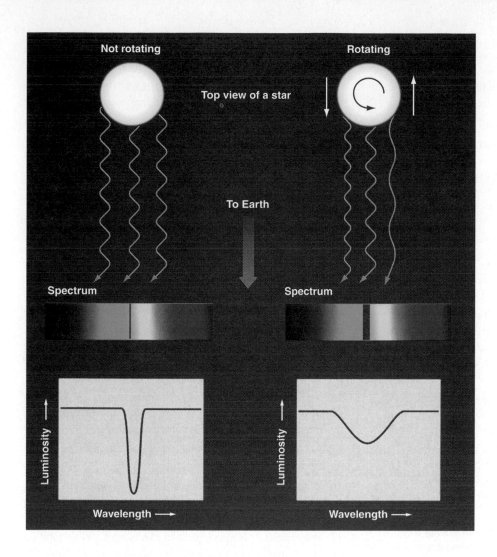

Figure 16.9
The rotation of a star broadens its spectral lines.

Summary

16.1 The total energy emitted per second by a star is called its **luminosity.** How bright a star looks to us is called its **apparent brightness.** For historical reasons, the apparent brightnesses of stars are often expressed in terms of **magnitudes.** If one star is five magnitudes brighter than another, it emits 100 times more energy. Since the apparent brightness of a star depends on its luminosity and distance, determination of apparent brightness and measurement of the distance to a star provide enough information to calculate its luminosity.

16.2 Stars have different colors, and these are indicators of temperature. The **color index** of a star is the difference in the magnitudes measured at any two different wavelengths. The difference between blue and visual magnitudes, B − V, is one frequently used color index; redder, cooler stars have more positive values of B − V.

16.3 The differences in the spectra of stars are principally due to differences in temperature, not composition. The spectra of stars are described in terms of seven **spectral classes.** In order of decreasing temperature, these spectral classes are O, B, A, F, G, K, and M. Spectra of stars of the same temperature but different atmospheric pressure have subtle differences, so spectra can be used to determine whether a star has a large radius and low atmospheric pressure (a **giant** star), or a small radius and high atmospheric pressure. Stellar spectra can also be used to determine the chemical composition of stars; hydrogen and helium make up most of the mass of all stars (just as they do in the Sun). Measurements of line shifts produced by the Doppler effect indicate the **radial velocity** of a star. Broadening of spectral lines by the Doppler effect is a measure of rotational velocity.

Review Questions

1. What two factors determine how bright a star appears to be in the sky?

2. Explain why color is a measure of a star's temperature.

3. What is the main reason that the spectra of stars are not all identical? Explain.

4. What elements are stars made of? How do we know this?

5. What two women astronomers made significant contributions to the understanding of stellar spectra? Discuss what each of them did.

6. Name at least three characteristics of a star that can be determined by measuring its spectrum. Explain how you would use a spectrum to determine these characteristics.

Thought Questions

7. If the star Sirius emits 23 times more energy than the Sun, why does the Sun appear brighter in the sky?

8. Draw a picture showing how two stars of equal intrinsic luminosity, one of which is blue and the other red, would appear on two images, one taken through a filter that passes mainly blue light, and the other through a filter that transmits mainly red light.

9. Table 16.1 lists the temperature ranges that correspond to various spectral types. What part of the star do these temperatures refer to?

10. Star A has lines of ionized helium in its spectrum, and star B has bands of titanium oxide. Which is hotter? Why? The spectrum of another star shows lines of ionized helium and also molecular bands of titanium oxide. What is strange about this spectrum? Can you suggest an explanation?

11. The spectrum of the Sun has hundreds of strong lines of non-ionized iron but only a few, very weak, lines of helium. A star of spectral type B has very strong lines of helium but very weak iron lines. Do these differences mean that the Sun contains more iron and less helium than the B star? Explain.

12. What are the approximate spectral classes of stars with the following characteristics?
   a. Balmer lines of hydrogen are very strong; some lines of ionized metals are present.
   b. Strongest lines are those of ionized helium.
   c. Lines of ionized calcium are the strongest in the spec-

trum; hydrogen lines show with only moderate strength; lines of neutral and ionized metals are present.
   d. Strongest lines are those of neutral metals and bands of titanium oxide.

13. Look at Appendix 13. Can you identify any relationship between the abundance of an element and its atomic weight? Are there any obvious exceptions to this relationship?

14. Appendix 10 lists the nearest stars. Are most of these stars hotter or cooler than the Sun? Do any of them emit more energy than the Sun? If so, which ones?

15. Look at Appendix 10. Which stars are the hottest? Which are the brightest? Are they the same?

16. What is the brightest star in the sky? The second brightest? What color is Betelgeuse? Use Appendix 11 to find the answers.

17. Why can only a lower limit to the rate of stellar rotation be determined from rotational broadening, rather than the actual rotation rate? (Refer to Figure 16.9.)

18. Most of the mass of every star (including the Sun) is made of hydrogen and helium. Why are these two elements the most common?

19. If you were so rich that you had $100 million to give away, what instrument or project in astronomy would you want to build or support? (This might be a good question to answer at the end of the semester.)

Problems

20. A fifth-magnitude star is about the faintest that can be seen without optical aid unless you have access to a very dark, unpolluted sky. How much fainter is it than a zero-magnitude star? A good pair of binoculars can reveal tenth-magnitude stars. How much fainter are these than stars of zero magnitude?

21. As seen from the Earth, the Sun has an apparent magnitude of about −26. What is the apparent magnitude of the Sun as seen from Saturn, about 10 AU away? (Remember that 1 AU is the distance from the Earth to the Sun.)

22. If a star has a color index of $B - V = 2.5$, how many times brighter does it appear in visual light than in blue light, relative to a standard star with $B - V = 0$?

23. What are the approximate spectral classes for stars whose wavelengths of maximum light have the following values? (See Section 4.2.)
   a. 290 nm
   b. 50 nm
   c. 600 nm
   d. 1200 nm
   e. 1500 nm

24. The following equation describes the quantitative relationship between the magnitudes and light flux received from two stars. If m_1 and m_2 are the magnitudes corresponding to stars from which we receive light flux in the amounts l_1 and l_2, the difference between m_1 and m_2 is defined by

$$m_1 - m_2 = 2.5 \log (l_2/l_1)$$

Use this equation to calculate the following:

a. The difference in magnitudes of two stars that differ in light flux received at the Earth by a factor of ten.

b. The difference in light flux received from two stars that differ by ten magnitudes.

c. The difference in magnitudes of two identical stars, one of which is ten times farther away than the other.

25. Appendix 11 lists the 20 brightest stars. How much more light flux do we receive from the brightest of these stars than from the faintest? Use the equation in Problem 24 to calculate the answer.

Suggestions for Further Reading

Fraknoi, A. and Freitag, R. "Women in Astronomy" in *Mercury*, Jan./Feb. 1992, p. 27. Part of an issue devoted to a discussion of women in astronomy.

Hearnshaw, J. "Origins of the Stellar Magnitude Scale" in *Sky & Telescope,* Nov. 1992, p. 494. A good history of how we have come to have this cumbersome system.

Hearnshaw, J. *The Analysis of Starlight.* 1986, Cambridge U. Press. A history of spectroscopy in astronomy.

Hirshfeld, A. "The Absolute Magnitude of Stars" in *Sky & Telescope*, Sep. 1994, p. 35.

Kaler, J. *Stars.* 1992, Scientific American Library/W. H. Freeman. Good modern review of our understanding of stars.

Kaler, J. *Stars and Their Spectra.* 1989, Cambridge U. Press. A detailed introduction to the field of spectroscopy and what it can tell us about the stars.

Kaler, J. "Origins of the Spectral Sequence" in *Sky & Telescope*, Feb. 1986, p. 129.

Kidwell, P. "Three Women of American Astronomy" in *American Scientists*, May/June 1990, p. 244. Focuses on Annie Cannon and Cecilia Payne.

Sneden, C. "Reading the Colors of the Stars" in *Astronomy*, Apr. 1989, p. 36. Discusses what we learn from spectroscopy.

Steffey, P. "The Truth about Star Colors" in *Sky & Telescope*, Sep. 1992, p. 266. About the color index, and how the eye and film "see" colors.

Welther, B. "Annie J. Cannon: Classifier of the Stars" in *Mercury*, Jan./Feb. 1984, p. 28.

Using REDSHIFT ™

1. Turn on only the stars and set the display to *Mercator projection* with a zoom factor to 0.2. Adjust the *Magnitude Limit* until only the brightest 10 stars appear.

Which star is brightest?

What is the brightest blue star?

What is the brightest red star?

Which star is most like the Sun?

How much brighter is the brightest star than the faintest?

2. Turn on the *Constellation Boundaries* (patterns off) and set the *Faint Magnitude* limit to 4.

Which star in Leo is closest?

Which of its stars is farthest?

Are the distances related to the stars' brightnesses?

Which star is hottest?

Which star is coolest?

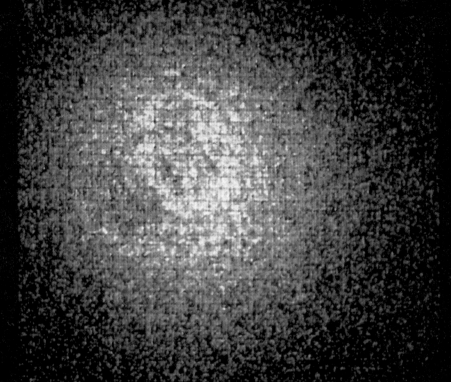

The determination of stellar masses depends on measuring the gravitational effects of one star on a companion star. With modern instruments and computer processing we can now measure binary star systems in which the stars are very close to each other. The top image shows a long time-exposure of the binary star Sigma Herculis. Atmospheric turbulence produces a blurred image of both stars that is about 2 arc-sec in diameter. New techniques can remove this blurring and reveal two stars separated by just 0.07 arcsec. (Anthony Readhead/Palomar Observatory, Caltech)

CHAPTER 17

The Stars: A Celestial Census

Thinking Ahead
A typical person might live seven decades or so. A typical star lives for many billions of years. No one (and no culture on Earth) has ever watched a star go through all the stages of its life cycle. How then can we begin to understand the evolution of a star from its birth to its death?

Imagine for a moment that you are the captain of a well-equipped starship, sent from a civilization that orbits another star to study life-forms elsewhere. (You happen to resemble an intelligent cauliflower with appendages, but you have some excellent scientists among your crew.) You have arrived on Earth, and your schedule gives your crew exactly one Earth-day to learn all they can about the life cycle of the dominant life form on this new planet.[1] What would be your advice to the landing party?

The crew could pursue many strategies, after all. One might be for each cauliflower to watch a single member of the human species for a whole day, to see if any of them went through important life-cycle changes. For a random sample of humans, this would typically turn out to be a disappointingly unproductive strategy. A better approach might be for the crew to make a widespread survey, cataloging as many different humans as possible

---

[1] An interesting question is whether a new visitor to Earth would immediately infer that humans are the dominant life-form here. Some commentators have suggested that a more reasonable conclusion would be that cars rule the Earth. After all, they seem to swallow and disgorge humans (and other animals) at will, and humans are clearly their servants—washing, feeding, and polishing them with domestic regularity.

and then trying to determine which of the differences were significant for understanding how humans develop.

If you don't know anything about human beings to begin with (being an intelligent cauliflower, after all), those human characteristics that are important (and those that are simply random variations from one specimen to the next) would not be at all obvious. For example, you might note that humans come in different colors and hypothesize that they start life a rich, dark brown, getting lighter and lighter with age. The humans with the darkest skin would be the youngest, according to this theory. It's an interesting idea, but it would turn out to be completely wrong. Or you might note that humans come with a bit of fuzzy stuff on top of their heads, and make the very reasonable suggestion that the length of this fuzz is a good measure of age. This also would not get you very far down the road toward the cauliflower Nobel Prize.

On the other hand, there are characteristics that (properly understood) might help you pin down the human life cycle. For the first part of a human life, body mass and height are reasonably good indicators of age, for example. And the smoothness of our skin tends to decrease as we get older. To understand such subtle indicators would require a broad sampling of human characteristics and many years of careful study (even though your data would be collected in a single day).

Astronomers faced a similar problem with stars (Figure 17.1). Stars live such a long time that nothing much can be gained from staring at one for a human lifetime. It was necessary to measure the characteristics of many stars (to take a celestial census, in effect) and then determine which characteristics would help us understand the stars' life stories. Like the cauliflowers of our example, astronomers tried a variety of hypotheses about stars until they came up with the right approach to comprehending their development. But the key was making a thorough census of the stars around us.

A Stellar Census

Before we can make our own survey, we need to agree on a unit of distance appropriate to the objects we are studying. The stars are all so far away that kilometers (and even astronomical units) would be very cumbersome to use; so—as discussed in the Prologue—astronomers use a much larger "measuring stick" called the *light year* (*LY*). A light year is the distance that light (the fastest signal we know) travels in one year. Since light covers an astounding 300,000 km per second, and since there are a lot of seconds in one year, a light year is a very large quantity: 9.5

(a)

(b)

Figure 17.1

(a) Stars, like people, come in a variety of colors. (Jeff Greenberg/Photo Researchers) (b) This image shows a technique invented by David Malin of the Anglo-Australian Observatory to demonstrate the colors of stars. We are looking at the constellation of Orion. The camera was held fixed, and the turning of the Earth caused the stars to drift across the image during the long exposure. As the picture was being taken, the photographer changed the focus of the lens to make each star into a wider and wider streak, thus bringing out the subtle colors. The three stars making the belt of Orion are in the center, the bright yellow-reddish star near the upper left is Betelgeuse, and the whitish-blue star at the lower right is Rigel. (© David Malin, Anglo-Australian Telescope Board)

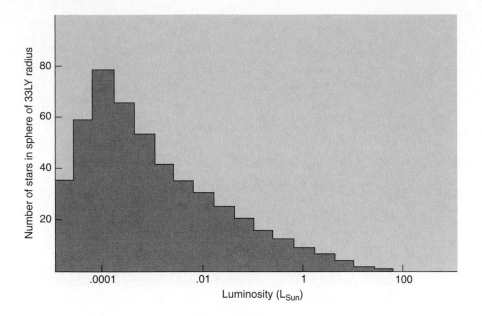

trillion (9.5×10^{12}) km to be exact. (Bear in mind that the light year is a unit of *distance* even though the term *year* appears in it.) If you drove at the legal U.S. speed limit without stopping for food or rest, you would not arrive at the end of a light year in space until roughly 12 million years had passed. And the closest star is more than 4 LY away.

Notice that we have not yet said much about how such enormous distances can be measured. That is a complicated question, to which we will return in later chapters. For now, let us assume that distances have been measured for stars in our cosmic vicinity so that we can proceed with our census.

The Luminosity Function

As a first step, let's examine the stars in our immediate neighborhood. For example, Appendix 10 lists the stars found within 16 LY of the Sun, and also gives their luminosities, a characteristic defined in Chapter 16. (Remember that in order to understand how stars "work," we are interested in how much energy they put out, not how bright they happen to appear from Earth.) With surveys in hand, we can ask how many stars in a given volume of space are intrinsically very luminous, and how many are faint.

Figure 17.2 shows the results of counting stars in a volume of space slightly larger than the one for which data are given in Appendix 10. The stars in this region range from about 100 times the luminosity of the Sun, or 100 L_{Sun}, to less than one ten-thousandth L_{Sun}. If we divide these stars into groups according to luminosity, we can count how many fall into each narrow range. The relationship, shown in Figure 17.2, is called the *luminosity function.* (You are probably familiar with such graphs from surveys of people—for example, the number of resi-

dents in a city who fit into various income or age brackets.) As you can see, the Sun is more luminous than the vast majority of stars in our vicinity.

But does the local neighborhood contain samples of all types of stars? Surely the neighborhood in which you live does not contain all the types of people—distinguished according to age, education, income, race, etc.— that live in the entire country. For example, a few people do live to be over 100 years old, but there may be no such individual within several miles of where you live. In order to sample the full range of the human population, you would have to extend your census to a much larger area.

Similarly, there are some types of stars not found in our immediate neighborhood. A clue that we are missing something in our stellar census comes from the fact that only three first-magnitude stars—Sirius, Alpha Centauri, and Procyon—are found within 16 LY of the Sun (see Appendix 10). Why are we missing most of the stars that appear brightest in the sky when we take our census of the local neighborhood?

The answer, interestingly enough, is that the stars that appear brightest are *not* the closest ones to us. These stars are bright because they emit a very large amount of energy—so much, in fact, that they do not have to be nearby to look brilliant. You can confirm this by looking at Appendix 11, which gives distances for the 20 stars that appear brightest from Earth. In fact, it turns out that *most* of the stars visible without a telescope are hundreds of light years away and many times more luminous than the Sun. Among the 6000 stars visible to the unaided eye, at most 50 are intrinsically fainter than the Sun. Figure 17.3 is a graph showing the distribution of the luminosities of the 30 brightest-appearing stars. Notice that the most luminous of these stars emits nearly 100,000 times more energy than does the Sun.

These highly luminous stars are missing from the solar neighborhood because they are very rare. None of

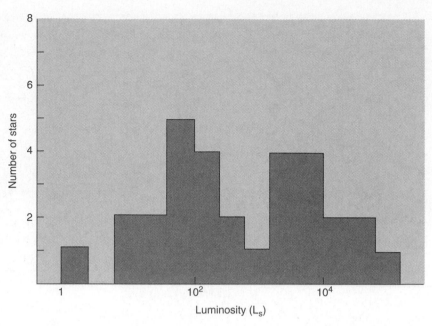

Figure 17.3
Distribution (bar graph) of the luminosity of the 30 brightest-appearing stars in our sky. We plot the number of stars in each range of luminosity.

them happens to be in the tiny volume of space immediately surrounding the Sun that was surveyed to get the data shown in Figure 17.2.

On the other hand, stars fainter than the Sun cannot be seen with the unaided eye unless they are *very* nearby. For example, stars with luminosities ranging from 10^{-2} to 10^{-4} L_{Sun} are very common, but a star with a luminosity of 10^{-2} L_{Sun} would have to be within 5 LY to be visible to the naked eye—and only three stars (all in one system) are this close to us. The nearest of these three stars, Proxima Centauri, cannot even be seen without a telescope because it has such a low luminosity. The common nearby stars typically do not send enough light across interstellar distances to be seen with the unaided eye, and just don't make it into a "top ten" list of stars that appear bright.

Let's again consider the highly luminous stars—those 100 or more times as luminous as the Sun. Although such stars are rare, they are visible to the unaided eye even when hundreds to thousands of light years away. For example, a star with a luminosity of 10^4 L_{Sun} can be seen without a telescope out to a distance of 5000 LY. Such stars are very rare, and we would not expect to find one nearby. The volume of space included within a distance of 5000 LY, however, is enormous; so even though they are intrinsically rare, many such highly luminous stars are readily visible to the unaided eye.

The contrast between these two samples of stars—those that are close to us and those that can be seen with the unaided eye—is an example of a *selection effect*. When a population of objects (stars in this example) includes a great variety of different types, we must be careful what conclusions we draw from an examination of any particular subgroup. Certainly we would be fooling ourselves if we assumed that the stars visible to the unaided eye are characteristic of the general stellar population; this subgroup is heavily weighted to the most luminous stars. It requires much more effort to assemble a complete data set for the nearest stars, since most are so faint that they can be observed only with a telescope. However, it is only by doing so that astronomers are able to work out the properties of the vast majority of the stars, which are actually much smaller and fainter than our own Sun.

The Density of Stars in Space

Now let's use our surveys to answer another question. What is the typical spacing between stars? There are at least (we may have missed a few) 59 stars within 16 LY of the Earth, counting the members of binary and multiple star systems, and the Sun itself. A sphere with a radius of 16 LY has a volume of about 17,000 cubic LY (LY^3). Since this volume of space contains at least 59 stars, the density of stars in space in the neighborhood of the Sun is about one star for every 300 LY^3. The average distance between stars is then about 7 LY (the cube root of 300). The Sun and Alpha Centauri, which are separated by 4.4 LY, are somewhat closer than the average. In any case, you can see that stars (and star systems) live in splendid isolation, with enormous gulfs of space among them.

Density is more typically expressed in g/cm^3, not stars per LY^3 as used in the preceding paragraph. To see how dense our neighborhood is in more common terms, assume the matter contained in stars could be spread out evenly over space. If a typical star has a mass 0.4 times that of the Sun, the average density of matter in the solar neighborhood is only about 3×10^{-24} g/cm^3—or about one hydrogen atom per cubic centimeter. (Notice that planets, if they exist elsewhere, have too little mass compared to stars for us to bother including them.) How does the density get to be so small? While stars have a lot of mass, the amount of space between the stars is so enormous that the average cosmic density of our vicinity is very low indeed.

17.2

Stellar Masses

We have seen that the Sun is more luminous than most stars. How does its mass compare with other stars? We will see as we study stars that mass is the prime indicator of what type and length of life a star will have. Yet it is not very easy to measure the mass of stars directly: somehow we need to put a star on the cosmic equivalent of a scale. Luckily, not all stars live like the Sun, in isolation from other stars. The best information about stellar masses comes from the study of **binary stars**—two stars that orbit around one another, bound together by gravity. We can use their pull on each other to measure their mass.

Many, but not all, stars are members of binary star systems. For example, among the 59 nearest stars, 27 (roughly one-half) are members of systems containing more than one star. Masses for stars can be calculated from measurements of their orbits, just as the mass of the Sun can be derived by measuring the orbits of the planets around it (see Chapter 2).

Binary Stars

Before we discuss in more detail how mass can be measured, we want to take a closer look at stars that come in pairs. The first binary star was discovered in 1650, less than half a century after Galileo began to observe the sky with a telescope. John Baptiste Riccioli, an Italian astronomer, noted that the star Mizar, in the middle of the Big Dipper's handle, appeared through his telescope as two stars. Since that discovery, thousands of binary stars have been cataloged. (Astronomers call any pair of stars close to each other in the sky *double stars,* but not all of these are two stars that are physically associated. Some are just chance alignments of stars that are actually at different distances from us.) Although stars most commonly come in pairs, there are also triple and quadruple systems.

One well-known binary star is Castor, located in the constellation of Gemini. By 1804 astronomer William Herschel, who also discovered the planet Uranus, had noted that the fainter component of Castor had slightly changed its position relative to the brighter component. (We use the term *component* to mean a member of a star system.) Here was evidence that one star was moving around another. It was the first evidence that gravitational influences exist outside the solar system. The orbital motion of the binary star Kruger 60 is shown in Figure 17.4. Binary star systems in which both of the stars can be seen with a telescope are called **visual binaries.**

A class of double stars in which both stars *cannot* be seen was discovered by Edward C. Pickering at Harvard in 1889. He was examining the spectrum of Mizar and found that the dark absorption lines in the brighter star's spectrum were usually double. Not only were there two lines where astronomers normally saw only one, but the spacing of the lines was constantly changing. At times the lines even became single. Pickering correctly deduced that the brighter component of Mizar, called Mizar A, is itself really two stars that revolve about each other in a period of 104 days. Stars like Mizar A, which appear as single stars when photographed or observed visually through the telescope, but which spectroscopy shows really to be double stars, are called **spectroscopic binaries.**

Mizar, by the way, is a good example of just how complex such star systems can be. Mizar has been known for centuries to have a faint companion called Alcor, which can be seen without a telescope. Mizar and Alcor form an *optical double*—a pair of stars that appear close together in the sky but do not orbit around each other. Through a telescope Mizar can be seen to have another, closer companion that does orbit around it; Mizar is thus a visual binary. The two components that make up this visual binary, known as Mizar A and Mizar B, are both spectroscopic binaries. So, Mizar is really a quadruple system of stars.

Strictly speaking, it is not correct to describe the motion of a binary star system by saying that one star orbits

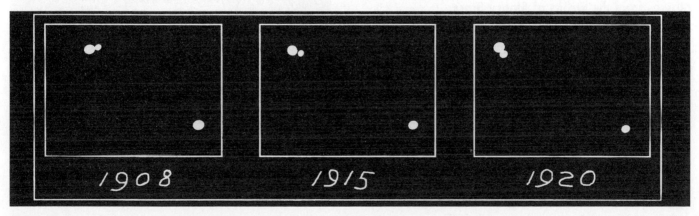

Figure 17.4
Revolution of a binary star. Three photographs covering a period of about 12 years show the mutual revolution of the two stars that make up the nearby star system Kruger 60. (Yerkes Observatory)

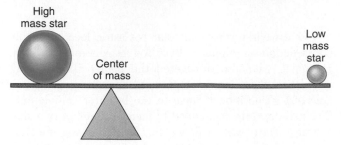

Figure 17.5
In a binary star system, both stars orbit their center of mass. A star with higher mass will be balanced closer to the center of mass, while a star with lower mass will be balanced farther from it.

the other. Gravitation is a *mutual* attraction. Each star exerts a gravitational force on the other, with the result that both stars orbit a point between them called the *center of mass*. Imagine that the two stars are seated, one at each end of a seesaw. The point at which the fulcrum would have to be located in order for the seesaw to balance is the center of mass, and it is always closer to the more massive star (Figure 17.5).

Figure 17.6 shows two stars moving around their center of mass, along with the spectrum we observe from the system at different times. When one star is approaching us relative to the center of mass, the other star is receding from us. In the top illustration, star A is moving toward us, so the lines in its spectrum are Doppler-shifted toward the blue end of the spectrum. Star B is moving away from us, so its spectrum shows a red shift. When we observe the composite spectrum of the two stars, each line appears double. When the two stars are both moving across our line of sight (neither away from nor toward us), however, they both have the same radial velocity (that of the pair's center of mass); hence the spectral lines of the two stars come together. This is shown in the second and fourth illustrations in Figure 17.6. A plot showing how the velocities of the stars change with time is called a *radial-velocity curve*; one for the binary system in Figure 17.6 is shown in Figure 17.7.

Masses from the Orbits of Binary Stars

We can estimate the masses of double star systems by using Newton's reformulation of Kepler's third law (discussed in Section 2.3). Kepler found that the time a planet takes to go around the Sun is related by a specific mathematical formula to its distance from the Sun. In our binary star situation, if two objects are in mutual revolution, then the period (P) with which they go around each other is related to the semimajor axis (D) of the orbit of one with respect to the other, according to the equation

$$D^3 = (M_1 + M_2)P^2$$

where D is in astronomical units, P is measured in years, and $M_1 + M_2$ is the sum of the masses of the two stars in units of the Sun's mass. Thus if we can observe the size of

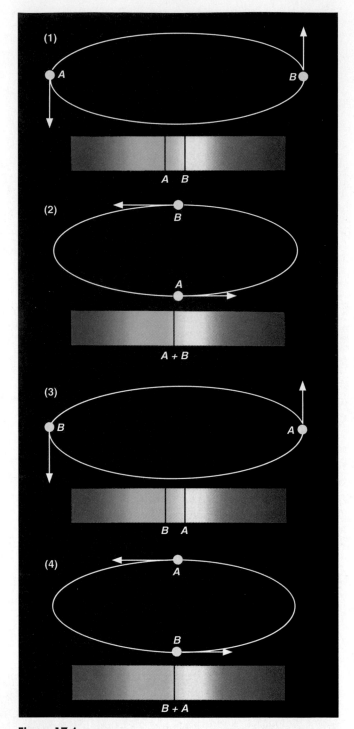

Figure 17.6
A schematic drawing of the motions of two stars orbiting each other. When one star is moving toward the Earth, the other is moving away; half a cycle later the situation is reversed. Doppler shifts cause the spectral lines to move back and forth. At times, lines from both stars can be seen well-separated from each other. When the two stars are moving perpendicular to our line of sight, the two lines are exactly superimposed, and so we see only a single spectral line.

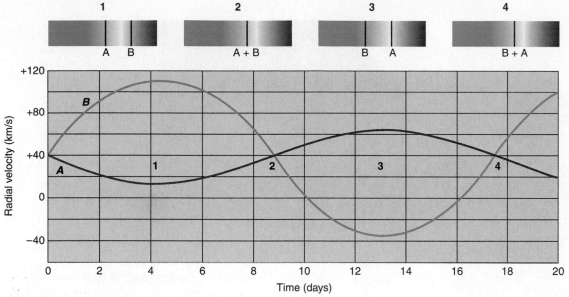

Figure 17.7

Radial-velocity curves for a spectroscopic binary system showing how the two components alternately approach and recede from the Earth. The positions on the curve corresponding to the illustrations in Figure 17.6 are marked.

the orbit and the period of mutual revolution of the stars in a binary system, we can calculate the sum of their masses.

Most spectroscopic binaries have periods ranging from a few days to a few months, with separations of usually less than 1 AU between their member stars. Recall that an AU is the distance from the Earth to the Sun, so this is a small separation and very hard to see at the distances of stars. This is why many of these systems are known to be double only through careful study of their spectra.

The mathematical analysis of a radial-velocity curve (such as the one in Figure 17.7) to determine the masses of the stars in a spectroscopic binary is complex in practice but not hard in principle. The speeds of the stars are measured from the Doppler effect. We then measure the period—how long the stars take to go through an orbital cycle—which can be determined from the velocity curve. Knowing how fast the stars are moving and how long they take to go around tells us the circumference of the orbit and hence the separation of the stars in kilometers or astronomical units. Using Kepler's law, the period and the separation allow us to calculate the sum of the stars' masses.

Of course, knowing the sum of the masses is not as useful as knowing the mass of each star separately. But the relative orbital speeds of the two stars can tell us how much of the total mass each star has. The more massive star has a smaller orbit and hence moves more slowly to get around in the same time. In practice we also need to know how the binary system is oriented in the sky to our line of sight; but if we do and the just-described steps are carried out carefully, the result is a measurement of the masses of each of the two stars in the system.

In short, a good understanding of the motion of two stars around a common center of mass, combined with the laws of gravity, allows us to measure the masses of stars in such systems. These measurements are absolutely crucial to developing a theory of how stars evolve. One of the best things about this method is that it is independent of the location of the binary system. It works as well for stars 1000 LY away as for those in our immediate neighborhood.

The Range of Stellar Masses

How large can the mass of a star be? We find that stars more massive than the Sun are rare. There are no stars within 30 LY of the Sun with masses greater than four times that of the Sun. Searches for massive stars at large distances from the Sun indicate that masses up to about 100 times that of the Sun do occur, but very few stars are this huge (Figure 17.8). There is no convincing evidence for the existence of any stars with masses significantly exceeding about 100 times the mass of the Sun.

According to theoretical calculations, the smallest mass that a true star can have is about 1/12 that of the Sun. By a true star, astronomers generally mean one that becomes hot enough to fuse hydrogen to form helium. Stars with masses smaller than the Sun's are very common, although they are difficult to detect if far away.

Objects with masses between 1/100 and 1/12 that of the Sun may produce energy for a brief time by means of nuclear reactions involving deuterium, but they do not become hot enough to force protons to combine to produce helium. Such objects are called **brown dwarfs,** and for

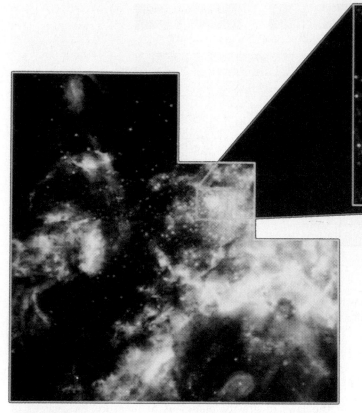

Figure 17.8

R 136, a cluster that contains many massive stars. Part of a giant region of cosmic raw material called the 30 Doradus nebula (larger image), this cluster of young hot stars (smaller image) contains some of the most massive stars known, including some with masses as large as 100 M_{Sun}. The nebula and cluster are 160,000 LY away in the Large Magellanic Cloud, a smaller galaxy that is a satellite of our Milky Way. The cluster is very difficult to make out clearly with telescopes on the ground; this image was one of the tests to which astronomers subjected the repaired Hubble Space Telescope. If it can distinguish stars in the R 136 cluster clearly (and see for yourself that it does), then the Space Telescope must be working pretty well. (NASA/STScI)

more than a decade astronomers have been trying to find examples of such "failed stars." This is not easy to do, since they are extremely dim and thus difficult to observe. Ingenious techniques are required to make sure that a very faint object is a true brown dwarf and not merely a very low-mass star.

In 1995 three groups of astronomers announced what appears to be convincing evidence for brown dwarfs in three different star systems. Two are located in a cluster of stars called the Pleiades, which is young enough so that any newly formed brown dwarfs have not had time to fade away to complete obscurity. One candidate, called PPL 15, appears to be just on the borderline between stars and brown dwarfs. The second, called Teide 1, seems less massive but still requires more investigation. However, the third candidate, in orbit around a nearby star called Gliese 229, appears to have somewhere between 20 and 40 times the mass of Jupiter, and is thus very likely a brown dwarf. Its spectrum, as observed with the giant Keck telescope, shows evidence of methane, a molecule that would be destroyed by heat in the outer layers of a star. Many groups around the world are continuing to hunt for other examples of these fascinating but elusive objects.

Still-smaller objects with masses less than 1/100 the mass of the Sun are called *planets*. They may radiate energy produced by the radioactive elements that they contain, and they may also radiate heat generated by slow gravitational contraction. However, their interiors will never reach temperatures high enough for nuclear reactions to take place. Jupiter, whose mass is about 1/1000 the mass of the Sun, is unquestionably a planet, for example. Until the 1990s we could only detect planets in our own solar system, but now we have begun to detect them elsewhere as well (see Chapter 20).

The Mass–Luminosity Relation

Now that we have measurements of the characteristics of many different types of stars, we can search for relationships among the characteristics. For example, we can ask whether the mass and luminosity of a star are related. It turns out that for most stars, they are: the more massive stars are generally also the more luminous. This relationship, known as the **mass–luminosity relation,** is shown graphically in Figure 17.9. Each point represents a star whose mass and luminosity are both known. Horizontal position on the graph shows the star's mass, given in units of the Sun's mass, and vertical position shows its luminosity in units of the Sun's luminosity.

Notice how good this relationship is: most stars fall along a line running from the lower left (low mass, low luminosity) corner of the diagram to the upper right (high mass, high luminosity) corner. About 90 percent of all stars obey the mass–luminosity relation illustrated in Figure 17.9. Later we will explore why such a relationship exists, and what we can learn from the roughly 10 percent of stars that "disobey."

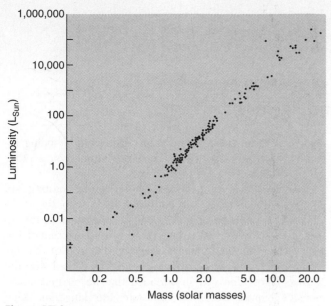

Figure 17.9

The mass–luminosity relation. The plotted points show the masses and luminosities of stars for which both of these quantities are known to an accuracy of 15 to 20 percent. The three points lying below the sequence of points are all white dwarf stars. (Adapted from data compiled by D. M. Popper)

17.3

Diameters of Stars

It is easy to measure the diameter of the Sun. Its angular diameter—that is, its apparent size on the sky—is about 1/2°. If we know the angle the Sun takes up in the sky and how far away it is, we can calculate its true (linear) diameter, which is 1.39 million km, or about 109 times the diameter of the Earth.

Unfortunately, the Sun is the only star whose angular diameter is easily resolved. All the other stars are so far away that they look like pinpoints of light through even the largest telescopes. (They often seem to be bigger, but that is merely distortion introduced by turbulence in the Earth's atmosphere.) Luckily, there are several techniques that astronomers can use to estimate the sizes of stars; we discuss two of them next.

Diameters of Stars Whose Light Is Blocked by the Moon *occultation*

One method, which gives very precise diameters but can be used for only a few stars, is to observe the dimming of light that occurs when the Moon passes in front of a star. What astronomers measure (with great precision) is the time required for the star's brightness to drop to zero as the edge of the Moon moves across it. Since we know how rapidly the Moon moves in its orbit around the Earth, it is possible to calculate the angular diameter of the star. If the distance to the star is also known, this gives us its diameter in kilometers. This method works only for fairly bright stars that happen to lie along the zodiac, where the Moon (or, much more rarely, a planet) can pass in front of them as seen from the Earth.

Eclipsing Binary Stars

A second technique for measuring diameters works for those stars that are members of **eclipsing binaries,** and so we must make a brief detour from our main story to examine this type of star system. Some double stars are lined up in such a way that when viewed from the Earth, each star passes in front of the other during every revolution (Figure 17.10). When one star blocks the light of the other, preventing it from reaching the Earth, the luminos-

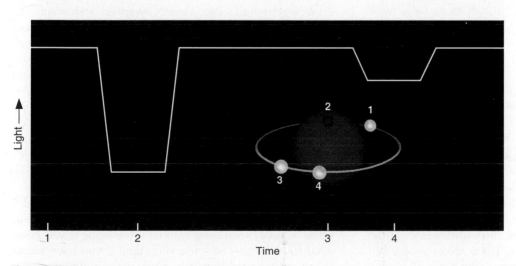

Figure 17.10

The light curve of a hypothetical eclipsing binary star with total eclipses (one star passes directly in front of and behind the other). The numbers indicate parts of the light curve corresponding to various positions of the smaller star in its orbit.

Astronomy and Mythology:
Algol the Demon Star and Perseus the Hero

The name Algol comes from the Arabic *Ras al Ghul,* meaning "the demon's head." The word *ghoul* in English has the same derivation. As discussed in Chapter 1, many of the bright stars have Arabic names because during the long dark ages in medieval Europe, it was Arabic astronomers who preserved and expanded the Greek and Roman knowledge of the skies. The reference to the demon is part of the ancient Greek legend of the hero Perseus, who is commemorated by the constellation in which we find Algol, and whose adventures involve many of the characters associated with the northern constellations.

Perseus was one of the many half-god heroes fathered by Zeus (Jupiter in the Roman version), the king of the gods in Greek mythology. Zeus had, to put it delicately, a roving eye and was always fathering somebody or other with a human maiden who caught his fancy. (Perseus derives from *Per Zeus,* meaning "fathered by Zeus.") Set adrift with his mother by an (understandably) upset stepfather, Perseus grew up on an island in the Aegean Sea. The king there, taking an interest in Perseus' mother, tried to get rid of the young man by assigning him an extremely difficult task.

In a moment of overarching pride, a beautiful young woman named Medusa had compared her golden hair with that of the goddess Minerva. The Greek gods did not take kindly to being compared to mere mortals, and Minerva turned Medusa into a "Gorgon," a hideous, evil creature with writhing snakes for hair and a face that turned anyone who looked at it into stone. (In these mythological stories, there were no second or third chances; one strike and you were out.) Perseus was given the task of slaying the demon, which seemed like a pretty sure way to get him out of the way forever.

But because Perseus had a god for a father, some of the other gods gave him tools for the job, including Minerva's reflective shield and the winged sandals of Hermes (Mercury in the Roman story). By flying over her and looking only at her reflection, Perseus was able to cut off Medusa's head without ever looking at her directly. Taking her head (which could still turn onlookers to stone even without being attached to her body) with him, Perseus continued on to other adventures.

He next came to a rocky seashore, where boasting had gotten another family into serious trouble with the gods. Queen Cassiopeia had dared to compare her own beauty to that of the Nereids, sea nymphs who were daughters of Poseidon (Neptune in Roman mythology), the god of the sea. (Today one of the planet Neptune's moons is named after the Nereids.) Poseidon was so offended that he created a sea-monster named Cetus to devastate the kingdom. King Cepheus, Cassiopeia's beleaguered husband, consulted the oracle, who told him that he must sacrifice his beautiful daughter Andromeda to the monster.

When Perseus came along and found Andromeda chained to a rock near the sea, awaiting her fate, he rescued her by turning the monster to stone. (Scholars of mythology actually trace the essence of this story back to far-older legends from ancient Mesopotamia, in which the god-hero Marduk vanquishes a monster named Tiamat. Symbolically, a hero like Perseus or Marduk is usually associated with the Sun, the monster with the power of night, and the beautiful maiden with the fragile beauty of dawn, which the Sun releases after its nightly struggle with darkness.)

All the characters in these Greek legends can be found as constellations in the sky, not necessarily resembling their namesakes, but serving as reminders of the story. (See the star map for October, after the Appendices.) For example, vain Cassiopeia is sentenced to be very close to the celestial pole, rotating perpetually around the sky and hanging upside down every winter. The ancients imagined Andromeda still chained to her rock (it is much easier to see the chain of stars than to recognize the beautiful maiden in this star grouping). Perseus is next to her with the head of Medusa swinging from his belt. Algol represents this gorgon head, and has long been associated with evil and bad fortune in such tales. Some commentators have speculated that the star's change in brightness (which can be observed with the unaided eye) may have contributed to its unpleasant reputation, with the ancients regarding such a change as an evil sort of "wink."

---

ity of the system decreases, and astronomers say that an *eclipse* has occurred.

The first eclipsing binary was discovered very shortly after Pickering found that Mizar A is a spectroscopic binary, and the discovery helped solve a long-standing puzzle in astronomy. The star Algol, in the constellation of Perseus, changes its brightness in an odd but regular way. Normally Algol is a second-magnitude star, but at intervals of 2^d, 20^h, 49^m it fades to one-third of its regular brightness. After a few hours it brightens to normal again. This effect is easily seen even without a telescope if you know what to look for—try observing Algol if you have access to clear skies without too much light pollution. (See "Making Connections" above.)

In 1783, more than a century before the spectroscopic observations of stars became possible, a young

English astronomer named John Goodricke (profiled in Chapter 18) made a careful study of Algol. Even though Goodricke could neither hear nor speak, he made a number of major discoveries in the 21 years of his brief life. He suggested that Algol's unusual brightness variations might be due to an invisible companion that regularly passes in front of the brighter star and blocks its light. Unfortunately, Goodricke had no way to test this idea, since the equipment available at that time was not good enough to measure Algol's spectrum.

In 1889 the German astronomer Hermann Vogel demonstrated that, like Mizar, Algol is a spectroscopic binary. The spectral lines of Algol were not observed to be double because the fainter star of the pair gives off too little light compared with the brighter for its lines to be conspicuous in the composite spectrum. Nevertheless, the periodic shifting back and forth of the brighter star's lines gave evidence that it was revolving about an unseen companion. (The lines of both components need not be visible for a star to be recognized as a spectroscopic binary.)

The discovery that Algol is a spectroscopic binary verified Goodricke's hypothesis. The plane in which the stars revolve is turned nearly edgewise to our line of sight, and each star passes in front of the other during every revolution. (The eclipse of the fainter star in the Algol system is not very noticeable because the part of it that is covered contributes little to the total light of the system. This second eclipse can, however, be detected by careful measurements.) Any binary star produces eclipses if viewed from the proper direction, near the plane of its orbit, so that one star passes in front of the other (see Figure 17.10). But from our vantage point on Earth, only a few of these stars are oriented in this way.

Diameters of Eclipsing Binary Stars

We now turn back to the main thread of our story to discuss how all this can be used to measure the sizes of stars. The technique involves making a careful *light curve* of an eclipsing binary, a graph that plots how the brightness changes with time. Let us consider a hypothetical binary system in which the stars are very different in size, like those illustrated in Figure 17.11. To make life easy, we will assume that the orbit is viewed exactly edge-on. Even though we cannot see the two stars separately in such a system, the light curve can tell us what is happening. When the small star just starts to pass behind the large star (a point we call *first contact*), the brightness begins to drop. The eclipse becomes total (the small star is completely hidden) at the point called second contact. At the end of the total eclipse (third contact), the small star begins to emerge. When the small star has reached last contact, the eclipse is completely over.

To see how this allows us to measure diameters, look carefully at Figure 17.11. During the time interval between first and second contacts, the small star has moved a distance equal to its own diameter. During the time in-

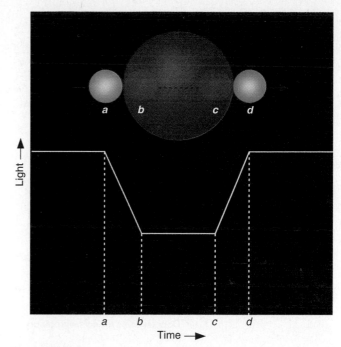

Figure 17.11
The light curve of a hypothetical eclipsing binary star, whose orbit we view exactly edge-on, in which the two stars fully eclipse each other. From the time intervals between contacts it is possible to estimate the diameters of the two stars.

terval from first to third contacts, the small star has moved a distance equal to the diameter of the large star. If the spectral lines of both stars are visible in the spectrum of the binary, the speed of the small star with respect to the large one can be measured from the Doppler shift. But knowing the speed with which the small star is moving and how long it took to cover some distance can tell us the span of that distance—in this case the diameters of the stars. The speed, multiplied by the time interval from first to second contact, gives the diameter of the small star. Multiply it by the time between first and third contact and you get the diameter of the large star.

In actuality, the orbits are generally not seen exactly edge-on, and the light from each star may be only partially blocked by the other. Furthermore, binary star orbits, just like the orbits of the planets, are ellipses, not circles. However, all these effects can be sorted out from very careful measurements of the light curve.

Stellar Diameters

The results of stellar size measurements confirm that most nearby stars are roughly the size of the Sun—with typical diameters of a million kilometers or so. Faint stars, as we might have expected, are generally smaller than more luminous stars. However, there are some dramatic exceptions to this simple generalization.

A few of the very luminous stars, those that are also red in color (indicating relatively low surface tempera-

Characteristic	Technique
Surface temperature	1. Determine the color (very rough).
	2. Measure the spectrum and get the spectral type.
Chemical composition	Determine which lines are present in spectrum.
Luminosity	Measure apparent brightness and compensate for distance.
Radial velocity	Measure the Doppler shift in the spectrum.
Rotation	Measure the width of spectral lines.
Mass	Measure the period and radial velocity curves of spectroscopic binary stars.
Diameter	1. Measure the way a star's light is blocked by the Moon.
	2. Measure the light curves and Doppler shifts for eclipsing binary stars.

TABLE 17.1
Measuring the Characteristics of Stars

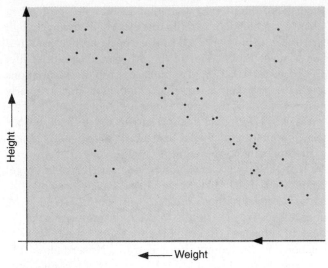

Figure 17.12
The plot of the height and weight of a representative group of human beings. Most lie along a "main sequence" representing normal people, but there are a few exceptions.

tures), turn out to be truly enormous. These stars are called, appropriately enough, *giants* or *supergiants*. An example is Betelgeuse, the second-brightest star in the constellation of Orion and one of the dozen brightest stars in our sky. Its diameter is greater than 10 AU, large enough to fill the entire inner solar system almost as far out as Jupiter. In Chapter 21 we will look in detail at the evolutionary process that leads to the formation of giant and supergiant stars.

17.4

The H–R Diagram

In this chapter and the previous one we have described some of the characteristics by which we might classify stars, and how those characteristics are measured. These ideas are summarized in Table 17.1. We have also given an example of a relationship between two of these characteristics in the mass–luminosity relation discussed earlier. When the characteristics of large numbers of stars were measured at the beginning of the 20th century, astronomers were able to begin a deeper search for patterns and relationships in these data.

To help understand what sorts of relationships might be found, let us return briefly to our intelligent cauliflowers who are trying to make sense of their data about human beings. Being good scientists, they might try plotting their data in different ways. Suppose they make a plot of the height of a good sampling of humans against their weight (which is a measure of their mass). Such a plot is shown in Figure 17.12 and has some interesting features. The way we have chosen to present our data, height in-

creases upward, while weight increases to the left. Notice that humans are not randomly distributed in that graph: most fall along a sequence that goes from the upper left to the lower right.

We can conclude from this graph that generally speaking, taller human beings weigh more, while shorter ones weigh less. This makes sense if you are familiar with the structure of human beings: typically, if we have bigger bones, we have more flesh to fill out our larger frame. It's not mathematically exact—there is a wide range of variation—but it's not a bad overall rule. And, of course, there are some dramatic exceptions. You occasionally see a short human who is very overweight and would thus be more to the bottom left of our diagram than the average sequence of people. Or you might have a very tall, skinny fashion model with great height but relatively small weight, who would be found near the upper right of the figure. A similar diagram has been found extremely useful for understanding the lives of stars.

In 1913 American astronomer Henry Norris Russell plotted the luminosities of stars against their spectral classes (a way of denoting their surface temperatures). This investigation, and a similar independent study in 1911 by Danish astronomer Ejnar Hertzsprung (Figure 17.13), led to an extremely important discovery concerning how these characteristics of stars are related.

Features of the H–R Diagram

Following Hertzsprung and Russell, let us plot the temperature (or spectral class) of a selected group of nearby stars against their luminosity and see what we find

(continued on page 354)

Henry Norris Russell

When Henry Norris Russell graduated from Princeton University, his work had been so brilliant that the faculty decided to create a new level of honors degree beyond "summa cum laude" for him. His students later remembered him as a man whose thinking was three times faster than just about anybody else's. His memory was so phenomenal, he could correctly quote an enormous number of poems and limericks, the entire Bible, tables of mathematical functions, and almost anything he had learned about astronomy. He was nervous, active, competitive, critical, and very articulate—he tended to dominate every meeting he attended. In outward appearance he was an old-fashioned product of the 19th century who wore high-top black shoes and high starched collars, and carried an umbrella every day of his life. His 264 papers were enormously influential in many areas of astronomy.

Born in 1877, the son of a Presbyterian minister, Russell showed early promise. When he was 12, his family sent him to live with an aunt in Princeton so he could attend a top preparatory school. He lived in the same house in that town until his death in 1957 (interrupted only by a brief stay in Europe for graduate work). He was fond of recounting that both his mother and his maternal grandmother had won prizes in mathematics, and that he probably inherited his talents in that field from their side of the family.

Before Russell, American astronomers devoted themselves mainly to surveying the stars and making impressive catalogs of their properties—especially their spectra (as described in Chapter 16). Russell began to see that interpreting the spectra of stars required a much more sophisticated understanding of the physics of the atom, a subject that was being developed by European physicists in the 1910s and 20s.

Russell embarked on a lifelong quest to ascertain the physical conditions inside stars from the clues in their spectra; his work inspired, and was continued by, a generation of astronomers, many trained by Russell and his collaborators.

Russell also made important contributions in the study of binary stars and the measurement of star masses, the origin of the solar system, the atmospheres of planets, and the measurement of distances in astronomy, among other fields. He was an influential teacher and popularizer of astronomy, writing a column on astronomical topics for *Scientific American* magazine for over 40 years. He and two colleagues wrote a textbook for college astronomy classes that helped train astronomers and astronomy enthusiasts over several decades. This book set the scene for the kind of textbook you are now reading, which not only lays out the facts of astronomy but explains how they fit together. Russell gave lectures around the country, often emphasizing the importance of understanding modern physics in order to grasp what was happening in astronomy.

Harlow Shapley, director of the Harvard College Observatory, called Russell "the dean of American astronomers." He was certainly regarded as the leader of the field for many years, and was consulted on many astronomical problems by colleagues from around the world. Today, one of the highest recognitions that an astronomer can receive is an award from the American Astronomical Society called the Henry Norris Russell Prize, set up in his memory.

Figure 17.13
Ejnar Hertzsprung (1873–1967) and Henry Norris Russell (1877–1957) independently discovered the relationship between the luminosity and surface temperature of stars that is summarized in what is now called the H–R diagram. (Sterrewacht Leiden and Princeton University Archives)

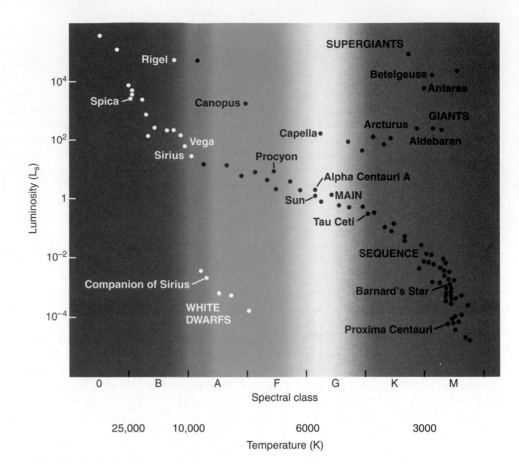

Figure 17.14
The H–R diagram for a selected sample of stars. Luminosity is plotted along the vertical axis. Along the horizontal axis we can plot either temperature or spectral type. Several of the brightest stars are identified by name.

(Figure 17.14). Such a plot is frequently called the **Hertzsprung–Russell diagram,** abbreviated **H–R diagram.** It is one of the most important and widely used diagrams in astronomy, with applications that extend far beyond the purposes for which it was originally developed nearly a century ago.

It is customary to plot H–R diagrams in such a way that temperature increases toward the left and luminosity toward the top. Notice the similarity to our plot of height and weight for people. Stars, like people, are not distributed over the diagram at random, as they would be if they exhibited all combinations of luminosity and temperature. Instead, we see that the stars cluster into certain parts of the H–R diagram. The great majority are aligned along a narrow sequence running from the upper left (hot, highly luminous) to the lower right (cool, less luminous). This band of points is called the **main sequence.** It represents a relationship between *temperature* and *luminosity* that is followed by most stars. Hotter stars are more luminous than cooler ones.

A number of stars, however, lie above the main sequence on the H–R diagram, in the upper right (cool, high luminosity) region. How can a star be at once cool, meaning each point on the star does not put out all that much energy, and yet very luminous? The only way is for the star to be enormous—to have so many points on its surface that the *total* energy output can still be large.

These stars must be giants or supergiants, the stars of huge diameter we discussed above.

The stars in the lower left (hot, low luminosity) corner of the diagram, on the other hand, have high surface temperatures, so that each point on a given star puts out a lot of energy. How then can the overall star be dim? It must be that it has a very small surface area; such stars are known as **white dwarfs** (white because the colors blend together to make them look bluish-white). We will say more about these puzzling objects in a moment. Figure 17.15 is a schematic H–R diagram for a large sample of stars, drawn to make the various types more apparent.

Now think back to our discussion of star surveys. It is difficult to plot an H–R diagram that is truly representative of all stars, because most stars are so faint that we cannot see those outside our immediate neighborhood. The stars plotted in Figure 17.14 were selected because their distances are known. This sample omits many intrinsically faint stars that are nearby but have not had their distances measured, so it shows fewer faint main-sequence stars than a "fair" diagram would. To be truly representative of the stellar population, an H–R diagram should be plotted for all stars within a certain distance. Unfortunately, our knowledge is reasonably complete only for stars within 10 to 20 LY of the Sun, among which there are no giants or supergiants. Still, from many surveys (and more can now be done with new, more powerful

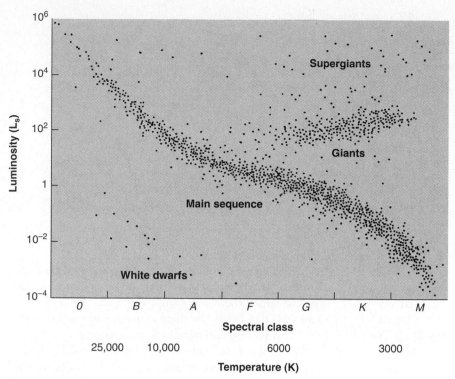

Figure 17.15
Schematic H–R diagram for many stars.

telescopes) we estimate that overall about 90 percent of the stars in our part of space are main-sequence stars, about 10 percent are white dwarfs, and fewer than 1 percent are giants or supergiants.

This result can be used directly to understand the lives of stars. Permit us another quick analogy with people. Suppose the intelligent cauliflowers return to your town and this time focus their attention on the location of young people, ages 6 to 18. Their survey teams fan out and take data about where such youngsters are found at all times during a 24-hour day. Some are found in the pizza parlor, others at home, others at the movies, many in school. After surveying a very large number of young people, the teams determine that, averaged over the course of the 24 hours, one-third of all youngsters are found in school.

How can we interpret this result? Does it mean that two-thirds of students are truants and the remaining one-third spend all their time in school? No, we must bear in mind that the survey teams counted youngsters throughout the full 24-hour day. Some survey teams worked at night, when most youngsters were at home asleep, and others in the late afternoon, when they were on their way home from school (and most likely to be enjoying a pizza). If the survey was truly representative, we *can* conclude, however, that if an average of one-third of all youngsters are found in school, then humans ages 6 to 18 must spend about one-third of their time in school.

We can do something similar for stars. We find that on average 90 percent of all stars are located on the main

sequence of the H–R diagram. If we can identify some activity or life stage with the main sequence, then it follows that stars must spend 90 percent of their lives in that activity or life stage.

Understanding the Main Sequence

In Chapter 15 we discussed the Sun as a representative star. We saw that what stars such as the Sun "do for a living" is to convert hydrogen to helium deep in their interiors via the process of nuclear fusion, thus producing energy. We found that fusion is the means by which a star can maintain itself in equilibrium and continue to shine. The fusion of hydrogen to helium is an excellent source of energy for a star, because the bulk of every star consists of hydrogen atoms.

Our theoretical models of how stars evolve over time show us that a typical star will spend about 90 percent of its life fusing the abundant hydrogen in its core into helium. This then is a good explanation of why 90 percent of all stars are found on the main sequence in the H–R diagram. But if all the stars on the main sequence are doing the same thing (fusing hydrogen), why are they distributed along a sequence of points? That is, why do they differ in luminosity and surface temperature (which is what we are plotting on the H–R diagram)?

To help us understand how main-sequence stars differ, we can use one of the most important results from our studies of model stars (constructed on computers as described in Chapter 15). Astrophysicists have been able to

TABLE 17.2
Characteristics of Main-Sequence Stars

Spectral Type	Mass	Luminosity	Temperature	Radius
	(Sun = 1)	(Sun = 1)		(Sun = 1)
O5	40	7×10^5	40,000 K	18
B0	16	27×10^4	28,000 K	7
A0	3.3	55	10,000 K	2.5
F0	1.7	5	7,500 K	1.4
G0	1.1	1.4	6,000 K	1.1
K0	0.8	0.35	5,000 K	0.8
M0	0.4	0.05	3,500 K	0.6

show that the structure of stars that are in equilibrium and that derive all their energy from nuclear fusion is completely and uniquely determined by just two quantities—*total mass* and *composition*. This fact provides an interpretation of many features of the H–R diagram.

Imagine a cluster of stars forming from a cloud of interstellar "raw material" whose chemical composition is similar to the Sun's. All condensations that become stars will then begin with the same chemical composition and will differ from one another only in mass. Now suppose that we compute a model of each of these stars for the time at which it becomes stable and derives its energy from nuclear reactions, but before it has time to alter its composition appreciably as a result of these reactions.

The models calculated for these stars allow us to determine their luminosities, temperatures, and sizes. If we plot the results from the models—one point for each model star—on the H–R diagram, we get something that looks just like the main sequence we saw for real stars.

And here is what we find when we do this. The model stars with the largest masses are the hottest and most luminous, and they are located at the upper left of the diagram. The least-massive model stars are the coolest and least luminous, and they are placed at the lower right of the plot. The other model stars all lie along a line running diagonally across the diagram. The main sequence turns out to be a sequence of stellar masses.

This makes sense if you think about it. The most massive stars have the strongest gravitational pull, and can thus compress their centers to the greatest degree. This means they are the hottest inside and the best at generating energy from nuclear reactions deep within—hence they shine with the greatest luminosity and have the hottest surface temperatures. The stars with lowest mass, in turn, are coolest and thus the least luminous. Our Sun lies somewhere in the middle of these extremes (as you can see in Figure 17.14). The characteristics of representative main-sequence stars are listed in Table 17.2.

Note that this is exactly what we found earlier, in Section 17.2, when we examined the mass–luminosity rela-

tion for stars whose mass and luminosity we know (see Figure 17.9). We observed that 90 percent of all stars seem to follow the relationship; these are the 90 percent of all stars that lie on the main sequence in our H–R diagram. Our models and our observations agree.

What about the other stars on the real H–R diagram—the giants and supergiants and the white dwarfs? As we will see in the next few chapters, these are what main-sequence stars turn into as they age; they are the later stages in a star's life. As a star consumes its nuclear fuel, its source of energy changes, as do its chemical composition and interior structure. These changes cause the star to alter its luminosity and surface temperature so that it no longer lies on the main sequence.

Extremes of Stellar Luminosities, Diameters, and Densities

We can use the H–R diagram to explore the extremes in size, luminosity, and density found among the stars. Such extreme stars are not just interesting to fans of the *Guinness Book of World Records,* but can teach us a lot about how stars work. For example, we saw that the most massive main-sequence stars are the most luminous ones. We know of a few extreme stars that are a million times more luminous than the Sun, with masses up to 100 times the Sun's mass. These superluminous stars, most of which are at the upper left of the H–R diagram, are exceedingly hot, very blue stars of spectral type O. These are the stars that would be the most conspicuous at vast distances in space.

The cool supergiants in the upper right corner of the H–R diagram are as much as ten thousand times as luminous as the Sun. These stars also have diameters very much larger than that of the Sun. As discussed above, some supergiants are so large that if the solar system could be centered in one, the star's surface would lie beyond the orbit of Mars. We will have to ask, in coming chapters, what process can make a star swell up to such an enormous size, and how long these "swollen" stars can last in their distended state.

In contrast, the very common red, cool, low-luminosity stars at the lower end of the main sequence are much smaller and more compact than the Sun. An example of such a red dwarf is Ross 614B, with a surface temperature of 2700 K and only 1/2000 of the Sun's luminosity. We call such a star a dwarf because its diameter is only 1/10 that of the Sun. A star with such a low luminosity also has a low mass (about 1/12 that of the Sun). This combination of mass and diameter means that the star has an average density about 80 times that of the Sun. Its density must be higher, in fact, than that of any known solid found on the surface of the Earth. (Despite this, the star is made of gas throughout, because its center is so hot.)

The faint red main-sequence stars are not the stars of the most extreme densities, however. The white dwarfs, at the lower left corner of the H–R diagram, have densities many times greater still.

The White Dwarfs

The first white dwarf star was detected in 1862, but its spectrum was not obtained until 1914. Called Sirius B, it forms a binary system with Sirius, the brightest-appearing star in the sky. It eluded discovery and analysis for a long time because it is very, very faint. Although only 8 LY away, the white dwarf companion of Sirius is quite difficult to see without a rather large telescope (Figure 17.16). (Since Sirius is often called the Dog Star–being located in the constellation of Canis Major, the big dog—Sirius B is sometimes nicknamed the Pup.)

We have now found hundreds of white dwarfs. A good example of a typical white dwarf is the nearby star 40 Eridani B. Its surface temperature is a relatively hot 12,000 K, but its luminosity is only $1/275 \, L_{Sun}$. Calculations show that its radius is only 1.4 percent of the Sun's, or about the same as that of the Earth, and its volume is 2.5×10^{-6} that of the Sun. Its mass, however, is 0.43 times the Sun's mass, just a little less than half. To fit such a substantial mass into so tiny a volume, the star's density must be about 170,000 times the density of the Sun, or more than 200,000 g/cm³. A teaspoonful of this material would have a mass of some 50 tons! At such densities, matter cannot exist in its usual state; we will examine the peculiar behavior of this type of matter in Chapter 22. For

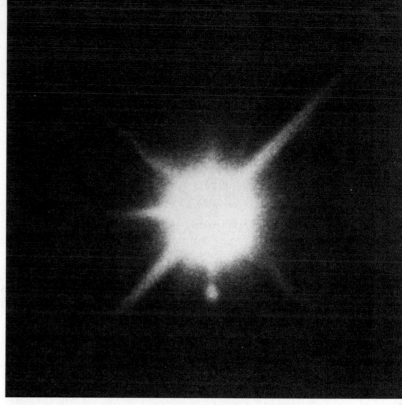

Figure 17.16
Sirius and its faint companion, the white dwarf star Sirius B. Sirius itself is overexposed on this image so that the fainter star can be seen. (Lick Observatory)

now, we should just note that white dwarfs are dying stars, reaching the end of their productive lives and ready for their stories to be over.

The British astrophysicist (and science popularizer) Arthur Eddington described the first known white dwarf this way: "The message of the companion of Sirius, when decoded, ran: 'I am composed of material three thousand times denser than anything you've ever come across. A ton of my material would be a little nugget you could put in a matchbox.' What reply could one make to something like that? Well, the reply most of us made in 1914 was, 'Shut up; don't talk nonsense.'" Today, however, astronomers not only accept that stars as dense as white dwarfs exist, but—as we will see—have found even denser objects in their quest to understand the evolution of different types of stars.

Summary

17.1 To understand the properties of stars, we must make wide-ranging surveys. We find the stars that appear brightest to our eyes are bright primarily because they are intrinsically very luminous, not because they are the closest to us. Most of the nearest stars are intrinsically so faint that they can be seen only with the aid of a telescope. The luminosity of stars ranges from about $10^{-4} \, L_{Sun}$ to more than $10^6 \, L_{Sun}$. Stars with low luminosity are much more common than stars with high luminosity.

17.2 The masses of stars can be determined by analysis of the orbits of **binary stars**. The three types of double stars are **visual binaries, spectroscopic binaries,** and **eclipsing binaries.** Stellar masses range from about 1/12 to (rarely) 100 times the mass of the Sun. Objects having less mass than stars do are called **brown dwarfs** and *planets.* The most massive stars are, in most cases, also the most luminous, and this correlation is known as the **mass–luminosity relation.**

17.3 The diameters of stars can be determined by measuring the time it takes an object (the Moon, a planet, or a companion star) to pass in front of it and block its light. Diameters of members of **eclipsing binary** systems can be determined through analysis of their orbital motions.

17.4 The **Hertzsprung–Russell diagram,** or **H–R diagram,** is a plot of stellar luminosity as a function of surface temperature. Most stars lie on the **main sequence,** which extends diagonally across the H–R diagram from high temperature and high luminosity to low temperature and low luminosity. The position of a star along the main sequence is determined by its mass: high-mass stars emit more energy and are hotter than low-mass stars on the main sequence. Main-sequence stars derive their energy from the fusion of hydrogen to helium. About 90 percent of the stars in the solar neighborhood lie on the main sequence. Only 10 percent of the stars are **white dwarfs,** and fewer than 1 percent are *giants* or *supergiants.*

Review Questions

1. How does the intrinsic luminosity of the Sun compare with that of the 30 brightest stars? With that of the stars within 15 LY? Refer to Appendices 10 and 11.

2. Name and describe the three types of binary systems.

3. Describe two ways of determining the diameter of a star.

4. What are the largest and smallest known values of the mass, luminosity, surface temperature, and diameter of stars?

5. You are able to take spectra of both stars in an eclipsing binary system. List all properties of the stars that can be measured from their spectra and light curves.

6. Sketch an H–R diagram. Label the axes. Show where cool supergiants, white dwarfs, the Sun, and main-sequence stars are found.

Thought Questions

7. Is the Sun an average star? Why or why not?

8. Suppose you want to determine the average education level of people throughout the nation. Since it would be a great deal of work to survey every citizen, you decide to make your task easier by asking only the people on campus. Will you get the right answer? Will your survey be distorted by a selection effect? Explain.

9. Why do most known visual binaries have relatively long periods, and most spectroscopic binaries relatively short periods?

10. Figure 17.11 shows the light curve of a hypothetical eclipsing binary star in which the light of one star is completely blocked by another. What would the light curve look like for a system in which the light of the smaller star is only partially blocked by the larger one? Assume the smaller star is the hotter one. Sketch the relative positions of the two stars that correspond to various portions of the light curve.

11. There are fewer eclipsing binaries than spectroscopic binaries. Explain why. Within 50 LY of the Sun, visual binaries outnumber eclipsing binaries. Why? Which is easier to observe at large distances—a spectroscopic binary or a visual binary?

12. The eclipsing binary Algol drops from maximum to minimum brightness in about 4 hours, remains at minimum brightness for 20 min, and then takes another 4 hours to return to maximum brightness. Assume that we view this system exactly edge-on, so that one star crosses directly in front of the other. Is one star much larger than the other, or are they fairly similar in size?

13. Consider the following data on five stars:

Star	Apparent Magnitude	Spectrum
1	12	G, main sequence
2	8	K, giant
3	12	K, main sequence
4	15	O, main sequence
5	5	M, main sequence

    a. Which is the hottest?
    b. Coolest?
    c. Most luminous?
    d. Least luminous?
    f. Most distant?

In each case, give your reasoning. (Recall that apparent magnitude is a measure of apparent brightness, where the *larger* the number, the *dimmer* the star appears to us.)

14. Which changes by the largest factor along the main sequence from spectral types O to M—mass, luminosity, or radius?

15. Suppose you want to search for main-sequence stars with very low mass using a space telescope. Will you design your telescope to detect light in the ultraviolet or the infrared part of the spectrum. Why?

16. An astronomer discovers a type-M star with a large luminosity. How is this possible? What kind of star is it?

17. Approximately 6000 stars are bright enough to be seen without a telescope. Are any of these white dwarfs? Use the information given in this chapter, and explain your reasoning.

Problems

18. Plot the luminosity functions of the nearest stars (see Appendix 10) and of the 20 brightest stars (Appendix 11). Explain how and why these two luminosity functions differ.

19. Find the combined mass of two stars in a binary system whose period of mutual revolution is two years, and for which the semimajor axis of the relative orbit is 2 AU.

20. We view a binary star exactly edge-on and observe eclipses. The star being eclipsed has an orbital velocity of 100 km/s. The time interval from first to second contact is $2^h\ 30^m$. The time from second to third contact is 10^h, and from third to fourth contact, $2^h\ 30^m$, again. What are the diameters of the two stars? How do these compare with the diameter of the Sun?

21. In Figure 17.7, is Star A or Star B more massive? Assume the orbit is viewed edge-on. What is the diameter of each star's orbit? If both stars are main-sequence stars, which is more luminous? Which is hotter?

22. Verify that a red dwarf with 1/12 the mass and 1/10 the radius of the Sun has a density 80 times that of the Sun. Calculate the density of Ross 614B, the red supergiant described in Section 17.4, which has 50 times the mass and 400 times the radius of the Sun. The outer parts of such a star would constitute an excellent laboratory vacuum.

23. Suppose you weigh 70 kg on the Earth. How much would you weigh on the surface of a white dwarf star the same size as the Earth but having a mass 300,000 times larger (nearly the mass of the Sun)?

Suggestions for Further Reading

Croswell, K. "The Grand Illusion: What We See Is Not Necessarily Representative of the Universe" in *Astronomy*, Nov. 1992, p. 44.

Croswell, K. "Visit the Nearest Stars" in *Astronomy*, Jan. 1987, p. 16. Explores the 19 nearest stars.

Davis, J. "Measuring the Stars" in *Sky & Telescope*, Oct. 1991, p. 361. Explains direct measurements of stellar diameters.

DeVorkin, D. "Henry Norris Russell" in *Scientific American*, May 1989.

Kaler, J. *Stars*. 1992, Scientific American Library/W. H. Freeman. Good modern introduction.

Kaler, J. "Journeys on the H–R Diagram" in *Sky & Telescope*, May 1988, p. 483.

Kopal, Z. "Eclipsing Binary Stars: Algol and Its Celestial Relations" in *Mercury*, May/June 1990, p. 88.

Moore, P. *Astronomers' Stars*. 1987, Norton. Focuses on 15 specific stars, including Mizar and Algol, that have been important in the development of astronomy.

Nielsen, A. "E. Hertzsprung—Measurer of Stars" in *Sky & Telescope*, Jan. 1968, p. 4.

Parker, B. "Those Amazing White Dwarfs" in *Astronomy*, July 1984, p. 15. Focuses on the history of their discovery.

Phillip, A. and Green, L. "Henry N. Russell and the H–R Diagram" in *Sky & Telescope*, April 1978, p. 306.

Using REDSHIFT™

1. Turn on only the stars and set the *Faint Magnitude Limit* to magnitude 2. Turn off the undefined and composite spectral types and undefined luminosities. Set the sky to *Mercator projection*, full sky view. Count how many stars are in each spectral type and luminosity class.

Which types are the most common?

Explain why this would be true for the brightest stars in the sky.

2. Using the full-sky *Mercator projection*, compare the colors of the stars of magnitudes 0 to 3 and the colors of stars of magnitudes 5 to 6. Is there a difference?

What do the colors say about the stars we see with the naked eye?

3. Set the *Faint Magnitude Limit* so about 10 stars appear in the *Mercator* all-sky map. Click on each star to obtain its visual magnitude and spectral type. Plot each star in a spectral type versus magnitude diagram.

Does the plot reveal any useful information? Explain why.

This beautiful image shows a giant cluster of stars called 47 Tucanae, visible from the Earth's Southern Hemisphere. Such crowded groups, which astronomers call *globular clusters*, contain hundreds of thousands of stars, including some of the RR Lyrae variables discussed in this chapter. (Photo by David Malin, courtesy of the Anglo-Australian Telescope Board)

CHAPTER 18

Celestial Distances

Thinking Ahead

When you are driving on a country road late at night, a point of light in the darkness can be a puzzling thing. Is it a nearby firefly, an oncoming motorcycle some distance away, or the porch light of a house much farther down the road? In the same way, astronomers are confronted with the question of how to tell the distances of stars when all that our eyes show us are faint points of light.

The determination of astronomical distances is central to understanding the nature of stars, but measuring such distances accurately is very difficult. After all, we cannot go out and lay a tape measure between the Sun and even the nearest star. Over the years, astronomers have developed a variety of clever techniques for estimating the vast distances that separate us from the stars. For nearby stars, we can use methods similar to the ones surveyors use here on Earth. For more distant objects, we have to apply some of the information about stellar luminosities and temperatures described in the previous two chapters, or use as a guidepost a special type of star that varies in brightness.

We now have a chain of methods for measuring cosmic distances, one that stretches from the Earth to the stars to the farthest reaches of the universe. One of the characteristics of that chain is that its links depend on one another: the measurement of distances to remote galaxies depends on the measurement of distances to the stars within our own Galaxy, which in turn depends on the accuracy of measurements within the solar system. The entire chain of cosmic distances is only as strong as its weakest

Y ou have made the universe too large, says she. I protest, said I . . . when the Heavens were a little blue arch, stuck with stars, I thought the universe was too strait and close, I was almost stifled for want of air. But now [that] it is enlarged in height and breadth . . . I begin to breathe with more freedom, and think the universe to be incomparably more magnificent than it was before.

Bernard de Fontenelle in *Conversations on the Plurality of Worlds* (1686).

link, and so it is important that every link be as accurate as possible.

In this chapter we begin with the fundamental definitions of distances on Earth, and then extend our reach outward to the distant stars.

18.1

Fundamental Units of Distance

The first measures of distances were based on human dimensions—the inch as the distance between knuckles on the finger, or the yard as the span from the extended index finger to the nose of the British king. Later, the requirements of commerce led to some standardization of such units, but each nation tended to set up its own definitions. It was not until the middle of the 18th century that any real efforts were made to establish a uniform, international set of standards.

The Metric System

One of the enduring legacies of the Napoleonic era was the establishment of the *metric system* of units, officially adopted in France in 1799 and now used in most countries around the world. The fundamental metric unit of length is the *meter*, originally defined as one ten-millionth of the distance along the Earth's surface from the equator to the pole. French astronomers of the 17th and 18th centuries were pioneers in determining the dimensions of the Earth, so it was logical to use their information as the foundation of the new system.

Practical problems exist with a definition expressed in terms of the size of the Earth, since anyone wishing to determine the distance from one place to another can hardly be expected to go out and remeasure the planet. Therefore an intermediate standard meter consisting of a bar of platinum–iridium metal, was set up in Paris. In 1889, by international agreement, this bar was defined to be exactly 1 m in length, and precise copies of the original meter bar were made to serve as standards for other nations.

Other units of length are derived from the meter. Thus 1 kilometer (km) equals 1000 m, 1 centimeter (cm) equals 1/100 m, and so on. Even the old British and American units, such as the inch and the mile, are now defined in terms of the metric system.

Modern Redefinitions of the Meter

In 1960 the official definition of the meter was changed again. As a result of improved technology for generating spectral lines of precisely known wavelength, the meter was redefined to equal 1,650,763.73 wavelengths of a particular atomic transition in krypton-86. The advantage of this redefinition is that anyone with a suitably equipped laboratory can reproduce a standard meter, without reference to any particular metal bar.

In 1983 the meter was redefined once more, this time in terms of the velocity of light. At this point, the length of the standard unit of time, the *second*, had been fixed by international agreement as 9,192,631,770 times the frequency of a cesium-133 atomic clock. The meter was measured to be the distance light travels in a vacuum in a time interval of 1/299,792,458.6 s. This then defines the speed of light in a vacuum to be 299,792,458.6 m/s. Today, therefore, light travel-time provides us our basic unit of length. Putting it another way, a distance of one *light second* (LS) (the amount of space light covers in one second) is defined to be 299,792,458.6 m. We could just as well use the light second as the fundamental unit of length, but for practical reasons (and to respect tradition), we have defined the meter as a small fraction of the light second.

Distances Within the Solar System

The work of Copernicus and Kepler (see Chapters 1 and 2) established the *relative* distances of the planets—that is, how far from the Sun one planet is compared to another. But their work could not establish the *absolute* distances (in light seconds or meters or other standard units of length). This is like knowing the height of all the students in your class only as compared to the height of your astronomy instructor, but not in inches or centimeters. Somebody's height has to be measured directly.

Similarly, to establish absolute distances, astronomers had to measure one distance in the solar system directly. Estimates of the distance to Venus were made as Venus crossed the face of the Sun in 1761 and 1769, and an international campaign was organized to estimate the distance to the Earth-approaching asteroid Eros in the early 1930s. But not until the past three decades could planetary distances be measured with extremely high precision.

The key to our modern determination of solar system dimensions is *radar*, a type of radio wave that can bounce off solid objects (Figure 18.1). As discussed in several earlier chapters, by timing how long a radar beam (traveling at the speed of light) takes to reach another world and return, we can measure the distance involved very accurately. In 1961, radar signals were bounced off Venus for the first time, providing a direct measurement of the distance from Earth to Venus in terms of light seconds (the round-trip travel time of the radar signal). Subsequently, radar has also been used to determine the distances to Mercury, Mars, the satellites of Jupiter, the rings of Saturn, and several asteroids.

The "measuring stick" of distance astronomers use within the solar system is the *astronomical unit* (AU), the average distance from the Earth to the Sun. We then express all the other distances in the solar system in terms of the astronomical unit. Note, by the way, that it is not possible to use radar to measure the distance to the Sun directly because the Sun does not reflect radar very efficiently. But we can measure the distance to many other

Figure 18.1
Radar telescope of the NASA Deep Space Network in California's Mojave Desert. This dish-shaped antenna can send and receive radar waves, and measure the distances to planets, satellites, and asteroids. (NASA/JPL)

solar system objects and use Kepler's laws to give us the length of the astronomical unit.

Years of painstaking analyses of radar measurements have led to a determination of the length of the astronomical unit to a precision of about one part in a billion. The length of 1 AU can be expressed in light travel time as 499.004854 LS, or about 8.3 light minutes (LM). If we use the definition of the meter given previously, this is equivalent to 1 AU = 149,597,892,000 m.

These distances are, of course, given here to a much higher level of precision than is normally needed. In this text, we are usually content to express numbers to a couple of significant places and leave it at that. For our purposes it will be sufficient to round off these numbers:

Speed of light: $c = 3.00 \times 10^8$ m/s $= 3.00 \times 10^5$ km/s

Length of light second: LS $= 3.00 \times 10^8$ m
$= 3.00 \times 10^5$ km

Astronomical unit: AU $= 1.50 \times 10^{11}$ m $= 1.50 \times 10^8$ km
$= 500$ LS

We now know the absolute distance scale within our own solar system with fantastic accuracy. This is the first link in the chain of cosmic distances.

Surveying the Stars

It is an enormous step to go from the planets to the stars. The nearest star is hundreds of thousands of astronomical units from the Earth. Yet in principle, we can survey distances to the stars using the same technique that a civil engineer employs to survey the distance to an inaccessible mountain or tree—the method of *triangulation.*

Triangulation

A practical example of triangulation is your own sense of depth perception. As you are pleased to discover every morning when you look in the mirror, your two eyes are located some distance apart and thus view the world from two different vantage points. This dual perspective allows you to get a general sense of how far away objects are.

To see what we mean, take a pen and hold it a few inches in front of your face. Look at it first with one eye (closing the other), and then switch eyes. Note how the pen seems to shift relative to objects across the room. Now hold the pen at arm's length: the shift is less. If you play with the pen for awhile, you will notice that the farther away you hold it, the less it seems to shift. Your brain automatically performs such comparisons and gives you a pretty good sense of how far away things in your immediate neighborhood are.

If your arms were made of rubber, you could stretch the pen far enough away from your eyes that the shift would become imperceptible. This is because our depth perception fails for objects more than a few tens of meters away. It would take a larger distance between viewing perspectives than the spacing between the eyes to see the shift of an object a city block or more from you.

Let's see how surveyors take advantage of this idea. Suppose you are trying to measure the distance to a tree across a deep river (Figure 18.2). You set up two observing stations some distance apart. That distance (AB in Figure 18.2) is called the *baseline.* Now the direction to the tree (C in the figure) in relation to the baseline is observed from each station. Note that C appears in different directions from the two stations. This apparent change in direction of the remote object due to a change in vantage point of the observer is called **parallax.** The parallax is also the angle that lines AC and BC make—in mathematical terms, the angle subtended by the baseline. A knowledge of the angles at A and B, and the length of the baseline, AB, allows the triangle ABC to be solved for any of its dimensions—say, the distance AC or BC. The solution could be accomplished by constructing a scale drawing, or by using the technique of trigonometry to make a numerical calculation. The greater the parallax, the nearer the object.

The farther away the tree (or astronomical object), the longer the baseline has to be to give us a reasonable measurement. Unfortunately, nearly all astronomical ob-

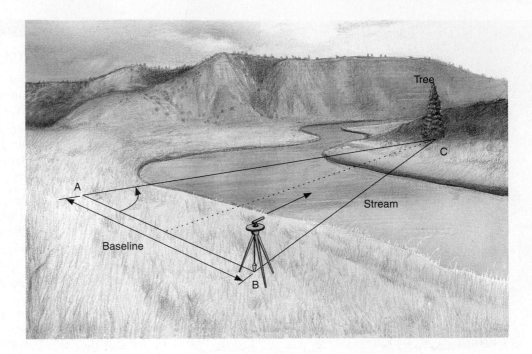

Figure 18.2
Triangulation allows us to measure distances to inaccessible objects. By getting the angle to a tree from two different vantage points, we can calculate the properties of the triangle they make, and thus the distance to the tree.

jects are very far away. To measure their distances requires either a very large baseline or highly precise angular measurements, or both. The Moon is the only object near enough that its distance can be found fairly accurately with measurements made without a telescope. Ptolemy determined the distance to the Moon correctly to within a few percent. He used the Earth itself as a baseline, measuring the position of the Moon relative to the stars at two different times of night.

With the aid of telescopes, later astronomers were able to measure the distances to the nearer planets or asteroids by using the Earth's diameter as a baseline. This is how the astronomical unit was first established. To reach for the stars, however, requires a much longer baseline for triangulation, and extremely sensitive measurements. Such a baseline is provided by the Earth's annual trip around the Sun.

Distances to Stars

As the Earth travels from one side of its orbit to the other, it graciously provides us with a baseline of 2 AU, or about 300 million km. Although this is a much bigger baseline than the diameter of the Earth, the stars are so far away that the resulting parallax shift is still not visible to the naked eye for even the closest stars.

In Chapter 1 we discussed how this perplexed the ancient Greeks, some of whom had suggested that the Sun might be the center of the solar system, with the Earth in motion around it. Aristotle and others argued, however, that the Earth could not be revolving about the Sun. If it were, they said, we would observe the parallax of the nearer stars against the background of more distant objects as we viewed the sky from different parts of the

Earth's orbit (Figure 18.3). Tycho Brahe advanced the same argument nearly 2000 years later, when his careful measurements of stellar positions with the unaided eye revealed no such shift. These early observers did not realize how truly distant the stars were, and did not have the tools to measure parallax shifts too small to be seen with the human eye.

By the 18th century, when there was no longer serious doubt about the Earth's revolution, it became clear that the stars must be extremely distant. Astronomers equipped with telescopes began to devise instruments capable of measuring the tiny shifts of nearby stars relative to the background of more distant (and thus unshifting) celestial objects. This was a significant technical challenge, since, even for the nearest stars, parallax angles are usually only a fraction of a second of arc. Recall that one second of arc is an angle of only 1/3600th of a degree (see Chapter 1). A coin the size of a quarter would appear to have a diameter of 1 arcsec if you were viewing it from a distance of about 3 miles, or 5 km. No wonder it took astronomers a while before they could measure such tiny shifts.

The first successful detections of stellar parallax were in the year 1838, when Friedrich Bessel (in Germany), Thomas Henderson (a Scottish astronomer working at the Cape of Good Hope), and Friedrich Struve (in Russia) independently measured the parallaxes of the stars 61 Cygni, Alpha Centauri, and Vega, respectively (Figure 18.4). Even the closest star, Alpha Centauri, showed a total displacement of only about 1.5 arcsec during the course of a year.

Figure 18.3 shows how such measurements work; seen from opposite sides of the Earth's orbit a star shifts positions when compared to a pattern of more distant

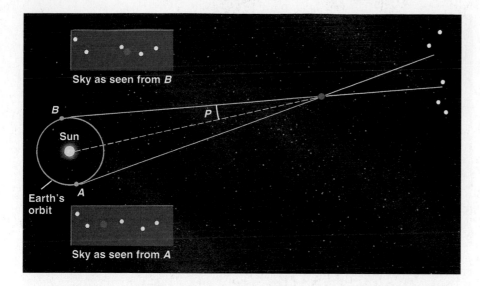

Figure 18.3
As the Earth revolves around the Sun, the direction in which we see a nearby star varies with respect to distant stars. We define the parallax of the nearby star to be one-half of the total change in direction, and usually measure it in arcseconds.

Sky as seen from *B*

B

Sun

P

Earth's orbit

A

Sky as seen from *A*

stars. Astronomers actually define parallax to be *one-half* the angle that a star shifts when seen from opposite sides of the Earth's orbit (the angle labeled *P* in Figure 18.3). The reason for this is just that they prefer to deal with a baseline of 1 AU instead of 2 AU.

Units of Stellar Distance

With a baseline of 1 AU, how far away would a star have to be to have a parallax of 1 arcsec? The answer turns out to be 206,265 AU, or 3.1×10^{13} km (in words, 31 million million kilometers). We give this unit a special name, the **parsec** (abbreviated **pc**)—derived from "the distance at which we have a **par**allax of one **sec**ond." The distance of

a star in parsecs (r) is just the reciprocal of its parallax (p) in arcseconds. That is,

$$r = \frac{1}{p}$$

Thus a star with a parallax of 0.1 arcsec would be found at a distance of 10 pc, and one with a parallax of 0.05 arcsec would be 20 pc away.

Back in the days when most of our distances came from parallax measurements, this was a useful unit of distance, but it is not as intuitive as the **light year,** which we defined in Section 17.1. Light years have the advantage of being related directly to both the definition of the meter, which is expressed in terms of the speed of light, and the radar measurements that determine the length of the astronomical unit. In this text, we will use light years as our unit of distance, but many astronomers still use parsecs when they write technical papers or talk with each other at meetings. Conversion between the two distance units is simple: 1 pc = 3.26 LY, and 1 LY = 0.31 pc.

Another advantage of the light year as a unit is that it emphasizes the fact that as we look out into space, we are also looking back into time. The light that we see from a star 100 LY away left that star 100 years ago. What we study is not the star as it is now, but rather as it was in the past. The light from a distant galaxy that reaches our telescopes today most likely left its source before the Earth even existed.

Figure 18.4
Friedrich Wilhelm Bessel (1784–1846) made the first authenticated measurement of the distance to a star (61 Cygni) in 1838, a feat that had eluded many dedicated astronomers for almost a century. (Yerkes Observatory)

ASTRONOMY BASICS

Naming Stars

If you've read this far, you have probably noticed that stars have a confusing assortment of names. Just look at the first three stars to have their parallax measured: 61 Cygni, Alpha Centauri, and Vega. Each of these names comes from a different tradition of designating stars.

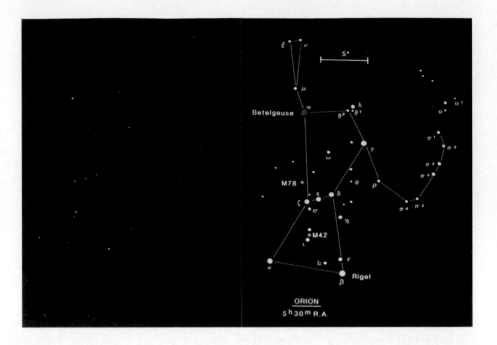

A photograph and diagram of the brightest objects in or near the star pattern of Orion the hunter (of Greek mythology) in the constellation of Orion. The Greek letters in Bayer's system are shown in yellow. The objects denoted M42 and M78 are not stars but *nebulae*—clouds of gas and dust; these numbers come from a list of "fuzzy objects" made by Charles Messier in 1781. (Richard Norton, Science Graphics)

The brightest stars have names that derive from the ancients. Some are from the Greek, such as Sirius, which means "the scorched one,"—a reference to its brilliance. A few are from Latin, but many of the best-known names are from Arabic because, as discussed in Chapter 1, much of Greek and Roman astronomy was "rediscovered" in Europe after the dark ages by means of Arabic translations. Vega, for example, means "swooping Eagle," and Betelgeuse (pronounced "Beetle-juice") means "right hand of the central one."

In 1603 the German astronomer Johann Bayer introduced a more systematic approach to naming stars. For each constellation, he assigned a Greek letter to the brightest stars, roughly in order of brightness. In the constellation of Orion, for example, Betelgeuse is the brightest star, so it got the first letter in the Greek alphabet—alpha—and is known as Alpha Orionis. (Orionis is the possessive form of Orion, so Alpha Orionis means "the first of Orion.") A star called Rigel, being the second brightest in that constellation, is called Beta Orionis. Since there are 24 letters in the Greek alphabet, this system allows the labeling of 24 stars in each constellation, hardly a complete sample.

In 1725 the English Astronomer Royal John Flamsteed introduced yet another system, in which the brighter stars eventually got a number in each constellation in order of their location or, more precisely, their right ascension. The system of sky coordinates that includes right ascension was discussed in Chapter 1. In this system Betelgeuse is called 58 Orionis and 61 Cygni is the 61st star in the constellation of Cygnus, the swan.

It gets worse. As astronomers began understanding more and more about stars, they drew up a series of specialized star catalogs, and fans of those catalogs began calling stars by their catalog numbers. If you look at Appendix 10—our list of the nearest stars (many of which are much too faint to get an ancient name, Bayer letter, or Flamsteed number)—you will see references to some of these catalogs. An example is a set of stars labeled with a BD number, for "Bonner Durchmusterung." This was a mammoth catalog of over 324,000 stars in a series of zones in the sky, organized at the Bonn Observatory in the 1850s and 1860s. Keep in mind that this catalog was made before photography or computers came into use, so the position of each star had to be measured (at least twice) by eye, a daunting undertaking.

There is also a completely different system for keeping track of stars whose luminosity varies, and another for stars that brighten explosively at unpredictable times. Astronomers have gotten used to the many different star-naming systems, but students often find them bewildering and wish astronomers would settle down to one. Don't hold your breath: in astronomy, as in many fields of human thought, tradition holds a powerful attraction. Still, with high-speed computer databases to aid human memory, names may become less and less necessary. Today, astronomers often refer to stars by their precise locations on the sky rather than by their names or catalog numbers.

The Nearest Stars

No known star (other than the Sun) is within 1 LY or even 1 pc of the Earth. The stellar neighbors nearest to the Sun are three stars that make up a multiple system in the constellation of Centaurus. To the unaided eye the system appears as a single bright star, called Alpha Centauri, only 30° from the south celestial pole and hence not visible from the mainland United States. Alpha Centauri itself is a binary star—two stars in mutual revolution, too close together to be distinguished without a telescope. These two stars are 4.4 LY from us. Nearby is the third member of the system, a faint star known as Proxima Centauri. Proxima, with a distance of 4.3 LY, is slightly closer to us than the other two stars. (By the way, a few astronomers

Parallax and Space Astronomy

One of the most difficult things about precisely measuring the tiny angles of parallax shifts from Earth is that you have to observe the stars through our planet's atmosphere. As we saw in Chapter 5, the effect of the atmosphere is to spread out the points of starlight into fuzzy disks, making exact measurements of their positions more difficult. Astronomers have long dreamed of being able to measure parallax from space, and two instruments have recently turned this dream into reality.

The name of the Hipparcos satellite, launched in 1989 by the European Space Agency, is both an abbreviation for **High Precision Par**allax **Co**llecting **S**atellite and a tribute to Hipparchus, the pioneering Greek astronomer whose work we discussed in Chapter 1. The satellite was designed to make the most accurate parallax measurements in history from 36,000 km above the Earth. However, the failure of its rocket motor meant it did not get the needed boost to reach the desired altitude, and it spent its four-year life in an elliptical orbit that varied from 500 to 36,000 km high. In this orbit, the satellite repeatedly passed through the Earth's radiation belts, which fi-nally took its toll on the sensitive electronics.

Nevertheless, the satellite instruments were able to measure the positions of about 100,000 stars to an accuracy of 0.001 to 0.002 arcsec, as well as to make very accurate measurements of star motions. When these observations have been processed and published, they will form the most accurate catalog of stellar distances yet available.

Artist's conception of the Hipparcos satellite. (ESA)

The Hubble Space Telescope, whose on-board sensors must pinpoint with great accuracy the direction in which the telescope is pointing, can also perform precise parallax measurements for relatively nearby stars. In the next decade, astronomers will thus be able to know the distances of the stars within our own "corner" of the Galaxy with much greater precision. But no instruments, on Earth or in space, can overcome the natural limitations of this technique. Beyond a certain distance, one simply cannot triangulate; the more remote stars guard the secret of their distances too jealously.

---

have started to question whether Proxima is actually bound to the Alpha Centauri pair; some lines of evidence show that it may simply be passing close to Alpha temporarily.)

The nearest star visible without a telescope from most parts of the United States is the brightest-appearing of all the stars, Sirius, which has a distance of 8 LY. As we saw in Chapter 17, it too is a binary system, composed of a faint white dwarf orbiting a bluish-white main-sequence star. It is interesting to note that light reaches us from the Sun in about 8 min, and from Sirius in about 8 years.

The Extent of Parallax Measurements

Even though the parallaxes of stars are very small, today we can measure them to within a few thousandths of an arcsecond with ground-based telescopes. So far, parallaxes have been measured for more than 10,000 stars. Only for a fraction of them, however, are the parallaxes large enough (about 0.05 arcsec or more) to be measured with a precision of 10 percent or better, which is what astronomers need to make effective use of the distance measurement. A parallax of 0.05 arcsec corresponds to a distance of 20 pc, or a little more than 60 LY, a very small distance in the cosmic scheme of things. Measurements from space will soon expand the reach of accurate parallaxes to 1000 LY or more (see "Making Connections" box). However, even 1000 LY is merely 1 percent the size of our Galaxy's main disk. If we are going to reach very far away from our own neighborhood with our chain of methods for measuring cosmic distances, we need some completely new techniques.

explained in ch 25

Variable Stars: One Key to Cosmic Distances

Standard Bulbs Revisited

Let's briefly review why measuring distances to the stars is such a struggle. As discussed in Section 16.1, our problem is that stars come in a bewildering variety of intrinsic luminosities. (If stars were light bulbs, we'd say they come in a wide range of wattages.) Suppose, instead, that all stars had the same wattage or luminosity; in that case the more distant ones would always look dimmer, and we could tell how far away a star was simply by how dim it appeared. In the real universe, however, when we look at a star in our sky (with eye or telescope) and measure its apparent brightness, we cannot know whether it looks dim because it's a low-wattage bulb, because it is far away, or perhaps some of each.

Astronomers need to discover something else about the star that allows us to "read off" its intrinsic luminosity—in effect, to know what the star's true wattage is. With this information, we can then attribute how dim it looks from Earth to its distance. Recall that the apparent brightness of an object decreases with the square of the distance to that object. Therefore, if we know the luminosity of a star and its apparent brightness, we can calculate how far away it is.

Thus astronomers have searched for techniques that allow us to determine the luminosity of a star, and it is to these techniques that we turn next.

Variable Stars

The breakthrough in measuring distances to remote parts of our own Galaxy, and to other galaxies as well, came from the study of *variable stars*. Most stars are constant in their luminosity, at least to within a percent or two. Like the Sun, they generate a steady flow of energy from their interiors. However, some stars are seen to vary in brightness and for this reason are called variable stars. Many such stars vary on a regular cycle, like the flashing bulbs that decorate stores and homes during the winter holidays.

Let's define some tools to help us keep track of how a star varies. A graph that shows how the brightness of a variable star changes with time is called a **light curve** (Figure 18.5). The *maximum* is the point of the light curve where the star has its greatest brightness; the *minimum* is the point where it is faintest. If the light variations repeat themselves periodically, the interval between the two maxima is called the *period* of the star.

Pulsating Variables

There are two special types of variable stars for which—as we will see—measurements of the light curve give us accurate distances. These are the **cepheids** and the **RR Lyrae variables,** both of which are **pulsating variable stars.** A pulsating variable star actually changes its diameter with time—periodically expanding and contracting, as your chest does when you breathe. We now understand that these stars are going through a brief unstable stage in their lives. Their expansion and contraction can be measured by using the Doppler effect. The spectral lines shift toward the blue as the surface of the star moves toward us, and then to the red as it shrinks back. As the star pulsates, it changes overall color, indicating that its temperature is also varying. And, most important for our purposes, the luminosity of the pulsating variable also changes in a regular way as it expands and contracts.

Cepheid Variables

Cepheids are large, yellow, pulsating stars named for the first-known star of the group, Delta Cephei. The variability of Delta Cephei was discovered in 1784 by the young English astronomer John Goodricke (see "Voyagers in Astronomy" box). The star rises rather rapidly to maximum light and then falls more slowly to minimum light, taking a total of 5.4 days for one cycle. The curve in Figure 18.5 represents the light variation of Delta Cephei.

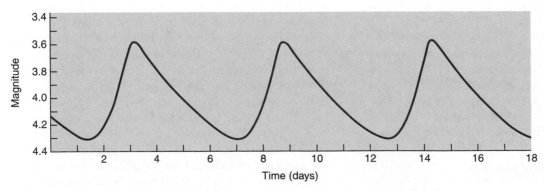

Figure 18.5
Light curve (plot of brightness as a function of time) of a typical cepheid variable.

John Goodricke (1764–1786)

The brief life of John Goodricke is a testament to the human spirit under adversity. Born deaf, and unable to speak, Goodricke nevertheless made a number of pioneering discoveries in astronomy through patient and careful observations of the heavens.

Born in Holland where his father was on a diplomatic mission, Goodricke was sent back to England at age eight to study at a special school for the deaf. He did sufficiently well to enter Warrington Academy, a secondary school that offered no special assistance for students with handicaps. His mathematics teacher there inspired an interest in astronomy, and in 1781, at age 17, Goodricke began observing the sky at his family home in York, England. Within a year he had discovered the brightness variations of the star Algol (discussed in Chapter 17), and suggested that an unseen companion star was causing the changes, a theory that waited over 100 years for proof. His paper on the subject was read before the Royal Society (the main British group of scientists) in 1783, and won him a medal from that distinguished group.

In the meantime, Goodricke had discovered two other stars that varied regularly, Beta Lyrae and Delta Cephei, both of which contin-

A portrait of John Goodricke by artist J. Scouler, which now hangs in the Royal Astronomical Society in London. There is some controversy about whether this is actually what Goodricke looked like, or whether the painting was much retouched to please his family. (Courtesy of the San Diego State University special collections library)

ued to interest astronomers for years to come. Goodricke shared his interest in observing with his older cousin, Edward Pigott, who went on to discover other variable stars during his much longer life. But Goodricke's time was quickly drawing to a close; at age 21, only two weeks after he was elected to the Royal Society, he caught a cold while making astronomical observations and never recovered.

Today the University of York has a building named Goodricke Hall, and a plaque that honors his contributions to science. Yet if you go to the churchyard cemetery where he is buried, an overgrown tombstone has only the initials "J.G." to show where he lies. Astronomer Zdenek Kopal, who has looked carefully into Goodricke's life, has speculated on why the marker is so modest: perhaps the rather staid Goodricke relatives were ashamed of having a "deaf-mute" in the family, and could not sufficiently appreciate how much a man who could not hear could nevertheless see.

Several hundred cepheid variables are known in our Galaxy. Most cepheids have periods in the range of 3 to 50 days, and luminosities that are about 1000 to 10,000 times greater than that of the Sun. Their variations in luminosity range from a few percent to a factor of 10. Polaris, the North Star, is a cepheid variable that for a long time varied by one-tenth of a magnitude, or by about 10 percent in visual luminosity, in a period of just under four days. Recent measurements indicate that the amount by which the brightness of Polaris changes is decreasing, and that sometime in the future this star will no longer be a pulsating variable. This is just one more piece of evidence that stars really do evolve and change in fundamental ways as they age.

The Period–Luminosity Relationship

The importance of cepheid variables lies in the fact that their periods and average luminosities are directly related. The longer the period (the longer the star takes to vary), the greater the luminosity. This **period–luminosity** relationship was a remarkable discovery, one for which astronomers still (pardon the expression) thank their lucky stars. The period of such a star is easy to measure—a good telescope and a good clock are all you need. Once you have the period, the relationship (which can be put into precise mathematical terms) will give you the luminosity (the actual wattage) of the star. Astronomers can then compare this intrinsic brightness with the apparent brightness: as we saw, the difference between the two allows them to calculate the distance.

The relation between period and luminosity was discovered in 1908 by Henrietta Leavitt, a staff member at the Harvard College Observatory (one of a number of women working for low wages assisting Edward Pickering, the Observatory's director—see the "Voyagers in Astronomy" box on Annie Cannon in Chapter 16). She did not publish the results in a form that attracted the attention of astronomers until four years later. Some hundreds of cepheid variables had been discovered in the Large and Small Magellanic Clouds (Figure 18.6), two great star systems that are actually neighboring galaxies (although they were not known to be galaxies then).

These systems presented a wonderful opportunity to study the behavior of variable stars independent of their distance. For all practical purposes, the Magellanic

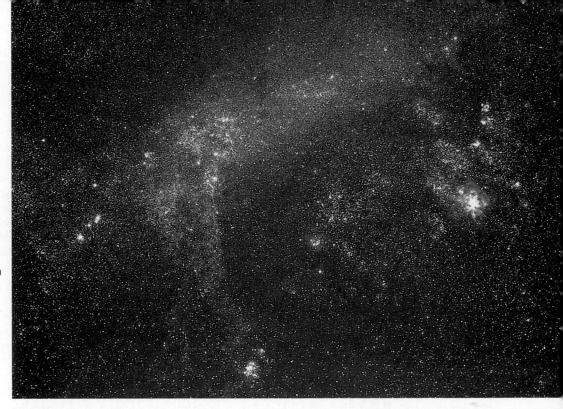

Figure 18.6
The Large Magellanic Cloud, a small irregular-shaped galaxy near our own Milky Way. It was in this galaxy that Henrietta Leavitt discovered the cepheid period–luminosity relationship. (National Optical Astronomy Observatories)

Clouds are so far away that astronomers can assume that all the stars in them are at roughly the same distance from us. (In the same way, all the suburbs of Los Angeles are roughly the same distance from New York City. Of course if you are *in* Los Angeles, you will notice distances between the suburbs; but compared to how far away New York City is, they seem small.) If all the variable stars in the Magellanic Clouds are at roughly the same distance, any difference in their apparent brightness must be a reflection of differences in their intrinsic luminosities.

Leavitt found that the brighter-appearing cepheids always have the longer periods of light variation. Thus, she reasoned, the period must be related to the luminosity of the stars. When Leavitt did this work, the distance to the Magellanic Clouds was not known, so she was only able to show differences in luminosities, not to calculate the actual cepheid luminosities themselves. To define the period–luminosity relationship with actual numbers (to *calibrate* it), astronomers first had to measure the actual distances to a few nearby cepheids in another way. (This was accomplished by finding cepheids associated in clusters with other stars whose distances could be estimated from their spectra, as discussed in the next section of this chapter.) Once the relationship was thus defined, it could give us the distance to any cepheid, wherever it might be located.

Here at last was the technique astronomers had been searching for to break the confines of distance that parallax imposed on them. Cepheids could be observed and monitored, it turned out, in many parts of our own Galaxy and in other galaxies as well. Astronomers, including Ejnar Hertzsprung and Harvard's Harlow Shapley, immedi-

ately saw the potential of the new technique; they and many others set to work exploring more-distant reaches of space using cepheids as signposts. As we will see, this work still continues, as the Hubble Space Telescope and other modern instruments try to identify and measure individual cepheids in galaxies farther and farther away (Figure 18.7).

RR Lyrae Stars

A related group of stars, whose nature was understood somewhat later than that of the cepheids, are called RR Lyrae variables, named for the star RR Lyrae, the best-known member of the group. More common than the cepheids but also less luminous, thousands of these pulsating variables are known in our Galaxy. The periods of RR Lyrae stars are always less than one day, and most of them change in brightness by less than one magnitude.

Astronomers have observed that the RR Lyrae stars occurring in any particular star cluster all have about the same apparent magnitude. Since stars in a cluster are all at approximately the same distance, it follows that RR Lyrae variables must all have nearly the same intrinsic luminosity—which turns out to be about 50 L_{Sun} (in this sense, RR Lyrae stars are a little bit like standard bulbs). Figure 18.8 displays the ranges of periods and absolute magnitudes for both the cepheids and the RR Lyrae stars.

RR Lyrae stars can be detected out to a distance of about 2 million LY, and cepheids to about 60 million LY. Compare these limits with parallaxes that even from space will probably not be measured for stars more distant than

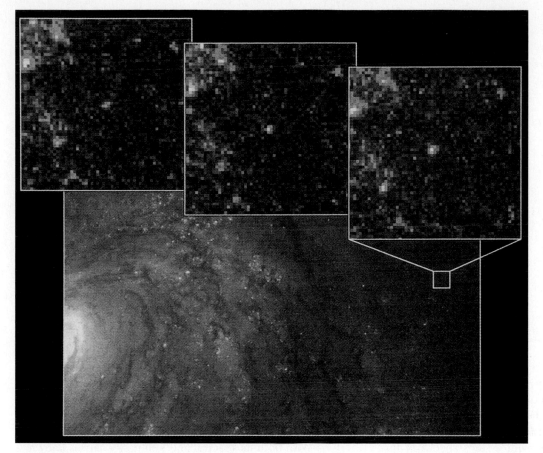

Figure 18.7
This image of part of the galaxy called M100 was taken with the Hubble Space Telescope in 1994. The insets show a single cepheid in the galaxy, going through its cycle of brightness variations. What makes this image remarkable is that M100 is 51 million LY away, the most distant galaxy in which individual cepheids had been identified and measured at that time. (Wendy Freedman, Carnegie Institution of Washington, and NASA)

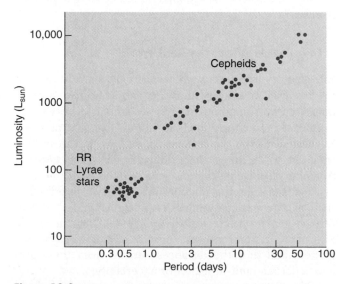

Figure 18.8
The period–luminosity relation for cepheid variables. Also shown are the period and luminosity for RR Lyrae stars.

about 1000 LY. You can see from this comparison just how important the discovery of the period–luminosity relationship was in enabling astronomers to extend their measurements of cosmic distances.

18.4

The H–R Diagram and Cosmic Distances

Distances from Spectral Types

As satisfying and productive as variable stars have been for distance measurement, their use still has many limitations. An obvious one is that you must first find a pulsating variable in a star group before this method will yield a distance. Suppose we need the distance to a star or stars where no variables are to be found nearby. In this case, the H–R diagram can come to our rescue.

If we can observe the spectrum of a star, we can estimate its distance from our understanding of the H–R diagram. As discussed in Chapter 16, a detailed examination of a stellar spectrum allows astronomers to classify the star into one of the *spectral types* indicating surface temperature. (The types are O, B, A, F, G, K, and M; each of these can be divided into numbered subgroups.) In general, however, the spectral type alone is not enough to al-

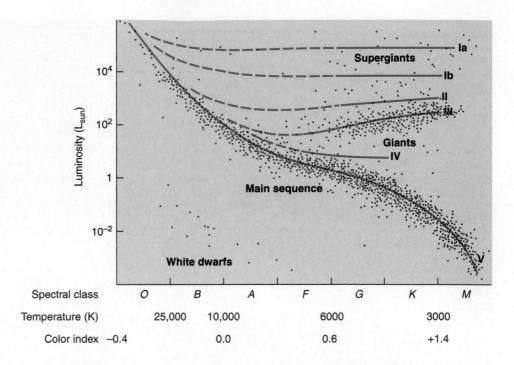

Figure 18.9
Luminosity classes for stars on the Hertzsprung–Russell diagram.

Spectral class	O	B	A	F	G	K	M
Temperature (K)	25,000	10,000		6000		3000	
Color index	−0.4		0.0		0.6		+1.4

low us to estimate luminosity. Look again at Figure 17.14. A G2 star could be a main-sequence star with a luminosity of 1 L_{Sun}, or it could be a giant with a luminosity of 100 L_{Sun}, or even a supergiant with a still-higher luminosity.

But we can learn more from a star's spectrum than just its temperature. Remember, for example, that we can tell pressure differences in stars from the details of the spectrum (see Section 16.3). This is very useful, because giant stars are larger (and have lower pressures) than main-sequence stars, and supergiants are still larger than giants. If we look in detail at the spectrum of a star, we can determine whether it is a main-sequence star, a giant, or a supergiant.

Suppose, for example, that the spectrum, color, and other properties of a distant G2 star match those of the Sun exactly. It is then reasonable to conclude that this distant star is likely to be a main-sequence star just like the Sun, and to have the same luminosity as the Sun.

The most widely used system of star classification divides stars of a given spectral class into six categories called **luminosity classes.** These luminosity classes are denoted by Roman numerals as follows:

Ia: Brightest supergiants
Ib: Less-luminous supergiants
II: Bright giants
III: Giants
IV: Subgiants (intermediate between giants and main-sequence stars)
V: Main-sequence stars

The full spectral specification of a star includes its luminosity class. For example, a main-sequence star with spectral class F3 is written as F3 V. For an M2 giant, the specification would be M2 III. Figure 18.9 illustrates the approximate positions of stars of various luminosity classes on the H–R diagram. The dashed portions of the lines represent regions with very few or no stars.

With both its spectral and luminosity classes known, a star's position on the H–R diagram is uniquely determined. Since the diagram plots luminosity versus temperature, this means we can now read off the star's intrinsic luminosity. As before, if we know how luminous the star really is, and see how dim it looks, the difference allows us to calculate its distance. (For historical reasons, astronomers sometimes call this method of distance determination *spectroscopic parallax;* we find this name misleading, however, since the method has nothing to do with parallax.)

A Few Words About the Real World

Introductory textbooks such as ours work hard to present the material in a straightforward and simplified way. In doing so, we sometimes do our students a disservice by making scientific techniques seem *too* clean and painless. In the real world, the techniques we have just described turn out to be messy and difficult, and often give astronomers headaches that last long into the day!

For example, the relationships we have described—such as the period–luminosity relationship for certain variable stars—aren't exactly straight lines on a graph. The points representing many stars scatter widely when plotted, and thus the distances derived from them also have a certain built-in scatter or uncertainty.

The distances we measure with the methods we have discussed are therefore only accurate to within a certain percentage of error—sometimes 10 percent, sometimes 25 percent, sometimes as much as 50 percent or more. A 25 percent error for a star estimated to be 10,000 LY away means it could be anywhere from 7500 to 12,500 LY away. This would be an unacceptable uncertainty if you were loading fuel into a spaceship for a trip to the star, but is not a bad first figure to work with if you are an astronomer stuck on planet Earth.

Nor is the construction of H–R diagrams as easy as you might think at first. To make a good diagram, you need to measure the characteristics of many stars, which can be a time-consuming task. If the process is to advance our knowledge, the stars you measure must be far away; hence spectra are difficult to obtain. Astronomers and their graduate students may have to spend many nights at the telescope (and many days back home working with their data) before they get their distance measurement.

Nevertheless, with these tools—parallaxes for the nearest stars, RR Lyrae variable stars and the H–R diagram for clusters of stars in our own and nearby galaxies, and cepheids out to distances of 50 million LY—we can measure distances throughout our own Galaxy and beyond to neighboring stellar systems. Such distances, combined with measurements of composition, luminosity, and temperature made with the techniques described in Chapters 16 and 17, are the arsenal of information we need to trace the evolution of stars from birth to death, the subject to which we turn in the chapters that follow.

Summary

18.1 Early measurements of length were based on human dimensions, but today we use worldwide standards that specify lengths in units such as the meter. Distances within the solar system are now determined by timing how long it takes radar signals to travel from the Earth to the surface of a planet or other body, and then return.

18.2 Half the shift in a nearby star's position relative to very distant background stars, as viewed from opposite sides of the Earth's orbit, is called the **parallax** of that star and is a measure of its distance. The units used to measure stellar distance are the **light year (LY),** the distance light travels in one year, and the **parsec (pc),** the distance of a star with a parallax of 1 arcsec. One parsec = 3.26 LY. The first successful measurements of stellar parallaxes were reported in 1838. The star nearest the Sun is Alpha Centauri, which is actually a system of three stars at a distance of about 4.4 LY. Parallax measurements are a fundamental link in the chain of cosmic distances.

18.3 **Cepheids** and **RR Lyrae** stars are two types of **pulsating variable stars.** *Light curves* of these stars show that their luminosities vary with a regularly repeating period. Both types of variables obey a **period–luminosity** relationship, so measuring their periods can tell us their luminosities. Then we can calculate their distances by comparing their luminosities with their apparent brightnesses.

18.4 Stars with identical temperatures but different pressures (and diameters) have slightly different spectra. Spectral classification can therefore be used to estimate the **luminosity class** of a star, as well as its temperature. By seeing where the star is located on an H–R diagram, we can establish its luminosity. This, with the star's apparent brightness, again yields its distance.

Review Questions

1. Explain how parallax measurements can be used to determine distances to stars. Why can we not make accurate measurements of parallax beyond a certain distance?

2. Make up a table relating the following units of astronomical distance: kilometer, Earth radius, solar radius, astronomical unit, light year, and parsec.

3. Suppose you have discovered a new RR Lyrae variable star. What steps would you take to determine its distance?

4. Explain how you would use the spectrum of a star to estimate its distance.

5. Which method would you use to obtain the distance to each of the following?
- **a.** An asteroid crossing the Earth's orbit.
- **b.** A star astronomers believe to be no more than 50 LY from the Sun.
- **c.** A group of stars in the Milky Way Galaxy that includes a significant number of variable stars.
- **d.** A star that is not variable, but for which you can obtain a clearly defined spectrum.

6. What would be the advantage of making parallax measurements from Pluto rather than from Earth? Would there be a disadvantage?

7. Parallaxes are measured in fractions of an arcsecond. One arcsecond equals 1/60 arcmin; an arcminute is in turn 1/60°. To get some idea of how big 1° is, go outside at night and find the Big Dipper. The two pointer stars at the end of the bowl are 5.5° apart. The two stars across the top of the bowl are 10° apart. (Ten degrees is also about the width of your fist when held at arm's length and projected against the sky.) Mizar, the second star from the end of the Big Dipper's handle, appears double. The fainter star, Alcor, is about 12 arcmin from Mizar. For comparison, the diameter of the full Moon is about 30 arcmin. The belt of Orion is about 3° long. Why did it take until 1838 to make parallax measurements for even the nearest stars?

8. For centuries, astronomers wondered whether comets were true celestial objects, like the planets and stars, or a phenomenon that occurred in the atmosphere of the Earth. Describe an experiment to determine which of these two possibilities is correct.

9. The Sun is much closer to the Earth than are the nearest stars, yet it is not possible to measure accurately the parallax of the Sun relative to the stars by measuring its position directly. Explain why.

10. Parallaxes of stars are sometimes measured relative to the positions of galaxies or distant objects called quasars. Why is this a good technique?

11. Figure 18.5 is the light curve for the prototype cepheid variable Delta Cephei. How does the luminosity of this star compare with that of the Sun?

12. Look at Appendices 10 and 11. What percentage of the stars in each list are main-sequence stars (remember the luminosity classes)? Why is this percentage so different for the two lists of stars?

13. Suppose you measure the temperature of a star to be identical to that of the Sun. Is this enough information to determine its distance? Explain.

14. Which of the following can you determine about a star without knowing its distance: radial velocity, temperature, apparent magnitude, luminosity.

Problems

15. A radar astronomer who is new at the job claims she beamed radio waves to Jupiter and received an echo exactly 48 min later. Do you believe her? Why?

16. Demonstrate that a light year contains 9.46×10^{12} km. Demonstrate that 1 pc equals 3.086×10^{13} km and that it also equals 3.26 LY. Show your calculations.

17. Give the distances to stars having the following parallaxes:
   a. 0.1 arcsec
   b. 0.5 arcsec
   c. 0.005 arcsec
   d. 0.001 arcsec

18. Give parallaxes of stars having the following distances:
   a. 10 pc
   b. 3.26 LY
   c. 326 LY
   d. 10,000 pc

19. Consider a cepheid with a luminosity 10,000 times that of the Sun. If a given telescope and detector can detect a solar-type star out to a distance of only 1000 LY, to what distance can this same telescope observe the cepheid?

20. In the text we say that all of the cepheids in the Large Magellanic Cloud are nearly at the same distance away, and therefore their apparent brightnesses correlate with their periods. Look up the diameter and distance of the Large Magellanic Cloud in Appendix 12. What is the percentage difference in the distances of two stars, one of which is located on the edge of the galaxy closest to us, and the other on the far side? Suppose these stars have identical periods and identical intrinsic luminosities. How much will their apparent brightnesses differ? How will this difference affect the correlation between period and apparent brightness?

Suggestions for Further Reading

Croswell, K. "Visit the Nearest Stars" in *Astronomy*, Jan. 1987, p. 16. On the 19 nearest stars.

Ferris, T. *Coming of Age in the Milky Way*. 1988, Morrow. A history of how we established the scale of the cosmos.

Hirshfeld, A. "The Absolute Magnitude of Stars" in *Sky & Telescope*, Sep. 1994, p. 35. Good review, with charts.

Hodge, P. "How Far Away Are the Hyades?" in *Sky & Telescope*, Feb. 1988, p. 138. A history of how we measure distance to this important cluster.

Marschall, L. et al. "Parallax You Can See" in *Sky & Telescope*, Dec. 1992, p. 626. On measuring parallax for yourself.

Reddy, F. "How Far the Stars" in *Astronomy,* June 1983, p. 6. Nice summary of the entire chain of distances.

Rowan-Robinson, M. *The Cosmological Distance Ladder.* 1985, W. H. Freeman. Somewhat technical introduction to measuring distances in the universe.

Struve, O. "The First Determinations of Stellar Parallax" in *Sky & Telescope.* Vol. 16, 1956, pp. 9, 69.

Upgren, A. "New Parallaxes for Old" in *Mercury,* Nov./Dec. 1980, p. 143.

Using **REDSHIFT** ™

1. Zoom in on Corona Borealis and set the *Faint Magnitude Limit* to 5. Which star is brightest? Which star is intrinsically most luminous? Which star is the supergiant? Which star is on the main sequence?

Comparing the spectral type and visual magnitude to that of other stars in the constellation, can you deduce the luminosity class of Beta Coronae Borealis? Comparing the characteristics of Gamma Coronae Borealis to Kappa Coronae Borealis, can you predict the luminosity class for Kappa?

2. Set the *Faint Magnitude Limit* to 6 and center on Mensa in the Southern Hemisphere. Comparing its spectral type, luminosity class, and visual magnitude to other stars in the constellation, estimate the distance to Delta Mensae.

3. The distance given for Beta Orionis is incorrect. Compare spectral types and luminosity classes of the brightest 10 stars in Orion and use the inverse-square law of radiation to estimate the distance to Beta.

4. Turn the *Constellation Boundaries* on and set the *Faint Magnitude Limit* to 5.

Count the number of stars in Gemini in each spectral type (O, B, A, F, G, K, and M). Do the same for luminosity type (I, II, III, IV, and V).

Do Gemini's bright stars follow the same distributions of spectral type and luminosity class as for stars as a whole?

This region contains some of the most colorful interstellar clouds ever photographed. The blue area at the top is a cloud of dust surrounding the star Rho Ophiuchi; the blue color is starlight reflected from grains of dust. At the lower right, the bright star Antares, a cool red supergiant, is embedded in a thick dust cloud of its own making. (Immediately to the left of Antares is M4, a much more distant cluster of stars.) The reddish nebulae glow with light emitted from hydrogen atoms. (Photo by David Malin, © Royal Observatory, Edinburgh)

CHAPTER 19

Between the Stars: Gas and Dust in Space

Thinking Ahead

Astronomers examining the sky with telescopes come across an area that is generally crowded with stars except for a small dark section that has only a few. How can we determine if this dark region is simply a lack of stars in that particular direction, or a place where a cloud of some dark material blocks our view of stars behind it?

To begin our discussion of the life story of the stars, we must first examine where they come from. One of the most exciting discoveries of 20th-century astronomy was that our Galaxy contains not only stars, but vast quantities of what we might call "raw material" for stars—atoms or molecules of gas, and tiny solid particles we call dust. As we will see, the existence of this raw material helps us understand how new stars can continue to form, and gives us important clues about our own origins billions of years ago.

By earthly standards the space between stars is empty. No laboratory on Earth can produce so complete a vacuum. Yet this "emptiness" contains vast clouds of gas and solid particles. Some of these clouds are visible and are called **nebulae** (Latin for "clouds"). Others emit energy only at infrared or radio wavelengths. Still others make their existence known because they absorb some of the light passing through them. Astronomers refer

[Photographs] showed the entire group of stars [the Pleiades cluster] with an entangling system of nebulous matter, which seemed to bind together the different stars with misty wreaths and streams of filmy light . . . all of which is entirely beyond the keenest vision and the most powerful telescope.

E. E. Barnard, writing in *Popular Astronomy*, Vol. 6, p. 439 (1898)

to all the material between stars as **interstellar matter;** the collection of interstellar matter is called the *interstellar medium*.

Interstellar clouds do not last for the lifetime of the universe, but instead collide, coalesce, and grow. Some form stars within them that then inject energy into the cloud material and disperse it. When stars die they, in turn, eject some of their material into interstellar space. This material can then form new clouds and begin the cycle over again.

The Interstellar Medium

About 99 percent of the material between the stars is in the form of a *gas*—individual atoms or molecules. The most abundant elements in this gas are hydrogen and helium (which we saw are also the most abundant elements in the stars). The remaining 1 percent is solid—frozen particles sometimes called **interstellar grains** (Figure 19.1).

If all the interstellar gas were spread out smoothly, there would be about one atom of gas per cubic centimeter in interstellar space. (In contrast, the air in the room where you are reading this book has roughly 10^{19} atoms per cubic centimeter.) The dust grains are even scarcer. A cubic kilometer of space would contain only a few hundred to a few thousand tiny grains, each typically less than one ten-thousandth of a millimeter in diameter. These numbers are just averages, however, because the gas and dust are not distributed smoothly. Instead, as the pictures in this chapter show, the distribution is patchy and irregular, with the denser regions being referred to as "clouds."

In some clouds the density of gas and dust may exceed the average by as much as a thousand times or more, but even this density is more nearly a vacuum than any attainable on Earth. To show what we mean, let's imagine a vertical tube in the Earth's atmosphere with a cross section of 1 m². Such a tube of air would contain more atoms than we would find if we could extend that same tube from the top of the atmosphere all the way to the edge of the observable universe, 10 to 15 billion LY away.

While the density of interstellar matter is very low, the volume of space in which such matter is found is huge, and so its total mass is substantial. To see why, we must bear in mind that stars occupy only a tiny fraction of the volume of the Milky Way Galaxy. For example, it takes light only about 2 s to travel a distance equal to the radius of the Sun, but more than four *years* to travel from the Sun to the nearest star. Even if the spaces among the stars are sparsely populated, there's just a lot of space out there.

Astronomers estimate that the total mass of gas and dust in the Milky Way Galaxy is equal to about 5 percent of the mass contained in stars. Don't let that 5 percent fool you: this means that the mass of the interstellar mat-

Figure 19.1
Clouds of luminous gas and opaque dust, such as the nebulosity called NGC 3603, are found between the stars. A bright cluster of stars (bright white region near the bottom) energizes the surrounding gas, causing it to glow. The reddish glow is interrupted by lanes and clouds of gas mixed with enough dust to block the light from behind them. (Anglo-Australian Telescope Board)

ter in our Galaxy amounts to several billion times the mass of the Sun. There is plenty of raw material in the Galaxy to make many generations of new stars and planets (and perhaps even astronomy students).

ASTRONOMY BASICS

Naming the Nebulae

As you look at the captions for some of the spectacular photographs in this and the next chapter, you will notice the variety of names given to the nebulae. A few, which in small telescopes look like something recognizable, are sometimes named after the creatures or objects they resemble. Examples include the Crab, Tarantula, and Keyhole Nebulae. But most only have numbers that are entries in a catalog of astronomical objects.

Perhaps the best-known catalog of nebulae (as well as star clusters and galaxies) was compiled by the French astronomer Charles Messier (1730–1817). Messier's passion was discovering comets, and his devotion to this cause earned him the nickname "The Comet Ferret" from King Louis XV. When comets are first seen coming toward the Sun, they look like little fuzzy patches of light; in small telescopes, they are easy to confuse with nebulae or with groupings of many stars so far away that their light is all blended together. Time and again, Messier's heart leapt as he thought he had discovered one of his treasured comets, only to find that he had "merely" observed a nebula or cluster.

In frustration, Messier set out to catalog the position and appearance of over 100 objects that could be mistaken for comets. For him, this list was merely a tool in the far more important work of comet hunting. He would be very surprised if he returned today to discover that no one recalls his comets anymore, but that his catalog of "fuzzy things that are not comets" is still widely used. When the opening image in this chapter refers to M4, it denotes the fourth entry in Messier's list.

A far more extensive listing was compiled under the title of the New General Catalog (NGC) of Nebulae and Star Clusters in 1888 by John Dreyer, working at the observatory in Armagh, Ireland. He based his compilation on the work of William Herschel and his son John, plus many other observers who followed them. With the addition of two further listings (called the Index Catalogs), Dreyer's compilation eventually included 13,000 objects. Astronomers today still use his NGC numbers when referring to most nebulae and star groups.

Figure 19.2

NGC 3576: a cloud of luminous gas surrounds a compact cluster of very hot stars. These hot stars ionize the hydrogen in the gas cloud, forming an H II region. When electrons then recombine with protons and move back down to the lowest energy orbit, emission lines are produced. The strongest hydrogen line that can be seen as visible light is in the red part of the spectrum, and is responsible for the color of the gas in the photograph. (Anglo-Australian Telescope Board)

19.2

Interstellar Gas

Some of the most spectacular astronomical photographs (Figure 19.2) show interstellar gas located near hot stars. This gas is heated to temperatures close to 10,000 K by the nearby stars; it then glows with the colors characteristic of the element hydrogen, which makes up the majority of the gas (about three-quarters by mass). The reddish color in the accompanying images is the telltale glow of hydrogen; ionized nitrogen also contributes. (The strongest of the lines in the visible spectrum of hydrogen is the red Balmer line; see Section 4.5.)

H II Regions—Gas Near Hot Stars

Hydrogen interstellar gas near very hot stars is ionized (the electron is stripped completely away) by the ultraviolet radiation from those stars. Since hydrogen is the main constituent of interstellar gas, we often characterize a region of space according to whether its hydrogen is neutral or ionized. A cloud of ionized hydrogen is called an **H II region.** (Spectroscopists use the Roman numeral I to indicate that an atom is neutral; successively higher Roman numerals are used for each higher stage of ionization. H II thus refers to hydrogen that has lost its one electron; Fe III is iron with two electrons missing.)

When the radiation from a hot star ionizes hydrogen in the surrounding gas, a neutral hydrogen atom is converted into a positive hydrogen ion (a proton) and a free electron. Such a detached proton won't remain alone forever when attractive electrons are around; it will capture a free electron, becoming neutral hydrogen once more. However, such a neutral atom can then absorb ultraviolet radiation again and start the cycle over. At a typical moment, most of the atoms are in the ionized state, which is why we call such areas H II regions.

The capture of the free electrons leads to the emission of light. The electrons cascade down through the various energy levels of the hydrogen atoms on their way to the lowest level, or ground state. During each transition downward, they give up energy in the form of light (see Chapter 4). This process of converting ultraviolet radiation into visible light is called *fluorescence*.

A fluorescent light works in basically the same way. Electrons within it are heated to high temperatures and then collide with atoms of mercury vapor in the tube. The mercury is excited to a high energy state because of these collisions. When the mercury atoms return to lower energy levels, they emit ultraviolet photons that in turn strike a phosphor-coated screen. The atoms in the screen absorb the ultraviolet photons and emit visible light by the process of fluorescence.

The interstellar gas contains other elements besides hydrogen. Many of them are also ionized in the vicinity of hot stars; they then capture electrons and emit light, just as hydrogen does, which allows them to be observed by astronomers. But generally, these atoms are outnumbered by the hydrogen atoms.

Neutral Hydrogen Clouds

Ionized hydrogen gas is the type of interstellar matter most often photographed, but observations show that most interstellar hydrogen is not ionized. The very hot stars required to produce H II regions are rare, and only a small fraction of interstellar matter is close enough to such hot stars to be ionized by them.

If interstellar gas located at large distances from stars is not ionized and cannot produce strong emission lines, how can we learn about such gas? One way is to search for the presence of atoms and molecules of interstellar matter in the spectra of stars. Cold clouds of gas can absorb some of the light from stars that lie behind them, producing dark spectral lines (as discussed in Chapter 4).

The first evidence for absorption by interstellar clouds came from the analysis of a spectroscopic binary star. While most of the lines in the spectrum of this binary shifted alternately from longer to shorter wavelengths and back again, as we would expect from the Doppler effect for one star in orbit around another, a few lines in the spectrum remained fixed in wavelength. Since both stars are moving in a binary system, lines that showed no motion puzzled astronomers. Subsequent work demonstrated that these lines were not formed in the star's atmosphere at all, but rather in a cold cloud of gas located between the Earth and the binary star.

The most conspicuous interstellar lines in the visible-light region of the spectrum are produced by atoms of sodium and calcium. We also see absorption in the spectrum produced by molecules (combinations of atoms) such as CN and CH. Ultraviolet observations made with orbiting telescopes have detected lines of carbon, hydrogen, oxygen, nitrogen, and other elements, as well as carbon monoxide and molecular hydrogen. We should note that although other elements may be more noticeable in such spectra, hydrogen is still the most abundant element in the colder regions (see next section).

The strengths of interstellar lines can be used to estimate how common each element is in the interstellar medium. For some elements, the interstellar abundances are about the same as their relative abundances in the Sun and other stars. For others, the relative abundance in interstellar space is noticeably lower. Interstellar gas has especially low abundances of elements that readily condense into solids (notably aluminum, calcium, and titanium, as well as iron, silicon, and magnesium). As we will see in Section 19.4, it is likely that many of the atoms of these elements have indeed combined to form tiny solid grains of interstellar dust. Thus they no longer show the spectral lines produced by atoms in the gaseous state.

Radio Observations of Cold Clouds: The 21-cm Line

By now you've probably gotten the message that hydrogen is the most common element out there. And most of the hydrogen in interstellar space is cold. Unfortunately, cold hydrogen—far from the radiation of hot stars—is not so easy to detect using visible light or ultraviolet radiation. Radio astronomers, however, discovered a way to learn about what turned out to be vast reservoirs of cold hydrogen gas in the Galaxy.

Cold clouds of interstellar gas emit spectral lines in the radio part of the electromagnetic spectrum. The most useful of these lines is produced by (no surprise) hydrogen at a wavelength of 21 cm. That's quite a long wavelength, implying that the wave has a low frequency and low energy. The energy is too low to come from electrons jumping between the energy levels discussed in Chapter 4. Where then does such a low-energy wave come from in hydrogen?

A hydrogen atom possesses a tiny amount of angular momentum because its electron is spinning on its axis and orbiting around the nucleus (proton). In addition, the proton has a spin of its own. If the proton and electron are spinning in opposite directions, the atom as a whole has a very slightly lower energy than if the two spins are aligned (Figure 19.3). If an atom in the lower energy state (spins opposed) acquires a small amount of energy, then the spins of the proton and electron can be aligned, leaving the atom in a slightly *excited state*. If the atom then loses that same amount of energy again, it returns to its ground state. The amount of energy involved corresponds to a wave of 21-cm wavelength.

Neutral hydrogen atoms can be excited in this way by collisions with electrons and other atoms. Such collisions are extremely rare in the sparse gases of interstellar space. An individual atom may wait many years before such an encounter aligns the spins of its proton and electron. Nevertheless, over many millions of years a significant fraction of the hydrogen atoms are excited by a collision. (Out there in cold space, that's about as much excitement as an atom typically experiences.) An excited atom can then lose its excess energy either by another collision or by giving off a radio wave with a wavelength of 21 cm. If there are no collisions, an excited hydrogen atom will wait an average of about 10 million years before emitting a photon and returning to its state of lowest energy.

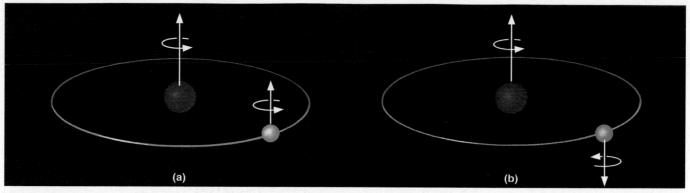

Figure 19.3
Formation of the 21-cm line. When the electron in a hydrogen atom is in the orbit closest to the nucleus, the proton and the electron may be spinning either in the same direction (left) or in opposite directions (right). When the electron flips over, the atom gains or loses a tiny bit of energy by either absorbing or emitting electromagnetic energy with a wavelength of 21 cm.

Equipment sensitive enough to detect the 21-cm line of neutral hydrogen became available in 1951. The mechanism for producing these waves was earlier described by Dutch astronomers, and they had actually prepared an experiment to detect 21-cm waves when a fire destroyed their equipment. As a result, two Harvard physicists, Harold Ewen and Edward Purcell, made the first detection (Figure 19.4), soon followed by confirmations from the Dutch and a group in Australia. Since

Figure 19.4
Harold Ewen in 1952, with the horn antenna that first detected 21-cm radiation, atop the Lyman Physics Laboratory at Harvard. The inset shows Edward Purcell, winner of the 1952 Nobel Prize in physics, a few years later. (Photos courtesy of E. M. Purcell and Harvard University)

that time many other radio lines produced by both atoms and molecules have been discovered (see following discussions).

Observations at 21 cm show that most of the neutral hydrogen in the Galaxy is confined to an extremely flat layer, less than 300 LY thick, that extends throughout the flat disk of the Milky Way. The 21-cm line is produced in cold hydrogen clouds that turn out to have temperatures of about 100 K. The diameters of these cold clouds range from about 3 to 30 LY; an imaginary spaceship traveling 1000 LY through space in the plane of the Galaxy would encounter, on average, only about two of these clouds. Overall, we can estimate that approximately 2 percent of interstellar space is filled with such cold clouds. The masses of the clouds are typically in the range 1 to 1000 times the mass of the Sun, with clouds of low mass being the most common.

Not all of interstellar hydrogen is found in these extremely cold clouds. Some hydrogen is warm, with temperatures of 3000 to 6000 K. At least 20 percent of the space between stars in the plane of the Milky Way is filled with such warm, neutral hydrogen. The warm hydrogen is not heated by nearby stars; what then warms it to such high temperatures? The answer came when even hotter interstellar gas was discovered by an orbiting observatory.

Ultra-Hot Interstellar Gas

The discovery of ultra-hot interstellar gas was a big surprise. Before the launch of astronomical observatories into space, models of the interstellar medium assumed that most of the region between stars was filled with cool hydrogen. But when astronomers were able to make ultraviolet observations from telescopes above the Earth's atmosphere, they discovered (for example) interstellar lines produced by oxygen atoms that had been ionized five times. To strip five electrons from their orbits around an oxygen nucleus requires a lot of energy. In fact, these observations imply that the temperature of the interstellar matter where these atoms occur must be approximately a *million* degrees.

It appears that the source of energy producing these remarkable temperatures is the explosion of massive stars at the ends of their lives. Such explosions, called *supernovae*, will be discussed in detail in Chapter 22. For now we'll just say that some stars, nearing the ends of their lives, become unstable and literally explode. These explosions send high-temperature gas, moving at velocities of thousands (and even tens of thousands) of kilometers per second, out into interstellar space, where it has a tremendous heating effect on the neighborhood gas.

Astronomers estimate that one supernova explodes roughly every 25 years somewhere in the Galaxy. On the average, the hot gas from a supernova will sweep through any given point in the Galaxy about once every 2 million years. At this rate, the sweeping action is continuous

enough to keep most of the space between clouds filled with gas at a temperature of a million degrees. When this ultra-hot gas comes into contact with a denser cloud, it heats the outer layers of the cloud to a temperature of a few thousand degrees.

Interstellar Molecules

As we just saw, visible-light and ultraviolet observations revealed the presence of a number of *simple* molecules among interstellar matter by absorbing some of the radiation of more distant stars. When more sophisticated equipment for obtaining spectra in radio and infrared wavelengths became available, astronomers found more complex molecules in interstellar clouds as well.

Just as atoms leave their "fingerprints" in the spectrum of visible light, so the vibration and rotation of atoms within molecules can leave spectral fingerprints in radio and infrared waves. If we spread out the radiation at such longer wavelengths, we can detect emission or absorption lines in the spectrum that are characteristic of specific molecules. Over the years, experiments in Earth laboratories have shown us the exact wavelengths associated with changes in the rotation and vibration of many molecules—giving us a template of possible lines against which we can now compare our observations of interstellar matter.

About a hundred kinds of molecules have been identified in interstellar space, most in giant clouds containing substantial amounts of gas and dust. These more complex molecules are mostly composed of combinations of hydrogen, oxygen, carbon, nitrogen, and sulfur atoms. Among them are molecules of H_2 (molecular hydrogen), water, ammonia, and hydrogen sulfide. Astronomers also identify a number of organic molecules (those associated with carbon chemistry on Earth), such as formaldehyde (used to preserve living tissues), hydrogen cyanide, and alcohol (see "Making Connections" box). Relatively heavy molecules such as HC_9N and $HC_{11}N$, which have long chains of carbon atoms, are found in some cold clouds. Other features in infrared spectra hint at complex rings of atoms; some of these are called *polycyclic aromatic hydrocarbons* (PAHs) by chemists because on Earth some of them have a strong smell or aroma. PAHs may have as many as 50 carbon atoms, and can be considered intermediate between molecules and solid grains.

The cold interstellar clouds also contain cyanoacetylene (HC_3N) and acetaldehyde (CH_3CHO), generally regarded as starting points for amino acid formation. These are the building blocks of proteins, and thus one of the fundamental chemicals from which living organisms on Earth are constructed. The presence of these organic molecules does not imply that life exists in space. But it does show that the chemical building blocks of life can form under a wide range of conditions in the universe. As we learn more about how complex molecules are pro-

Cocktails in Space

Among the molecules astronomers have identified in interstellar clouds is alcohol, which comes in two varieties: methyl (or wood) alcohol, and ethyl alcohol (the kind you find in cocktails). Ethyl alcohol is a pretty complex molecule, written by chemists as C_2H_5OH. It is quite plentiful (relatively speaking)—in clouds where it has been identified we detect up to one molecule for every cubic meter. The largest of the cold clouds (which can be several hundred light years across) have enough ethyl alcohol to make 10^{28} fifths of liquor.

Spouses of future interstellar astronauts, however, need not fear that their wives or husbands will become interstellar alcoholics. Even if a spaceship were equipped with a giant funnel 1 km across, and could scoop it through such a cloud at the speed of light, it would take about a thousand years to gather up enough alcohol for one standard martini!

Furthermore, the very same clouds also contain water (H_2O) molecules. Your scoop would gather them up as well, and there are a lot more of them because they are simpler and thus easier to form. For the fun of it, one astronomical paper actually calculated the proof of a typical cloud. Proof is the ratio of alcohol to water in a drink, where 0 proof means all water, 100 proof means half alcohol and half water, and 200 proof means all alcohol. The proof of the interstellar cloud was only 0.2, hardly enough to blur your memory before a big astronomy exam.

duced in interstellar clouds, we gain an increased understanding of the kinds of processes that preceded the beginnings of life on Earth billions of years ago.

How do such fragile molecules survive in the harsh environment of interstellar space? Most of the more complex molecules would be dissociated (torn apart into individual atoms) by the short-wavelength radiation that comes from stars. Therefore these molecules can survive only where they are shielded from ultraviolet starlight. There are dense, dark, giant clouds that contain dust, which acts like a thick blanket of interstellar smog and keeps ultraviolet starlight from penetrating the interior of the cloud (although the radio waves that signal the presence of the molecules readily get out). It is in these clouds that we find molecules, in fact, these giant clumps of interstellar matter are called *molecular clouds*.

Some scientists also speculate that the surfaces of the dust grains (see Section 19.4)—which would seem very large if you were an atom—provide "nooks and crannies" where atoms can stick long enough to join with other atoms and form molecules. (Think of the dust grains as "interstellar social clubs" where lonely atoms can meet and form meaningful relationships.) Dusty clouds not only protect molecules but provide an environment that encourages their formation.

As we will see in the next chapter, the giant clouds of interstellar material where molecules are most common are among the most interesting structures in the Galaxy. It is here that we are most likely to see new stars forming from raw material, and beginning the life cycle that is our subject in the next several chapters.

19.3

A Model of the Interstellar Gas

Table 19.1 summarizes the characteristics of the various types of clouds that populate interstellar space. The contrasts are fascinating. There are cold, dense clouds in which hydrogen is not ionized. There are clouds so hot that atoms are mainly ionized and molecules cannot sur-

TABLE 19.1
Interstellar Gas

Type of Region	Temperature (K)	Density (number/cm^3)	Description
H I: cold clouds	10^2	50	Hydrogen atoms in clouds typically 3–30 LY across.
H I: warm clouds	3–6×10^3	0.3	Hydrogen is warmer but not ionized; found in clouds.
Hot gas	10^5–10^6	10^{-3}	Found throughout the Galaxy; hydrogen is ionized; heated by supernova explosions.
H II regions	10^4	10^2–10^4	Found near hot stars; hydrogen mostly ionized.
Giant molecular clouds	10	as high as 10^5	50–200 LY across; have dense clumps.

vive. The densities differ by a factor of a million, with the coolest clouds having the highest density. What do these data tell us about the interstellar medium? Where might we expect to find the various types of clouds? What is an individual cloud like? How do the clouds change as time passes?

Structure and Distribution of Interstellar Clouds

One important requirement for a model of the interstellar medium is that the clouds and the million-degree gas between them must be at approximately the same pressure. Suppose they were not. If the cloud pressure were higher, the cloud would expand until its pressure matched that of its environment. If, on the other hand, the pressure of the hot gas were greater than that of a cloud embedded in it, the hot gas would compress the cloud and force it to shrink until its pressure became high enough to resist further compression. Imagine a balloon as an analogy. If you add air to it, thereby increasing its pressure, the balloon will expand. If you squeeze it with your hands, you can make the balloon smaller because of the greater external pressure.

The pressure in a gas depends on both the temperature and the number of particles per cubic centimeter (its density). If the gas particles are hotter, their energy of motion is greater, and they collide harder with one another. If the density is greater, there are more collisions. For the gas pressures in two different regions next to each other to be the same, the region at higher temperature must have a lower density. A region with lower temperature, such as a cool cloud, must have a higher density. (In the same way, you can fit more people into a room when

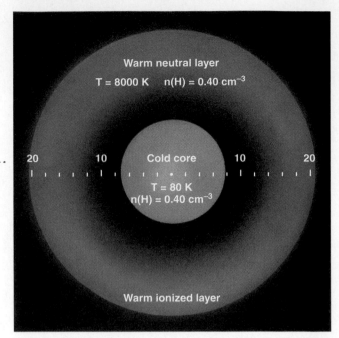

Figure 19.5
A typical interstellar cloud consists of a dense, cold core surrounded by a warm envelope. The horizontal scale shows the distance from the center in light years.

they are not moving around very much, say at a formal cocktail party. But if they want to do the waltz, fewer of them will fit.)

Figures 19.5 and 19.6 show in a schematic way what interstellar clouds look like based on the requirement that gas pressures must be nearly the same everywhere. Individual clouds are scattered at random throughout the Galaxy. The typical cloud may be a few tens of light years

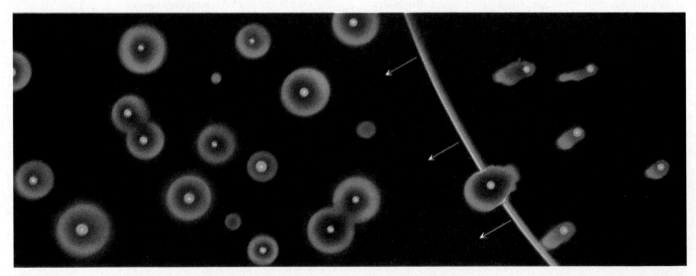

Figure 19.6
Interstellar clouds are embedded in hot, low-density gas heated to temperatures as high as a million degrees K by supernova explosions. In the upper right, a supernova remnant is shown sweeping through interstellar space. (Diagram adapted from work published in *The Astrophysical Journal* by C. McKee and J. Ostriker)

in diameter. The clouds are embedded in gas with a temperature of a million degrees or so, the legacy of exploding stars. The outer portions of the clouds are heated by this gas via conduction—the direct transfer of energy from the hot gas to the outer atoms of the clouds. The temperature of the outer portion of a typical cloud is thus raised to a few thousand degrees.

If a cloud is large enough, it can shield its innermost core from being heated, and the core may have a temperature 100 times lower and a density correspondingly 100 times higher. Typical values for a cloud core are a temperature of 80 K and a density of 40 hydrogen atoms per cubic centimeter.

Interstellar Matter Around the Sun

Let's take a brief aside to look at the distribution of interstellar matter in our own neighborhood. The Sun is located in a region where the density of interstellar matter is unusually low. The temperature of interstellar matter in the vicinity of the Sun is about 10^6 K, and its density is only 5×10^{-3} hydrogen atoms per cubic centimeter. This region of low-density gas, called the *Local Bubble*, extends to a distance of at least 300 LY from the Sun.

If interstellar space near the Sun contained the normal number of clouds, we would expect to have detected approximately 2000 of them within the Local Bubble. We have not: clouds of the type shown in Figure 19.5 are conspicuously absent. One possible explanation is that a supernova explosion in the past 100,000 to ten million years swept the region we now see as the Local Bubble nearly clean of interstellar clouds. Such an explosion would have heated the small amount of remaining gas to very high temperatures.

While typical cold hydrogen clouds are very rare within the Local Bubble, a few clouds do exist. The Sun itself seems to be inside a cloud with a density of 0.1 hydrogen atom per cubic centimeter and a temperature of 10,000 K. This cloud is so tenuous that it is referred to as Local Fluff. We do see one sizable warm cloud in the direction toward the center of our Galaxy but within 60 LY of the Sun. It may be that the Local Fluff is the warm, partially ionized edge of a denser, cooler cloud (see Figure 19.5) and that the Sun is just entering this cloud as it moves through the Galaxy.

Evolution of Interstellar Clouds

The model of interstellar gas described here presents a picture of the interstellar medium as it appears, on the average. In reality, interstellar gas pressures are not always balanced. Clouds collide and coalesce; they are torn apart by supernovae explosions and disrupted by stellar winds and radiation pressure from hot stars. The interstellar medium is constantly replenished with gas ejected by supernovae and hot stars. Initially, this gas forms relatively small clouds, then grow through collisions and mergers

with other clouds. Ultimately the process leads to the formation of giant clouds with diameters as large as 200 LY, and masses that exceed 100,000 times that of the Sun. It is in these giant clouds that the most vigorous star formation occurs. The newly formed stars then go through their lives, and some (as we will see) become supernovae. As they explode they eject gaseous material, thus starting the cycle over again.

19.4

Cosmic Dust

Figure 19.7 shows a striking example of what is actually a common sight—a dark region on the sky that appears nearly empty of stars. For a long time astronomers debated whether these dark regions were empty "tunnels" through which we looked beyond the stars of the Milky Way Galaxy into intergalactic space, or clouds of some dark material that blocked the light of the stars beyond. American astronomer E. E. Barnard is generally credited with showing from his extensive series of nebulae photographs that the latter interpretation is the correct one (see "Voyagers in Astronomy" box).

Dark Nebulae

The dark cloud (called Barnard 86) seen in Figure 19.7 blocks the light of the many stars that lie behind it—note

Figure 19.7
A star cluster (NGC 6520) next to a dark cloud of interstellar matter (Barnard 86). Old stars in our Galaxy are yellowish in color and form the brightest part of the Milky Way. The dark cloud is seen in front of these background stars, visible only because it blocks out the light from the stars beyond. The small cluster of young blue stars may have begun, millions of years ago, as just such a dark cloud. (Anglo-Australian Observatory)

Edward Emerson Barnard

Born in 1857 in Nashville, Tennessee, two months after his father died, Edward Barnard grew up in such poor circumstances that he had to drop out of school at age nine to help support his ailing mother. He soon became assistant to a local photographer, where he learned to love both photography and astronomy, destined to become the dual passions of his life. He worked as a photographer's aide for 17 years, studying astronomy on his own. In 1883 he obtained a job as an assistant at the Vanderbilt University Observatory, which enabled him at last to take some astronomy courses.

Married in 1881, Barnard built a house for his family that he could ill afford. But as it happened, a patent medicine manufacturer offered a $200 prize (a lot of money in those days) for the discovery of any new comet. With the determination that became characteristic of him, Barnard spent every clear night searching for comets. He discovered seven of them between 1881 and 1887, earning enough money to make the payments on his home; this "Comet House" later became a local attraction. (By the end of his life, he had found 17 comets through diligent observation.)

In 1887 Barnard got a position at the newly founded Lick Observatory, where he soon locked horns with the director, Edward Holden, a blustering administrator who made Barnard's life miserable. (To be fair, Barnard soon tried to do the same for him.) Despite being denied the telescope time that he needed for his photographic work, Barnard in 1892 managed to discover the first new satellite found around Jupiter since Galileo's day, a stunning observational feat that earned him world renown. Now in a position to demand more telescope time, he perfected his photographic techniques and soon began to publish the best images of the Milky Way taken up to that time. It was during the course of this work that he began to examine the dark regions among the crowded star lanes of the Galaxy, and to realize that they must be vast clouds of obscuring material (rather than "holes" in the distribution of stars).

Astronomer/historian Donald Osterbrock has called Barnard an "observaholic": his daily mood seemed to depend entirely on how clear the sky promised to be for his night of observing. He was a driven, neurotic man, concerned about his lack of formal training, fearful of being scorned, and afraid that he might somehow slip back into the poverty of his younger days. He had difficulty taking vacations and lived for his work: only serious illness could deter him from making astronomical observations.

In 1895 Barnard, having had enough of the political battles at Lick, accepted a job at the Yerkes Observatory near Chicago, where he remained until his death in 1923. He continued his photographic work, publishing compilations of his images that became classic photographic atlases, and investigating the varieties of nebulae revealed in his photographs. He also made measurements of the sizes and features of planets, participated in observations of solar eclipses, and carefully cataloged dark nebulae (see Figure 19.7). In 1916 he discovered the star with the largest proper motion (see Chapter 16), the second-closest star system to our own, now called Barnard's Star in his honor.

E. E. Barnard at the Lick Observatory 36-in refractor.
(Mary Lea Shane Archives of the Lick Observatory)

---

how the regions in other parts of the photograph are crowded with stars. The only stars visible in the direction of Barnard 86 are those that happen to lie in front of it from our perspective. Barnard 86 is an example of a relatively dense cloud or *dark nebula* of tiny solid grains, which are often referred to as interstellar dust. Opaque clouds are conspicuous on any photograph of the Milky Way (see the figures in Chapter 24). The "dark rift," which runs lengthwise down a long part of the summer Milky Way and appears to split it in two, is produced by a collection of such obscuring clouds.

Dust clouds are invisible in the visible light region of the spectrum; we find them only when denser clouds block the light from stars behind them (Figure 19.8).

However, they glow brightly in the infrared. Small dust grains absorb visible light and ultraviolet radiation very efficiently. The grains are heated by the absorbed radiation, typically to temperatures from 20 to about 500 K, and reradiate this heat at infrared wavelengths. We can use Wien's law (see Section 4.2) to estimate where in the electromagnetic spectrum this radiation falls. For a temperature of 100 K, the maximum is at about 30 micrometers (μm; 1 μm $= 10^{-4}$ cm), while grains as cold as 20 K will radiate most strongly near 150 μm. The Earth's atmosphere is opaque to radiation at these wavelengths, so emission by interstellar dust is best measured from space.

Observations from above the Earth's atmosphere by IRAS (the Infrared Astronomical Satellite) showed that

dust clouds are present throughout the plane of the Milky Way (Figure 19.9). The bright patches of emission have been given the name **infrared cirrus.** The closest infrared cirrus clouds are about 300 LY away.

Reflection Nebulae

The tiny interstellar dust grains absorb only a portion of the starlight they intercept. At least half of the starlight that interacts with a grain is merely scattered—that is, it is redirected helter skelter in all directions. Since neither the absorbed nor the scattered starlight reaches us directly, both absorption and scattering make stars look dimmer. The effects of both processes are called **interstellar extinction;** something astronomers must especially take into account when they observe stars in dusty or distant regions of our Galaxy.

Some dense clouds of dust are close to luminous stars, and scatter enough starlight to become visible. Such a cloud of dust, illuminated by starlight, is called a *reflection nebula* since the light we see is starlight reflected off the grains of dust. One of the best-known examples is the nebulosity around each of the brightest stars in the Pleiades cluster (Figure 19.10). The dust grains are small, and such small particles turn out to scatter light with blue wavelengths more efficiently than light at red wavelengths (Figure 19.11). A reflection nebula, therefore, usually appears bluer than its illuminating star.

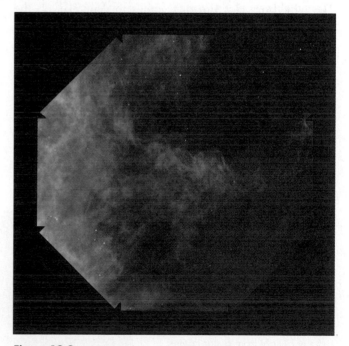

Figure 19.9
The Infrared Astronomical Satellite discovered patchy clouds of dust throughout the plane of the Milky Way Galaxy. These have been nicknamed "infrared cirrus" because of their resemblance to high clouds in the Earth's atmosphere. In this infrared image, the cirrus clouds are at the left, while three dusty nebulae can be seen to the right. (IPAC/JPL/NASA)

Figure 19.10

The Pleiades open star cluster. This cluster contains hundreds of stars (only a few are visible here) and is located about 400 LY from the Sun. The blue nebulosity is starlight reflected by interstellar dust in a cloud that the cluster happens to be passing through at the present time. (Royal Observatory Edinburgh/Anglo-Australian Telescope Board)

A similar scattering process explains why the Earth's atmosphere looks blue even though the gases that make up our air are transparent. As sunlight comes in, it scatters from the molecules of air. Again, the small size of the molecules means that the blue colors scatter much more efficiently than the greens, yellow, and reds. Thus the blue in sunlight is scattered out of the beam and all over the sky. The light from the Sun that comes to your eye, on the other hand, is missing some of its blue, so the Sun looks a bit yellower than it would from space.

Gas and dust are generally intermixed in space, although the proportions are not exactly the same everywhere. The presence of dust is apparent on many photographs of emission nebulae; Figure 19.12 shows a beautiful image of the Trifid Nebula in the constellation of Sagittarius, where we see an H II region surrounded by a blue reflection nebula. Which type of nebula appears brighter depends on the kind of stars that cause the gas

and dust to glow. Stars cooler than about 25,000 K have so little ultraviolet radiation of wavelengths shorter than 91.2 nm (required to ionize hydrogen) that the reflection nebulae around such stars outshine the emission nebulae. Stars hotter than 25,000 K emit enough ultraviolet energy that the emission nebulae produced around them generally outshine the reflection nebulae.

Interstellar Reddening

Seventy years ago, astronomers were puzzled by the existence of stars whose spectral lines indicated that they were intrinsically hot and blue, although their overall appearance showed the reddish colors of much cooler stars. Today we understand that the light from these stars is not only dimmed but also **reddened** by interstellar dust. Most of their violet, blue, and green light has been scattered or absorbed, but some of their orange and red light,

Figure 19.11

Interstellar dust scatters blue light more efficiently than red light, thereby making distant stars appear redder, and giving clouds of dust near stars a bluish hue. Here a red ray of light from a star comes straight through to the observer, while a blue ray is shown scattering. A similar scattering process makes the Earth's sky look blue.

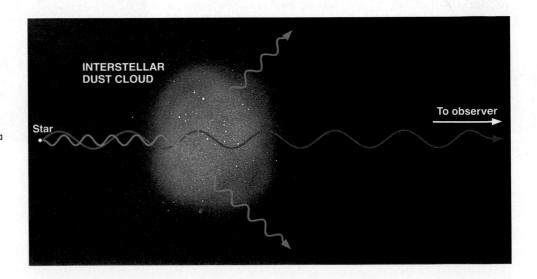

INTERSTELLAR DUST CLOUD

Star

To observer

Figure 19.12
The Trifid Nebula (M20) in the constellation Sagittarius. In the reddish region, the hydrogen is ionized by nearby hot stars and glows through the process called fluorescence. The blue region is a reflection nebula. The Trifid Nebula is about 30 LY in diameter and about 3000 LY from the Sun. This beautiful image was processed by David Malin using special photographic techniques to bring out faint details. The bluish nebulosity can be seen surrounding the H II region. (Anglo-Australian Telescope Board)

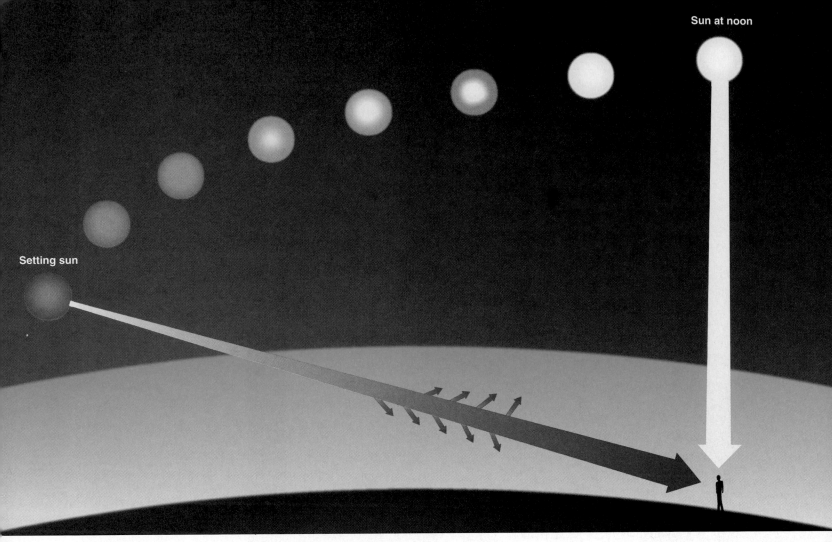

Figure 19.13
When light from the setting Sun traverses a long path through the Earth's atmosphere, blue light is scattered out of the direct light path, thereby making the Sun look redder.

with longer wavelengths, penetrates the obscuring dust and reaches Earth-based telescopes. (So, strictly speaking, *reddening* is not the most accurate term for this process, since no red color is added; instead, blues and related colors are subtracted, so it should more properly be called "deblueing.")

We have all seen an example of reddening. The Sun appears much redder at sunset than it does at noon. The lower the Sun is in the sky, the farther its light must travel through the atmosphere. Over this greater distance there is a higher probability that sunlight will be scattered. Since red light is less likely to be scattered than blue light, the Sun appears more and more red as it approaches the horizon (Figure 19.13).

The fact that light is reddened by interstellar dust means that long-wavelength radiation is transmitted more efficiently than short-wavelength radiation. Consequently, if we wish to see farther in a direction with considerable interstellar material, we should look at long wavelengths. This simple fact provides one of the motivations for the development of infrared astronomy. In the infrared region at 2 μm (2000 nm), for example, the obscuration is only one-sixth as great as in the visible region (500 nm), and we can therefore study stars that are more than twice as distant before their light is blocked by interstellar dust. This ability to see farther by observing in the infrared portion of the spectrum represents a major gain for astronomers trying to understand the structure of our Galaxy or probing its puzzling center (see Chapter 24).

Interstellar Grains

The preceding paragraphs have described the reddening and dimming of starlight by interstellar dust. But what exactly is this dust? Observations can give us some vital clues about the nature of the dust grains.

First of all, they tell us that the absorption of light is done by *solid particles* and not by interstellar gas. Except for specific spectral lines, atomic or molecular gas is almost transparent. Consider the Earth's atmosphere. Despite its incredibly high density compared with that of interstellar gas, it is so transparent as to be practically invisible (except when humans produce smog by filling it with solid particles of smoke and dirt; volcanoes, too, sometimes inject enough dust to make the sky look hazy). The quantity of *gas* required to produce the observed absorption of light in interstellar space would have to be enormous. The gravitational attraction of so great a mass of gas would produce effects upon the motions of stars that would be easily detected. Such effects are not observed, and thus the interstellar absorption cannot be the result of gases: the culprit must be something else.

Although gas does not absorb much light, we know from everyday experience that tiny solid or liquid particles can be very efficient absorbers. Water vapor in the air is quite invisible. When some of that vapor condenses into tiny water droplets, however, the resulting cloud is opaque. And dust storms, smoke, and smog offer familiar examples of the efficiency with which solid particles absorb light. On the basis of arguments like these, astronomers have concluded that widely scattered *solid* particles in interstellar space are responsible for the observed dimming of starlight. What are these particles made of? And how did they form?

Our observations show that a great deal of this dust exists; hence it must be primarily composed of elements that are abundant in the universe (and in interstellar matter). After hydrogen and helium, the most abundant elements are oxygen, carbon, and nitrogen. These three elements, along with magnesium, silicon, iron, and perhaps hydrogen itself, are thought to be the most important components of interstellar dust.

Observations support this hypothesis. Many heavy elements, including iron, magnesium, and silicon, are less abundant in interstellar gas than in the Sun and young stars. These heavy elements are assumed to be missing from the interstellar *gas* because they are condensed into solid particles of interstellar *dust*.

We also know from measurements of both total extinction and reddening that typical individual grains must be just slightly smaller than the wavelength of visual light. If the grains were a lot smaller, they would not block the light efficiently. Similarly, a wall of bowling balls, which are considerably smaller than the wavelength of radio waves, would not keep radio signals from reaching us inside a bowling alley.

On the other hand, if the dust grains were much larger than the wavelength of light, then starlight would not be reddened. Again, a bowling ball, which *is* much larger than the wavelength of light, would block both blue and red light with equal efficiency. In this way we can deduce that a characteristic interstellar dust grain contains 10^6 to 10^9 atoms and has a diameter of 10^{-7} to 10^{-8} m (10 to 100 nm). This is actually more like the specks of solid matter in cigarette smoke than the larger grains of dust you find under your desk when you are too busy studying astronomy to clean properly.

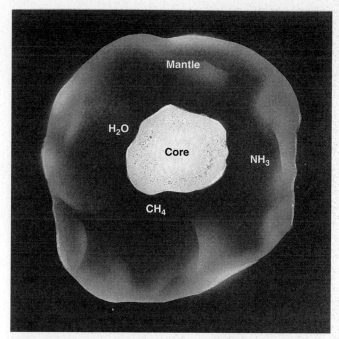

Figure 19.14
A typical interstellar grain is thought to consist of a core of rocky material (silicates), graphite, or possibly iron surrounded by a mantle of ices. Typical grain sizes are 10^{-7} to 10^{-8} m.

Observations of absorption features in spectra produced by interstellar material indicate that there must be many types of solid particles. Some interstellar grains apparently consist of a core of rock-like material (silicates); other grains appear to be nearly pure carbon (graphite). The nuclei of the grains are probably formed in shells of cooling gas ejected by red giants and other stars nearing the ends of their lives (these will be discussed in Chapter 22).

The grain nuclei may subsequently be incorporated into an interstellar cloud, where they can grow by gathering other atoms. The most widely accepted model pictures the grains with rocky cores and icy mantles (Figure 19.14). The most common ices in the grains are water (H_2O), methane (CH_4), and ammonia (NH_3).

19.5

Cosmic Rays

In addition to gas and dust, a third class of particles, noteworthy for the high speeds with which they travel, is found in interstellar space. They were first discovered in 1911 by the Austrian physicist Victor Hess, who flew simple instruments aboard balloons and demonstrated that high-speed particles are indeed coming to Earth from space (Figure 19.15). These particles, which became known as **cosmic rays** (an awkward name that is retained for historical reasons), resemble ordinary interstellar gas

Figure 19.15
Victor Hess returns from a 1912 balloon flight that reached an altitude of 5 1/3 km. It was on such balloon flights that Hess discovered cosmic rays. (Photo courtesy of Martin Pomerantz)

in terms of composition. In their behavior, however, cosmic rays differ radically from ordinary interstellar gas.

Composition of Cosmic Rays

Cosmic rays are high-speed atomic nuclei, electrons, and positrons. Velocities equal to 90 percent of the speed of light are typical. Most cosmic rays are hydrogen nuclei (protons) stripped of their accompanying electron. Helium and heavier nuclei constitute about 9 percent of the cosmic ray particles. The number of cosmic ray electrons is only about 2 percent the number of protons. Ten to 20 percent of the cosmic ray particles with masses equal to the mass of the electron carry positive charge rather than the negative charge that characterizes electrons. Such particles are called **positrons** and are a form of antimatter (see Section 15.2).

The abundances of various atomic nuclei in cosmic rays mirror the abundances in stars and interstellar gas, with one important exception. The light elements lithium, beryllium, and boron are far more abundant in cosmic rays than in the Sun and stars. These light elements are formed when high-speed cosmic ray nuclei of carbon, nitrogen, and oxygen collide with protons in interstellar space and break apart. (By the way, if you, like most readers, have not memorized all the elements and want to see how any of those we mention fit into the sequence of ele-

ments, you will find them all listed in Appendix 13 in order of the number of protons they contain.)

Cosmic rays reach the Earth in substantial numbers, and we can determine their properties either by capturing them directly, or by observing the reactions that occur when they collide with atoms in our atmosphere. The total energy deposited by cosmic rays in the Earth's atmosphere is only about one-billionth the energy received from the Sun, but it is comparable to the total energy received in the form of starlight. Some of the cosmic rays come to the Earth from the surface of the Sun. But most—as we will see—come from outside the solar system.

Origin of Cosmic Rays

There is a serious problem in identifying the source of cosmic rays. Since light travels in straight lines, we can tell where it comes from simply by looking. Cosmic rays are charged particles, and their direction of motion can be changed by magnetic fields. The paths of cosmic rays are curved both by magnetic fields in interstellar space, and by the Earth's own field. Calculations show that low-energy cosmic rays may spiral many times around the Earth before entering the atmosphere where we can detect them. If an airplane circles an airport many times before landing, it is impossible to determine the direction and city from which it came. So, too, after a cosmic ray circles the Earth several times, it is impossible to know where its journey began.

There are a few clues, however, about where cosmic rays might be generated. We know, for example, that magnetic fields in interstellar space are strong enough to keep all but the most energetic cosmic rays from escaping the Galaxy. It therefore seems likely that they are produced somewhere inside the Galaxy. The only likely exceptions are those with the very highest energy. Such cosmic rays move so rapidly that they are not significantly influenced by interstellar magnetic fields, and thus they could escape our Galaxy. By analogy, they could escape other galaxies as well, so some of the highest-energy cosmic rays that we detect may have been created in some distant galaxy. Still, most cosmic rays must have their source inside the Milky Way.

We can also estimate how far typical cosmic rays travel before striking the Earth. The light elements lithium, beryllium, and boron hold the key. Since they are formed when carbon, nitrogen, and oxygen strike interstellar protons, we can calculate how long, on the average, cosmic rays must travel through space in order to experience enough collisions to account for the relative abundances of lithium and the other light elements that they contain. It turns out that the required distance is about 30 times around the Galaxy. At speeds near the speed of light, it takes perhaps 3 to 10 million years for the average cosmic ray to travel this distance. This is only a small fraction of the age of the Galaxy or the universe, so cosmic rays must have been created fairly recently on a cosmic time scale.

The best candidates for a source of cosmic rays are the supernova explosions, which mark the deaths of massive stars. There are enough explosions, and the explosions generate enough energy, to account for the observed number of cosmic rays. What we do not yet know is what precise mechanism in these explosions accelerates pro-

tons and other atomic nuclei to the fast speeds we see. Some collapsed stars (including star remnants left over from supernova explosions) may, under the right circumstances, also serve as accelerators of particles. In any case, we again find that the raw material of the Galaxy is enriched by the life cycle of stars. In the next chapters we will see that as stars live and die, their material can be recycled in complex and intriguing ways.

Summary

19.1 About 5 percent of the visible matter in the Galaxy is in the form of gas and dust, which serves as the raw material for new stars. The material making up **interstellar matter** is distributed in a patchy way, sometimes becoming visible as glowing clouds of gas called **nebulae.** The most abundant elements in the interstellar gas are hydrogen and helium. About 1 percent of the interstellar matter is in the form of solid **interstellar grains.**

19.2 Interstellar gas near hot stars emits light by *fluorescence;* that is, light is emitted when an electron is captured by an ion and cascades down to lower energy levels. Glowing clouds of ionized hydrogen are called **H II regions** and have temperatures of about 10,000 K. Most hydrogen in interstellar space is not ionized and can best be studied by radio measurements of the 21-cm line. About 2 percent of interstellar space is filled with cold (100 K) clouds with densities of 50 atoms per cubic centimeter. Another 20 percent of the space between stars is filled with warm (3000 to 6000 K) neutral hydrogen clouds at a density of 0.3 atom per cubic centimeter. Some gas has been heated by supernova explosions to temperatures of 10^6 K.

19.3 Observations of interstellar gas are consistent with a model in which this gas is distributed in the form of clouds with cold, high-density cores surrounded by warm, lower-density envelopes, all embedded in a hot gas of very

low density. The pressure in these three types of regions is approximately the same. The Sun is located at the edge of a low-density (0.1 atom per cubic centimeter; $T = 10^4$ K) cloud called the Local Fluff. The Sun and this cloud are located within the Local Bubble, a region extending to at least 300 LY from the Sun, within which the density of interstellar material is extremely low.

19.4 Interstellar dust grains absorb and scatter starlight. The effects of both processes are referred to as **interstellar extinction** and cause distant stars to appear fainter than they would if no dust grains lay along the path traversed by the starlight. Much of the dust is found in clouds called **infrared cirrus.** Since typical temperatures of the dust are 20 to 500 K, interstellar grains are best observed in the infrared. Because dust grains scatter blue light more efficiently than red light, stars seen through dust appear **reddened.** Interstellar grains typically have sizes comparable to the wavelength of light. They probably have a core of silicates or carbon surrounded by a mantle of such ices as water, ammonia, and methane.

19.5 Cosmic rays are particles that travel through interstellar space at a typical speed of 90 percent the speed of light. The most abundant elements in cosmic rays are the nuclei of hydrogen and helium, but **positrons** are also found. It is likely that many cosmic rays are produced in supernova explosions.

Review Questions

1. Identify several dark nebulae in photographs in this book (don't restrict yourself to this chapter alone). Give the figure numbers of the photographs, and specify where the dark nebulae are to be found on them.

2. Why do nebulae near hot stars look red? Why do dust clouds near stars look blue?

3. Describe the characteristics of the various types of interstellar gas clouds.

4. Prepare a table listing the different ways in which (a) dust and (b) gas can be detected in interstellar space.

5. Describe how the 21-cm line of hydrogen is formed. Why is this line an important tool for understanding the interstellar medium?

6. Describe the properties of the dust grains found in the space between stars.

7. Why is it difficult to determine where cosmic rays come from?

8. What causes reddening of starlight? Explain how the reddish color of the Sun's disk at sunset is caused by the same process.

9. Suppose a bright reflection nebula appears yellow. What kind of star is probably producing it?

10. Describe the spectrum of each of the following:
  a. Starlight reflected by dust
  b. A star behind invisible interstellar gas
  c. An emission nebula

11. According to the text, a star must be hotter than about 25,000 K to produce an H II region. Both white dwarfs and main-sequence O stars have temperatures hotter than 25,000 K. Which type of star can ionize more hydrogen? Why?

12. From the comments in the text about which kinds of stars produce emission nebulae and which kinds are associated with reflection nebulae, what can you say about the temperatures of the stars in the Pleiades (see Figure 19.10)?

13. One way to calculate the size and shape of the Galaxy is to estimate the distances to faint stars from their observed apparent magnitudes, and to note the distance at which stars are no longer observable. The first astronomers to try this experiment did not know that starlight is dimmed by interstellar dust. Their estimates of the size of the Galaxy were much too small. Explain why.

14. Short-wavelength light is *absorbed* more efficiently by the Earth's atmosphere than is long-wavelength light. Sunburns are caused primarily by sunlight with wavelengths between 280 and 320 nm. The heat that we feel, however, is produced mainly by infrared radiation. Use this information to explain in general terms why it is easier to get sunburned at noon than in the late afternoon, even though the Sun feels nearly as hot at, say, 4:00 P.M. as it does at noon.

15. New stars form in regions where the density of gas and dust is high. Suppose you wanted to search for some recently formed stars. Would you more likely be successful if you observed at visible wavelengths or at infrared wavelengths? Why?

16. In big cites, you can see much farther on days without smog. Why?

17. The Sun is located in a region where the density of interstellar matter is low. Suppose that instead it were located in a dense cloud that dimmed visible light significantly. How would this have affected the development of civilization on Earth? For example, would it have presented a problem for early navigators?

18. Suppose that the average density of hydrogen gas in our Galaxy is one atom per cubic centimeter. If the Galaxy is a sphere with a diameter of 100,000 LY, how many hydrogen atoms are in the interstellar gas? What is the mass of this quantity of hydrogen?

19. H II regions can exist only if there is a nearby star hot enough to ionize hydrogen. Hydrogen is ionized only by radiation with wavelengths shorter than 91.2 nm. What is the temperature of a star that emits its maximum energy at 91.2 nm (use Wien's law)? Based on this result, what are the spectral types of those stars likely to provide enough energy to produce H II regions?

20. According to the text, there is about one atom of hydrogen in every cubic centimeter of interstellar space, and per-

haps 10^3 dust grains per 10^9 m^3. If approximately 1 percent of the total mass of the interstellar medium is in the form of dust grains, what is the typical mass of a dust grain?

21. Suppose the density of a typical dust grain is 3 g/cm^3, and its mass is the value found in Problem 20; what is its radius? (Assume the grain is spherical, and remember that the volume of a sphere is given by the formula $V = (4/3)\pi R^3$.)

22. Suppose a star is behind a cloud of dust that dims its brightness by a factor of 100. Suppose you do not realize the dust is there. How much in error will your distance estimate be? Can you think of any measurement you might make to detect the dust?

Suggestions for Further Reading

Blitz, L. "Giant Molecular Cloud Complexes in the Galaxy" in *Scientific American*, Apr. 1982, p. 84.

Bowyer, S. et al. "Observing a Partly Cloudy Universe" in *Sky & Telescope*, Dec. 1994, p. 36. Review of the Extreme Ultraviolet Explorer mission and what it has shown us about the ISM.

Friedlander, M. *Cosmic Rays.* 1989, Harvard U. Press. The definitive introduction.

Helfand, D. "Fleet Messengers from the Cosmos" in *Sky & Telescope,* Mar. 1988, p. 265. Excellent summary of history and current understanding of cosmic rays.

Malin, D. *A View of the Universe.* 1993, Sky Publishing and Cambridge U. Press. Album by the skilled Australian astronomical photographer; includes some of the most dramatic color images of nebulae ever taken.

Marschall, L. "The Secrets of Interstellar Clouds" in *Astronomy*, Mar. 1982, p. 6.

Shore, L. and Shore, S. "The Chaotic Material Between the Stars" in *Astronomy*, June 1988, p. 6.

Smith, D. "Reflection Nebulae: Celestial Veils" in *Sky & Telescope*, Sep. 1985, p. 207.

Teske, R. "The Star That Blew a Hole in Space" in *Astronomy*, Dec. 1993, p. 31. On the Local Bubble.

Verschuur, G. "Interstellar Molecules" in *Sky & Telescope*, Apr. 1992, p. 379.

Verschuur, G. "Barnard's Dark Dilemma" in *Astronomy*, Feb. 1989, p. 30.

Verschuur, G. *Interstellar Matters*. 1989, Springer Verlag. Essays on dust and the interstellar medium; combines good historical material and reviews of modern topics.

Wynn-Williams, G. "Bubbles, Tunnels, Onions, Sheets: The Diffuse Interstellar Medium" in *Mercury*, Jan./Feb. 1993, p. 2.

Wynn-Williams, G. *The Fullness of Space: Nebulae, Stardust, and the Interstellar Medium*. 1992, Cambridge U. Press. A thorough introduction to the matter between stars.

Using REDSHIFT™

1. Check the images in the *Photo Gallery* under Galactic, Gaseous Nebulae. Identify the following nebulae as principally bright, dark, or reflection. Also record secondary features of the nebulae that belong in a different class.

a. M8   d. NGC 2024
b. M16   e. NGC 2237
c. M42   f. NGC 6589-90

A portion of the Orion Nebula, about 1.6 LY across diagonally, recorded by the Hubble Space Telescope. It combines images taken in the red light of nitrogen gas, the green light of hydrogen, and the blue light of oxygen. The gases in the nebula are set to glow by the strong ionizing radiation from a brilliant O-type star that lies outside the region of this image, off to the bottom left. Visible throughout the region are a series of small, elongated "cocoons" of gas and dust, inside which stars are forming. Several of these can be seen in the inset, which shows a field only 0.14 LY across. Images such as this are good demonstrations of the capabilities of the repaired Hubble. (C. R. O'Dell/NASA)

CHAPTER 20

The Birth of Stars and the Search for Planets

Thinking Ahead

Although they had long suspected that other stars should have planets, astronomers were not able to discover a planet around a star like the Sun until 1995. Why did it take them so long?

Having examined the raw material from which stars form, we are ready to take a look at the process by which new stars emerge from interstellar matter. The mechanisms that make stars can sometimes also lead to one or more planets around those stars, so in this chapter we also examine the evidence for planets outside the solar system.

As we begin our study of the birth of stars, it may be useful to remember some of the key things we've learned so far. Table 20.1 on page 398 summarizes several important ideas about stars that were established in previous chapters.

The formation of stars is not merely a topic of historical interest. As we will see in this chapter, star formation is a continuous process that is going on *right now*. Stars of all masses, low as well as high, are being formed in the interiors of dust and gas clouds, throughout our Galaxy and others. In the last few years, new telescopes and instruments have enabled astronomers to look deep within nearby regions of star formation, and to obtain tantalizing glimpses of stars in the very earliest stages of their lives.

All these illustrious worlds,
and many more,
Which by the tube
astronomers explore . . .
Are suns, are centers,
whose superior sway,
Planets of various
magnitudes obey.

Sir Richard Blackmore in *The Creation*, Book II (quoted by A. Meadows in *The High Firmament*)

Figure 20.1
Stars form in clouds of gas and dust. M16 (the Eagle Nebula) is a large region of such cosmic raw material, visible because about 2 million years ago it produced a cluster of bright stars whose light ionizes the gas lying nearby. The cluster can be seen at the upper right of this image. The "elephant trunks" of dusty material visible near the center of the image are seen in much more detail on the cover of our text.
(Anglo–Australian Observatory Board)

20.1

Star Formation

If we want to find the youngest stars—those still in the process of formation—we must look in places with plenty of the raw material required to make stars. Since stars are made of gas, we focus our attention (and our telescopes) on the dense clouds of gas that dot the Milky Way (Figure 20.1).

Molecular Clouds: Stellar Nurseries

As we saw in Chapter 19, the most massive clouds—and some of the most massive objects in the Milky Way Galaxy—are the **giant molecular clouds.** These enormous reservoirs of gas and dust contain enough mass to make anywhere from a hundred to a million Suns, and the diameter of a typical cloud is 50 to 200 LY. Their name reflects the coldness of their interiors; with characteristic temperatures of about 10 K, most of their gas atoms are bound into molecules. Observations show that these clouds are the birthplaces of most stars in our Galaxy.

Molecular clouds are not smooth, but rather contain clumps or *dense cores* of material with very low temperatures and densities much higher (10^4 to 10^5 atoms per cubic centimeter) than is typical of most of the rest of the cloud. Both of these conditions—low temperature and high density—are just what is required to make new

stars. In order to form a star—that is, a dense, hot ball of matter capable of starting nuclear reactions deep within—we need a massive clump of atoms and molecules to shrink in radius and increase in density by a factor of nearly 10^{20}. Such a drastic collapse is brought about by the force of gravity.

TABLE 20.1
Basics About Stars from Earlier Chapters

- Stable (main-sequence) stars such as the Sun maintain equilibrium by producing energy through nuclear fusion in their cores. The ability to generate energy by fusion defines a star. (Sections 15.2, 15.3)
- Each second in the Sun approximately 600 million tons of hydrogen undergo fusion into helium, with about 4 million tons turning to energy in the process. This rate of hydrogen use means that eventually the Sun (and all other stars) will run out of central fuel. (Section 15.2)
- Stars come with many different masses, ranging from 1/12 M_{Sun} to roughly 100 M_{Sun}. There are far more low-mass than high-mass stars. (Section 17.2)
- The most massive main-sequence stars (spectral type O) are also the most luminous and have the highest surface temperature. The lowest-mass stars on the main sequence (spectral type M) are the least luminous and the coolest. (Section 17.4)
- A galaxy of stars such as the Milky Way contains enormous amounts of gas and dust—enough to make billions of stars like the Sun. (Section 19.1)

Figure 20.2
The star group named after Orion, the hunter of Greek mythology. Three stars close to each other in a line mark Orion's belt. The ancients imagined a sword hanging from the belt; one of the objects in this sword is the Orion Nebula. (Andrea Dupree, Harvard-Smithsonian CA, and Ronald Gilliland, NASA)

As we saw in Chapter 15 (and will continue to see in future chapters), the essence of the stars' life story is the ongoing competition between two forces: *gravity* and *pressure*. The force of gravity, pulling inward, tries to make a star collapse. Internal pressure produced by the motions of the gas atoms, pushing outward, tries to force the star to expand. When these two forces are in balance, the star is stable. Major changes in the structure of a star occur when one or the other of these two forces gains the upper hand. When a star is first forming, low temperature and high density both work to give gravity the advantage. Let's begin our examination of what happens by looking at a specific region of star formation.

The Orion Molecular Cloud

The best-studied of the stellar nurseries is in the constellation of Orion, the hunter, about 1500 LY away (Figure 20.2). A luminous cloud of dust and gas, called the Orion Nebula, can easily be seen with binoculars in the middle of the sword hanging from Orion's belt (Figure 20.3). This luminous material, however, is a mere "pimple" of light on the surface of a much larger molecular cloud. Most of the cloud does not glow with visible light, but betrays its presence by the radiation it gives off at infrared and radio wavelengths (Figure 20.4).

The Orion Nebula itself is a fascinating place, where a young cluster of hot stars is shining with great energy. Its radiation is opening up the cocoon of dust inside

Figure 20.3
A wide-angle view of some of the star-forming regions of Orion. The Orion Nebula is the fuzzy white object near the bottom. A cloud of dust that resembles a horsehead can be seen against the reddish glow of hydrogen in the upper left. The bright, over-exposed star above the horsehead and the one at the very top center of the image are two of the stars in Orion's belt. All the visible features of gas and dust are evidence of the much larger molecular cloud that lies behind them and that (at radio wavelengths) would fill most of the field shown here. (Anglo–Australian Observatory Board)

which it formed, exposing the stars to our view (Figure 20.5). Only a small number of stars in this Trapezium cluster can be seen with visible light, but infrared images—which penetrate the dust better—actually reveal over a thousand stars (Figure 20.6). The image that opens this chapter shows some of the wonderfully complex structure of gas, dust, and newly formed stars that characterizes this region.

(text cont. on page 401)

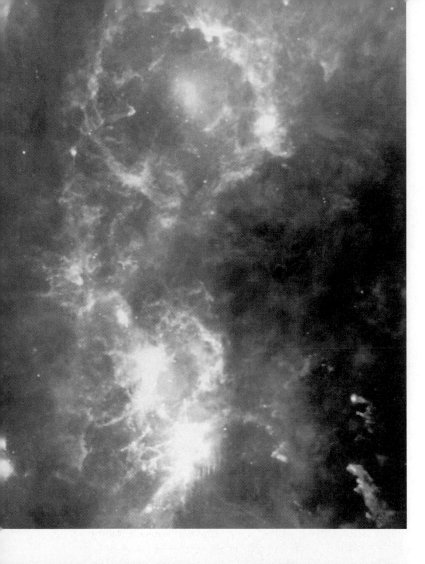

◄Figure 20.4
A wide-field view of Orion at three infrared wavelengths, taken with the Infrared Astronomical Satellite. The emission at 12 μm (typically coming from stars) is colored blue; emission at 60 μm is colored green; and emission at 100 μm (which comes from interstellar dust) is colored red. The two colorful regions seen in Figure 20.3 are the two bright yellow splotches at bottom right; the lower one is the Orion Nebula, and the higher one is the region of the horsehead, which can be seen as a small yellow finger poking out to the right. The horsehead area marks the region of Orion's belt. The red giant Betelgeuse is an intense point in the upper center of the image. To its right is a large ring that appears to be the remnant of an exploding star. Note how the infrared image reveals some of the vast quantity of dust in this region, and betrays the presence of the giant Orion Molecular Cloud. (Infrared Processing and Analysis Center/NASA)

Figure 20.5
In this wonderful image of the Orion Nebula, Australian photographer David Malin has used a special photographic technique called unsharp-masking to bring out subtle details. You can clearly see how the energy from the Trapezium cluster (the highly exposed white region) is opening up the gas and dust cloud within which it formed. The energy is ionizing the gas and reflecting from the dust to show the expanding bubble of the nebula. (Anglo–Australian Telescope Board)

◄Figure 20.6
The inner part of the Orion Nebula is seen in the infrared in this image taken with an electronic infrared camera on the 3.8-m United Kingdom Infrared Telescope on Mauna Kea. Blue colors represent radiation at 1.2 μm, green at 1.65 μm, and red at 2.2 μm. The hot O- and B-type stars of the Trapezium cluster are seen as blue in the center of the image. Dusty regions are seen as red and purple. The image covers a region of about 2 LY by 2 LY. About 500 stars can be found on this image, most belonging to the Trapezium cluster. This makes the region one of the densest groupings of stars astronomers have yet found. (© Mark McCaughrean 1989/NASA)

Interesting as the nebula may be, the full Orion molecular cloud is the truly impressive structure. In its long dimension it stretches over a distance of about 100 LY. The total quantity of molecular gas is about 200,000 times the mass of the Sun. A wave of star formation that began about 12 million years ago at one edge of this molecular cloud, near the western shoulder of the Orion star figure, has slowly moved through the cloud, leaving behind groups of newly formed stars. The stars in Orion's belt are about 8 million years old, and the stars near the Trapezium range in age from 300,000 to a million years.

Why is there so much material left around the newly formed stars in the Trapezium cluster? We have come to understand that star formation is not a very efficient process, using typically only a few percent—at most perhaps 25 percent—of the gas in a molecular cloud. The leftover material is heated either by the radiation of hot stars or by the explosions of the most massive stars that form. (We will see in later chapters that the most massive stars go through their lives very quickly, and end them by exploding.) Whether gently or explosively, the material in the neighborhood of the new stars is blown away into interstellar space. Older groups of stars can therefore be easily observed in visible light because they are no longer shrouded in dust and gas.

Because of the correlation between stellar ages and position in the Orion region, we know that star formation has been moving progressively through this molecular cloud. While we do not know what initially caused stars to begin forming in Orion, there is good evidence that the first generation of stars triggered the formation of additional stars, which in turn led to the formation of still more stars (Figure 20.7).

The basic idea is this: When a massive star is formed, it emits a large amount of ultraviolet radiation that heats the surrounding gas in the molecular cloud. This heating increases the pressure in the gas and causes it to expand.

When massive stars exhaust their supply of fuel, they explode, and the energy of the explosion also heats the gas. The hot gases burst into the surrounding cold cloud, compressing the material in it until the cold gas is at the same pressure as the expanding hot gas.

For the regions of cooler gas this can be a dramatic change. In the conditions typical of molecular clouds, the compression is enough to increase the gas density by a factor of 100. At densities this high, stars can begin to form in the compressed gas. Such a chain reaction—where the brightest and hottest stars of one area become the cause of star formation "next door"—seems to have occurred not only in Orion but also in many other molecular clouds.

The Birth of a Star

Although regions such as Orion can give us insight into how star formation might begin, its subsequent stages are still shrouded in mystery (and a lot of dust). There is almost a factor of 10^{20} difference between the density of a molecular cloud core and that of the youngest stars that can be detected. So far, we have been unable to get any *direct* evidence about what happens within a cloud as material comes together through the action of gravity and collapses through this range of densities to form a star.

Observations of this stage of stellar evolution are nearly impossible for several reasons. First, the dust-shrouded interiors of molecular clouds where stellar births take place cannot be observed with visible light. It is only with the new techniques of infrared and millimeter radio astronomy that we are able to make any measurements at all. In addition, the time scale for the initial collapse, measured in thousands of years, is very short, astronomically speaking. Since each star spends such a tiny fraction of its life in this stage, relatively few stars are going through the collapse process at any given time.

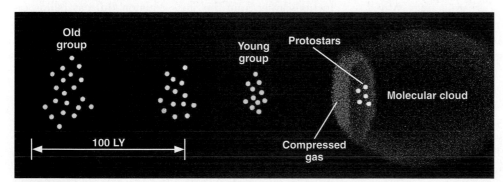

Figure 20.7
Schematic diagram showing how star formation can move progressively through a molecular cloud. The oldest group of stars lies to the left of the diagram and has expanded because of the motions of individual stars. Eventually the stars in the group will disperse and no longer be recognizable as a cluster. The youngest group of stars lies to the right, next to the molecular cloud. This group of stars is only 1 to 2 million years old. The pressure of the hot, ionized gas surrounding these stars compresses the material in the nearby edge of the molecular cloud and initiates the gravitational collapse that will lead to the formation of more stars.

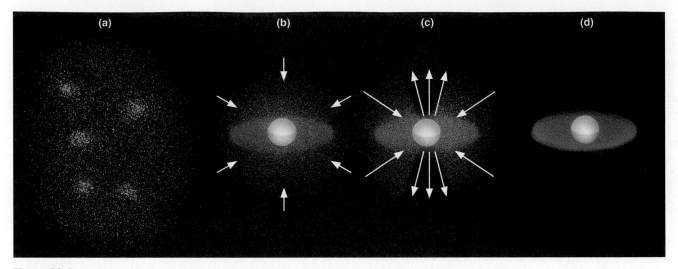

Figure 20.8
The formation of a star. (a) Dense cores form within a molecular cloud. (b) A protostar with a surrounding disk of material forms at the center of a dense core, accumulating additional material from the molecular cloud through gravitational attraction. (c) A stellar wind breaks out along the two poles of the star. (d) Eventually this wind sweeps away the cloud material and halts the accumulation of additional material, and a newly formed star surrounded by a disk becomes observable. (Based on drawings by F. Shu, F. Adams, and S. Lizano)

Furthermore, the collapse of a new star occurs in a region so small (0.3 LY) that in most cases we cannot resolve it with existing techniques. For all these reasons we have yet to catch a star in the act, so to speak, of its initial collapse. Nevertheless, through a combination of theoretical calculations and the limited observations available, astronomers have pieced together a picture of what the earliest stages of stellar evolution are likely to be.

During the period when a clump of matter in a molecular cloud is contracting to become a true star, we call it a **protostar.** Since a molecular cloud is more massive than a typical star by a factor of 100,000 or more, many protostars must form within each cloud.

The first step in the process of creating stars is the formation within the cloud—through a process we do not fully understand—of the dense cores of material discussed earlier (Figure 20.8a). These dense cores then attract additional matter because of the gravitational force they exert on the cloud material surrounding them. Eventually, the gravitational force of the infalling gas becomes strong enough to overwhelm the pressure exerted by the cold material that forms the dense cores. The material then undergoes a rapid collapse, the density of the core increases greatly as a result, and a protostar is formed.

According to the conservation of angular momentum (discussed in Chapter 2), a rotating body spins more rapidly as it decreases in size. In other words, if the object can turn its material around a smaller circle, it can move that material more quickly. This is exactly what happens when a protostar is forming: as it shrinks, its rate of spin goes up. Since the natural turbulence inside a molecular

cloud tends to give any portion of it some initial spinning motion (even if it is very slow), all collapsing cores are expected to spin as they become protostars.

Rapid spin can begin to affect the shape of a ball of collapsing material. Rotation can keep material from falling inward at the equator of the spinning ball of gas (even while it continues falling in at the poles). You may have observed this same effect on the amusement park ride in which you stand with your back to a cylinder that is spun faster and faster. As you spin really fast, you are pushed against the wall so strongly that you could not possibly fall toward the center of the cylinder. So, too, the collapsing gas around a protostar can fall easily from directions away from the star's equator. But gas falling inward toward the equator is held back by the rotation, and forms an extended disk around the equator (as shown in Figure 20.8b).

The protostar and disk at this stage are embedded in an envelope of dust and gas, from which material is still falling onto the protostar. This dusty envelope blocks visible light, but infrared radiation can get through. As a result, a protostar itself in this phase of evolution is observable only in the infrared region of the spectrum. However, the dusty clouds of material around the star can sometimes be glimpsed when they are illuminated by the light of nearby stars that formed some time earlier, as in the Orion Nebula. The inset for the image that opens this chapter shows some beautiful examples of elongated dust clouds observed with the Hubble Space Telescope. Infrared observations reveal stars in the process of forming inside most of these clouds.

Winds and Jets

Observations show that a protostar eventually goes through a stage involving a powerful outflow of particles from its surface. This **stellar wind** consists mainly of protons (hydrogen nuclei) and electrons streaming away from the star at speeds of about 200 km/s (about 440,000 mi/h). Let us examine the effect of this wind on the material around the forming star.

When the wind first starts up, the disk of material around the star's equator blocks the wind in this direction. Where the wind particles *can* escape most effectively is in the direction of the star's poles (Figure 20.8c). If this model is correct, we should see beams of particles shooting out in opposite directions from the polar regions of new stars during this stage of their formation.

Such double beams (or *jets*) are precisely what astronomers observe around some protostars. Sometimes the material flowing out can be quite broad (Figure 20.9), but on other images, we see a remarkably narrow set of jets. On occasion, the jets collide with a somewhat denser lump of material nearby and excite its atoms to glow. These glowing regions, called *Herbig–Haro* (or *HH*) *objects* after the two astronomers who first identified them, allow us to trace the progress of the jets quite a distance from the star that produces them.

Figure 20.10 shows a series of recent images of Herbig–Haro objects, taken with the Hubble Space Telescope, that confirm our general picture of star formation but that show a considerable amount of complex structure in the jets and the clouds of material they energize. It may be that great clumps of material from the star's inner disk fall toward the star and then are caught up by the wind and blown outward, much as a clump of leaves can be

(text cont. on page 405)

Figure 20.9
An infrared image of the protostar S106. Gas flows outward from this object in directions perpendicular to the dense disk that surrounds it. The disk is located where the outward flows appear to pinch together. (National Optical Astronomy Observatories)

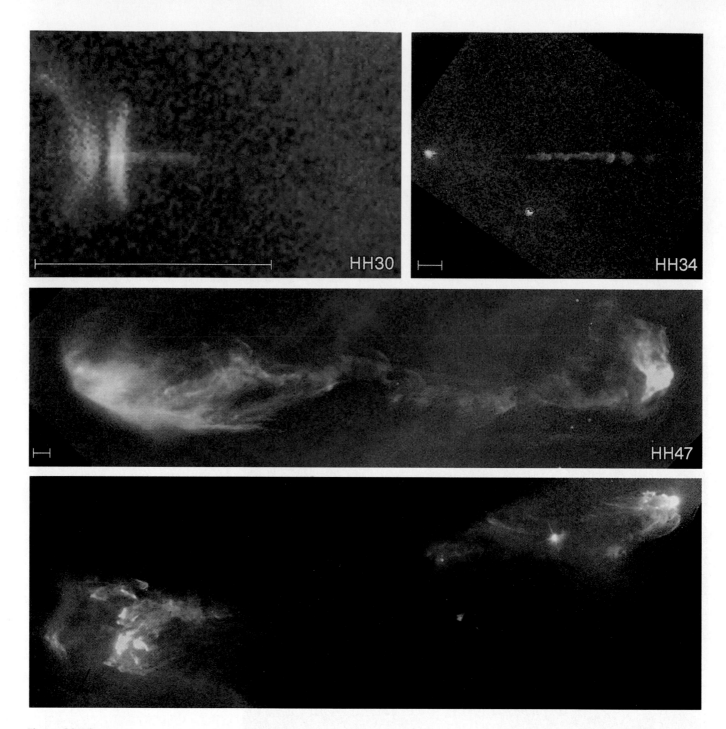

Figure 20.10

A collection of recent images taken with the Hubble Space Telescope, showing disks and jets from protostars. On the first three images the white bar at the bottom is the size of 1 AU. (a) Upper left: HH 30, an edge-on disk around a protostar in the constellation of Taurus about 450 LY away. The reddish jet is amazingly narrow and shows knots of material that can be monitored to measure the rate at which material in the jet is moving. The speed turns out to be about a third of a million km/h. (C. Burrows and NASA) (b) Upper right: HH 34, a protostar in the vicinity of the Orion Nebula, displays a jet with beaded structure that gets wider as it gets farther from its star. (J. Hester and NASA) (c) Middle: HH 47, a protostar 1500 LY away (invisible inside a dusty disk at the left edge of the image), produces a very complicated jet. The star may actually be wobbling, perhaps because it has a companion. Light from the star illuminates the white region at the left because light can emerge perpendicular to the disk (just as the jet does). At right the jet is plowing into existing clumps of interstellar gas, producing a shock wave that resembles an arrowhead. (J. Morse and NASA) (d) Bottom: This view of HH 1 and 2 shows a classic double-beam jet emanating from a protostar (hidden in a dust disk in the center) in the constellation of Orion. Tip to tip, these jets are more than 1 LY wide. The bright regions (first identified by Herbig and Haro) are places where the jet is slamming into a clump of interstellar gas. (C. Burrows, J. Morse, J. Hester, and NASA)

blown away together in a windstorm on Earth. This might explain the clumpy structure we are beginning to see in many of these jets, but such ideas are still very speculative. Clearly, we have a lot more to learn about the details of these intriguing objects.

The wind from a forming star will ultimately sweep away its obscuring envelope of dust and gas, leaving behind the naked disk and protostar, which can now be seen with visible light (Figure 20.8c and 20.8d). We should note that at this point, the protostar itself is still contracting slowly and has not yet reached the main-sequence stage on the H–R diagram (a concept introduced in Chapter 17). But the disk can be more easily detected at this point, especially when observed at infrared wavelengths. In the last decade astronomers have accumulated a great deal of evidence for the presence of such disks around many newly forming stars.

This description of a protostar surrounded by a rotating disk of gas and dust sounds very much like what happened when the Sun and planets formed (see Section 13.3). Do the disks around protostars also form planets? We will return to this question at the end of the chapter.

The H–R Diagram and the Study of Stellar Evolution

One of the best ways to summarize all of these details about how a star or protostar changes with time is to use an H–R diagram. As a star consumes its nuclear fuel, its luminosity and temperature change. Thus its position on the H–R diagram, in which luminosity is plotted against temperature, also changes. As a star ages, we must replot it in different places on the diagram. Therefore astronomers often speak of a star *moving* on the H–R diagram, or of its evolution *tracing out a path* on the diagram. Of course, in this context, "tracing out a path" has nothing to do with the star's motion through space; this is just a shorthand way of saying that its temperature and luminosity change as it evolves.

To estimate just how much the luminosity and temperature of a star change as it ages, we must resort to calculations. In a theoretical study of stellar evolution, we compute a series of *models* for a star, each successive model representing a later point in time. Stars may change for a variety of reasons. Protostars, for example, change in size because they are contracting, and their temperature and luminosity change as they do so. After nuclear fusion begins in the star's core (see Chapter 15), main-sequence stars change because they are using up their nuclear fuel.

Given a model that represents a star at one stage of its evolution, we can calculate what it will be like at a slightly later time. At each step, the model predicts the lu-

minosity and size of the star, and from these we can figure out its surface temperature. A series of points on an H–R diagram, calculated in this way, allows us to follow the life changes of a star and hence is called its *evolutionary track*.

Evolutionary Tracks

Let's now use these ideas to follow the evolution of protostars that are on their way to becoming main-sequence stars. The evolutionary tracks of newly forming stars with a range of stellar masses are shown in Figure 20.11. These young stellar objects are not yet producing energy by nuclear reactions, but derive their energy from gravitational contraction—by the sort of process proposed for the Sun by Helmholtz and Kelvin in the last century (see Chapter 15).

Initially, a protostar remains fairly cool, with a very large radius and a very low density. It is transparent to infrared radiation, and the heat generated by gravitational contraction can be radiated away freely into space. Because heat builds up slowly inside the protostar, the gas pressure remains low, and the outer layers fall almost unhindered toward the center. Thus the protostar undergoes very rapid collapse, corresponding to the roughly vertical lines at the right of Figure 20.11. As the star shrinks its surface area gets smaller, and so its total luminosity decreases. The contraction stops only when the protostar becomes dense and opaque enough to trap the heat released by gravitational contraction.

Stars first become visible only after the stellar wind described earlier clears away the surrounding dust and gas, and this occurs near the point at which the evolutionary tracks in Figure 20.11 change from being nearly vertical to being nearly horizontal (although, as you can see, the details are not quite the same for stars of different mass). At this point, changes inside the contracting star keep the luminosity of stars like our Sun roughly constant, but start building up the surface temperature. Thus the star "moves" to the left in the H–R diagram.

To help you keep track of the various stages that stars undergo during their lives, it can be useful to compare the development of a star to that of a human being. (Clearly, you will not find an exact correspondence, but thinking through the stages in human terms may help you remember some of the ideas we are trying to emphasize.) Protostars might be compared with human embryos, as yet unable to sustain themselves, but drawing resources from their environment as they grow. Just as the birth of a child is the moment it is called upon to produce its own energy (through eating and breathing), so astronomers say that a star is born when it is able to sustain itself through nuclear reactions.

When the central temperature becomes high enough (about 10 million K) to fuse hydrogen into helium, we say that the star has *reached the main sequence* on the H–R diagram. (Astronomers call such a star a main-sequence

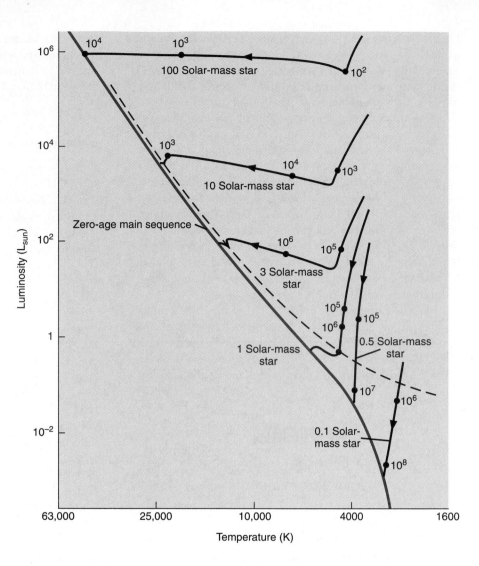

Figure 20.11

Evolutionary tracks for contracting protostars of different masses seen on an H–R diagram. The numbers next to each dark point on a track are the rough number of years it takes an embryo star to reach that stage in its life. You can see that the more mass a star has, the shorter the time it takes to go through each stage. Stars that lie above the dashed line would typically still be surrounded by infalling material and would be hidden by it.

star even when the discussion does not concern the H–R diagram.) It is now a full-fledged star, more or less in equilibrium, and its rate of change slows dramatically. Only the gradual depletion of hydrogen as it is transformed into helium in the core slowly changes the star's properties. The mass of a star determines exactly where it falls on the main sequence. As Figure 20.11 shows, massive stars on the main sequence have high temperatures and high luminosities. Low-mass stars have low temperatures and low luminosities.

Objects of extremely low mass never achieve high enough central temperatures to ignite nuclear reactions. The lower end of the main sequence stops where stars have a mass just barely great enough to sustain nuclear reactions at a sufficient rate to stop gravitational contraction. This critical mass is calculated to be about 1/12 the mass of the Sun. Objects below this critical mass are called either brown dwarfs or planets. Exactly where the dividing line should be drawn between brown dwarfs and planets is still a controversial subject among astronomers. Increasingly, "planet" is used to refer to an object formed in a disk surrounding a protostar, while a "brown dwarf," like a true star, forms at the center of the protostellar disk.

At the other extreme, the upper end of the main sequence terminates at the point where the energy radiated by the newly forming massive star becomes so great that it halts the accretion of additional matter. The upper limit of stellar mass is thought to be about 100 solar masses.

Evolutionary Time Scales

In general, a star's pre-main-sequence evolution slows down as it moves along its evolutionary track toward the main sequence. The numbers labeling the points on each track in Figure 20.11 are the times, in years, required for the embryo stars to reach those stages of contraction. The time for the whole evolutionary process, as the figure shows, depends on the mass of the star. Stars of mass much higher than the Sun's reach the main sequence in a few thousand to a million years. The Sun required millions of years before it was born. Tens of millions of years are required for stars of lower mass to evolve to the lower main sequence. (We will see that massive stars go through *all* stages of evolution faster than do low-mass stars.)

We will take up the subsequent stages in the life of a star in the next chapter, examining what happens after stars arrive on the main sequence and begin a "prolonged adolescence" of fusing hydrogen to form helium. But now we want to examine the connection between the formation of stars and planets.

Evidence That Planets Form Around Other Stars

Having developed on one, and finding it essential to our existence, we have a special interest in planets. Yet planets outside the solar system are extremely difficult to detect. The amount of light a planet reflects from its star is a depressingly tiny fraction of the light that star gives off. Furthermore, planets are lost in the glare of their much brighter parent stars. You might compare a planet orbiting a star to a mosquito flying around one of those giant spotlights at shopping center openings. Imagine viewing the scene from some distance away—say from an airplane. You could see the spotlight just fine, but what are your chances of catching the light reflected from the mosquito?

Despite such daunting difficulties, we would very much like to know whether planets are common around other stars. We can search for planets elsewhere *indirectly*—by observing their effects on the parent star, or by detecting a disk of material from which planets might be condensing.

Disks Around Protostars: Planetary Systems in Formation?

It is a lot easier to detect the raw material from which planets might be assembled than to detect planets after they are fully formed. The reasoning goes like this: As we saw in Chapter 13, planets form by the agglomeration of gas and dust particles in orbit around a newly created star. Each dust particle is heated by the young protostar and radiates in the infrared region of the spectrum. Before any planets form, we can detect such radiation from all of the individual dust particles that are destined to become part of the planets.

However, once those particles gather together and form a few planets, the overwhelming majority are hidden in the interiors of the planets where we cannot see them. All we can now detect is the radiation from the outside surfaces of the planets, which cover a drastically smaller area than the dusty disk from which they formed. The amount of infrared radiation is therefore greatest *before* the dust particles combine into planets. For this reason, our search for planets begins with a search for radiation from the material required to make them.

As we have seen in this chapter, the presence of a disk of gas and dust appears to be an essential part of star formation. Observations show that at least 50 percent of all protostars are surrounded by disks (Figure 20.12) that range in size from 10 to 1000 AU. For comparison, the average diameter of the orbit of Pluto—that is, the size of our own planetary system—is 80 AU. The mass contained in these disks is typically 1 to 10 percent of the mass of

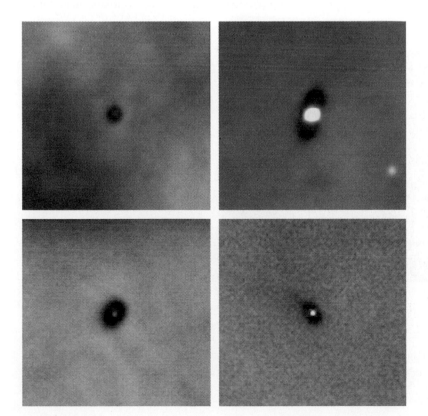

Figure 20.12
Hubble Space Telescope images of four disks around young stars in the Orion Nebula. The dark, dusty disks are seen silhouetted against the bright backdrop of the glowing gas in the nebula. The size of each image is about 30 times the diameter of our planetary system; this means the disks we see here range in size from two to eight times the orbit of Pluto. The red glow at the center of each disk is a young star, no more than a million years old. See this chapter's opening image for some disks in which stars are not easily visible.
(M. McCaughrean, C. R. O'Dell, and NASA)

our own Sun, which is more than the mass of all the planets in our solar system put together. Observations therefore demonstrate that a large fraction of stars begin their lives with enough material in the right place to form a planetary system.

Evolution of Proto-Planetary Disks

This result is encouraging, but the existence of rotating disks of raw material is still not enough. Do planets actually form? There is indirect evidence that they do. If we measure the temperature and luminosity of a protostar, we can place it in an H–R diagram like the one shown in Figure 20.11. By comparing the real star to our models of how protostars should evolve with time, we can estimate its age. What astronomers have discovered is that the properties of the disks change systematically with the ages of the encircled protostars.

In order to understand what the changes mean, we need to bear two things in mind. First, when we study the disks with infrared detectors, what we are actually able to measure is the heat radiation from dust. The dust absorbs energy radiated by the central protostar, and then the dust re-emits that energy as infrared radiation.

Second, because the protostar is the primary source of energy, the dust closest to the protostar is hottest. Temperatures in the disk at 0.1 AU from the star are about 1000 K; at 1 AU, the temperature is a few hundred degrees; and at their outer boundaries the disks are very, very cold—only a few tens of degrees. Therefore, if we observe some dust with temperatures of 1000 K or more in a given disk, we know that the dust extends nearly all

the way to the surface of the star. If no hot matter is detectable, then the regions close to the star must be very nearly empty of dust.

What the observations show is that if a protostar is less than 3 million years old, its disk does contain hot dust. The disk, which is opaque to visible light, must therefore extend all the way from very close to the surface of the star out to tens or hundreds of astronomical units away. But by the time protostars reach the "advanced age" of 30 million years, their disks have thinned out and become transparent.

Most interestingly, around 3- to 30-million-year-old stars we find disks with outer parts that are still opaque but with so little hot material that we cannot detect the inner disk at all. That is, the inner regions close to the star are seriously depleted of dust. In these objects, the disk looks like a donut with the protostar centered in its hole.

How is it possible to have two rather sharply separated zones in the disk—one with a high density of dusty material and the other with almost none? This separation is especially puzzling since calculations show that matter from the outer, dense region should spiral inward quickly (in less than about 50,000 years) and fill the inner hole. This time span is pretty short compared to how much time the star spends in this general stage of its life. The fact that we see such holes around a large fraction of the protostars with ages between 3 and 30 million years indicates that there must be some mechanism to prevent the matter from flowing inward.

Calculations show that the formation of a Jupiter-sized planet would halt the inward flow. Suppose such a planet does form at a distance of a few astronomical units

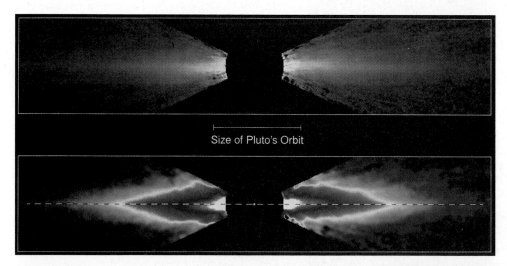

Figure 20.13
The disk around the star Beta Pictoris, as seen with the Hubble Space Telescope. The star itself is blocked out so that the much fainter disk can be observed. The cleared-out area discussed in the text is also within the blocked-out area. (Top) A visible light image of the disk, shining with reflected starlight. Note the size of Pluto's orbit, given by the line below. (Bottom) Image-processing techniques have been used to assign different colors to each level of brightness in the disk. The pink-white inner part of the disk is slightly tilted relative to the plane of its outer parts (red-yellow-green). (C. Burrows, J. Krist, ESA/NASA)

from the protostar, presumably due to the progressive agglomeration of matter from the disk. This very agglomeration can contribute to the creation of dust-free regions. Calculations also show that any small dust particles and gas that were initially located in the region between the protostar and the planet, and that are not swept up by the planet, will accrete onto the star in about 50,000 years.

Matter lying outside the planet's orbit, on the other hand, is prevented from moving into the hole by the gravitational forces exerted by the planet. (We saw something similar in Saturn's rings, where the action of the shepherd moons keeps the material near the edge of the rings from spreading out.) The formation of giant planets can, according to the calculations, thus produce and sustain the holes in the disks surrounding very young stars.

This phenomenon can actually be seen in a star that is no longer a protostar, but has maintained a significant disk even as it has reached the main sequence. Called Beta Pictoris, it is located about 50 LY away in the southern constellation of Pictor. What distinguishes this star is that its disk (at least 2000 AU wide) can be detected with visible as well as with infrared light. At the heart of the disk is a clear zone about 50 AU wide—roughly the size of our own planetary system.

Figure 20.13 shows a recent Hubble Space Telescope image of the Beta Pictoris disk, where the star and the clear zone have been blocked out to reveal the structure of the disk in unprecedented detail. As the bottom image (in exaggerated color) shows, the inner part just outside the clear zone is slightly tilted relative to the outer parts (the dotted line shows the orientation of most of the disk). One possible interpretation of this inner tilt (which should long ago have been smoothed out in such a system) is that it is maintained by the presence of one or more planets in the clear area inside the disk. (If this interpretation is correct, then the Beta Pictoris disk may well be a dustier, thicker version of our solar system's Kuiper belt, discussed in Chapter 12.) Much more work will have to be done with this star before we can be sure about our conclusion, but it is intriguing to consider that more mature disks might also hold clues about the presence of planets around other stars.

20.4

Planets Beyond the Solar System: Search and Discovery

Once again, the fact that the formation of giant planets *could* explain such observations does not *prove* that such planets actually exist. A variety of techniques have been used to search for better evidence of planets, with some dramatic recent successes. Since planets are so difficult to image, the kinds of experiments possible with current technology look for the effects of surrounding planets on the central star—in particular, for changes in the star's motion through space caused by the orbital motion of surrounding planets.

Search for Orbital Motion

To understand how this approach works, consider a single Jupiter-like planet in orbit about a star. As described in Chapter 17, both the planet and the star in such systems actually revolve about their common center of mass. The sizes of their orbits are inversely proportional to their masses. Suppose the planet has a mass about one-thousandth that of the star; the size of the star's orbit is then one-thousandth that of the planet.

In other words, while the star will be much easier to see than the planet, its motion will be quite small. To get a sense of how difficult such an observation might be, let's first see how hard Jupiter (the most massive planet in our solar system) would be to detect from the distance of a nearby star. Consider an alien astronomer trying to observe our own system from Alpha Centauri, the closest star system to our own.

The diameter of Jupiter's apparent orbit viewed from this distance is 10 arcsec, and that of the Sun's orbit is 0.010 arcsec. Remember, even with the best telescope *we* presently have, these alien astronomers could not detect faint Jupiter directly. But if they could measure the apparent position of the Sun to this precision, they would see it describe an orbit of diameter 0.010 arcsec with a period equal to that of Jupiter, which is 12 years. In other words, if they watched it for 12 years, they would see the star "wiggle" back and forth in the sky by this minuscule fraction of a degree. From the observed motion and the period, they could deduce the mass of Jupiter and its distance using Kepler's laws.

It is just possible (but very difficult) to make astronomical measurements from the Earth's surface today with enough precision to detect Jupiter-mass companions around the nearer low-mass stars in this way. In the 1960s, after many decades of positional measurements with refracting telescopes, several astronomers thought they had discovered planets, but all of these claims have since been found to be in error. More recent applications of this technique with higher-precision, ground-based instruments have so far turned up nothing. With improved equipment, however, it would be possible to extend this search to several hundred nearby stars.

The technique just described relies on measuring the changes in the *position* of the star as its orbit moves it back and forth across the sky. But as the star and planet orbit each other, part of their motion will be in our line of sight. Such motion (as discussed in Chapter 4) can be measured using the Doppler effect and the star's spectrum.

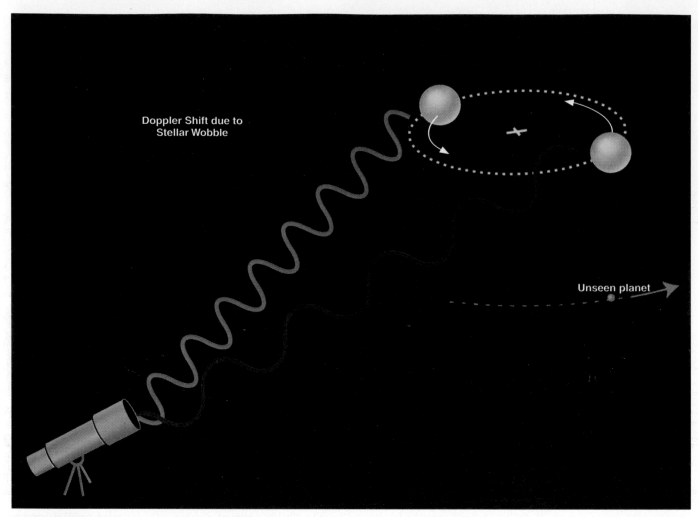

Figure 20.14
The motion of a star and a planetary companion around a common center of mass can be detected as a cyclical change in the Doppler shift of the star. When the star is moving away from us, the lines in its spectrum show a tiny redshift; when it is moving toward us they show a tiny blueshift. The change in color (wavelength) has been exaggerated here for educational purposes. In reality, the Doppler shifts we measure are extremely small and require sophisticated equipment to be detected. (Diagram courtesy of G. Marcy, San Francisco State University)

The Discovery of Planets

It was this method that finally allowed astronomers to establish the existence of planets around other stars like our own in 1995. With modern technology, astronomers can measure very tiny changes in the radial velocity of a star (its motion toward or away from us). As the star moves back and forth in orbit around the system's center of mass in response to the gravitational tug of an orbiting planet, the lines in its spectrum will shift back and forth (Figure 20.14).

Let's again consider the example of the Sun. Its radial velocity changes by about 13 m/s with a period of 12 years because of the gravitational force of Jupiter, and slightly more (15 m/s) if the effects of Saturn are also included. This corresponds to about 30 mi/h, roughly the speed at which many of us drive around town.

Detecting motion at this level in a star's spectrum presents an enormous technical challenge, but not an insurmountable one. Several groups of astronomers around the world have been actively engaged in searching for planets using specialized spectrographs designed for this purpose. Observers engaged in such programs must be patient as well as skillful, since the regular cycle of changes for a planet in an orbit like Jupiter's take a decade or so to become apparent. (Note that this means that the planets closest to their stars—and thus taking the shortest time to revolve around the center of mass—are the ones we are most likely to find at the beginning of such searches.)

The first planet discovered using this technique was found in 1995 by Michel Mayor and Didier Queloz of the Geneva Observatory, using an advanced-design spectrograph at the 1.9-m telescope of the Haute-Provence Ob-

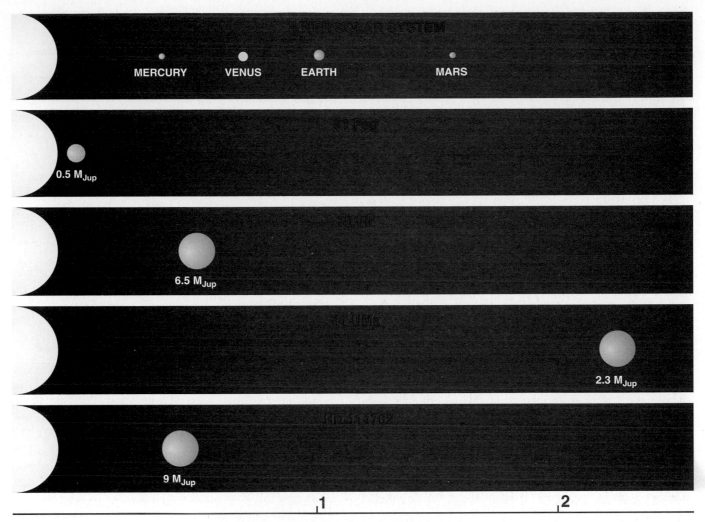

Figure 20.15
The relative positions of the planets around the first five normal stars known to have planetary companions. Other systems will probably have been found by the time you read this. The masses indicated for each planet are minimum masses. The sizes of objects are not to scale.

servatory in France. This planet orbits a main-sequence star resembling our Sun called 51 Pegasi, about 40 LY away. (The star can be found in the sky near the great square of Pegasus, the flying horse of Greek mythology, one of the best-known star patterns.) To everyone's surprise, the planet takes a mere 4.2 days to orbit around the star. Remember that Mercury, the innermost planet in our solar system, takes 88 days to go once around.

This means the planet must be very close to 51 Pegasi, circling it about 7 million km away (Figure 20.15). At that distance, the energy of the star should heat the planet's surface to a temperature close to 1000°C (a bit hot for future tourism). Since we do not know the angle at which we are seeing the planet's orbit, we can only calculate the minimum mass it could have, which is about half the mass of Jupiter (or more than 150 times that of the Earth). The actual mass may be somewhat larger than this; but, in any case, this is clearly a jovian and not a ter-

restrial-type planet. How a self-respecting jovian planet got so close to its star (it probably formed farther out in the system), whether or not the star's heat is continuously evaporating some of its outer material, and how frequently such close-in planets develop around stars are questions future observations and theoretical modeling will have to address.

With a "year" lasting only four Earth days, the effects of the new planet (for which some astronomers have already suggested the name Bellerophon, after the hero who rode Pegasus in Greek mythology) become evident in the spectrum of its star after only a few nights of observation. This enabled Geoffrey Marcy and Paul Butler of San Francisco State University to confirm its existence using the 3-m reflector at the Lick Observatory in California only one week after the Swiss team made their announcement. The two California astronomers and their students had been observing the spectra of about 120 nearby Sun-

TABLE 20.2
The First Planets Discovered Around Other Stars

Name of Star	Estimated Distance	Minimum Planet Mass	Period of Planet's Orbit	Semimajor Axis
	(LY)	(Jupiters)	(days)	(AU)
51 Pegasi	40–60	0.5	4.2	0.05
70 Virginis	30	6.5	116	0.5
47 Ursae Majoris	40	2.3	1100	2
HR 3522	44	0.8	14.76	0.1
HD 114762	100	9	84	0.38
HR 5185 (Tau Boötes)	62	3.9	3.3	0.05
PSR 1257 + 12	1600	A. 0.00005	25.3	0.19
		B. 0.011	66.5	0.36
		C. 0.009	98.2	0.47

like stars for over eight years with a spectrograph that can measure radial velocity changes to within 3 m/s. (Ironically, they had not included 51 Pegasi in their initial search program because it had been mistakenly listed as not quite resembling the Sun in a standard star catalog.)

In early 1996 Marcy and Butler were able to announce the discovery of planets around two other nearby stars, 70 Virginis and 47 Ursae Majoris. The two new planets are farther from their stars, taking 116 days and 1100 days, respectively, to go around. This puts them in orbits that are less shocking to our solar system prejudices—the first would be between the orbits of Mercury and Venus, the second between those of Mars and Jupiter. Their minimum masses are estimated around 6.5 and 2.3 Jupiters, making them quite substantial but still definitely planets and not brown dwarfs.

Two other stars, HR 3522 and HD 114762 (their names are merely entries in a lengthy star catalog), were soon added to the list of those with planetary companions. The planet around HD 3522 is similar to the one around 51 Pegasi in that its period is only a few days and its orbit is quite close to its star. The planet around HD 114762 takes 84 days to orbit its star, which means its orbit is just a bit smaller than Mercury's. It has a minimum mass of 9 Jupiters, the largest so far.

It is important to bear in mind that our current techniques can only detect planets with masses comparable to that of Jupiter. Planets with masses like Earth's might exist around these or other stars, but they would cause much smaller (and, at present, unobservable) Doppler shifts in their stars' spectra. Table 20.2 lists the planets we know about at the time this text is going to press, although others may have been discovered by the time you read this.

The discovery of planets around a number of Sun-like stars has ended centuries of conjecture and debate about the uniqueness of our solar system. It appears, as theorists have been suggesting for some time, that the formation of planets around the Sun was not a unique event, but a process that accompanies the formation of many stars. To the degree that planets are the only sort of habitats where we can imagine life as we know it, the discovery of these planets lends a new spirit of optimism to the search for life elsewhere, a subject to which we will return in the book's Epilogue.

Planets and Pulsars

To be absolutely fair, we should point out that the first planetary-mass objects discovered (in 1992) were not found around living stars like the ones discussed in the previous section, but around a kind of stellar "corpse" called a neutron star (to be described in Chapter 22). This particular neutron star (whose catalog designation is PSR 1257+12) has three and possibly four planets orbiting around it (also shown in Figure 20.15). Many such dead stars give off sharp pulses of radio waves and are thus called *pulsars*. (If you are intrigued by how dead stars can emit sharp pulses of radiation, you now have another reason to read Chapter 22.)

In any case, astronomers can identify the motion of the neutron star around its center of mass by regular changes in the arrival of the pulses (in the same way that the motion of a living star can be seen in the Doppler shift of its spectrum). But there may be an important difference between the planets found around such stellar corpses and the planets discussed earlier. Because the formation of such a dead star generally involves the explosion of most of the original star, astronomers are convinced that the planet-mass companions of PSR 1257+12 are not "original" planets—in the sense that they existed during the star's life. Instead, they are probably "second generation planets" consisting of the remains of a nearby companion star that did not survive the explosion. Scoured by intense radiation from the spinning neutron star, such companions would be nothing like the planets in our own solar system. This is why more interest has greeted the discovery of planets around stars like the Sun.

Imaging Planets

What about looking for planets directly rather than just studying their parent stars? After all, "seeing is believing" is a very human prejudice. One future possibility might be to detect the (very slight) decrease in a star's light when one of its planets passes in front of it. But if *seeing* a planet means taking an image of it, then no extra-solar-system planet—even one as large as those we have discovered—could be seen with current equipment. Let's discuss the problems outlined at the start of Section 20.3 in a bit more detail.

Suppose, for example, that you were a great distance away and wished to detect reflected light from the planet Jupiter. Jupiter intercepts and reflects just about one-billionth of the Sun's radiation, so its apparent brightness in visible light is only a billionth that of the Sun. Smaller and less-reflective planets would be even more difficult to see. But the faintness of potential planets is not the biggest problem.

The real difficulty is that the faint light from a planet is swamped by the blaze of radiation from its parent star. If you are near-sighted, try looking at streetlights at night with your glasses off. You will see a halo of light surrounding every light. Bright stars seen through a telescope also appear to be surrounded by a bright halo of light (see the picture of Supernova 1987A, which does not appear as a sharp point of light, in Figure 22.10). In this case the problem is not that the telescope is near-sighted, but rather slight imperfections in its optics and atmospheric blurring prevent the light from coming to focus in a completely sharp point. Planets, if any, would lie within this halo, and their faint light could not be seen in the glare.

Astronomers are now trying to devise ways to suppress this glare. One of the best approaches appears to be observing in space in order to escape the effects of the Earth's atmosphere. The optimum wavelength range in which to observe is the infrared, since planets get brighter in the infrared while stars get fainter, thereby reducing the brightness contrast.

Even so, it may not be possible to obtain such images from the Earth's neighborhood. We are still deep within the Sun's *zodiacal cloud*—a thin cloud of dust left over from the formation of the planets—which tends to scatter and spread out starlight. It may be necessary to make the observations far from the Sun—say at the orbit of Jupiter—to minimize the scattered light from the star around which the planet orbits. It will also be necessary to use special techniques to artificially suppress the light from the central star. NASA is actively exploring the feasibility of such a mission sometime in the next century.

Even better than imaging would be a spectrum of a planet. For example, we have discussed that the oxygen in the Earth's atmosphere is unstable, combining easily with other elements. This means it has to be replenished through the action of the life-forms we call plants, which produce oxygen during photosynthesis. If we ever obtain a spectrum of a planet containing substantial amounts of oxygen in its atmosphere, we can conclude that the presence of life is very likely. It is not at all far-fetched to think that sometime in the next century such an experiment could be performed.

The search for planets around other stars seems to be a project whose time has come. If numerous large planets are out there, it is now within our capability to find them. The discovery of smaller planets like our Earth may still be some time away, but perhaps not impossible. Once we have found a representative number of planets, astronomers will for the first time be able to consider the formation of our own planetary system within a broader cosmic context.

Summary

20.1 Most stars form in **giant molecular clouds** that have masses as large as 10^6 times the mass of the Sun and typical diameters of 50 to 200 LY. The best-studied molecular cloud is Orion, where star formation began about 12 million years ago and is moving progressively through the cloud. Recently formed hot stars are exposing many stages of the process of star formation to our view in Orion. The formation of a star inside a molecular cloud begins with a dense core of material, which accretes matter and collapses due to gravity. The accumulation of material halts when the **protostar** develops a strong **stellar wind.** A turbulent cloud will form a rotating star with an equatorial disk of material. The wind tends to emerge more easily in the direction of a protostar's poles, leading to jets of material. These can collide with the material around the star and produce Herbig–Haro objects.

20.2 The evolution of a star can be described in terms of its temperature and luminosity changes, which can best be followed by plotting them on an H–R diagram. Protostars generate energy through gravitational contraction. The initial gravitational collapse takes several thousand years, after which a slow contraction typically continues for millions of years, until the star reaches the **main sequence** and nuclear reactions begin. The higher the mass of a star, the shorter the time it spends in each stage of evolution. Stars range from about 1/12 to 100 times the mass of the Sun.

20.3 There is observational evidence that most protostars are surrounded by a disk with a large enough diameter and enough mass (as much as 10 percent that of the Sun) to form a system of planets. Furthermore, some of these disks are observed to change with time. Initially, an opaque disk extends all the way to the surface of the protostar. After a few million years the inner part of the disk is cleared of dust, and the disk is then shaped like a donut with the protostar centered in the hole. The development of the hole can be explained if a large planet has formed at its outer boundary.

20.4 At present we can search for planets around nearby stars only by looking for the star's motion around the star–planet system's common center of mass. This can be done by looking either for changes in the star's position on the sky over time, or in its radial velocity (as seen in the Doppler shift of the lines in its spectrum). So far, planets have been detected around six Sun-like stars using this method (but more are being found regularly). Ambitious space experiments are now being planned that might make it possible to image extra-solar-system planets and even to obtain spectra of them. Oxygen would be an unmistakable indicator of biological activity. These experiments are not yet feasible but will probably be carried out sometime during the next century.

Review Questions

1. Give several reasons that the Orion Molecular Cloud is such a useful "laboratory" for studying the stages of star formation.

2. Why is star formation more likely to occur in cold molecular clouds than in regions where the temperature of the interstellar medium is several hundred thousand degrees?

3. Why have we learned a lot more about star formation since the invention of detectors sensitive to infrared radiation?

4. Describe what happens when a star forms. Begin with a dense core of material in a molecular cloud, and trace the evolution up to the point at which the newly formed star reaches the main sequence.

5. Why is it so hard to see planets around other stars, and so easy to see them around our own?

6. What techniques have been used to search for planets around other stars?

7. How were the first planets around stars like the Sun discovered?

8. Explain why taking an image of a planet around another star is so difficult, and how astronomers hope to overcome these difficulties in the future.

Thought Questions

9. A friend of yours, who did not do well in her astronomy class, tells you that she believes all stars are old and that none could possibly be born today. What arguments would you use to persuade her that stars are being born somewhere in the Galaxy during our lifetimes?

10. Look at the four stages in the birth of a star shown in Figure 20.8. In which stage(s) is the star visible in optical radiation? In infrared radiation? In which stage(s) is it generating energy by converting hydrogen to helium?

11. An enormously wealthy donor has just given a university in your state a large sum of money to "unlock the secrets of star birth." You are on a committee to decide how to spend the money. What kind of instruments would you buy or build? What kind of studies would you fund? Explain each answer.

12. Observations suggest that it takes more than 3 million years for the dust to begin clearing out of the inner regions of the disk surrounding protostars. Suppose this is the minimum time required to form a planet. Would you expect to find a planet around a ten-solar-mass star? (Refer to Figure 20.11.)

13. The evolutionary track for a star with one solar mass remains nearly vertical in the H–R diagram for awhile (see Figure 20.11). How is its luminosity changing during this time? Its temperature? Its radius? What is its source of energy?

14. Suppose you wanted to image a planet around another star. Would you try to observe in the optical or in the infrared? Why? Would the planet be easier to see if it were at 1 AU or 5 AU?

Problems

15. If a giant molecular cloud is 100 LY across and has a mass equivalent to 10,000 Suns, what is its average internal density? Suppose that half of the cloud's mass is made up of dust grains, each with a mass of 10^{-13} kg; how many such grains per cubic centimeter are in the cloud?

16. Use Figure 20.11 to determine how long, compared to the Sun, a star with ten times the Sun's mass would spend in the pre-main-sequence stages of its life. Express as a fraction.

17. An astronomer has just found a new planet around a Sun-like star. If its orbit has a period of 2.4 years, use Kepler's law to determine the size in astronomical units of its semimajor axis.

Caillault, J. et al. "The New Stars of M42" in *Astronomy*, Nov. 1994, p. 40. On studies of circumstellar disks in Orion.

Cohen, M. *In Darkness Born: The Story of Star Formation*. 1988, Cambridge U. Press. A fine introduction to the topic.

Fienberg, R. "Pulsars, Planets, Pathos" in *Sky & Telescope*, May 1992, p. 493. (Update May 1994, p. 10.)

Goldsmith, D. *The Astronomers*. 1991, St. Martin's Press. Chapter 7 includes a section on the birth of stars.

Lada, C. "Deciphering the Mysteries of Stellar Origins" in *Sky & Telescope*, May 1993, p. 18. Good review; fine beginning article.

Lemonick, M. "Searching for Other Worlds" in *Time*, Feb. 5, 1996, p. 52. On the discovery of more planets.

MacRobert, A. and Roth, J. "The Planet of 51 Pegasi" in *Sky & Telescope*, Jan. 1996, p. 38.

O'Dell, C. R. "Exploring the Orion Nebula" in *Sky & Telescope*, Dec. 1994, p. 20. Good review with recent Hubble results.

Stahler, S. "The Early Life of Stars" in *Scientific American*, July 1991, p. 48.

Appendix 1 lists some World Wide Web sites with information about searches for planets orbiting other stars, and for nebulae with protostars.

Using REDSHIFT ™

1. Display the 10 brightest stars in the constellation Boötes. Which is (are) most likely to have planets?

2. Use Figure 20.11 to estimate how long it took for each of these stars to reach the main sequence.

a. Beta Canes Venaticorum

b. Epsilon Eridani

c. Alpha Pictoris

d. Gamma Cassiopeiae

e. Procyon, Alpha Canis Minoris

f. Theta Ursae Majoris

g. Menkalinan, Beta Aurigae

h. Tau Herculis

i. Eta Cephei

j. Mu Herculis

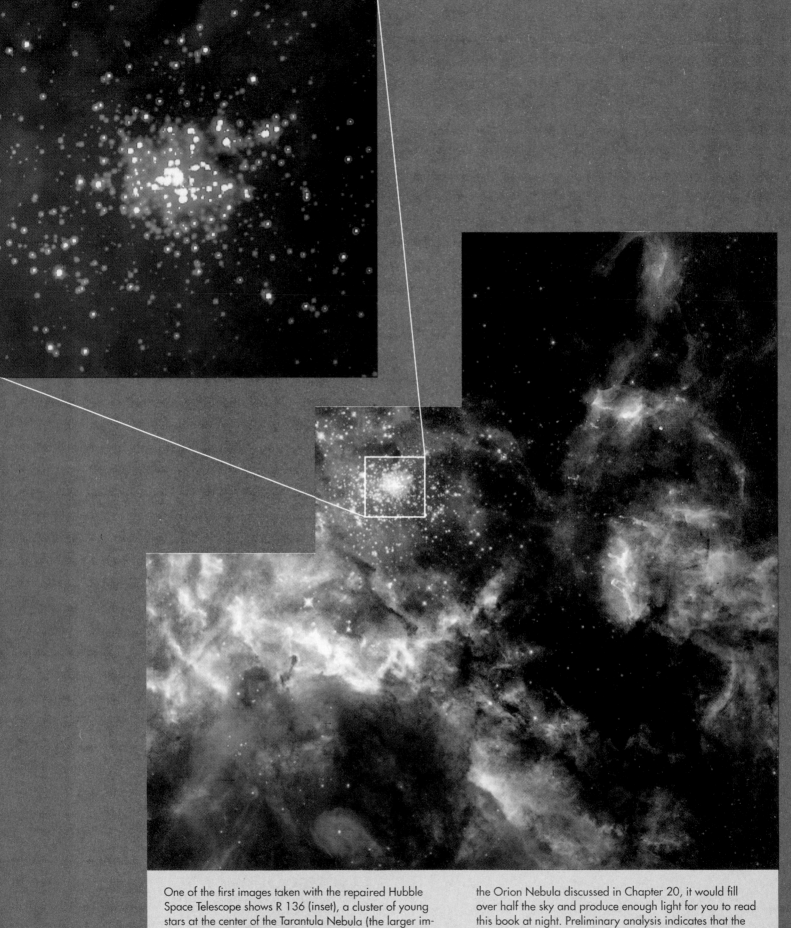

One of the first images taken with the repaired Hubble Space Telescope shows R 136 (inset), a cluster of young stars at the center of the Tarantula Nebula (the larger image). The nebula is an enormous region of gas and dust located in one of our nearest neighbor galaxies, the Large Magellanic Cloud. If the Tarantula were at the distance of the Orion Nebula discussed in Chapter 20, it would fill over half the sky and produce enough light for you to read this book at night. Preliminary analysis indicates that the cluster has several thousand closely packed stars in it. (STScI/NASA)

CHAPTER 21

Stars: From Adolescence to Old Age

Thinking Ahead

While the Sun's output of heat and light seems steady and reliable now, the fact is that we humans have only been around for a tiny fraction of the Sun's life. We have good evidence that as billions of years go by, stars like the Sun change drastically. How will the evolution of our star affect the habitability of planet Earth?

The universe at large would suffer as little, in its splendor and variety, by the destruction of our planet, as the verdure and sublime magnitude of a forest would suffer by the fall of a single leaf.

Thomas Chalmers in *Discourses on the Christian Revelation Viewed in Connection with the Modern Astronomy* (1817)

We now turn from the birth of stars and planets to the rest of their life stages. This is not an easy task, since, as we saw in Chapter 17, stars live much longer than astronomers. Thus we cannot hope to see the life story of any single star unfold before our eyes or telescopes. Like the rushed crew of our imaginary starship in Chapter 17 (with only one day to study the lives of the Earth's human inhabitants), we must survey as many of the stellar inhabitants in the Galaxy as possible. If we are lucky (and thorough), we can catch at least a few of them in each possible stage of their lives.

Surveys of stars in our own neighborhood and beyond do reveal a variety of stars with different characteristics (Figure 21.1). Some of the differences come about because stars have different masses and thus different temperatures and luminosities. But others are the result of changes that occur as a star ages. Through a combination of observation, theory, and clever detective work, we can use these differences to piece together the life story of a star.

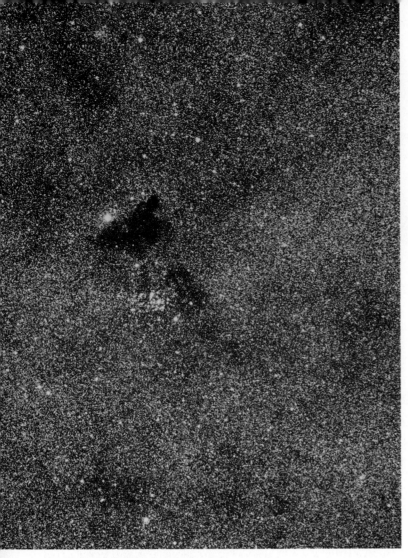

Figure 21.1
This snapshot of a crowded region of stars in the constellation of Sagittarius shows a yellowish population of older stars with a small cluster of younger bluish stars (NGC 6520), as well as a dark cloud (Barnard 86) inside which new protostars are forming. (© Anglo–Australian Telescope Board)

21.1

Evolution from the Main Sequence to Giants

One of the best ways to represent a "snapshot" of a group of stars is by plotting their properties on an H–R diagram. We have already used the H–R diagram to follow the evolution of protostars up to the time they reach the main sequence. Now let's see what happens next.

Once a star has reached the main-sequence stage of its life, it derives its energy almost entirely from the conversion of hydrogen to helium via the process of nuclear fusion (see Chapter 15). Since hydrogen is the most abundant element in stars, this process can help maintain the star's equilibrium for a long time. Thus all stars remain on the main sequence for most of their lives. Some as-

tronomers like to call this the star's "prolonged adolescence" or "adulthood" (continuing our analogy with the stages in a human life).

Since only 0.7 percent of the hydrogen used in fusion reactions is converted to energy, the star does not change its *total* mass appreciably during this long period. It does, however, change the chemical composition in its central regions, where the nuclear reactions occur: hydrogen is gradually depleted and helium accumulates. This change of composition forces the star to alter its structure, including its luminosity and size.

The original main sequence, corresponding to stars of different mass but roughly the same chemical composition, is sometimes called the **zero-age main sequence.** We use the term "zero-age" to note that each star reaches the main sequence when its hydrogen fusion reactions begin (the star's birth). Eventually, though, the point that represents any star on the H–R diagram will evolve away from the main sequence. That is, each star's luminosity and surface temperature (the characteristics recorded on the H–R diagram) will begin to change.

As helium accumulates in the center of a star, calculations show that the temperature and density in the inner region slowly grow, increasing the rate at which hydrogen fuses into helium. When it is hotter, each proton has more energy of motion on average; this means it will be more likely to interact with other protons and undergo fusion. (For the proton–proton cycle described in Chapter 15, the rate of fusion goes up roughly as the temperature to the fourth power. If the temperature were to double, the fusion would increase by a factor of 2^4, or 16 times.)

Consequently, the rate of nuclear energy generation increases with time, and the luminosity of the star gradually rises. Initially, however, these changes are small, and stars remain close to the zero-age main sequence for most of their lifetimes.

Lifetimes on the Main Sequence

How many *years* a star remains in the main-sequence stage depends on its mass. You might think that a more-massive star, having more fuel, would last longer, but it's not as simple as that. The lifetime of a star in a particular stage involving nuclear fusion depends on how much fuel it has, and on *how fast* it uses up that fuel. More-massive stars use up their fuel much more quickly than stars of low mass.

As we just saw, the rate of fusion is very strongly dependent on the star's inner temperature. And what determines how hot a star's central regions get? It is the mass of the star—the weight of the overlying layers—that controls how compressed and hot the star becomes at its core. The more massive the star, then, the faster it races through its storehouse of central hydrogen. Although massive stars *have* more fuel, they burn it so prodigiously that their lifetimes are much shorter than those of their low-mass counterparts.

TABLE 21.1
Lifetimes of Main-Sequence Stars

Spectral Type	Mass	Lifetime on Main Sequence
	(Mass of Sun = 1)	
O5	40	1 million years
B0	16	10 million years
A0	3.3	500 million years
F0	1.7	2.7 billion years
G0	1.1	9 billion years
K0	0.8	14 billion years
M0	0.4	200 billion years

The main-sequence lifetimes of stars of different masses are listed in Table 21.1. As this table shows, the most-massive stars spend only a few million years on the main sequence. A star of 1 solar mass remains there for about 10 billion years, while a star of about 0.4 solar mass has a main-sequence life of some 200 billion years, a value that is longer than the current age of the universe. (Bear in mind, however, that every star spends *most* of its total lifetime on the main sequence; stars spend an average of 90 percent of their lives peacefully fusing hydrogen into helium.)

Note that these results are not merely of academic interest. Human beings developed on a planet around a G-type star whose stable main-sequence lifetime was so long that it afforded life on Earth plenty of time to evolve. If we were to search for intelligent life like our own on planets around other stars, it would be a pretty big waste of time to search around O- or B-type stars. These stars remain stable for so short a time that the development of creatures complicated enough to take astronomy courses is very unlikely.

From Main-Sequence Star to Red Giant

Eventually all the hydrogen hot enough to undergo fusion is used up inside a star. The core (where nuclear reactions take place) now contains only helium, "contaminated" by whatever small percentage of heavier elements the star had to begin with. Energy can no longer be generated by hydrogen fusion in this helium core and, as we will see, helium requires far greater temperatures to undergo fusion. The helium in the core can be thought of as the accumulated "ash" from the nuclear "burning" of hydrogen during the main-sequence stage.

With nothing more to supply heat to the central region of the star, gravity again takes over: after a long period of stability, the core begins to contract. Once more the star's energy is partially supplied by gravitational energy, in the way described by Kelvin and Helmholtz (see

Section 15.1). As the star's core shrinks, the energy of the inward-falling material is converted to heat.

The heat generated in this way has an effect on the hydrogen that spent the whole long main-sequence time just outside the core. Like an understudy waiting in the wings of a hit play for a chance at fame and glory, this hydrogen was almost (but not quite) hot enough to undergo fusion and take part in the main action that sustains the star. Now, the additional heat produced by the shrinking core puts this hydrogen "over the limit," and a shell of hydrogen nuclei just outside the core becomes hot enough for hydrogen fusion to begin.

Energy now pours outward from this shell, flowing, as heat always does, to the cooler outer regions. This begins to heat up layers of the star farther out, causing them to expand a bit. Meanwhile the helium core continues to contract, producing more heat right around it, which leads to more fusion in the shell of fresh hydrogen. The fusion produces more energy, which also flows out into the upper layers of the star.

These changes result in a substantial and rather rapid readjustment of the star's entire structure, so that the star leaves the main sequence altogether. Most stars actually generate more energy in this stage than they did when hydrogen fusion was confined to the core; thus they increase in luminosity. The outer layers of the star begin to expand, and the star grows to enormous proportions—we say it becomes a *giant*.

Depending on their mass, these giant stars can become so large that if we were to replace the Sun with one of them, its outer atmosphere could extend to the orbit of Mars or even beyond. This is the next stage in the life of a star, as it moves (to continue our analogy with human lives) from its long period of youth into middle age. (After all, many human beings today also see their outer layers expand a bit during middle age.)

The expansion of the outer layers causes the temperature at the stellar surface to decrease. As it cools, the star's overall color becomes redder (we saw in Chapter 4 that in nature red corresponds to cooler temperature). So the star becomes simultaneously more luminous and cooler; on the H–R diagram this means it moves upward and to the right. The star becomes one of the *red giants* first discussed in Chapter 17 (Figure 21.2). You might say that these stars have "split personalities": their cores are contracting while their outer layers are expanding.

Models for Evolution to Red Giants

As discussed earlier, astronomers can construct computer models of stars with different masses and compositions to see how stars change throughout their lives. Figure 21.3, based on theoretical calculations by University of Illinois astronomer Icko Iben, shows an H–R diagram with several tracks of evolution from the main sequence to the red-giant stage. Tracks are shown for stars of several

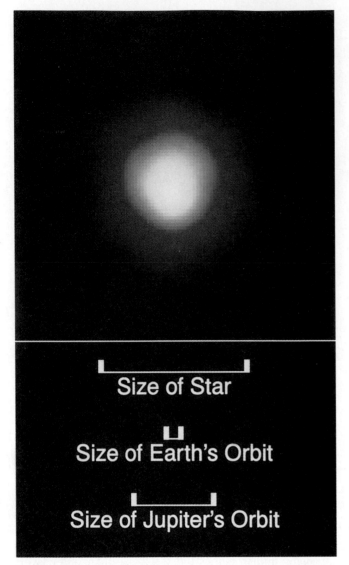

Figure 21.2
An easily observed example of a red giant star is Betelgeuse in the constellation of Orion (see Figure 20.2). Here we see an image taken in ultraviolet light with the Hubble Space Telescope—the first direct image ever made of the surface of another star. As shown by the bars at the bottom, Betelgeuse has an extended atmosphere so large that, if it were at the center of our solar system, it would stretch past the orbit of Jupiter. (A. Dupree, R. Gilliland, and NASA)

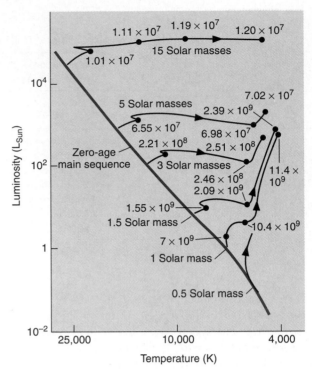

Figure 21.3
The predicted evolution of stars of different masses from the main sequence through the red-giant stage on the H−R diagram. The numbers show how many years the stars take to become red giants after leaving the main sequence. (Based on calculations by I. Iben)

masses, with chemical compositions similar to that of the Sun. The red line is the initial or zero-age main sequence.

You can see how the changes inside the star we have been discussing force it to "move" on the H−R diagram. The numbers along the tracks in Figure 21.3 indicate the times, in years, required for the stars to reach those points in their evolution after leaving the main sequence. Once again, you can see that the more massive a star, the more quickly it goes through each stage in its life.

21.2

Star Clusters

The description of stellar evolution just given is based on calculations. No star completes its main-sequence lifetime or its evolution to a red giant quickly enough for us to observe these structural changes as they happen. Fortunately, nature has provided us with a way to test our calculations.

Instead of observing the evolution of a single star, we can look at a group or *cluster* of stars. If a group of stars is very close together in space, held together by gravity, it is reasonable to assume that the individual stars in the group all formed at nearly the same time, from the same cloud, and with the same composition. We therefore expect that these stars will differ from one another only in mass and thus in how quickly they go through each stage of their lives.

Since stars with higher masses evolve more quickly, we can find clusters in which massive stars have already completed their main-sequence phase of evolution and become red giants, while stars of lower mass in the same cluster are still on the main sequence, or even undergoing

TABLE 21.2
Characteristics of Star Clusters

	Globular Clusters	Open Clusters	Associations
Number in Galaxy	150	Thousands	Thousands
Location in Galaxy	Halo and nuclear bulge	Disk (and spiral arms)	Spiral arms
Diameter (LY)	50–300	<30	100–500
Mass (solar masses)	10^4–10^6	10^2–10^3	10^2–10^3
Number of stars	10^4–10^6	50–1000	5–50
Color of brightest stars	Red	Red or blue	Blue
Luminosity of cluster (L_{Sun})	10^4–10^6	10^2–10^6	10^4–10^7

pre-main-sequence gravitational contraction. This way, we can see many stages of stellar evolution among the members of a single cluster.

There are three types of star clusters: globular clusters, open clusters, and stellar associations. Their properties are summarized in Table 21.2. As we will see, globular clusters contain only very old stars, while open clusters and associations contain young stars.

Globular Clusters

The **globular clusters** were given their name because of appearance. One of the most famous, M13 in the constellation of Hercules, passes nearly overhead on a summer evening at most places in the United States. As Figure 21.4 shows, M13 is a nearly symmetrical round system of many, many stars, with the highest concentration near the

Figure 21.4
The globular cluster M13. (U.S. Naval Observatory)

center. (The opening photo for Chapter 18, a wonderful image of another globular cluster, shows this type of concentration really well.) The brightest stars in globular clusters are red giants.

Most globular clusters contain hundreds of thousands of member stars. What would it be like to live inside such a cluster? In the densest part of M13, stars would be roughly a million times more concentrated than in our neighborhood. There would still be plenty of space between the stars, however. If the Earth orbited one of M13's inner stars, the nearest stars to our own would be *light months* away. They would still appear as points of light but would be brighter than any of the stars we see in our own sky. The Milky Way would probably be impossible to see through the bright haze of starlight produced by the cluster.

About 150 globular clusters are known in our Galaxy. Most of them are in a spherical halo (or cloud) surrounding the flat disk formed by the majority of the Galaxy's stars and interstellar matter. All the globular clusters are very far from the Sun, and some are found at distances of 60,000 LY or more from the galactic plane. The linear diameters of globular star clusters range from 50 LY to more than 300 LY. The most massive known globular cluster, called Omega Centauri (Figure 21.5), was recently found to contain enough mass to make 5 million stars like our Sun.

Open Clusters

Open clusters, on the other hand, are found in the disk of the Galaxy, often associated with interstellar matter. Open clusters contain far fewer stars than do globular clusters (Figure 21.6). The stars in open clusters usually appear well-separated from one another, even in the central regions, which explains the name we give this type of cluster. The Galaxy contains thousands of open clusters, but we can see only a small fraction of them. Interstellar dust (see Chapter 19) dims the light of more-distant clusters so much that they are undetectable.

Several open clusters are visible to the unaided eye. Most famous among them is the Pleiades (Figure 19.10),

Figure 21.5
The most massive known globular cluster is Omega Centauri, whose central region is shown here. The field of view, about 90 LY across, contains hundreds of thousands of stars. The full cluster is actually significantly larger. (European Southern Observatory)

Figure 21.6
The Jewel Box (NGC 4755) is an open cluster of young, bright stars at a distance of about 8000 LY from the Sun. The name comes from its description by John Herschel as "a casket of variously colored precious stones." (Photo by David Malin; © Anglo–Australian Telescope Board)

which appears as a tiny group of six stars (some people can see even more) arranged like a dipper in the constellation of Taurus, the bull. A good pair of binoculars shows dozens of stars in the cluster, and a telescope reveals hundreds. The Hyades is another famous open cluster in Taurus. To the naked eye, it appears as a V-shaped group of faint stars marking the face of the bull. Telescopes show that the Hyades actually contains more than 200 stars.

Typical open clusters contain several dozen to several hundred member stars. Compared with globular clusters, open clusters are small, usually having diameters of less than 30 LY. A few brilliant stars in some open clusters, however, may cause them to outshine the far richer globular clusters.

Stellar Associations

An **association** appears as a group containing 5 to 50 hot, bright O and B stars scattered over a region of space some 100 to 500 LY in diameter. (Associations contain low-mass stars as well, but these are much fainter and less conspicuous.) Such a concentration of hot, luminous stars indicates that star formation in the association has occurred in the last million years or so. Since O and B stars go through their lives so quickly, they would not be around any more if they had formed much earlier. It is therefore not surprising that associations are typically found in regions rich in gas and dust.

Because associations, like ordinary open clusters, lie in regions occupied by interstellar matter, most are hidden from our view. There are probably several thousand undiscovered associations in our Galaxy.

21.3

Checking Out the Theory

Open clusters are younger than globular clusters, and associations are typically somewhat younger still. We know this because the stars in these different types of clusters are found in different places in the H–R diagram, and we can use their locations in combination with theoretical calculations to estimate their ages. The fact that stellar evolution theory can explain why the H–R diagrams of

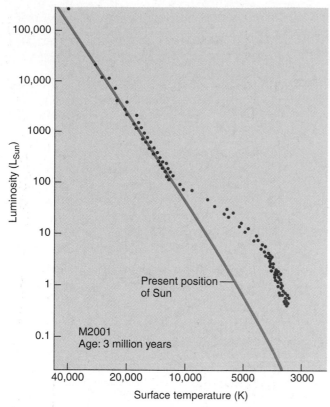

Figure 21.7
The H−R diagram of a hypothetical cluster at an age of 3 million years. Note that the high-mass (high-luminosity) stars have already arrived at the main-sequence stage of their lives, while the lower-mass (lower-luminosity) stars are still to the right of the zero-age main sequence, not yet hot enough to begin fusion of hydrogen.

Figure 21.8
The cluster NGC 2264, at a distance of 2500 LY. This region of newly formed stars is a complex mixture of red hydrogen gas ionized by hot embedded stars, dark obscuring dust lanes, and brilliant young stars. (Photo by David Malin; © Anglo–Australian Telescope Board)

globular clusters appear to be so different from those of open clusters and associations is one of the best arguments that the theory is right.

H−R Diagrams of Young Clusters

What does theory predict for the H−R diagram of a cluster whose stars have recently condensed from an interstellar cloud? After a few million years ("recently" for astronomers), the most-massive stars should have completed their contraction phase and be on the main sequence, while the less-massive ones should be off to the right, still on their way to the main sequence. These ideas are illustrated in Figure 21.7, which shows the H−R diagram calculated by R. Kippenhahn and his associates at Munich for a hypothetical cluster with an age of 3 million years.

There are real star clusters that fit this description. The first to be studied (in about 1950) was NGC 2264, which is still associated with the region of gas and dust from which it was born (Figure 21.8). Its H−R diagram is shown on Figure 21.9. Among the other star clusters in

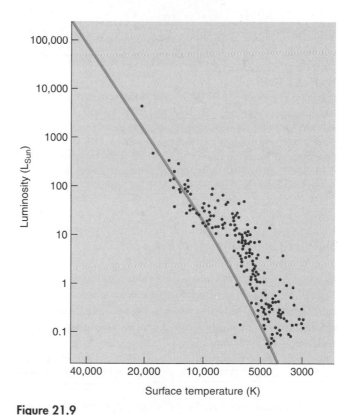

Figure 21.9
The H−R diagram for cluster NGC 2264. Compare to Figure 21.7; although the points scatter a bit more here, the theoretical and observational diagram are remarkably—and satisfyingly—similar. (Data by M. Walker)

Figure 21.10
The open star cluster NGC 3293. All the stars in such clusters form at about the same time. The most massive stars, however, exhaust their nuclear fuel more rapidly and hence evolve more quickly than stars of low mass. As stars evolve, they become redder. The bright orange star in NGC 3293 is the member of the cluster that has evolved most rapidly. (Photo by David Malin; © Anglo–Australian Telescope Board)

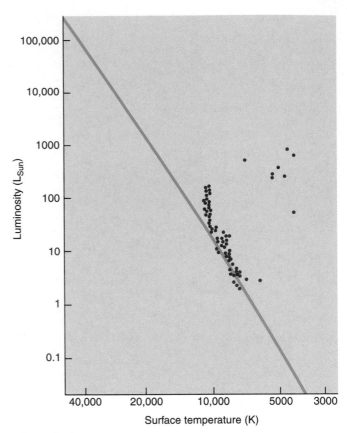

Figure 21.11
The H–R diagram for M41, a cluster older than NGC 2264. Note that M41 has several red giants and that some of its more-massive stars are no longer close to the zero-age main sequence (blue line).

such an early stage is the one in the middle of the Orion Nebula (shown in Figures 20.5 and 20.6).

As clusters get older, their H–R diagrams begin to change. After a short time—less than a million years after reaching the main sequence—the *most*-massive stars use up the hydrogen in their cores and evolve off the main sequence to become red giants. As more time goes on, stars of lower and lower mass begin to leave the main sequence and make their way to the upper right of the H–R diagram. Figure 21.10 is a photograph of NGC 3293, a somewhat older open cluster. One massive star has evolved to be a red giant and stands out as an especially bright orange member of the cluster.

Figure 21.11 shows the H–R diagram of the open cluster M41, which is roughly 100 million years old; a number of red giants have moved off to the right. Note the gap that appears in this H–R diagram between the stars near the main sequence and the red giants. In the snapshot of stellar evolution represented by the H–R diagram, a gap does not necessarily represent a region of temperatures and luminosities that stars avoid. It simply represents a domain of temperature and luminosity through which a star moves very quickly as it evolves. We see a gap for M41 because at this particular moment we have not caught a star in the process of scurrying across this part of the diagram.

H–R Diagrams of Older Clusters

After 4 billion years have passed, the H–R diagram of a cluster will look quite a bit different. By this time, many more stars—including stars only a few times more massive than the Sun—have begun to leave the main sequence (Figure 21.12). This means that no stars are left near the top of the main sequence; only the low-mass stars near the bottom remain. The older the cluster, the lower the point on the main sequence where stars begin to move toward the red giant region. Astronomers actually use this "turn-off point" as a measure of the age of the cluster (compare Figures 21.9 and 21.12). To get actual

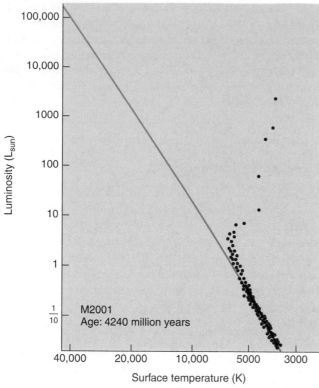

Figure 21.12
The H–R diagram of a hypothetical cluster at an age of 4.24 billion years. Note that most of the stars on the upper part of the main sequence have now turned off toward the red-giant region.

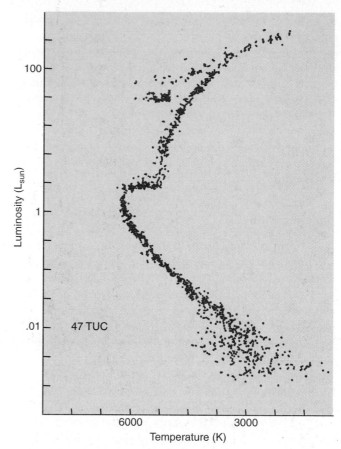

Figure 21.13
The H–R diagram of the globular cluster 47 Tucanae. Note the scale of luminosity is different from those of the other H–R diagrams in this chapter. We are focusing on the lower portion of the main sequence, the only part where stars still remain in this old cluster. (Data by J. Hesser and collaborators)

ages (in years) rather than relative ages, we must compare the appearance of our calculated H–R diagrams for different ages to actual H–R diagrams. Since the stars in a cluster formed at roughly the same time, this tells us the ages of the many individual stars that make up the cluster.

The oldest clusters are the globulars, typically about 15 billion years old. The H–R diagram of the globular cluster 47 Tucanae (pictured in the opening image for Chapter 18) is shown in Figure 21.13. Notice that its scale is different from the other H–R diagrams in this chapter. In Figure 21.12, for example, the luminosity scale on the left side of the diagram goes from 1/10 to 100,000 times the Sun's luminosity. But in Figure 21.13, the luminosity scale has been significantly reduced in extent. So many stars in this old cluster have had time to turn off the main sequence, only the very bottom of the main sequence remains.

Note that globular clusters have main sequences that turn off at a luminosity close to that of the Sun. Star formation in these systems evidently ceased billions of years ago, so no new stars are coming on to the main sequence to replace the ones that have turned off. Open clusters, on the other hand, are often found in regions of interstellar matter, where star formation can still take place. Apparently, new open clusters are still forming in our Galaxy. There are open clusters of all ages from less than 1 million

to several billion years, although no open cluster is as old as the globular clusters.

The globular clusters are the oldest structures in our Galaxy (and in other galaxies as well). Their study provides important clues about the age and early evolution of the Galaxy, and gives us a crucial reference point for understanding the age of the entire universe. We will return to their properties in future chapters.

21.4

Further Evolution of Stars

The stellar life story related so far applies to all stars: every one of them moves through the stages of contracting as a protostar, living most of its life as a stable main-sequence star, and eventually moving off the main sequence toward the red-giant region. (The pace at which the star goes through the stages depends, of course, on its mass, with more-massive stars evolving more quickly.) But after this point, the life stories of stars of different masses diverge, with a wider range of behavior possible according to mass, composition, and the presence of any nearby companion stars.

Because we have written this book for non-science students taking their first astronomy course, we will recount a somewhat simplified version of what happens to stars as they move toward the final stages in their lives. We will (perhaps to your heartfelt relief) not delve into all the possible ways stars can behave. Instead, we will focus on the key stages in the evolution of single stars, and show how the evolution of high-mass stars differs from that of low-mass stars (such as our Sun).

Helium Fusion

Let's begin by considering stars whose initial masses are comparatively low, no more than about two to three times the mass of our Sun. (That may not sound all that low, but we know that stars far more massive than this exist; we will see what happens to them in the next section.) Because there are many more of these low-mass stars than high-mass stars (see Chapter 17), the vast majority of stars—including our Sun—follow the scenario we are about to relate. By the way, we carefully used the term "initial masses" of stars because, as we will see, stars can lose quite a bit of mass in the process of aging and dying.

Remember that red giants have a helium core where no energy generation is taking place, surrounded by a shell where hydrogen is undergoing fusion. The core, however, is shrinking and growing hotter. Our studies of nuclear reactions in stars tell us that as the helium in the core heats up, no fusion reactions producing a stable end product occur until temperatures reach 100 million K. But once the dense core of the star reaches this extremely high temperature, three helium atoms can fuse to form a single carbon nucleus. (This process is called the **triple alpha process,** so named because nuclear physicists call the nucleus of the helium atom an alpha particle.) When the triple alpha process begins in the low-mass stars, our calculations show that the entire core is ignited in a quick burst of fusion called the **helium flash.**

Stars in Your Little Finger

Stop reading for a moment and look at your little finger. It's full of carbon atoms because carbon is a fundamental chemical building block for life on Earth. Every one of those carbon atoms was once inside a red giant star, and was fused from helium nuclei in the triple alpha process. All the carbon on Earth—in you, in the charcoal you use for barbecuing, and in the diamonds you might exchange with your sweetheart—was "cooked" by previous generations of stars. How the carbon atoms (and other elements) made their way from

inside some of those stars to become part of the Earth is something we will discuss in the next chapter. For now, we want to emphasize that our description of stellar evolution is not merely of interest to specialists in astronomy, but is, in a very real sense, the story of our own cosmic "roots"—the history of how our atoms originated among the stars. We are made of star-stuff.

Becoming a Giant Again

After the helium flash occurs, the star continues to fuse the helium in its core for awhile, returning to the kind of equilibrium between pressure and gravity that characterized the main-sequence stage. Following the helium flash, the star's surface temperature increases and its overall luminosity decreases. The point representing the star on the H–R diagram thus moves to a new position to the left of and somewhat below its place as a red giant (Figure 21.14).

During this period, the core of the star is stable, fusing helium to form carbon. A newly formed carbon nucleus can sometimes be joined by another helium nucleus to produce a nucleus of oxygen.

At a temperature of 100 million degrees, the inner core is converting its helium fuel to carbon (and oxygen) at a rapid rate. Thus, the new period of stability cannot last very long: it is far shorter than the main-sequence stage. Soon all the helium hot enough for fusion will be used up, and again the inner core will not be able to generate energy via fusion. Once more, gravity will take over.

The situation is analogous to the end of the main-sequence stage when the star's central hydrogen was used up, but the star now has a somewhat more complicated structure. Again the star's core begins to collapse under its own weight. Heat released by the shrinking of the carbon and oxygen core flows into a shell of helium just above the core. This helium, which had not been hot enough for fusion into carbon earlier, is heated just enough for fusion to begin and to generate a flow of energy.

Farther out in the star is another shell where fresh hydrogen is fusing to form helium. Once again the outer regions of the star expand. Its brief period of stability over, the star moves back to the red giant domain on the H–R diagram for a brief time (Figure 21.14). In this stage, the star actually becomes somewhat brighter than it was during its first term as a giant, but it is merely a brief and final burst of glory.

Recall that the last time the star was in this predicament, helium fusion came to its rescue. The temperature at the star's center eventually became hot enough for the *product* of the previous step of fusion (helium) to become the *fuel* for the next step (helium fusing into carbon). But the step after the fusion of helium nuclei requires so hot a temperature that the kinds of lower-mass stars we are dis-

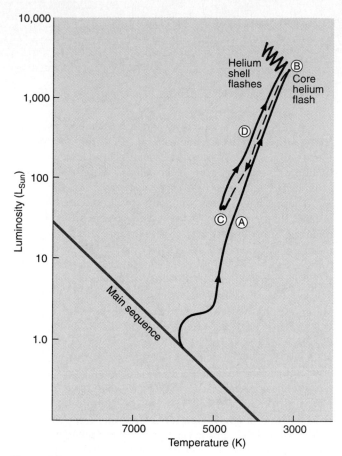

Figure 21.14
The evolution of a star like the Sun on an H–R diagram. Each stage is labeled with a letter: (A) the star evolves from the main sequence to be a red giant, decreasing in surface temperature and increasing in luminosity; (B) a helium flash occurs at this point, leading to a readjustment of the star's internal structure and to (C) a brief period of stability during which helium is fused to carbon and oxygen in the core (in the process the star becomes hotter and less luminous than it was as a red giant). (D) After the central helium is exhausted, the star becomes a giant again and moves to higher luminosity and lower temperature. By this time, however, the star has exhausted its inner resources and will soon begin to die. Where the evolutionary track becomes a dashed line, the changes are so rapid, they are difficult to model. (After calculations by Sackmann, Boothroyd, and Kraemer)

cussing simply cannot compress their cores to reach it. No further types of fusion are possible for such a star.

In stars with masses similar to that of the Sun, the formation of a carbon-oxygen core marks the end of the generation of nuclear energy at the center of the star. The star must now confront the fact that its death is near. We will discuss the death of stars in the next chapter, but in the meantime Table 21.3 summarizes the stages discussed so far in the life of a star with the mass of the Sun. One of the things that gives us confidence in our calculations of stellar evolution is that when we make H–R diagrams of

older clusters, we see stars in each of the stages we have been discussing.

Mass Loss from Giant Stars

When stars become giants and their envelopes expand, they can begin to lose a substantial fraction of their mass into space. Astronomers have observed a strong wind of material flowing away from a number of giant stars, but the mechanism for generating the outflow is not very well understood. We estimate that by the time a star like the Sun reaches the point of the helium flash, for example, it can lose as much as 25 percent of its mass.

This is our first example of a kind of cosmic stellar recycling scheme. We saw that stars form from vast clouds of raw material (gas and dust). As they age, stars begin to return part of themselves to the galactic reservoirs of raw material. Eventually, some of the expelled material from aging stars may participate in the formation of new star systems.

However, the atoms returned to the Galaxy by an aging star are not always the same ones it received initially. The star has fused new elements over the course of its life, and during the red-giant stage material from the star's central regions is dredged up and mixed with its outer layers. As a result, the wind blows outward some particles that were "newly minted" inside the star. (As we will see, this mechanism is even more effective for high-mass stars, but it does work for stars like the Sun.)

21.5

The Evolution of More-Massive Stars

If what we have described so far were the whole story of the evolution of stars and elements, we would have a big problem on our hands. We will see in Chapter 28 that in our best models of the first few minutes of the universe, everything starts with the two simplest elements—hydrogen and helium (plus a tiny bit of lithium). All the predictions of the model imply that no heavier elements were produced at the beginning. This certainly fits with the observation—discussed in several earlier chapters—that most of the universe is made of hydrogen and helium even today.

Yet when we look around us on Earth, we see lots of other elements besides hydrogen and helium. These elements must have been made somewhere in the universe, *and the only place hot enough to make them is inside stars.* One of the fundamental discoveries of 20th-century astronomy is that the stars are the source of all the chemical richness that characterizes our world and our lives.

(text cont. on page 429)

The Red Giant Sun and the Fate of the Earth

How will the evolution of the Sun affect conditions on Earth? While the Sun has appeared reasonably steady in size and luminosity over recorded human history, that brief span means nothing compared to the kinds of time scales we have been discussing. Let's examine the long-term prospects for our planet.

The Sun took its place as a main-sequence star approximately 4.5 billion years ago. At that time it emitted only about 70 percent of the energy that it radiates today. As a result of the changes in its core caused by the buildup of helium, the Sun will continue to increase in luminosity as it ages. More and more radiation will reach each of the planets. Sooner or later—atmospheric models are not yet good enough to say exactly when, but estimates range from 500 million to 2 billion years—this increased heating of the Earth will cause a runaway greenhouse effect. The oceans will start to evaporate, resulting in more clouds and water vapor to hold in the heat. The Earth will start to resemble the Venus of today, and temperatures will become much too high for life as we know it.

The Sun will also expand as it ages, at first gradually as it completes the main-sequence stage, and then more quickly when it moves toward becoming a red giant. Our Sun will begin its climb on the H–R diagram to become a red giant about 6.5 billion years from now. In 7.8 billion years, it will have grown enough in size to engulf the planet Mercury. The surface of Venus will be molten, and temperatures on Earth will be greater than 750°C. Frozen water on the Galilean satellites of Jupiter will start to thaw out.

Then the Sun will experience a helium flash and shrink somewhat, but only temporarily. When it again becomes a giant, it will expand to a radius of 1 AU—the size of the Earth's orbit today. Mercury will again be inside the Sun, and friction with our star's outer atmosphere will make the planet spiral inward until it is completely vaporized.

The Earth and Venus will not be swallowed up, however, because they will no longer be in their present orbits. By this time, the Sun will have lost about 40 percent of its mass by means of its powerful red-giant winds. The gravitational force it exerts will therefore be smaller, and the size of the Earth's orbit (and those of the other planets) will increase accordingly. The Earth will be orbiting at about 1.7 AU. This is not enough to preserve our environment, however! The Sun will be about 5000 times brighter than it is now, and will heat the Earth to a temperature of 1300°C—hot enough to melt all the rocky material that makes up the continents.

From a personal perspective, such predictions are not especially worrisome. After all, no one alive today needs to take out an insurance policy against the Sun becoming very large and very bright billions of years from now. Some commentators feel that, given our inability to get along with each other, humans will have destroyed themselves long before the evolving Sun makes the Earth unlivable.

But if by some miracle our species manages to survive into that remote epoch, chances are that our technology and ability to manipulate the environment will have improved considerably. (Just think of how far we have come in ten thousand years. Imagine dropping off someone from ancient Egypt in a modern airport terminal!) If we survive a hundred million years, our species will probably leave the Earth, and it may become, at best, a museum world to which youngsters from many other worlds return to learn about their origins.

TABLE 21.3
The Evolution of a Star with the Sun's Mass

Stage	Time in This Stage	Surface Temperature	Luminosity	Diameter
	(yrs)	(K)	(L_s)	(Diameter of Sun = 1)
Main Sequence	11 billion	6000	1	1
Becomes Red Giant	1.3 billion	down to 3100	up to 2300	165
Helium Fusion	100 million	4800	50	10
Giant Again	20 million	3100	5200	180

In coming to understand the life stages of lower-mass stars, we have already seen where carbon and oxygen are produced: they are the results of fusion inside stars that become red giants. But where do the heavier elements we know and love (such as the silicon and iron inside the Earth, and the gold and silver in our jewelry) come from? Such heavier elements are formed in the later evolution of more-massive stars.

Making New Elements in Massive Stars

Massive stars evolve in much the same way that the Sun does (but always more quickly)—up to the formation of a carbon-oxygen core. One difference is that for stars with more than about twice the mass of the Sun, helium begins fusion more gradually, rather than with a sudden flash. Also, when more-massive stars become red giants, they become so bright and large that we call them *supergiants*. Such stars can expand until their outer regions become as large as the orbit of Jupiter, which is precisely what the Hubble Space Telescope has shown for the star Betelgeuse (Figure 21.2). They also lose mass very effectively, producing more dramatic winds and outbursts as they age. Figure 21.15 shows a wonderful image of the massive star Eta Carinae, with a great deal of ejected material clearly visible.

But the crucial way that massive stars diverge from the story we have outlined is that they can start further kinds of fusion in their centers. The outer layers of a star whose mass is greater than about eight solar masses weigh enough to compress the carbon-oxygen core until it becomes hot enough to ignite carbon. Carbon can fuse into neon, still more oxygen, and finally silicon. After each of the possible sources of nuclear fuel is exhausted, the core contracts until it reaches a temperature high enough to lead to the fusion of still heavier nuclei.

Theorists have now found mechanisms whereby virtually all chemical elements of atomic weights up to that of iron can be built up by this **nucleosynthesis** (the making of new atomic nuclei) in the centers of the more-massive red giant stars. Moreover, our theories are able to predict the relative abundances with which they occur in nature; that is, the way stars build up elements during various fusion reactions can explain why some elements are common and others quite rare.

Our discussion has barely skimmed the surface of the many different nuclear pathways that can build up elements. Many of these reactions were worked out by nuclear physicists who became interested in applying their understanding of nuclear reactions to the world of astronomy. Primary among them was a group led by the late William Fowler of the California Institute of Technology, who received the 1983 Nobel Prize in physics for his work (Figure 21.16). While the details of these reactions are beyond the scope of this book, they account remarkably well for where the various chemical elements come from.

Figure 21.15
With a mass at least 100 times that of the Sun, the hot supergiant Eta Carinae is one of the most massive stars known. The Hubble Space Telescope took this image of some of the material it has ejected in the course of its evolution. The red outer region is material ejected in an outburst seen in 1841. Moving away from the star at a speed of about 1000 km/s, it is rich in nitrogen and other elements formed in the interiors of such massive stars. The inner blue-white region is material ejected at lower speeds and thus still closer to the star. It appears blue-white because it contains dust and reflects the light of Eta Carinae. (J. Hester and NASA)

Differences in Chemical Composition of Different Clusters

The fact that elements are synthesized in stars over time explains an otherwise puzzling fact about globular and open clusters. The two kinds of clusters do not have the same chemical composition. Hydrogen and helium, the most abundant elements in stars in the solar neighborhood, are also the most abundant constituents of stars in all kinds of clusters. The exact abundances of the elements heavier than helium, however, are not the same in different types of clusters.

In the Sun and most of its neighboring stars, the combined abundance (by mass) of the heavy elements is between 1 and 4 percent of the star's mass. The strengths of the heavy-element lines in the spectra of most open-

Figure 21.16
This 1981 photo shows the team of physicists and astronomers who in a famous 1957 paper first worked out the nuclear pathways by which the heavier elements formed. From left to right, they are Margaret Burbidge, William Fowler, Fred Hoyle, and Geoffrey Burbidge. (Photo courtesy of Caltech)

cluster stars show that they, too, have 1 to 4 percent of their matter in the form of heavy elements. Globular clusters, however, are a different story. The heavy-element abundance of stars in typical globular clusters is found to be only 1/10 to 1/100 that of the Sun.

Differences in chemical composition are related to when stars were formed. The very first generation of stars initially contained only hydrogen and helium. We have seen that these stars, to generate energy, created heavier elements in their interiors. In the last stages of their lives they ejected matter, now enriched in heavy elements, into the reservoirs of raw material between the stars. Such matter was then incorporated into a new generation of stars.

This means that the relative abundance of the heavy elements must be less and less as we look farther into the past. We saw that the globular clusters are much older than the open clusters and associations. Since globular-cluster stars formed much earlier than those in open clusters, they have only a relatively small abundance of elements heavier than hydrogen and helium.

As the Galaxy gets older, the proportion of heavier elements increases. This means that the first generation of stars that formed in our Galaxy was unlikely to have been accompanied by a planet like the Earth, full of silicon, iron, and many other heavy elements. Such a planet (and the astronomy students who live on it) were only possible after stars had a chance to make and recycle their heavier elements.

Approaching Death

Compared with the main-sequence lifetimes of stars, the events that characterize the last stages of stellar evolution pass very quickly. As the star's luminosity increases, its rate of nuclear fuel consumption goes up rapidly—just at that point in its life when its fuel supply is beginning to run down. It is as if a person suddenly did everything possible to hasten death, by overeating, overdrinking, smoking like a chimney, and so on.

After the prime fuel, hydrogen, is exhausted in a star's core, other sources of nuclear energy are, as we have seen, available to the star—in the fusion first of helium and then of other, more complex elements. But the energy yield of these reactions is much less than that of the fusion of hydrogen to helium. And to trigger these reactions, the central temperature must be higher than that required for the fusion of hydrogen to helium, leading to even more rapid consumption of fuel, and faster change. Clearly this is a losing game, and very quickly the star reaches its end. In doing so, however, some remarkable things can happen—as we will see in Chapter 22.

Summary

21.1 When stars are born, they lie on the **zero-age main sequence.** The amount of time a star spends in the main-sequence stage depends on its mass. The fusion of hydrogen to form helium changes the interior composition of a star, which in turn results in changes in temperature, luminosity, and radius. Eventually, as stars age, they evolve away from the main sequence to become red giants. The core of a red giant is contracting, but the outer layers are expanding as a result of fusion in a shell outside the core.

21.2 Calculations that show what happens as stars age can be checked by measuring the properties of stars in clusters. The members of a given cluster were formed at about the same time and have the same composition, so the comparison of theory and observations is fairly straightforward. There are three types of star clusters. **Globular clusters** have diameters of 50 to 300 LY, contain hundreds of thousands of old stars, and are distributed in a halo around the Galaxy. **Open clusters** typically contain hundreds of young to middle-aged stars, are located in the plane of the Galaxy, and have diameters of less than 30 LY. **Associations** are found in regions of gas and dust, and contain extremely young stars.

21.3 The H–R diagram of stars in a cluster changes systematically as the cluster grows older. The most-massive stars evolve the most rapidly. In the youngest clusters and associations, highly luminous blue stars are on the main sequence; the stars with the lowest masses lie to the right of the main sequence and are still contracting toward it. With passing time, stars of progressively lower mass evolve away from (turn off) the main sequence. In globular clusters, which have typical ages of about 15 billion years, there are no luminous blue stars at all. Astronomers can use the turn-off point of the main sequence to determine the age of a cluster.

21.4 After stars become red giants, their cores eventually become hot enough to produce energy by fusing helium to form carbon and oxygen. The fusion of three helium nuclei produces carbon through the **triple alpha process.** The rapid onset of helium fusion in the core of a star is called the **helium flash.** After this, the star becomes stable and reduces its luminosity and size briefly. In stars with masses similar to or less than that of the Sun, fusion stops after the helium in the core has been exhausted. Fusion of hydrogen and helium in shells around the contracting core makes the star a bright giant again, but only temporarily.

21.5 In stars with masses greater than about eight solar masses, nuclear reactions involving carbon, oxygen, and still-heavier elements can build up nuclei as heavy as iron. This creation of elements is called **nucleosynthesis.** These late stages of evolution occur very quickly. Ultimately all stars must use up all of their available energy supplies. In the process of dying, most stars eject some matter, enriched in heavy elements, into interstellar space where it can be used to form new stars. Each succeeding generation of stars contains a larger proportion of elements heavier than hydrogen and helium. This progressive enrichment explains why the stars in open clusters contain more heavy elements than do those in globular clusters, and tells us where most of the atoms on Earth and in our bodies come from.

Review Questions

1. Compare the following stages in the lives of a human being and a star: pre-natal, birth, prolonged adolescence, middle age, old age. What does a star with the mass of our Sun do in each of these stages?

2. What is the main factor that determines where a star falls along the main sequence?

3. What happens when a star exhausts hydrogen in its core, stopping the generation of energy by nuclear fusion of hydrogen to helium?

4. Describe the evolution of a star with a mass similar to that of the Sun, from the protostar stage to the time it becomes a red giant. First give the description in words and then sketch the evolution on an H–R diagram.

5. Describe the evolution of a star with a mass similar to that of the Sun, from just after it first becomes a red giant to the time it exhausts the last type of fuel its core is capable of fusing. After describing the stages in words, sketch them on an H–R diagram.

6. Suppose you have discovered a new star cluster. How would you go about determining whether it is an open or a globular cluster? List several characteristics that might help you decide.

7. Explain how an H–R diagram can be used to determine the age of a cluster of stars.

8. Where did the carbon atoms in the trunk of a tree on your college campus come from originally? Where did the neon in the fabled "neon lights of Broadway" come from originally?

Thought Questions

9. Use star charts to identify at least one open cluster visible at this time of year. (Such charts can be found in *Sky & Telescope* and *Astronomy* magazines each month.) The Pleiades and Hyades are good autumn subjects, and Praesepe is good for springtime viewing. Go out and look at these clusters with binoculars and describe what you see.

10. Would you expect to find an Earth-like planet around a very low-mass star that formed right at the beginning of a globular cluster's life? Explain.

11. In the H–R diagrams for some young clusters, stars of *both* very low and very high luminosity are off to the right of the main sequence, whereas those of intermediate luminosity are on the main sequence. Can you offer an explanation? Sketch an H–R diagram for such a cluster.

12. If the Sun were a member of the cluster NGC 2264, would it be on the main sequence yet? Why?

13. Explain how you might decide whether stars in the red-giant region of the H–R diagram of a star cluster had evolved *away* from the main sequence, or were still evolving *toward* the main sequence.

14. If all the stars in a cluster have nearly the *same age*, why are clusters useful in studying evolutionary effects?

15. Suppose a star cluster were at such a large distance that it appeared as an unresolved spot of light through the telescope. What would you expect the overall color of the spot to be if it were the image of the cluster immediately after it was formed? How would the color differ after 10^{10} years? Why?

16. Suppose an astronomer told you she had found a type O main-sequence star that contained no elements heavier than helium. Would you believe her? Why?

Problems

17. Is the average mass of the stars in an association likely to be larger or smaller than the average mass in a globular cluster? Use the data in Table 21.2 to estimate the average mass of the stars in a globular cluster and in an association. Does this answer agree with your prediction?

18. What is the density in solar masses per cubic light year of the following clusters:

    a. A globular cluster 50 LY in diameter containing 10^5 stars

    b. A stellar association of 100 solar masses and 60 LY in radius

19. During the course of its main-sequence lifetime, a star converts about 10 percent of the hydrogen initially present into helium. How much does the mass of the star change as a result of fusion?

20. Stars exist that are as much as a million times more luminous than the Sun. Consider a star of mass 2×10^{32} kg and luminosity 4×10^{32} W. Assume that the star is 100 percent hydrogen, all of which can be converted to helium, and calculate how long it can shine at its present luminosity. (There are about 3×10^7 s in a year.)

21. Perform a similar computation for a typical star less massive than the Sun, such as one whose mass is 1×10^{30} kg and whose luminosity is 4×10^{25} W.

Suggestions for Further Reading

Darling, D. "Breezes, Bangs, and Blowouts: Stellar Evolution Through Mass Loss" in *Astronomy*, Sep. 1985, p. 78; Nov. 1985, p. 94.

Fortier, E. "Touring the Stellar Cycle" in *Astronomy*, Mar. 1987, p. 49. Observing objects that show the different stages of stellar evolution.

Kaler, J. "Giants in the Sky: The Fate of the Sun" in *Mercury*, Mar./Apr. 1993, p. 34.

Kaler, J. *Stars*. 1992, Scientific American Library/W. H. Freeman. Good modern introduction to stellar evolution.

Kaler, J. "The Largest Stars in the Galaxy" in *Astronomy*, Oct. 1990, p. 30. On red supergiants.

Mullan, D. "Caution! High Winds Beyond This Point" in *Astronomy*, Jan. 1982, p. 74. On stellar winds and mass loss.

Whitmire, D. and Reynolds, R. "The Fiery Fate of the Solar System" in *Astronomy*, Apr. 1990, p. 20. What will happen when the Sun becomes a red giant.

1. Use *RedShift* to determine the spectral type of each star listed below. Then use Table 21.1 to estimate how long each star will live on the main sequence; estimate the mass of each star; predict which stars will end their life after burning helium in their core and which will burn heavier elements in their core.

a. Beta Canes Venaticorum
b. Epsilon Eridani
c. Alpha Pictoris
d. Gamma Cassiopeiae
e. Procyon, Alpha Canis Minoris
f. Theta Ursae Majoris
g. Menkalinan, Beta Aurigae
h. Tau Herculis
i. Eta Cephei
j. Mu Herculis

2. To find which open and globular clusters are visible tonight to the naked eye or through binoculars, set the *Stellar Limiting Magnitude* to 5 and the *Deep-Sky Filter* to a faint magnitude of 5, with all objects except for star clusters turned off.

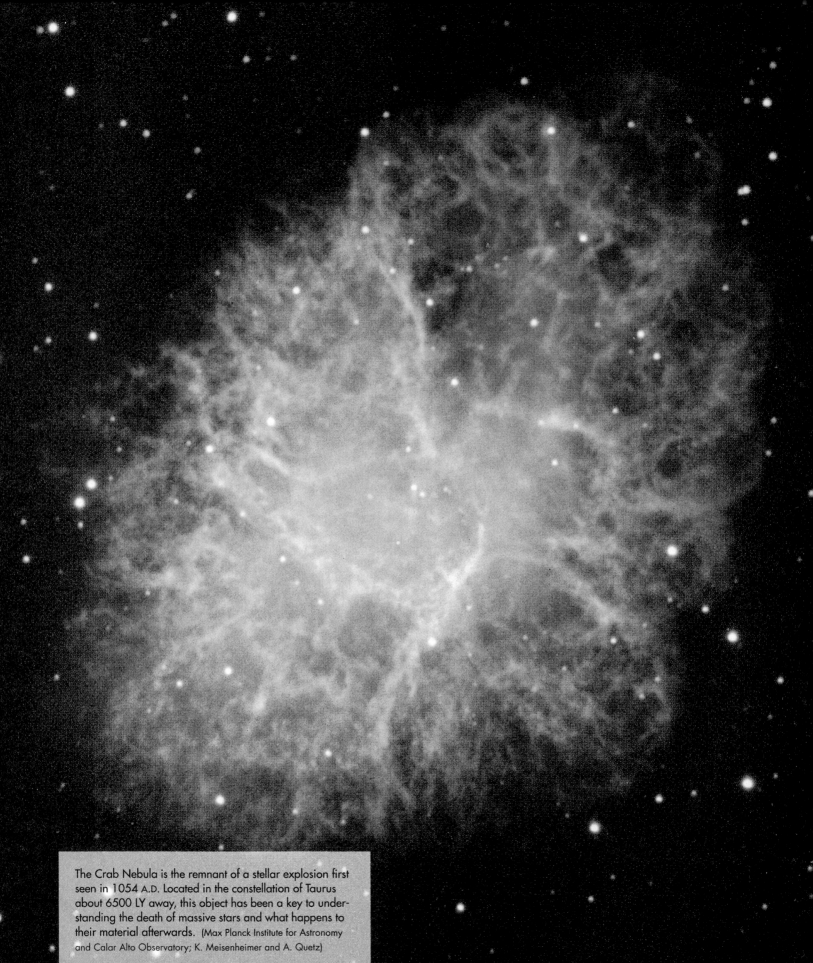

The Crab Nebula is the remnant of a stellar explosion first seen in 1054 A.D. Located in the constellation of Taurus about 6500 LY away, this object has been a key to understanding the death of massive stars and what happens to their material afterwards. (Max Planck Institute for Astronomy and Calar Alto Observatory; K. Meisenheimer and A. Quetz)

CHAPTER 22

The Death of Stars

Thinking Ahead

Several times in recorded history humanity awakened to find a new star in the sky, one so brilliant it was actually visible during the daytime. What do you think the reaction was to such a sight in ancient times? What would the public response be today if such a celestial event were to present itself with no warning?

He told me to look at my hand, for a part of it came from a star that exploded too long ago to imagine.

Paul Zindel in *The Effect of Gamma Rays on Man-in-the-Moon Marigolds.*

In the preceding two chapters we have followed the life story of stars, from the process of birth to the brink of death. Now we are ready to explore the ways that stars end their lives. Sooner or later every star exhausts its store of nuclear energy. Without an internal energy source, a star eventually gives way to the inexorable pull of gravity and collapses under its own weight. How that collapse takes place, and what kind of stellar "corpse" results from the process, depends on the star's mass at the time it collapses.

Bear in mind that a star's mass at the moment it is ready to die may be quite a bit smaller than it was at birth or during its long youth. As we noted in the last chapter, stars can lose a significant amount of mass in their middle and old ages. In this chapter we examine further ways that stars can lose mass in the last stages of their lives. Following the rough distinction made in the last chapter, we again discuss the stars of lower and higher mass separately.

Figure 22.1
Antares, found in the constellation of Scorpius, is one of the brightest stars in our sky. It is a supergiant with a large quantity of expelled material around it. Most of the energy from the star actually comes in the infrared region of the spectrum and cannot be seen on this visible-light image.
(© ROE/Anglo–Australian Telescope Board)

22.1

The Death of Low-Mass Stars

Let's begin with those stars whose final mass just before death is less than about 1.4 times the mass of the Sun. (We will explain why this is the crucial dividing line in a moment.) Note that most stars in the universe fall into this category. As we saw, the number of stars increases as mass decreases; as in the music business, only a few stars get to be superstars. Furthermore, many stars with an initial mass much greater than 1.4 solar masses (M_{Sun}) will be reduced to that level by the time they die. Our theoretical models indicate that stars starting with 5 M_{Sun} or less will surely be down to 1.4 M_{Sun} by the end; it may even be that stars with 10 M_{Sun} manage to lose enough mass to fit into this category.

Mass Loss

In Chapter 21 we left the life story of a star with a mass like the Sun's just as it was climbing upwards toward the giant region in the H–R diagram for the second time. Powered by fusion from both of its shells (one with helium fusion and one with hydrogen fusion), the star's outer layers expand and cool down again.

During this period, the star loses mass at an impressive rate. Over the tens of thousands of years that a star like the Sun spends in this stage, it can lose a substantial fraction of its mass. As the ejected material expands and cools, its atoms can join together until they form molecules and, eventually, tiny particles of dust. We saw in Chapter 19 that this is one important source of the interstellar dust that is now prevalent throughout the star-forming regions of our Galaxy.

The shroud of dust makes the giant stars in this stage of their lives difficult to observe. In fact, only recently, through observations with infrared and microwave telescopes, has some of this expanding material been identified (Figure 22.1). Also, this period is a relatively brief one, and thus it is hard to catch stars "in the act." Still, we have now been able to confirm that they do pass through such a stage. And while the outside of the star is expanding and being expelled, the seeds for the star's death are being sown in its core.

The Contracting Core

Recall from Chapter 21 that during this time the core of the star was undergoing another crisis. Earlier in its life, during a brief stable period, helium in the core had gotten hot enough to fuse into carbon (and oxygen). But as the helium hot enough to do this had become exhausted, the star's core had once more found itself without a source of pressure to balance gravity, and so had begun to contract.

This collapse is the most devastating (and final) event in the life of the core. Because the star's mass is relatively low, it cannot ever get the temperature inside hot enough to begin another round of fusion (the way larger-mass stars can). The core thus continues to contract until it becomes so dense that a new and different way for matter to behave helps it achieve a final state of equilibrium. But this equilibrium is established only after the core has reached an enormous density, nearly a million times the density of water! In the process, what remains of the star (after all the mass loss has taken place) becomes one of the strange *white dwarfs* that we first met in Chapter 17.

Degenerate Stars

Since white dwarfs are far more dense than any substance on Earth, the behavior of matter inside them is different from anything we know from everyday experience. Under such compressed conditions, electrons actually resist any further compression and set up a powerful pressure inside the core. This pressure is the result of the fundamental rules that govern the behavior of electrons. According to these rules, which have been verified by studying how electrons behave under laboratory conditions, no two electrons can be in the same place at the same time doing the same thing. We specify the *place* of an electron by its precise position in space, and specify what it is doing by its motion and the way it is spinning.

The temperature in the interior of a star is always so high that the atoms are stripped of virtually all their electrons. In normal stars the density of matter is also relatively low, and the electrons are moving rapidly. It is very unlikely that any two of them will be in the same place moving in exactly the same way at the same time. But this all changes when a star exhausts its store of nuclear energy and begins its final collapse.

As the star's core contracts, electrons are squeezed closer and closer together. Eventually a star like the Sun becomes so dense that further contraction would require two or more electrons to violate the rule against occupying the same place and moving in the same way. Such a hot, dense gas is said to be **degenerate** (a term coined by physicists and not related to the electron's moral character). The electrons in a degenerate gas resist further crowding with overwhelming pressure. (It's as if the electrons said, "You can press inward all you want, but there is simply no room for any other electrons to squeeze in here without violating the rules of our existence.")

White Dwarfs

White dwarfs, then, are stars with degenerate electron cores that have stabilized the star against further collapse. Calculations showing that stars can do this, and what happens for stars of different masses, were first carried out by the Indian-American astrophysicist S. Chandrasekhar (see "Voyagers in Astronomy" box). He was able to show how

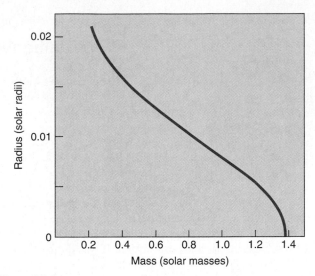

Figure 22.2
Theoretical relation between the masses and the radii of white dwarf stars. As you can see, the theory predicts that as the mass of the star increases (toward the right), its size gets smaller and smaller.

much a star will shrink before the degenerate electrons stop its contraction and hence what its final diameter will be (see Figure 22.2). Note that the larger the mass of the star, the *smaller* its radius.

A white dwarf with a mass like that of the Sun has a diameter about the same as that of the Earth. This is a remarkably small size for a star, and means that its material is very compressed. A teaspoonful of white dwarf material (if it could be brought to Earth in its compressed state) would weigh more than a full garbage truck. And, as we will see, such stars are quite hot on the outside—much hotter than the surface of the Sun.

Beyond the White Dwarf

When Chandrasekhar made his calculations about white dwarfs, he found something very surprising, which is clear on Figure 22.2 as well. According to the best theoretical models, a white dwarf with a mass of about 1.4 M_{Sun} or larger would have a radius of zero! What the calculations are telling us is that even the force of degenerate electrons cannot stop the collapse of a star with more mass than this. The maximum mass that a star can have and still become a white dwarf—1.4 M_{Sun}—is called the **Chandrasekhar limit.** Stars with masses exceeding this limit have a different kind of end in store—one that we will explore in the next section.

Planetary Nebulae

Before we describe the ultimate fate of white dwarfs, let's briefly return to the outer layers of our dying star. When the star was still in its giant state, some of its outer mate-

Subrahmanyan Chandrasekhar

Born in 1910 in Lahore, India, S. Chandrasekhar (known as Chandra to his friends and colleagues) grew up in a home that encouraged scholarship and an interest in science. His uncle, C. V. Raman, was a physicist who won the 1930 Nobel Prize. A precocious student, Chandra tried to read as much as he could about the latest ideas in physics and astronomy, although obtaining technical books was not easy in India at the time. He finished college at age 19 and won a scholarship to study in England. It was during the long boat voyage to get to graduate school that he first began doing calculations about the structure of white dwarf stars.

Chandra developed his ideas during and after his studies as a graduate student, showing—as we have discussed—that white dwarfs with masses greater than 1.4 times the mass of the Sun cannot exist, and that the theory predicts the existence of other kinds of stellar corpses as well. He wrote later that he felt very shy and lonely during this period, isolated from other students, afraid to assert himself, and sometimes waiting for hours to speak with some of the famous professors he had read about in India. His calculations soon brought him into conflict with certain distinguished astronomers, including Sir Arthur Eddington, who publicly ridiculed Chandra's ideas. At a number of meetings of astronomers, such leaders in the field as Henry Norris Russell refused to give Chandra the opportunity to defend his ideas, while allowing his more senior critics lots of time to criticize them.

S. Chandrasekhar (1910–1995)
(Courtesy of Emilio Segrè Visual Archives, Physics Today Collection)

Yet Chandra persevered, writing books and articles elucidating his theories, which turned out not only to be correct, but to lay the foundation for much of our modern understanding of the death of stars. (In 1983, he received the Nobel Prize in physics for this early work.)

In 1937 Chandra came to America and joined the faculty at the University of Chicago, where he remained for the rest of his life. There he devoted himself to research and teaching, making major contributions to many fields of astronomy—from our understanding of the motions of stars through the Galaxy to the behavior of the bizarre objects called black holes (see Chapter 23).

Perhaps because he remembered how alienated he had felt, Chandra spent a great deal of time with his graduate students, supervising the research of more than 50 PhD's during his life. He took his teaching responsibilities very seriously: during the 1940s, while based at the Yerkes Observatory, he willingly drove the more than 100-mile trip to the university each week to teach a class of only two students. As the university later reported, "Any concern about the cost-effectiveness of such a commitment was erased in 1957, when that entire class—consisting of T. D. Lee and C. N. Yang—won the Nobel Prize in physics."

Chandra also had a deep devotion to music, art, and philosophy, writing articles and books about the relationship between the humanities and science. He emphasized that "one can learn science the way one enjoys music or art. . . . Heisenberg had a marvelous phrase 'shuddering before the beautiful' . . . that is the kind of feeling I have."

---

rial was actually able to lift off. As a result, our star is now surrounded by one or more expanding shells of gas, each containing as much as 0.1 or 0.2 M_{Sun} of material (Figure 22.3).

This process strips our star of its outer layers, exposing the hotter regions underneath for the first time. As the star's core is heated by its compression, any remaining hydrogen is quickly fused into helium. As the core evolves to become a white dwarf, it becomes (as we saw) very hot—reaching surface temperatures of 100,000 K. Such stars are very strong sources of ultraviolet radiation, which pours out directly into the shell of recently expelled material.

As a result, this shell is heated, ionized, and set aglow (as are the H II regions around young hot stars described in Chapter 19). These glowing shells are among the most spectacular objects in the sky (Figure 22.4). They were given an extremely misleading name when first found: **planetary nebulae.** The name is derived from the fact that a few planetary nebulae, when viewed through a small telescope, bear a superficial resemblance to planets. Actually, they are thousands of times larger than the solar system and have nothing to do with planets, but once names are put into regular use in astronomy, it is extremely difficult to change them. There are tens of thousands of planetary nebulae in our own Galaxy, but many are hidden from view because their light is absorbed by interstellar dust.

Sometimes, as in Figure 22.5, a planetary nebula appears to be a ring. This is an illusion, caused by the fact that we are looking through more layers along the sides of the shell than in the middle. In the same way, the center of a soap bubble is often more transparent than its rim.

Figure 22.3
With the use of modern CCDs, the Ring Nebula (perhaps the best-known of all the planetary nebulae) reveals several shells of ejection. (George Jacoby and Bruce Balick, NOAO)

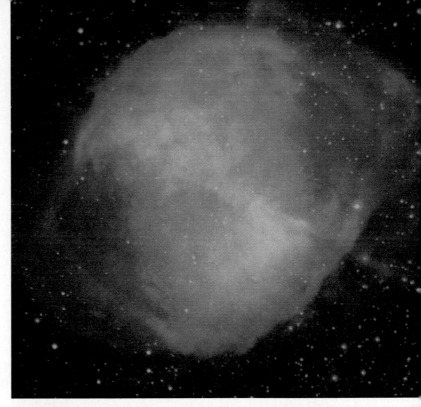

Figure 22.4
The Dumbbell Nebula is one of the most beautiful planetary nebulae in the sky. It is a shell of ejected material from a dying star, glowing because the hot star is giving off large amounts of ultraviolet radiation. (Max Planck Institute for Astronomy and Calar Alto Observatory; K. Meisenheimer and A. Quetz)

When the central star has had several episodes of ejection, or when the system contains more than one star, planetary nebulae can take on wonderfully complex structures, as shown in Figure 22.6.

Planetary nebula shells usually expand at speeds of 20 to 30 km/s, and a typical planetary nebula has a diameter of about 1 LY. If we assume that the gas shell has expanded at constant speed, we can calculate that the shells of all the planetary nebulae visible to us were ejected within the past 50,000 years or so. Still-older shells have expanded so much that they are simply too thin and tenuous to be seen. When we consider the relatively short time that each planetary nebula can be observed, and the number of such nebulae we see, we must conclude that a large fraction of all stars evolve through the planetary nebula phase. This confirms our view of planetary nebulae as a sort of "last gasp" of low-mass star evolution.

Figure 22.5
At a distance of about 400 LY, the Helix Nebula, NGC 7293, is the planetary nebula nearest the Sun. On photographs the Helix has a diameter about the same as that of the full Moon. The greenish color is produced by emission lines of ionized oxygen; the red is due to nitrogen and hydrogen. (Anglo-Australian Telescope Board)

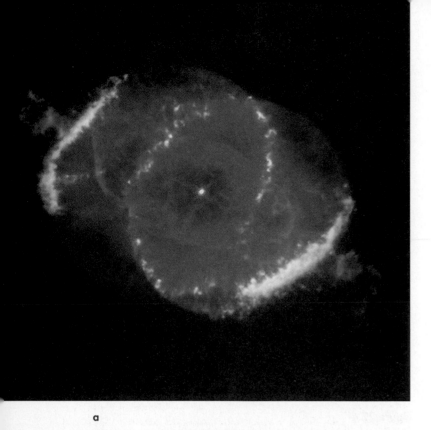

a

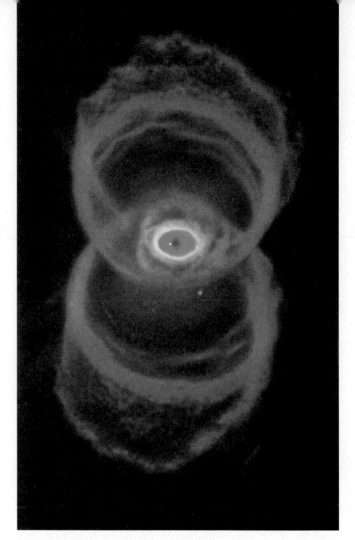

b

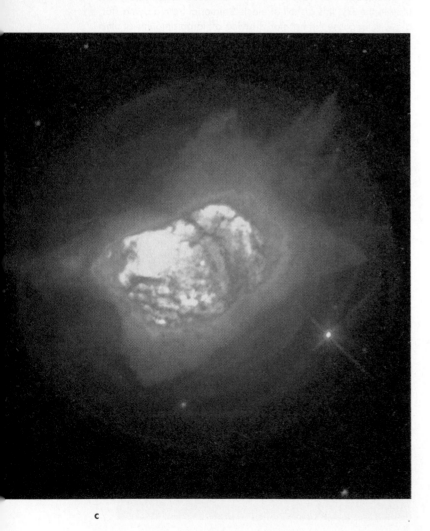

c

Figure 22.6

Three intriguing planetary nebulae as seen with the Hubble Space Telescope. (a) The Cat's Eye Nebula (NGC 6543) in the constellation of Draco shows several concentric shells of gas, jets of high-speed gas, and regions (in green) where the high-speed material is piling into previously ejected gas shells. The complexity of the features may mean that there are two stars at the center, both of which have ejected or are ejecting material in shells, rings around their equators, and/or jets from their polar regions. (P. Harrington, K. Borkowski, and NASA) (b) The Hourglass Nebula (MyCn 18), about 8000 LY away, also shows complicated structure. This picture is composed of three separate images taken in the light emitted by ionized nitrogen (red), hydrogen (green), and doubly ionized oxygen (blue). Notice that the star that gives rise to the material of the nebula is not in the center of the green region. Jets, rings, and companion stars may also be required to explain this complicated object. (R. Sahai, J. Trauger, NASA) (c) NGC 7027 is a planetary nebula located about 3000 LY from us in the constellation of Cygnus. This image combines both visible light and infrared information to bring out several episodes of ejection in the object. The initial (outer) blue shells were ejected much more symmetrically than the inner shells of material, which glow with infrared radiation from dust grains that have condensed within it. (H. Bond and NASA)

The Ultimate Fate of White Dwarfs

As a planetary nebula expands and fades away, the star inside—hot, dense, luminous, and dying—is now clearly revealed to our view. If the birth of a star is the onset of fusion reactions, then we must consider the end of all fusion reactions to be the time of a star's death. As the core is stabilized by degeneracy pressure, a last shudder of fusion passes through the outside of the star, consuming the little hydrogen still remaining. Now the star is a true white dwarf: it has no further source of energy, and so begins to cool. (Figure 22.7 shows the path of a star like the Sun on the H–R diagram during its final stages.)

The electrons in a degenerate gas move about, as do particles in any gas, but not with freedom. A particular electron cannot change position or momentum until another electron in an adjacent state gets out of the way. The situation is much like that in the parking lot after a big football game. Vehicles are closely packed, and a given car cannot move until the one in front of it moves, leaving an empty space to be filled. If one car leaves the lot, another can move ahead, and still others can follow behind it, producing a flow of cars toward the exit. It turns out

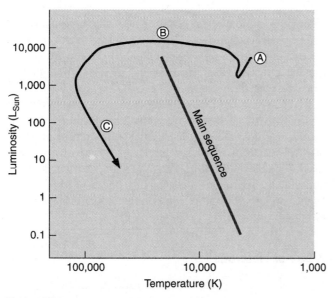

Figure 22.7
The evolutionary track on the H–R diagram for a star with a mass like the mass of the Sun which is nearing the end of its life. After the star becomes a giant again (point A on the diagram), it will lose more and more mass as its core begins to collapse. The mass loss will expose the hot inner core, which will appear at the center of a planetary nebula. In this stage the star moves across the diagram to the left as it becomes hotter and hotter during its collapse (point B on the diagram). At first the luminosity remains constant, but as the star begins to shrink significantly, it becomes less and less bright (point C). It is now a white dwarf and will slowly cool until all of its remaining store of energy is radiated away. (Assumes the Sun will lose about 46 percent of its mass during the giant stages; based on calculations by Sackmann, Boothroyd, and Kraemer)

that heat can flow through an electron-degenerate gas, and the star can slowly lose heat into space.

Since the white dwarf can no longer contract (or produce energy through fusion), its only energy source is the heat represented by the motions of the atomic nuclei in its interior. The light it emits comes from this internal stored heat, which is substantial. Gradually, however, the white dwarf radiates away its heat and fades out. After many billions of years, the heat will be gone, the nuclei will cease their motion, and the white dwarf will no longer shine. It will then be a *black dwarf*—a cold stellar corpse with the mass of a star and the size of a planet. It will be composed mostly of carbon and oxygen, the product of the most advanced fusion reaction of which the star was capable.

We have one final surprise as we leave our low-mass star in the stellar graveyard. Calculations of what happens inside a degenerate star as it cools indicate that the atoms in essence "solidify" into a giant, highly compact lattice (organized rows of atoms just like in a crystal). When carbon is compressed and crystallized in this way, it becomes a diamond. A white dwarf star is the most impressive engagement present you could ever see, although any attempt to mine the diamond-like material inside would crush an ardent lover instantly!

Evidence for Significant Mass Loss

Our model of the evolution of low-mass stars into white dwarfs assumes there is significant mass loss in the red giant phase of stellar evolution; this is required to bring down the masses of many stars below the Chandrasekhar limit of 1.4 M_{Sun}. What evidence do we have that stars starting out significantly more massive than this limit actually do shed so much of their mass?

White dwarfs have been found in young, open clusters—clusters so young that only stars with masses greater than 5 M_{Sun} have had time to exhaust their supplies of nuclear energy and complete their evolution to the white dwarf stage. In the Pleiades, for example, stars with masses of 4 to 5 M_{Sun} are still on the main sequence. Yet this cluster also has at least one white dwarf. The star that turned into this dwarf must have had a main-sequence mass exceeding 5 M_{Sun}, since stars with lower masses have not yet had time to exhaust their stores of nuclear energy. It must have gotten rid of enough matter so that its mass at the time nuclear energy generation ceased was less than 1.4 M_{Sun}.

Another convincing piece of evidence can be seen in the presence of a white dwarf star around the bright star Sirius (see Chapter 17). Sirius and its white dwarf companion are part of a binary star system, and we presume both stars formed at the same time. The mass of Sirius is a little more than 2 M_{Sun} and it is still on the main sequence. The mass of the white dwarf is about the mass of the Sun (half of Sirius' mass), yet it has already died and become a white dwarf.

Since more-massive stars evolve more quickly, the only way for Sirius' companion to be a white dwarf today is for it to have been *the more-massive star* when the pair formed. Thus it must have lost a substantial amount of its mass between the time it formed and now, when we see it as a white dwarf. Today astronomers believe that mass loss is an important part of the evolution of all stars.

Evolution of Massive Stars: An Explosive Finish

Thanks to mass loss, stars with masses up to at least 5 (and perhaps even 10 M_{Sun}) probably end their lives as white dwarfs. But we know stars can have masses as large as 100 M_{Sun}. What is the ultimate fate of the higher-mass stars?

Nuclear Fusion of Heavy Elements

As discussed in Section 21.5, only after the helium in its core is exhausted does the evolution of a massive star take a significantly different course from that of lower-mass stars. In a massive star, the weight of the outer layers is sufficient to force the carbon and oxygen core to contract until it becomes hot enough to fuse carbon into neon. This cycle of contraction, heating, and the ignition of another nuclear fuel is repeated several more times. After each of the possible nuclear fuels is exhausted, the core contracts until it reaches a temperature high enough to fuse still-heavier nuclei. Massive stars go through these stages very, very quickly—far more quickly than the main-sequence and red giant stages. In really massive stars, some fusion stages toward the very end can take only months or even days!

How long can this process of peacefully building up elements by fusion go on? It turns out that there is a limit: the fusion of silicon into iron is the last step in the sequence of peaceful element production. The fusion of iron begins a rapid sequence of dramatic events that can actually force the star to explode.

Up to this point, a key result of each fusion reaction has been to *release* energy. This is because the nucleus of each fusion product has been a bit more stable than the nuclei that formed it. As discussed in Chapter 15, light nuclei give up some of their binding energy in the process of fusing into more tightly bound, heavier nuclei. It is this released energy that helps keep a star in balance. But of all the nuclei known, iron is the most tightly bound and thus the most stable.

You might think of the situation like this: all small nuclei want to "grow up" to be like iron, and are willing to pay (release energy) to move toward that goal. But iron is a mature nucleus with good self-esteem, perfectly content being iron; it *requires* payment (must absorb energy) to

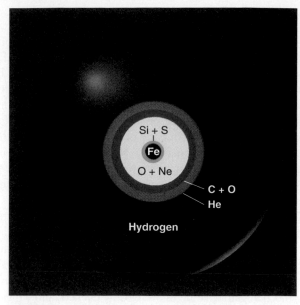

Figure 22.8
Just before its final gravitational collapse, a massive star resembles an onion. The iron core is surrounded by layers of silicon and sulfur, oxygen, neon, carbon mixed with some oxygen, helium, and finally hydrogen.

change its stable nuclear structure. This is the exact opposite of what has happened in each nuclear reaction so far: instead of providing energy to balance the inward pull of gravity, any fusion of iron would *remove* some energy from the core of the star.

At this stage of its evolution, a massive star resembles an onion with an iron core. As we get farther from the center, we find shells of decreasing temperature in which nuclear reactions involving nuclei of progressively lower mass—silicon and sulfur, oxygen, neon, carbon, helium, and finally hydrogen—are taking place (Figure 22.8). What happens when the iron core is no longer a source of energy for the star?

A Ball of Neutrons

In effect, a massive star builds a white dwarf in its center where no nuclear reactions are taking place. For stars that begin their evolution with masses of at least 10 M_{Sun}, this white dwarf is made of iron. For stars with initial masses in the range 8 to 10 M_{Sun}, the white dwarf that forms the core is made of oxygen, neon, and magnesium because the star never gets hot enough to form elements as heavy as iron. Whatever its composition, the white dwarf embedded in the center of the star is supported against further gravitational collapse by degenerate electrons.

While no energy is being generated within the white-dwarf core of the star, fusion does still occur in shells surrounding the core. As the shells finish their fusion reactions and stop producing energy, the ashes of the last reaction fall onto the white-dwarf core, increasing its mass. Ultimately, the core is pushed over the Chandrasekhar limit of 1.4 M_{Sun}. That is, it becomes so massive that the force exerted by degenerate electrons is no longer great enough to resist gravity. What happens next is completely out of the realm of ordinary experience.

A stellar core containing degenerate electrons is still mostly empty space. Electrons and atomic nuclei are, after all, extremely small. The electrons and nuclei in a stellar core may be crowded compared to the air in your room, but there is still lots of space between them. As mass is added to the core, its density begins to increase. The electrons at first resist being crowded closer together, but if the mass grows larger than 1.4 M_{Sun}, they can no longer do so.

When the density reaches 4×10^{11} g/cm^3 (400 billion times the density of water), some electrons are actually *squeezed into the atomic nuclei*, where they combine with protons to form neutrons. (This can only happen at the outrageous densities found in collapsing stars.) Some of the electrons are now gone, and so the core's ability to resist the crushing mass of the star's overlying layers is reduced even more. The collapse becomes catastrophic. The atomic nuclei absorb more and more electrons and ultimately become so saturated with neutrons that they cannot hold onto them.

At this point the neutrons are squeezed out of the nuclei, now exerting a new force. As is true for electrons, it turns out that the neutrons strongly resist being in the same place and moving in the same way. The force that can be exerted by such *degenerate neutrons* is much greater than that produced by degenerate electrons, so they can ultimately resist the collapse. Calculations show that the upper limit on the mass of stars made only of neutrons is about 3 M_{Sun}.

In other words, if the collapsing core has less mass than this, it will reach a stable state as a crushed ball made mainly of neutrons, which astronomers call a **neutron star.** However, if the mass of the core is greater than this limit, then even neutron degeneracy cannot stop the core from collapsing, and the star becomes something unbelievably compressed called a *black hole*. Black holes are the subject of Chapter 23; for now, we restrict ourselves to those stars in which the collapse of the core is in fact halted by degenerate neutrons.

Collapse and Explosion

The collapse that takes place when electrons are absorbed into the nuclei is very rapid. In less than a second, the core, which originally was approximately the same diameter as the Earth, collapses to a diameter of less than 20 km. The speed with which material falls inward reaches one-fourth the speed of light. The collapse halts only when the density of the core exceeds the density of an atomic nucleus (which is the densest form of matter we know). A typical neutron star is so compressed that to duplicate its density we would have to squeeze all the people

in the world into a single raindrop. That would give us one raindrop's worth of a neutron star.

Because of neutron degeneracy, this dense core strongly resists further compression, abruptly halting the collapse. The shock of the abrupt jolt generates waves throughout the outer layers of the star, causing those outer layers to blow off in a violent explosion called a **supernova.** Although the analogy is not exact, you might think of what happens as the parking lots at a local stadium fill up for a very popular rock concert. Suddenly, no further cars can squeeze into the parking structures. This causes a big chain reaction in the adjoining streets, where long lines of cars have formed.

Recently, both our models and observations of supernovae have led us to realize that the ghostly subatomic particles called *neutrinos,* introduced in Chapter 15, play a crucial role at this point. Each time an electron and a proton merge to make a neutron in the collapsing core, the merger releases a neutrino. Thus we expect vast numbers of neutrinos to be generated very rapidly as the neutron star forms.

While neutrinos ordinarily do not interact very much with ordinary matter (we accused them of being downright antisocial in Chapter 15), matter in the core of a collapsing star is far more compressed and thus more likely to interact with an enormous burst of neutrinos. We now believe that up to 90 percent of the energy of a supernova explosion may be carried by the neutrinos. They may also be responsible for getting the rest of the star to blow outward even while gravity is pulling everything inward.

Whatever the detailed mechanism responsible for the explosion of the star, we know such supernova explosions must occur because we can observe them happening in our Galaxy and in other galaxies as well. When they happen close-by, they can be among the most spectacular and important celestial events, as we will discuss in the next section. (Note that there are at least two different types of supernova explosions; the kind we have been describing are called, for historical reasons, Type II supernovae. We will describe how the types differ in Section 22.4.)

For now, Table 22.1 summarizes the discussion so far about what happens to stars of different initial masses at the ends of their lives. Like so much of our scientific understanding, this list represents a progress report: it is the best we can do with our present models and observations. The mass limits corresponding to various outcomes may change somewhat as models improve. There is much we do not yet understand about the details of what happens when stars die.

Supernova Observations

Supernovae were discovered long before astronomers realized that these spectacular cataclysms mark the death of stars (see "Making Connections" box). The word *nova* means "new" in Latin; before telescopes, when a star too dim to be seen with the unaided eye suddenly flared up in

(text cont. on page 445)

Supernovae in History

While many supernova explosions in our own Galaxy have gone unnoticed, a few were so spectacular that they were clearly seen and recorded by sky-watchers and historians at the time. We can use these records, going back two millennia, to help us pinpoint where the exploding stars were and thus where to look for their remnants today.

The most dramatic supernova was observed in the year 1006 A.D. It appeared in May as a brilliant point of light visible during the daytime, perhaps 100 times brighter than the planet Venus. It was bright enough to cast shadows on the ground during the night, and was recorded with awe and fear by observers all over Europe and Asia. No one had seen anything like it before; Chinese astronomers, noting that it was a temporary spectacle, called it a "guest star."

Astronomers David Clark and Richard Stephenson have scoured records from around the world to find over 20 reports of the 1006 supernova. This has allowed them to determine with some accuracy where in the sky the explosion occurred. They place it in the modern constellation of Lupus; at roughly the position they have determined we do find a supernova remnant, now quite faint. From the way its filaments are expanding, it indeed appears to be about 1000 years old.

Another guest star was clearly recorded in Chinese records in July 1054 A.D. The remnant of that star, called the Crab Nebula, is shown in the opening image for this chapter. It is a marvelously complex object, whose study has been a key to understanding the death of massive stars. We estimate that it was about as bright as the planet Jupiter, nowhere near as dazzling as the 1006 event but still quite dra-

matic to anyone who kept track of objects in the sky. There is some evidence that Native Americans in New Mexico, for example, recorded the new star with the crescent moon nearby in a cave painting at a ceremonially important location. Another fainter supernova was seen in 1181 A.D.

The next supernova became visible in November 1572, and, being brighter than the planet Venus, was quickly spotted by a number of observers, including the young Tycho Brahe (see Chapter 2). His careful measurements of the star over a year and a half showed that it was not a comet or something in the Earth's atmosphere, since it did not move relative to the stars. He correctly deduced that it must be a phenomenon belonging to the realm of the stars, not of the solar system. The remnant of Tycho's Supernova (as it is now called) can still be detected in many different bands of the electromagnetic spectrum.

Not to be outdone, Johannes Kepler, Tycho Brahe's scientific heir, found his own supernova in 1604. Fainter than Tycho's, it nevertheless remained visible for about a year. Kepler wrote a book about his observations and conclusions that was read by many with an interest in the heavens, including Galileo.

No supernova has been spotted in our Galaxy for the past 300 years. Since the explosion of a visible supernova is a chance event, there is no way to say when the next one might occur. Around the world, dozens of professional and amateur astronomers keep a sharp lookout for "new" stars that appear overnight, hoping to be the first to spot the next guest star in our sky and make a little history themselves.

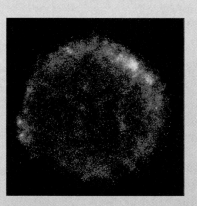

Several views of Tycho Brahe's 1572 supernova. (a) A broadsheet about the "new star" from 1573. (b) A modern map of the radio emission from the expanding shell of gas ejected by the supernova explosion. The different colors correspond to different intensities of radio emission, produced by extremely energetic electrons gyrating in a magnetic field. No central source has been found; apparently nothing remains of the star that exploded. (National Radio Astronomy Observatory/AUI) (c) As the debris from the explosion expands, it barrels into the surrounding interstellar gas, producing very high temperatures. The shocked material glows in x rays, as shown here in an image recorded with the Rosat x-ray satellite. (Max Planck Institute for Extraterrestrial Physics)

TABLE 22.1
The Ultimate Fate of Stars with Different Masses

Initial Mass (Mass of Sun = 1)	Final State at End of Its Life
<0.01	Planet
0.01 to 0.08	Brown dwarf
0.08 to 0.25	White dwarf made mostly of helium
0.25 to 8–10	White dwarf made mostly of carbon and oxygen
8–10 to 12	White dwarf made of oxygen–neon–magnesium*
12 to 40	Supernova explosion that leaves a neutron star
>40	Supernova explosion that leaves a black hole

* Stars in this mass range may produce a type of supernova different from the one we have discussed.

a brilliant explosion, observers concluded it must be a brand-new star. Twentieth-century astronomers reclassified the explosions with the greatest absolute luminosity as *super*novae.

From the historical records, from studies of the remnants of supernova explosions in our own Galaxy, and from analyses of supernovae in other galaxies, we estimate that, on average, one supernova explosion occurs somewhere in the Milky Way Galaxy every 25 to 100 years. Unfortunately, however, no supernova explosion has been detected in our Galaxy since the invention of the telescope. Either we have been exceptionally unlucky or, more likely, more-recent explosions have taken place in parts of the Galaxy where light is blocked from reaching us by interstellar dust.

At their maximum brightness, the most luminous supernovae have about 10 billion times the luminosity of the Sun. For a brief time, a supernova may outshine the entire galaxy in which it appears. After maximum brightness, the star fades in light and disappears from telescopic visibility within a few months or years. At the time of their outbursts, supernovae can eject material at typical velocities of 10,000 km/s (and speeds twice that high have been observed). A speed of 20,000 km/s corresponds to about 44 million mi/h, truly an indication of unimaginably great cosmic violence.

The Supernova Giveth and the Supernova Taketh Away

As most of the material that was once part of the star is blown into space by the supernova explosion, the life of a massive star comes to an end. But the death of each massive star is an important event in the history of its galaxy. The elements built up by fusion during the star's life are now "recycled" into space by the explosion, making them available (later) to form new stars and planets. Both because more of the star's material is lost into space, and because the ejection is so much more violent, the death of a massive star is a much more efficient recycler of newly minted starstuff than the more gentle death of a low-mass star. (To be fair, however, we should recall that there are a lot more low-mass stars than the kind of high-mass stars that make supernovae.)

The idea that supernova material becomes an important part of later generations of stars is not merely poetic speculation. We saw in Chapter 20 that the outward-moving shock of a supernova explosion compresses interstellar clouds and thus can help start the star-forming process. There is evidence in the detailed composition of ancient meteorites that the formation of our own solar system was triggered by a supernova in our neighborhood about 5 billion years ago.

But the supernova explosion has one further creative contribution to make, one we alluded to in the preceding chapter when we asked where the atoms in your jewelry came from. The supernova explosion produces a flood of energetic neutrons that barrel through the expanding material. These neutrons can be absorbed by iron and other nuclei, where they can turn into protons. Thus they build up the elements that are more massive than iron, including such terrestrial favorites as gold and silver. (This is the only place we know where such atoms as gold or uranium can be made.)

When supernovae explode, these elements (as well as the ones the star made during more stable times) are ejected into the existing gas between the stars, and mixed with it. Along with planetary nebulae, supernovae play a major role in building up the supply of chemical elements in the universe (Figure 22.9). Without them neither the authors nor the readers of this book would exist.

Furthermore, supernovae are thought to be the source of the high-energy *cosmic-ray* particles discussed in Section 19.5. Trapped by the magnetic field of the Galaxy, the particles from exploded stars continue to circulate around the vast spiral of the Milky Way. Scientists speculate that high-speed cosmic rays hitting the genetic material of Earth organisms over billions of years may have contributed to the steady *mutations*—subtle changes in the genetic code—that drive the evolution of life on our planet.

But, like so many creative events, supernovae also have a dark side. Suppose a life-form has the misfortune to develop around a star that happens to be close to a massive star destined to become a supernova. Such life-forms may find themselves snuffed out when the harsh radiation and high-energy particles from the explosion reach their world. If, as some astronomers speculate, life can develop on many planets around long-lived (lower-mass) stars, the suitability of that life's own star and planet may not be all that matters for its long-term evolution and survival. Life may well have formed around a number of

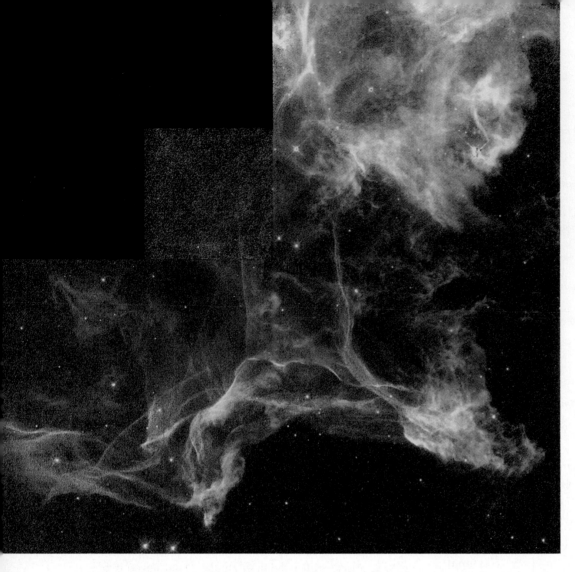

Figure 22.9
An image of a small portion of the old supernova remnant known as the Cygnus Loop, taken with the Hubble Space Telescope. The glowing gases mark the edge of a bubble-like blast wave produced by a stellar explosion that occurred about 15,000 years ago and is now approximately 130 LY wide. Here we see a region where the blast, moving from left to right, has hit a slightly denser cloud of interstellar gas and is causing it to glow. The image was taken through three different color filters: blue light shows doubly ionized oxygen atoms that glow with the energy of the blast-wave's passing; red light shows singly ionized sulfur atoms that have cooled off a bit since the shock passed them; and green shows light emitted by hydrogen atoms immediately behind the shock. In this way the supernova blast is mixing its atoms with those already in interstellar space, and compressing gas to help the formation of new generations of stars. (J. Hester/NASA)

pleasantly stable stars only to be wiped out because a nearby star went supernova.

What is a safe distance to be from a supernova explosion? A lot depends on the violence of the particular explosion, what type of supernova it is (see Section 22.4), and what level of destruction we are willing to accept. Calculations suggest that a supernova less than 50 LY away from us would certainly end all life on Earth, and that even one 100 LY away would have drastic consequences for the radiation levels here.

The good news is that there are at present no massive stars that promise to become supernovae within 50 LY of the Sun. (This is in part because the kinds of massive stars that become supernovae are overall quite rare.) The closest massive star to us, Spica (in the constellation of Virgo), is about 260 LY away, probably a safe distance.

Our ideas about supernovae are provocative, awe-inspiring, and occasionally even disturbing. What evidence do we have that supernova explosions really do have these consequences? Our best information about the vast energies involved comes from an event that was observed in 1987.

Supernova 1987A

Before dawn on February 24, 1987, Ian Shelton, a Canadian astronomer working at an observatory in Chile, pulled a photographic plate from the developer. Two nights earlier he had begun a survey of the Large Magellanic Cloud, a small galaxy that is one of the Milky Way's nearest neighbors in space. On his plate Shelton examined the Tarantula Nebula, a region of bright glowing gas where star formation is occurring. Nearby, where there should have been only faint stars, he saw a large bright spot. Concerned at first that his photograph was flawed, Shelton went outside to look at the Large Magellanic Cloud—and saw that a new object had indeed appeared in the sky (Figure 22.10). He soon realized he had discovered a supernova—one that could be seen with the unaided eye, despite the fact that it was 160,000 LY away.

Now known as SN 1987A, since it was the first supernova discovered in 1987, this brilliant newcomer to the southern sky gave astronomers their first opportunity to study the death of a relatively nearby star with modern instruments. Soon the world of astronomy was abuzz with

K, and the central density was about 5 g/cm^3, or about five times the density of water.

By the time hydrogen was exhausted at the star's center, a helium core of about six times the mass of the Sun had developed, and hydrogen fusion was proceeding in a shell surrounding this core. The core contracted and grew hotter until it reached a temperature of 170 million K and a density of 900 g/cm^3, at which time helium began to fuse to form carbon and oxygen. The surface of the star expanded to a radius of about 100 million km, a bit less than the distance from the Earth to the Sun. The star's luminosity nearly doubled, to 100,000 L$_{Sun}$, and it became a red supergiant. While in this stage the star lost some of its mass. This material has been detected by observations with the Hubble Space Telescope (Figure 22.11) because the light of the supernova flash is reflected from it.

Helium fusion lasted for only about 1 million years, forming a core of carbon and oxygen with a mass about four times that of the Sun. When helium was exhausted at the center of the star, the core contracted again, the surface also decreased in radius, and the star became a blue

Figure 22.10
Before-and-after pictures of the field around Supernova 1987A in the Large Magellanic Cloud. An arrow points to the star that exploded. It's easy to put such an arrow there *after* the explosion; astronomers can only wish such arrows appeared ahead of time so we could know which star was next. The difference in image quality between these pictures is an effect of the Earth's atmosphere, which was steadier when the plates used to make the pre-supernova picture were taken. (Anglo-Australian Telescope Board)

the news, and all available telescopes on Earth and in space were turned to the supernova. Because Shelton (and several other observers that same evening) had caught the star just as it was going off, theorists could at last test their detailed calculations of how massive stars die with data gathered at many different wavelengths.

By combining theory and observation, astronomers have reconstructed the life story of the star that became SN 1987A. Formed about 10 million years ago, it originally had a mass of about 20 M$_{Sun}$. For 90 percent of its life it lived quietly on the main sequence, converting hydrogen to helium. At this time its luminosity was about 60,000 times that of the Sun (L$_{Sun}$), and its spectral type was O. The temperature in its core was about 40 million

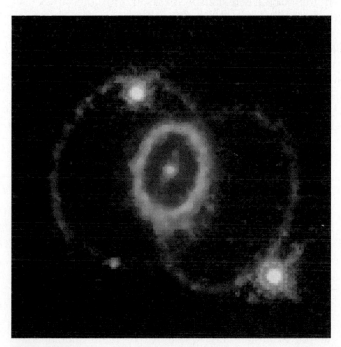

Figure 22.11
This image, taken with the repaired Hubble Space Telescope, shows (in yellow) a ring of stellar material ejected by the star that ultimately became SN 1987A. This material was ejected long before the supernova explosion and is now visible because the light of the supernova is reflected from it. Within 100 more years the expanding debris from the supernova, the elongated yellow blob in the center of the image, will plow into the ring and tear it apart. The outer rings (which are probably not in the same plane as the inner one) may or may not be earlier material released by the star; their origin is a source of great debate among astronomers. The blue stars are not associated with the supernova. (C. Burrows and NASA)

supergiant with a luminosity still about equal to 100,000 L_{Sun}. This is what it still looked like on the outside when it exploded. When the contracting core reached a temperature of 700 million K and a density of 150,000 g/cm^3, nuclear reactions began to convert carbon to neon, sodium, and magnesium. This phase lasted only about 1000 years.

The core, having exhausted carbon as a fuel, again contracted and heated—this time to a temperature of 1.5 billion K and a density of 10 million g/cm^3. The next stages in fusion, the conversion of neon to oxygen and magnesium, and then oxygen to silicon and sulfur, lasted for only a few years. At a temperature of 3.5 billion K and a density of 10^8 g/cm^3, iron began to form. Once iron was created, and the mass of the core exceeded 1.4 times that of the Sun, the collapse began. It was a catastrophic collapse, lasting only a few tenths of a second; the speed of infall in the outer portion of the iron core reached 70,000 km/s, about a fourth of the speed of light.

In the meantime, the outer shells of neon, helium, and hydrogen in the star did not yet know about the collapse. Information about the physical movement of different layers travels through a star at the speed of sound and cannot reach the surface in the few tenths of a second required for the core collapse to occur. Thus the surface layers of our star hung briefly suspended, much like a cartoon character that dashes off the edge of a cliff and hangs momentarily in space before realizing that he is no longer held up by anything.

The collapse of the core continued until the densities rose to several times that of an atomic nucleus. The resistance to further collapse then became very great, and the core rebounded. Infalling material ran into the "brick wall" of the rebounding core and was thrown outward with a great shock wave. Neutrinos poured out of the core, helping the shock wave blow the star apart. The shock reached the surface of the star a few hours later, and the star began to brighten into the supernova Ian Shelton observed.

Testing Our Theories

If this scenario (and all we have said about supernovae in general) is correct, we can make some predictions about the behavior of SN 1987A. If neutrinos play a crucial role in the explosion, vast quantities of them should come pouring out of the supernova even before the light emerges. We will examine the dramatic success of this prediction in the next section. In addition, the explosion should be rich in heavy elements just produced by the neutron-enrichment process described earlier. Even radioactive elements, ones that are unstable and decay pretty quickly into other elements, should be detectable in a supernova, since they will have *just formed*. (See Section 6.3 for a discussion of radioactivity.)

Figure 22.12 shows how the brightness of SN 1987A changed with time. In a single day the star soared in brightness by a factor of about 1000 and became just visi-

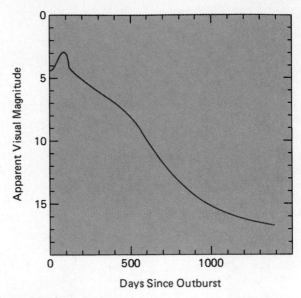

Figure 22.12
A graph that shows how the brightness of SN 1987A changed with time. (Courtesy N. Suntzeff/CTIO)

ble without a telescope. The star then continued to increase slowly in brightness until it was about the same apparent magnitude as the stars in the Little Dipper. Up until about day 40 after the outburst, the energy being radiated away was produced by the explosion itself. But what happened after that helped confirm our ideas about heavy element production. As the energy of the supernova explosion is radiated into space, we might expect the light from the explosion to fade away. But instead, SN 1987A remained bright, as a new source of energy from newly created radioactive elements came into play.

One of the elements formed in a supernova explosion is radioactive nickel, with an atomic mass of 56 (that is, the total number of protons plus neutrons in its nucleus is 56). Nickel-56 is unstable and changes spontaneously, with a half-life of about 6 days, to cobalt-56. It in turn decays with a half-life of about 77 days to iron-56, which is stable. Energetic gamma rays are emitted when these radioactive nuclei decay; they then serve as a new source of energy for the expanding layers. The gamma rays are absorbed in the overlying gas and re-emitted at visible wavelengths, keeping the remains of the star bright.

As you can see in Figure 22.12, astronomers did observe brightening due to radioactive nuclei in the first few months following the supernova's outburst, and then saw the extra light die away as more and more of the radioactive nuclei decayed to stable iron. The gamma-ray heating was responsible for virtually all of the radiation detected from SN 1987A after day 40. Some gamma rays also escaped directly without being absorbed. These were detected by Earth-orbiting telescopes at the wavelengths expected for the decay of radioactive nickel and cobalt, confirming our theory that new elements were formed in the crucible of the supernova.

Neutrinos from SN 1987A

If there had been any human observers in the Large Magellanic Cloud 160,000 years ago, the explosion we call SN 1987A would have been a brilliant spectacle in their skies. Yet we now know that less than 1/10 of 1 percent of the energy of the explosion appeared as visible light. About 1 percent of the energy was required to disrupt the star, and the rest was carried away by neutrinos. The overall energy in these neutrinos was truly astounding. In the first second, their total luminosity was 10^{46} W, which exceeds the luminosity of all the stars in all the galaxies in the part of the universe that we can observe. And the supernova generated this energy in a volume less than 50 km in diameter! Supernovae are by far the most violent events in the universe.

One of the most exciting results from observations of SN 1987A is that physicists actually detected these supernova neutrinos, thereby obtaining strong confirmation that the theoretical calculations of what happens when a star explodes are actually correct. The neutrinos were detected by two instruments, which might be called "neutrino telescopes," about 3 hours before the brightening of the star was first observed. (This is because the neutrinos get out of the exploding star more easily than light does.) Both neutrino telescopes, one in a deep mine in Japan and the other under Lake Erie, consist of several thousand tons of purified water surrounded by several hundred light-sensitive detectors. Incoming neutrinos interact with the water to produce positrons and electrons, which move rapidly through the water and emit deep blue light (Figure 22.13).

The Japanese system detected 11 neutrino events over an interval of 13 s, and the instrument beneath Lake Erie measured 8 events at the same time. Since the neutrino telescopes were in the Northern Hemisphere, and the supernova occurred in the Southern Hemisphere, the detected neutrinos had already passed through the Earth and were on their way back out into space when they were captured!

Only a few neutrinos were detected because the probability that they will interact with ordinary matter is very, very low. It is estimated that the supernova actually released 10^{58} neutrinos. About 50 billion of these passed through every square centimeter on the Earth, and about a million people experienced a neutrino interaction within their bodies. This interaction happened to only a single nucleus in each person, and thus had absolutely no biological effect; it went completely unnoticed by everyone concerned.

Since the neutrinos come directly from the heart of the supernova, their energies provide a measure of the temperature of the core as the star was exploding. The central temperature was about 200 billion K, a stunning figure to which no earthly analogy can bring much meaning. With neutrino telescopes we are peering into the final moment in the life stories of massive stars, and observing conditions beyond all human experience. Yet we are also seeing the unmistakable hints of our own origins.

22.3

Pulsars and the Discovery of Neutron Stars

After the supernova explosion fades away, all that is left behind is the neutron star. Neutron stars are the densest

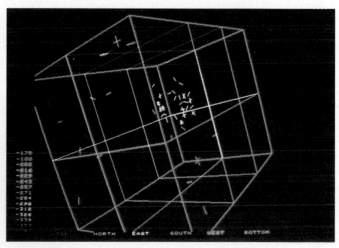

Figure 22.13

(Left) The neutrino detector under Lake Erie has 2048 light-sensitive tubes distributed around a tank that holds 8000 tons of pure water. (Right) A computer-generated display shows one of the neutrino detections on February 23, 1987. The yellow crosses and slashes near the center show which tubes were triggered by the passage of the neutrino. (F. Reines and J. C. van der Velde, IMB Collaboration)

TABLE 22.2
Properties of a Typical White Dwarf and Neutron Star

Property	White Dwarf	Neutron Star
Mass (Sun = 1)	1.0 (always <1.4)	Always > 1.4 and < 3
Radius	5000 km	10 km
Density	5×10^5 g/cm^3	10^{14} g/cm^3

objects in the universe; the force of gravity at their surface is 10^{11} times greater than what we experience at the Earth's surface. The interior of a neutron star is composed of about 95 percent neutrons, with a small number of protons and electrons mixed in. In effect, a neutron star is a giant atomic nucleus, with a mass about 10^{57} times the mass of a proton. Its diameter is more like the size of a small town or an asteroid than a star. (Table 22.2 compares the properties of neutron stars and white dwarfs.) From this description, a neutron star probably strikes you as the object least likely to be observed from hundreds or thousands of light years away.

Yet, to the surprise of astronomers, neutron stars manage to signal their presence across vast gulfs of space. We have found hundreds of them around the Galaxy, and have thus been able to confirm some of the bizarre properties of neutron stars predicted by our theories.

The Discovery of Neutron Stars

In 1967 Jocelyn Bell, a research student at Cambridge University, was studying distant radio sources with a special detector that had been designed and built by her advisor Antony Hewish to find rapid variations in radio signals (Figure 22.14). The project computers spewed out reams of paper, showing where the telescope had surveyed the sky, and it was the job of Hewish's graduate students to go through it all, searching for interesting phenomena. In September 1967 Bell discovered what she called "a bit of scruff"—a strange radio signal unlike anything seen before.

What Bell had found, in the constellation of Vulpecula, was a source of rapid, sharp, intense, and extremely regular pulses of radio radiation. Like the regular ticking of a clock, the pulses arrive precisely every 1.33728 s. Such exactness led the scientists to speculate that perhaps they had found signals from an intelligent civilization. Radio astronomers even half-jokingly dubbed the source "LGM" for "little green men." Soon, however, three similar sources were discovered in widely separated directions in the sky.

When it became apparent that this type of source was fairly common, astronomers concluded that they were highly unlikely to be signals from other civilizations. By today hundreds of such sources have been discovered; they are now called **pulsars,** short for pulsating radio sources. Antony Hewish won the 1974 Nobel Prize in physics for this work and other projects in radio astronomy.

The pulse periods of different pulsars range from a little longer than 1/1000 s to nearly 10 s. At first, the pulsars seemed particularly mysterious because nothing could be seen at their location on visible-light photographs. But then a pulsar was discovered right in the center of one of the best-known supernova remnants—the Crab Nebula. In addition to pulses of radio energy, we can observe pulses of visible light and x rays from the Crab as well (Figure 22.15).

The Crab Nebula is a fascinating object in many ways. Quite separate from the sharp pulses, the whole nebula glows with radiation at many wavelengths, and its overall energy output is over 100,000 times that of the Sun—not a bad trick for the remnant of a supernova that exploded in 1054 A.D., almost a thousand years ago. Can there be a connection between the pulsar and the large energy output of its host nebula?

Figure 22.14
Antony Hewish and Jocelyn Bell. (Courtesy, AIP Emilio Segrè Visual Archives, Weber Collection)

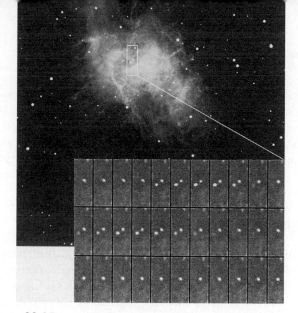

Figure 22.15
A series of photographs of the central part of the Crab Nebula taken by Nigel Sharp at Kitt Peak National Observatory. Note the star that seems to blink on and off: it is the pulsar, with a period of 0.033 s. (National Optical Astronomy Observatories)

A Spinning Lighthouse Model

The Crab pulsar itself emits considerably more energy than the Sun does. Yet that energy arrives in sharp bursts occuring 30 times each second. What type of object can emit such bursts of energy with a regularity that would be the envy of a Swiss watchmaker? As you might imagine, many ideas were proposed to account for the mysterious pulses. But most of them could not come up with a plausi-

ble physical mechanism to explain all the characteristics of the pulses, especially how rapid they are.

By applying a combination of theory and observation, astronomers have now determined that pulsars must be *spinning neutron stars.* Our model for how the pulses are generated involves something like a lighthouse on a rocky coast. To warn ships in all directions and yet not cost too much to operate, the light in a modern lighthouse turns, sweeping its beam across the dark sea. From the vantage point of a ship, you see a pulse of light each time the beam points in your direction. In the same way, something on a neutron star could sweep across the oceans of space, giving us a pulse of radiation each time the beam points at the Earth.

Neutron stars are ideal candidates for such a job because the collapse has made them so small that they can turn very rapidly. Recall the principle of the conservation of angular momentum. Even if the star was rotating very slowly when it was on the main sequence, its rotation had to speed up as it collapsed. Because the star's core shrinks until it is only 10 to 20 km across, it spins in only a fraction of a second. This is just the sort of time period we observe between pulsar pulses.

Any magnetic field that existed in the original star will be highly compressed when the core collapses to a neutron star. At the surface of the neutron star, protons and electrons are caught up in this spinning field and accelerated nearly to the speed of light. In only two places—at the north and south magnetic poles—can the trapped particles escape the strong hold of the magnetic field (Figure 22.16). Note that the magnetic north and south

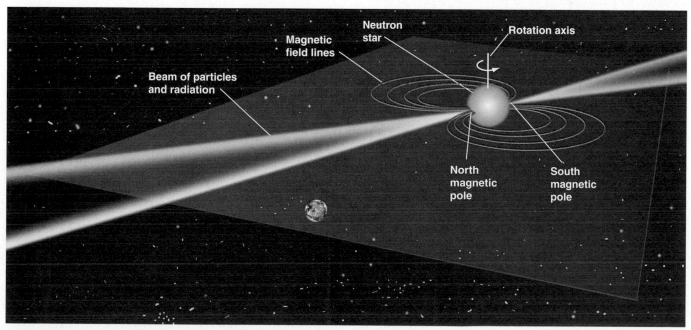

Figure 22.16
A diagram showing how emission at the magnetic poles of a neutron star can give rise to pulses of emission as the star rotates. This model requires that the magnetic poles be located in different places from the rotation poles.

poles do not have to be anywhere close to the north and south poles defined by the star's rotation. (We saw just such a difference between the two kinds of poles in the magnetic fields of Uranus and Neptune.)

At the two magnetic poles, the particles from the neutron star are focused into a narrow beam and come streaming out at enormous speeds, emitting energy over a broad range of the electromagnetic spectrum. The radiation itself is also confined to a narrow beam, explaining why the pulsar acts like a lighthouse. As the rotation carries first one and then the other magnetic pole of the star into our view, we see a pulse of radiation each time.

Proof of the Model

This model of a pulsar caused by beams of radiation from a highly magnetic and rapidly spinning neutron star is a very clever idea. But what evidence do we have that it is the correct model? A few pulsars are members of binary star systems for which enough information is available to calculate their masses using Kepler's laws. These pulsars have masses in the range of 1.4 to 1.8 times that of the Sun, just the sorts of masses that theorists predict neutron stars should have.

But there is an even better line of confirming argument, which brings us back to the Crab Nebula and its vast energy output. When the high-energy charged particles from the neutron star pulsar emerge along the poles of the magnetic field, they eventually hit the slower-moving material from the supernova. They energize this material and cause it to "glow" at many different wavelengths—just what we observe from the Crab Nebula. In a sense, the pulsar beams are a power source that renews the nebula long after the initial explosion of the star that made it.

Who pays the bills for all the energy we see coming out of a remnant like the Crab Nebula? After all, when energy emerges from one place, it must be depleted in another. The ultimate energy source in our model is the rotation of the neutron star, which propels charged particles outward and spins its magnetic field at enormous speeds. If our idea is correct, then as its energy is used to excite the Crab Nebula year after year, the pulsar inside the nebula will have to slow down because of lost energy. As it slows, the pulses eventually come a little less often; more time will elapse before the slower neutron star brings its beam back around.

While the Crab pulsar at first seemed rock-steady in the pacing of its pulses, several decades of careful observations have now clearly shown that they are in fact slowing down. Having measured how much the pulsar is slowing down, we can calculate how much rotation energy the neutron star is losing. It turns out to be the same as that emerging from the Crab Nebula in all its different forms of radiation. In other words, the rotating neutron star model can explain precisely why the Crab Nebula is glowing with the amount of energy we observe.

The Evolution of Pulsars

From observations of the several hundred pulsars discovered so far, astronomers have concluded that one new pulsar is born somewhere in the Galaxy every 25 to 100 years, the same rate at which supernovae are estimated to occur. Calculations suggest that the typical lifetime of a pulsar is about 10 million years; after that the neutron star no longer rotates fast enough to produce significant beams of particles and energy, and it is then no longer observable. We estimate that there are about 100 million neutron stars in our Galaxy.

According to present ideas, the Crab pulsar is rather young (only about 900 years old) and has a short period, while the other, older pulsars have already slowed to longer periods. Pulsars thousands of years old have lost too much energy to emit appreciably in the visible and x-ray wavelengths, and are observed only as radio pulsars; their periods are a second or more.

There is one other reason that we can see only a fraction of the pulsars in the Galaxy. Consider our lighthouse model again. On Earth, all ships approach on the same plane—the surface of the ocean—so the lighthouse can be built to sweep its beam over that surface. But in space, objects can be anywhere in three dimensions. As a given pulsar's beam sweeps over a circle in space, there is absolutely no guarantee that this circle will include the direction of Earth. In fact, if you think about it, many more circles in space will *not* include the Earth. Thus we estimate that we are unable to observe a large number of neutron stars whose beams miss us entirely.

At the same time, it turns out that only 3 of more than 400 pulsars discovered so far are embedded in the visible clouds of gas that mark the remnant of a supernova (Figure 22.17). This might at first seem mysterious, since we know that supernovae give rise to neutron stars and that we should expect each pulsar to have begun its life in a supernova explosion. But the lifetime of a pulsar turns out to be about 100 times longer than the length of time required for the expanding gas of a supernova remnant to disperse into interstellar space. Thus most pulsars are found with no other trace left of the explosion that produced them.

22.4

The Evolution of Binary Star Systems

The discussion of the life stories of stars presented so far has suffered from a bias—what we might call single-star chauvinism. Because the human race developed around a star that goes through life alone, we often tend to think of most stars in isolation. But as we saw in Chapter 17, it now appears that as many as half of all stars may develop in binary systems—those in which two stars are born enveloped in each other's gravitational embrace and go through life orbiting a common center of mass.

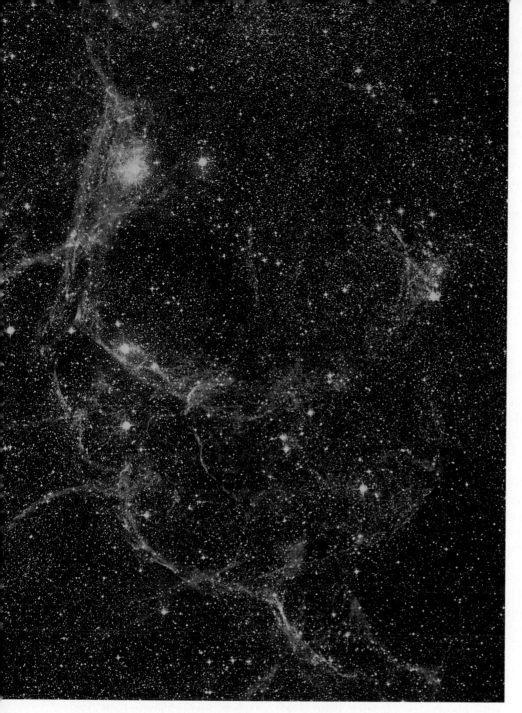

Figure 22.17
Here we see a small part of another (older) supernova remnant at whose center we can observe a pulsar. Located in the constellation of Vela, the supernova exploded over 10,000 years ago, and as a result the pulsar spins more slowly than the one in the Crab. The full remnant, most of which is not seen even on this wide-angle image, covers over 10° on the sky and is quite faint; special image-enhancement techniques were required to reveal the filaments we see. (© 1979 Royal Observatory, Anglo–Australian Telescope Board)

For these stars, the presence of a close-by companion can have a profound influence on their evolution. Stars can, under the right circumstances, exchange material, especially during the stages when one of them swells up into a giant or supergiant, or has a strong wind. When this happens and the companion stars are sufficiently close, material can flow from one star to another, decreasing the mass of the donor and increasing the mass of the recipient. Such *mass transfer* can be especially dramatic when the recipient is a stellar remnant such as a white dwarf or neutron star. While the detailed story of how such binary stars evolve is beyond the scope of our book, we do want to mention a few examples of how the stages described in this chapter may change when there are two stars in a system.

White Dwarf Explosions: The Mild Kind

Let's consider a system of two stars, in which one has become a white dwarf and the other is gradually transferring material onto it. As fresh hydrogen from the outer layers of its companion accumulates on the surface of the hot white dwarf, it begins to build up a layer of hydrogen. As more and more hydrogen accumulates and heats up on the surface of the degenerate star, the new layer eventually reaches a temperature that causes fusion to begin in a sudden, explosive way, blasting much of the new material away. The white dwarf quickly (but only briefly) becomes quite bright. To observers before the invention of the telescope, it seemed that a new star suddenly appeared, and they called it a **nova.** Novae fade away in a few months to a few years.

Hundreds of novae have been observed, each of them occurring in a binary star system, and each later showing a shell of expelled material. A number of stars have more than one nova episode, as more material accumulates on the white dwarf and the whole process repeats. As long as the episodes do not increase the mass of the white dwarf beyond the Chandrasekhar limit (by transferring too much mass), the white dwarf itself remains pretty much unaffected by the explosions on its surface.

White Dwarf Explosions: The Violent Kind

If a white dwarf accumulates matter from a companion star at a much faster rate, it can be pushed over the limit. With its mass now exceeding 1.4 times the mass of the Sun, such an object can no longer support itself as a white dwarf, and it begins to collapse. As it does so it heats up, and new nuclear reactions begin in the degenerate core. In less than one second an enormous amount of fusion takes place; the energy released is so great that it completely destroys the white dwarf. Gases are blown out into space at velocities of several thousand kilometers per second, and no central star remains behind.

Such an explosion is also called a supernova, since, like the destruction of a high-mass star, it can produce a huge amount of energy in a very short time, destroying the star in the process. We call these Type I supernovae (distinguished from the Type II's discussed earlier). Tycho's Supernova (see "Making Connections" box) appears to have been caused by such a white dwarf with an overly generous companion. (Very recent observations suggest that some mechanism allows white dwarfs with masses less than 1.4 M_{Sun} in binary systems to explode violently as well, although astronomers do not yet have a detailed model of how this happens.)

Neutron Stars with Companions

It is possible that a binary system can survive the explosion of one of its members. In that case, an ordinary star can share a system with a neutron star. If material is transferred from the "living" star to its "dead" (and highly compressed) companion, this material will be pulled in by the strong gravity of the neutron star. Such infalling gas will be compressed and heated to incredible temperatures. It will quickly become so hot that it will experience an explosive burst of fusion. The energies involved are so great that we would expect much of the radiation from such a burst to emerge as x rays and gamma rays.

And indeed, astronomers using high-energy observatories above the Earth's atmosphere (see Chapter 5) have recorded many objects that undergo just these types of x-ray and gamma-ray bursts. While there is much debate about the nature of the most energetic *bursters,* astronomers are reasonably confident that at least some of the x-ray bursts do come from material that falls onto the surface of a neutron star.

If the neutron star and its companion are positioned the right way, a significant amount of material can be transferred to the neutron star. If the neutron star is a pulsar, the new material can actually set it spinning faster (as spin energy is also transferred). Astronomers have found several pulsars spinning at a rate of almost *1000 times per second!* Such a rapid spin must have been externally caused. (Recall that the Crab Nebula pulsar, one of the youngest pulsars known, was spinning 30 times per second.) Indeed, some of the fast pulsars are clearly part of binary systems, while others may be alone only because they have consumed their former companions completely sometime in the past.

With this gruesome example, we have reached the end of our description of the final states of stars. Yet one piece of the story remains to be filled in. We saw that stars whose masses are less than 1.4 M_{Sun} at the time they run out of fuel end their lives as white dwarfs. Dying stars with masses between 1.4 and about 3 M_{Sun} become neutron stars. But, there are stars whose masses are greater than 3 M_{Sun} when they exhaust their fuel supplies; what becomes of them? The truly bizarre result of the death of such massive stellar cores is the subject of our next chapter.

Summary

22.1 During the course of their evolution, stars shed their outer layers and lose a significant fraction of their initial mass. Stars with masses up to about 10 M_{Sun} can lose enough mass to become *white dwarfs*, which have masses less than the **Chandrasekhar limit** (about 1.4 M_{Sun}). A typical white dwarf has a mass about the same as that of the Sun, and a diameter comparable to that of the Earth. The pressure exerted by **degenerate electrons** keeps white dwarfs from contracting to still smaller diameters. **Planetary nebulae** are shells of gas ejected by such stars, set to glowing by the ultraviolet radiation of the collapsing core. Eventually, white dwarfs cool off to become black dwarfs, stellar remnants made mainly of carbon and oxygen.

22.2 In a massive star, hydrogen fusion in the core is followed by several other fusion reactions involving heavier elements. Just before it exhausts all sources of energy, a massive star has an iron core surrounded by shells of silicon and sulfur, oxygen, neon, carbon, helium, and hydrogen. The fusion of iron requires energy (rather than releasing it). If the mass of a star's iron core exceeds the Chandrasekhar limit (but is less than 3 M_{Sun}), the core collapses until its density exceeds that of an atomic nucleus, forming a **neutron star** with a typical diameter of 20 km. The core rebounds and transfers energy outward, blowing off the outer layers of the star in a Type II **supernova** explosion. Studies of Supernova 1987A, including the detection of neutrinos, have confirmed theoretical calculations of what happens during such explosions.

22.3 At least some supernovae leave behind a highly magnetic, rapidly rotating neutron star, which can be observed as a **pulsar** if its beam of escaping particles and focused radiation is pointing toward us. Pulsars emit rapid pulses of radiation at regular intervals; their periods are in the range of 0.001 to 10 s. The rotating neutron star acts like a lighthouse, sweeping its beam in a circle. Pulsars lose energy as they age, the rotation slows, and their periods increase.

22.4 When a white dwarf or neutron star is a member of a close binary star system, its companion star can transfer mass to it. Material falling gradually onto a white dwarf can explode in a sudden burst of fusion and make a **nova.** If material falls rapidly onto a white dwarf, it can push it over the Chandrasekhar limit and cause it to explode completely as a Type I supernova. Material falling onto a neutron star can cause powerful bursts of x-ray and gamma-ray radiation.

Review Questions

1. How does a white dwarf differ from a neutron star? How does each one form? What keeps each from collapsing under its own weight?

2. Describe the evolution of a star with a mass like that of the Sun, from the main-sequence phase of its evolution until it becomes a white dwarf.

3. Describe the evolution of a massive star (say 20 times the mass of the Sun) up to the point at which it becomes a supernova. How does the evolution of a massive star differ from that of the Sun? Why?

4. How do the two types of supernovae discussed in this chapter differ? What kind of star gives rise to each type?

5. A star begins its life with a mass of 5 M_{Sun} but ends its life as a white dwarf with a mass of 0.8 M_{Sun}. List the stages in the star's life during which it most likely lost some of the mass it started with. How did mass loss occur in each stage?

6. If the formation of a neutron star leads to a supernova explosion, explain why only three out of the hundreds of known pulsars are found in supernova remnants.

7. How can the Crab Nebula shine with the energy of something like 100,000 Suns when the star that formed the nebula exploded almost 1000 years ago? Who pays the bills for much of the radiation we see coming from the nebula?

8. How is a nova different from a Type I supernova? How does it differ from a Type II supernova?

Thought Questions

9. You observe an expanding shell of gas through a telescope. What measurements would you make to determine whether you have discovered a planetary nebula or the remnant of a supernova explosion?

10. Arrange the following stars in order of age:
    a. A star with no nuclear reactions going on in the core, which is made primarily of carbon and oxygen.

b. A star of uniform composition from center to surface; it contains hydrogen but has no nuclear reactions going on in the core.

c. A star that is fusing hydrogen to form helium in its core.

d. A star that is fusing helium to carbon in the core, and hydrogen to helium in a shell around the core.

e. A star that has no nuclear reactions going on in the core, but is fusing hydrogen to form helium in a shell around the core.

11. Would you expect to find any white dwarfs in the Orion Nebula? (See Chapter 20 to remind yourself of its characteristics.) Why or why not?

12. Suppose no stars more massive than about 2 M_{Sun} had ever formed. Would life as we know it have been able to develop?

13. Would you be more likely to observe a Type II supernova (the explosion of a massive star) in a globular cluster or in an open cluster? Why?

14. Astronomers believe there are something like 100 million neutron stars in the Galaxy. Yet we have only found about 400 or so pulsars. Give several reasons that these numbers are so different. Explain each reason.

15. Would you expect to observe *every* supernova in our own Galaxy? Why or why not?

16. The Large Magellanic Cloud has about one-tenth the number of stars found in our own Galaxy. Suppose the mix of high- and low-mass stars is exactly the same in both galaxies. Approximately how often does a supernova occur in the Large Magellanic Cloud?

17. Look at the list of the nearest stars in Appendix 10. Would you expect any of these to become supernovae?

18. If most stars become white dwarfs at the ends of their lives, and the formation of white dwarfs is accompanied by the production of a planetary nebula, why don't we see many more planetary nebulae in the Galaxy than we do?

Problems

19. The gas shell of a particular planetary nebula is expanding at the rate of 20 km/s. Its diameter is 1 LY. Find its age. For this calculation, assume that there are 3×10^7 s/yr and 10^{13} km/LY.

20. Prepare a chart or diagram that exhibits the relative sizes of a typical red giant, the Sun, a typical white dwarf, and a neutron star of mass equal to the Sun's. You may have to be clever to devise such a diagram.

21. Suppose the central star of a planetary nebula is 16 times as luminous and 20 times as hot (about 110,000 K) as the Sun. Find its radius in terms of the Sun's. Compare this radius with that of a typical white dwarf.

22. The luminosity of a supernova at maximum light is about 10^{10} L_{Sun}. The Sun would be just barely visible without a telescope at a distance of 10 parsecs (32.6 LY). How far away is it possible to see a supernova with the unaided eye?

23. Suppose the radius of the pulsar in the Crab Nebula is 10 km/s, and its rotation period is 0.033 s. What is the rotational velocity of the star at its equator?

Suggestions for Further Reading

See some of the references for the last chapter, plus:

Balick, B. et al. "The Shaping of Planetary Nebulae" in *Sky & Telescope*, Feb. 1987, p. 125.

Bethe, H. and Brown, G. "How a Supernova Explodes" in *Scientific American*, May 1985, p. 60.

Chapman, C. and Morrison, D. *Cosmic Catastrophes*. 1989, Plenum Press. Chapter 18 is on supernovae.

Filippenko, A. "A Supernova with an Identity Crisis" in *Sky & Telescope*, Dec. 1993, p. 30. Good review of supernovae in general, and of 1993J in M81.

Graham-Smith, F. "Pulsars Today" in *Sky & Telescope*, Sep. 1990, p. 240.

Greenstein, G. "Neutron Stars and the Discovery of Pulsars" in *Mercury*, Mar./Apr. 1985, p. 34; May/June 1985, p. 66.

Kaler, J. "The Smallest Stars in the Universe" in *Astronomy*, Nov. 1991, p. 50. On white dwarfs, neutron stars, and pulsars.

Kaler, J. "Realm of the Hottest Stars" in *Astronomy*, Feb. 1990, p. 22. On planetaries and their central stars.

Kawaler, S. and Winget, D. "White Dwarfs: Fossil Stars" in *Sky & Telescope*, Aug. 1987, p. 132.

Kirshner, R. "Supernova: The Death of a Star" in *National Geographic*, May 1988, p. 618. Excellent introduction for beginners on SN 1987A.

Marschall, L. *The Supernova Story*, 2nd ed. 1994, Princeton U. Press. The introduction of choice to supernovae and SN 1987A.

Soker, N. "Planetary Nebulae" in *Scientific American*, May 1992, p. 78.

Tierney, J. "The Quest for Order: Profile of S. Chandrasekhar" in *Science '82*, Sep. 1982, p. 68.

Wallerstein, G. and Wolff, S. "The Next Supernova" in *Mercury*, Mar./Apr. 1981, p. 44.

1. To see which planetary nebulae might be visible tonight through binoculars or a small telescope, set the *Stellar Limiting Magnitude* to 5 and the *Deep-Sky Filter* to a faint magnitude of 10, with all objects off except for planetary nebulae.

2. Set the *Planetary Nebula* faint magnitude to 9 and count the number of nebulae. The number of stars visible to this limiting magnitude is roughly 250,000. Because most of these stars will go through the planetary nebula phase, the ratio of these two numbers gives an idea of the length of the planetary nebula phase.

Assuming the average star has a lifetime of 5 billion years, what is the length of the planetary nebula phase?

3. Using Table 22.1 and the masses for the stars found in *RedShift* exercise 21.1, determine which of these stars will end their lives as white dwarfs and which may end as neutron stars:

a. Beta Canes Venaticorum
b. Epsilon Eridani
c. Alpha Pictoris
d. Gamma Cassiopeiae
e. Procyon, Alpha Canis Minoris
f. Theta Ursae Majoris
g. Menkalinan, Beta Aurigae
h. Tau Herculis
i. Eta Cephei
j. Mu Herculis

The Hubble Space Telescope discovered this spiral-shaped disk of hot gas in the core of the galaxy known as M87. Measurements of the Doppler shift of the lines in its spectrum show that the material in the disk must be rotating at speeds of 550 km/s (about 2 million km/h). Such a high orbital velocity indicates that there must be about three billion solar masses concentrated at the center of the disk. So much mass in a very small volume must almost certainly be in the form of a black hole. The image also shows a jet of high-speed electrons emitted from the vicinity of the black hole. (H. Ford et al. and NASA)

CHAPTER 23

Black Holes and Curved Spacetime

Thinking Ahead

Most of us think that time is like the flow of a river with strong current. Much like twigs in the water, we seem caught in the flow, moving together from the past into the future. Einstein showed, however, that the flow of time is not uniform for every observer in the universe; depending on the strength of gravity, the pace can vary. What is the flow of time like near the collapsed remnants of the most massive stars, where gravity is very strong?

One thing is certain,

and the rest debate—

Light-rays, when near the Sun,

do not go straight.

Arthur S. Eddington (1920)

Most stars end their lives as white dwarfs and neutron stars. When a very massive star collapses at the end of its life, however, not even neutron degeneracy can support the core against its own weight. For a core whose mass is more than about three times that of the Sun (M_{Sun}), our theories predict that *no known force can stop it from collapsing forever!* Gravity simply overwhelms all other forces and crushes the core until it is infinitely small.

A star in which this occurs may become one of the strangest objects ever predicted by theory—a black hole. To understand what a black hole is, we need a theory that can describe the action of gravity under such extreme circumstances. Our best theory of gravity was put forward in 1916 by Albert Einstein and is called the theory of **general relativity.**

Figure 23.1
Albert Einstein has become a symbol for intellect in popular culture. The caption of this Italian ad translates to "Instinct says beer, reason says Carlsberg." (Photo courtesy of the archives, Caltech)

l'istinto dice birra.

la ragione dice Carlsberg.

Carlsberg Birra di Danimarca

General relativity is one of the major intellectual achievements of the 20th century; if it were music, we would compare it to the great symphonies of Beethoven or Mahler. Until recently, however, scientists had little need for a better theory of gravity; Isaac Newton's ideas (see Chapter 2) are perfectly sufficient for most of the objects we deal with. In the past three decades, however, general relativity has become more than just a beautiful idea; it is now essential in understanding pulsars, quasars (which will be discussed in Chapter 26), and many other astronomical objects and events, including black holes.

We should perhaps mention that this is the point in an astronomy course when many students start to feel a little nervous (and perhaps wish they had taken botany or some other earthbound course to satisfy the science requirement). This is because in popular culture, Einstein has become a symbol for mathematical brilliance that is simply beyond the reach of most people (Figure 23.1). So when we mentioned that the theory of general relativity was Einstein's work, you, like many other students, may have shuddered just a bit, convinced that anything Einstein did was beyond your understanding. This popular view is unfortunate and mistaken. While the detailed calculations of general relativity do involve a good deal of higher mathematics, the basic ideas are not difficult to understand (and are, in fact, almost poetic in the way they give us a new perspective on the world).

23.1

Principle of Equivalence

The fundamental insight that led to the formulation of the theory of general relativity starts with a very simple thought: If you were able to jump off a high building and fall freely, you would not feel your own weight. Einstein built on this idea to reach sweeping conclusions about the very fabric of space and time itself. He called it the "happiest idea of my life."

Einstein himself pointed out an everyday example that illustrates this effect. Notice how your weight seems to be reduced in a high-speed elevator when it accelerates from a stop to a rapid descent. Similarly, your weight seems to increase in an elevator that starts to move quickly upward. This effect is not just a feeling you have: if you stood on a scale in such an elevator, you could measure your weight changing (you can actually perform this experiment in some science museums).

In a *freely falling* elevator, with no air friction, you would lose your weight altogether. Near weightlessness can be achieved by taking an airplane to high altitude and then dropping rapidly for a while. This is how NASA trains its astronauts for the experience of free-fall in space; the scenes of weightlessness in the movie *Apollo 13* were filmed in the same way.

A more formal way to state Einstein's idea is the following. Suppose we have a spaceship that contains a windowless laboratory equipped with all the tools needed to perform scientific experiments. One future day, imagine that an astronomer wakes up after a long night celebrating some scientific breakthrough and finds herself sealed into this laboratory. She has no idea how it happened, but notices that she is weightless. This could be because she is at rest, or moving at some steady speed through space, far away from any source of gravity (in which case she has plenty of time to wake up). But it could also be because she is falling freely toward a planet like the Earth (in which case she might first want to check her distance from the surface before making coffee).

What Einstein postulated is that there is *no* experiment she can perform inside the sealed laboratory to determine whether she is floating in space or falling freely in

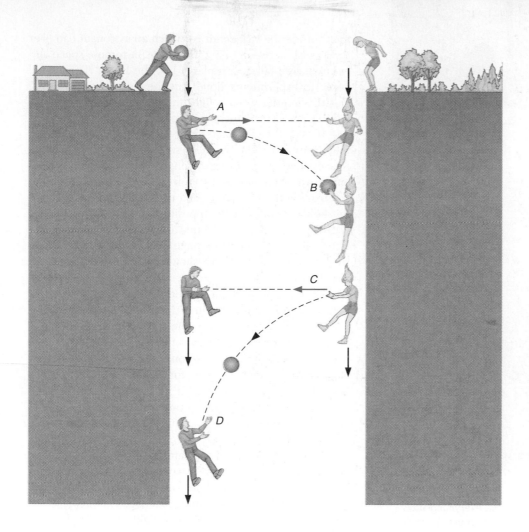

Figure 23.2
A brave couple playing catch as they descend into a bottomless abyss. Since the boy, girl, and ball all fall at the same speed, it appears to them that they can play catch by throwing the ball in a straight line between them. Within their world, there appears to be no gravity.

a gravitational field. As far as she is concerned, the two situations are completely *equivalent*. The idea that life in a freely falling laboratory is indistinguishable from, and hence equivalent to, life with no gravity is called the **equivalence principle.**

Gravity or Acceleration?

To explore the profound implications of this simple idea, let's consider as an example a foolhardy boy and girl who simultaneously jump from opposite banks into a bottomless chasm (Figure 23.2). If we ignore air friction, then we can say that while they fall, they both accelerate downward at the same rate and feel no external force acting on them. They can throw a ball back and forth, always aiming it straight at each other, as if there were no gravity. The ball falls at the same rate they do, so it can always remain in a line between them.

Such a game of catch is very different on the surface of the Earth. Everyone who grows up feeling gravity knows that a ball, once thrown, falls to the ground. Thus, in order to play catch with someone, you must aim the ball upward so that it follows an arc—rising and then falling as it moves forward—until it is caught at the other end.

Now suppose we isolate our freely falling boy, girl, and ball inside in a large box falling with them. No one inside the box is aware of any gravitational force. If the children let go of the ball, it doesn't fall to the bottom of the box or anywhere else, but merely sits there or moves in a straight line, depending on whether it is given any motion.

Astronauts in the Space Shuttle orbiting the Earth live in just such an accelerating environment (Figure 23.3). The Shuttle in orbit is falling freely around the Earth. While in free-fall, the astronauts live in a magical world where there seem to be no gravitational forces. One can give a wrench a shove, and it moves at constant speed across the orbiting laboratory. A pencil set in midair remains there, as if no force were acting on it.

Appearances are misleading however. There *is* a force in this situation. Neither the Shuttle nor the astronauts are *really* weightless, for they continually fall around the Earth, pulled by its gravity. But since all fall together—Shuttle, astronauts, wrench, and pencil—inside the Shuttle all gravitational forces appear to be absent.

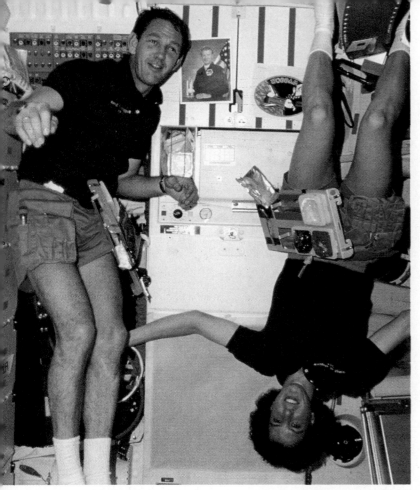

Figure 23.3
When the Space Shuttle is in free-fall in orbit, everything stays put or moves uniformly because there is no apparent gravitation acting inside the spacecraft. (NASA)

Thus the Shuttle provides an excellent example of the principle of equivalence—how local effects of gravity can be completely compensated by the right acceleration. To the astronauts, falling around the Earth creates the same effects as being far off in space, remote from all gravitational influences.

The Paths of Light and Matter

Einstein postulated that the equivalence principle is a fundamental fact of nature, and that there is *no* experi-

ment inside the spacecraft by which an astronaut can ever distinguish between being weightless in remote space and being in free-fall near a planet like the Earth. This would apply to experiments done with beams of light as well. But the minute we use light in our experiments, we are led to some very disturbing conclusions.

It is a fundamental observation from everyday life that beams of light travel in straight lines. Imagine the Space Shuttle is moving through empty space far from any gravity. Send a laser beam from the back of the ship to the front and it will travel in a nice straight line and land on the front wall exactly opposite the point from which it left the rear wall. If the equivalence principle really applies universally, then this same experiment performed in free-fall around the Earth should give us the exact same result.

Imagine that the astronauts again shine a beam of light along the length of their ship. But, as shown in Figure 23.4, when the Shuttle is in free-fall, it falls a bit between the time the light leaves the back wall and the time it hits the front wall. (The amount of the fall is grossly exaggerated in Figure 23.4 for educational purposes.) Therefore, if the beam of light follows a straight line but the ship's path curves downward, then the light should strike the front wall at a point higher than the point from which it left.

However, this would violate the principle of equivalence—the two experiments would give different results. We are thus faced with giving up one of our two assumptions. Either the principle of equivalence is not correct, or light does not always travel in straight lines. As Einstein did, let's select the second one, always keeping in mind that such theoretical choices in science must be validated by testing their predictions against experimental results.

If we say that the principle of equivalence is right, the beam must arrive directly opposite the point from which it started in the ship. Then the light, like the ball the children were throwing back and forth, *must fall with the ship* if that ship is in orbit around the Earth (Figure 23.4). This would make its path curve downward, like the path of the ball, and thus the light would hit the front wall exactly opposite the spot from which it came.

Thinking this over, you might well conclude that it doesn't seem like such a big problem; why can't light fall the way balls do. But, as discussed in Chapter 4, light is profoundly different from balls. Balls have mass, while light does not.

Figure 23.4
In a spaceship moving to the left (in this figure) in its orbit about a planet, light is beamed from the rear, A, toward the front, B. Meanwhile the ship is falling out of its straight path (the amount it falls is exaggerated here). We might therefore expect the light to strike at B', above the target in the ship. Instead, the light, bent by gravity, follows a curved path and strikes at C.

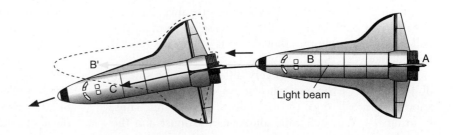

Here is where Einstein's intuition and genius allowed him to make a profound leap. Einstein suggested that the light curves down to meet the front of the Shuttle because the Earth's gravity actually bends the fabric of space and time. This radical idea—which we will explain below—keeps the behavior of light the same in empty space and free-fall, but changes some of our most basic and cherished ideas about space and time. The reason we take Einstein's suggestion seriously is that, as we will see, experiments now clearly show his intuitive leap was correct.

Spacetime and Gravity

Is light actually bent from its straight-line path by Earth's gravity? Einstein preferred to think that it is space and time that are affected by gravity; light, and everything else that travels through space and time, then finds its path affected. As we will see, light always follows the shortest path—but that path may not always be straight. For example, on the curved surface of planet Earth, if you want to fly from Chicago to Rome, the shortest distance is not a straight line but the arc of a *great circle* (which we defined at the beginning of Chapter 3).

To show what Einstein's insight really means, let's first consider how we locate an event in space and time. For example, imagine you have to describe to worried school officials the fire that broke out in your room when your roommate tried making shish kebob in the fireplace. You explain that your dorm is at 6400 College Avenue, which locates the room in a left-right direction; you are on the 5th floor, which tells where you are in the up-down direction; and you are the 6th room back from the elevator, which tells where you are in the forward-backward direction. Then you explain that the fire broke out at 6:23 P.M. (but was soon brought under control). Any event in the universe can thus be pinpointed using the three dimensions of space and the one dimension of time.

Newton considered space and time completely independent, and that continued to be the accepted view until the beginning of the 20th century. But Einstein showed that there was an intimate connection between space and time, and that only by considering the two together—in what we call **spacetime**—can we build up a correct picture of the physical world. We examine spacetime a bit more in the next section.

The gist of Einstein's general theory is that the presence of mass (gravity) curves or warps the fabric of spacetime. When something else—a beam of light, an electron, or the Starship *Enterprise*—enters such a region of distorted spacetime, the path of that something else will be different from what it would have been in the absence of the mass. This idea is often summarized in the following way: matter tells spacetime how to curve, and the curvature of spacetime tells other matter how to move.

The amount of distortion in spacetime depends on the mass involved. Terrestrial objects, such as the book you are reading, have far too little mass to introduce any noticeable distortion. (If, like most people, you have never heard of distorted spacetime, that's the reason; it plays no role in life on Earth. Newton's view of gravity is just fine for building bridges, skyscrapers, or amusement-park rides.) It takes a mass like a star to produce measurable distortions, and a massive star is best. So you see, we *are* eventually going to talk about collapsing stars again, but not before discussing Einstein's ideas (and the evidence for them) in a bit more detail.

Spacetime Examples

How can we understand the distortion of spacetime by the presence of some (significant) amount of mass? Let's try the following analogy. You may have seen maps of New York City that squeeze the full three dimensions of this towering metropolis onto a flat sheet of paper, and still have enough information so tourists will not get lost. Let's do something similar with diagrams of spacetime.

Figure 23.5, for example, shows the progress of a motorist driving to the east on a stretch of road in Kansas where the countryside is absolutely flat. Since our motorist is traveling only in the east-west direction, and the terrain is flat, we can ignore the other two dimensions of space. The amount of time elapsed since he left home is shown on the vertical axis, and the distance traveled eastward is shown on the horizontal axis. From A to B he drove at a uniform speed; unfortunately, it was too fast a uniform speed and a police car spotted him. From B to C

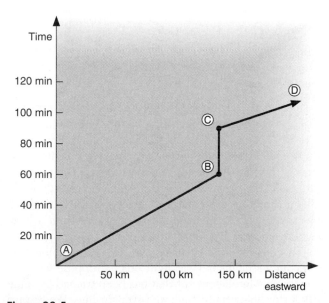

Figure 23.5
The progress of a motorist traveling across the Kansas landscape. Distance traveled is plotted along the horizontal axis. The time elapsed since the motorist left his starting point is plotted along the vertical axis.

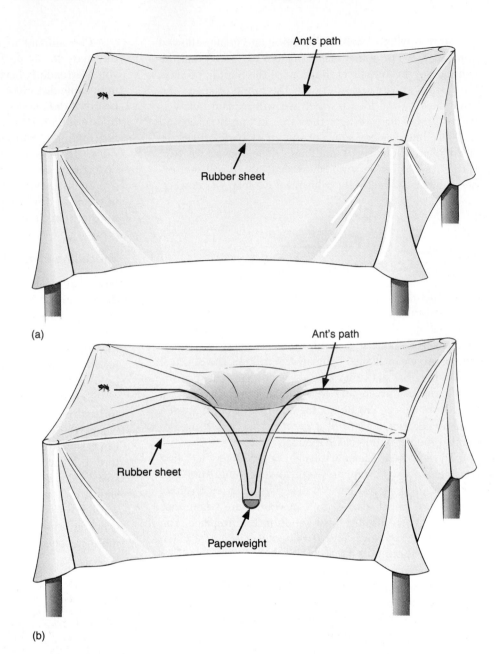

Figure 23.6
On a flat rubber sheet, trained ants have no trouble walking in straight lines. However, when a massive object puts a big depression into the sheet, then ants, who must walk where the sheet takes them, find their paths changed (warped) dramatically.

he stopped to receive his ticket and made no progress through space, only through time. From C to D he drove more slowly because the police car was behind him.

In the same way, we might (in our imaginations) squeeze the four dimensions of spacetime in the universe onto a flat sheet. To show the distortion, however, we will make it a rubber sheet that can stretch or warp if we put objects on it. Let's imagine we borrow one from a local hospital and stretch it taut on four posts.

To complete the analogy, we need something that travels in a straight line (as light does). Suppose we have an extremely intelligent ant that has been trained by some skilled psychologists to walk in a straight line. We reward the ant with a piece of sugar at the end of each run to reinforce the desired behavior.

We begin with just the rubber sheet and the ant, simulating empty space with no mass in it. We put the ant on one side of the sheet and it walks in a beautiful straight line over to the other side (Figure 23.6a). We next put a small grain of sand on the rubber sheet. While the sand does distort the sheet a tiny bit, this is not a distortion that we or the ant can measure. If we send the ant so it goes close to, but not on top of, the sand grain, it has little trouble walking in a straight line and earning its sugary reward.

Now we grab something with a little more mass—say a small pebble. It bends or distorts the sheet just a bit right around its position. If we send the ant into this region, it finds its path slightly altered by the distortion of the sheet. The distortion is not large, but if we follow the ant's path carefully, we notice it deviating slightly from a straight line.

The effect gets more noticeable as we increase the mass of the object we put on the sheet. Let's say we now use a massive paperweight. Such a heavy object distorts or warps the rubber sheet very effectively, putting good sag in it. We again send the ant so its path goes close to, but not on top of, the paperweight (Figure 23.6b). Far away from the paperweight, the ant has no trouble doing its walk.

But as the poor ant gets closer to the paperweight, it finds that the sheet is no longer straight! It has a big sag in it, down into which the ant is forced to go. Even worse, the ant must then climb up the other side of the sag, expending a lot of energy before it can return to walking on an undistorted part of the sheet. All this while, the ant is trying to move in a straight line, knowing that only straight-line walking will earn the reward. But through no fault of its own (after all, ants can't fly, so it has to stay on the sheet) the ant finds its path curved by the distortion of the sheet itself.

In the same way, according to Einstein's theory, we could say that light always "tries" to travel in a straight line through spacetime. But as the masses distorting spacetime get larger and larger, the shortest, most direct paths are no longer straight lines, but curves.

How large does a mass have to be before we can measure the distortion of spacetime? In 1916, when Einstein first proposed his theory, the grain of sand in our analogy might have been the entire Earth. There was, at that time, no way to measure the distortion of spacetime caused by a mass the size of our planet. To get the kind of small but measurable distortion caused by the pebble in our analogy, we needed the mass of the Sun. (We discuss how this effect was measured using the Sun in the next section.)

The paperweight in our analogy might be a white dwarf or neutron star. Around these compact, massive objects the distortion of spacetime is much more noticeable. And when, to return to the situation described at the beginning of the chapter, a star core with more than three times the mass of the Sun collapses forever, the distortions of spacetime can become truly mind-boggling!

23.3
Tests of General Relativity

What Einstein proposed was nothing less than a major revolution in our understanding of space and time. It was a new model of gravity, in which gravity was not seen as a force, but as something that changes the geometry of spacetime. Like all new ideas in science, no matter who advances them, Einstein's theory had to be tested by comparing its predictions against the experimental evidence. This was no easy matter, because the effects of the new view of gravity were apparent only when the mass was quite large. (For smaller masses, it required measuring techniques that would not become available until decades later.)

When the distorting mass is small, the predictions of general relativity must agree with those of Newton's the-

ory, which, after all, has served us admirably in our technology and in guiding space probes to the other planets. In familiar territory, therefore, the differences between predictions of the two theories are subtle and difficult to detect. Nevertheless, Einstein was able to demonstrate one proof of his theory that could be found in existing data, and to suggest another one that would be tested only a few years later.

The Motion of Mercury

Of the planets in our solar system, Mercury orbits closest to the Sun, and is thus located closest to where the Sun's gravity distorts spacetime. Einstein wondered if the distortion might produce a noticeable difference in the motion of Mercury that was not predicted by Newton's theory. It turned out that the difference was subtle but it was definitely there. Most important, it had already been measured.

Mercury has a highly elliptical orbit, so that it is only about two-thirds as far from the Sun at perihelion as it is at aphelion. It can be calculated that the gravitational effects (perturbations) of the other planets on Mercury should produce an advance of its perihelion. What this means is that each successive perihelion occurs in a slightly different direction as seen from the Sun (Figure 23.7).

According to Newtonian theory, the gravitational forces exerted by the planets will cause Mercury's perihelion to advance by about 531 arcsec per century. In the last century, however, it was observed that the actual advance is 574 arcsec per century. The discrepancy was first

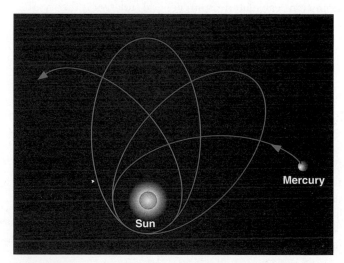

Figure 23.7
The major axis of the orbit of a planet, such as Mercury, rotates in space because of various perturbations. In Mercury's case the amount of rotation is larger than can be accounted for by the gravitational forces exerted by other planets; this difference is precisely explained by general relativity theory. The change from orbit to orbit has been significantly exaggerated on this diagram.

pointed out by Urban Leverrier, the co-discoverer of Neptune. Just as discrepancies in the motion of Uranus allowed astronomers to discover the presence of Neptune, so it was thought the discrepancy in the motion of Mercury could mean the presence of an undiscovered inner planet. Astronomers searched for this planet near the Sun, even giving it a name—Vulcan, after the Roman god of fire. (The name would later be used for the home planet of a famous character on a popular television show about future space travel.)

But no planet has ever been found near the Sun, and the discrepancy was still bothering astronomers when Einstein was doing his calculations. General relativity, however, predicts that due to the curvature of spacetime there should be a tiny additional push on Mercury, over and above that predicted by Newtonian theory, at each perihelion. The result is to make the major axis of Mercury's orbit rotate slowly in space. The prediction of relativity is that the direction of perihelion should change by 43 arcsec per century. This is remarkably close to the observed discrepancy, and it gave Einstein a lot of confidence as he advanced his theory. The relativistic advance of perihelion was later also observed in the orbits of several asteroids that come close to the Sun.

Deflection of Starlight

Einstein's second test was something that had not been observed before and would thus provide an excellent proof of his theory. Since spacetime is more curved in regions where the gravitational field is strong, we would expect light passing very near the Sun to appear to follow a curved path (Figure 23.8), just like the ant in our analogy. Einstein calculated from general relativity theory that starlight just grazing the Sun's surface should be deflected by an angle of 1.75 arcsec.

We encounter a small "technical problem" when photographing starlight coming very close to the Sun: the Sun is an outrageously bright source of starlight itself! But during a total solar eclipse much of the Sun's light is blocked out, allowing the stars to be photographed. In a

paper published during World War I, Einstein (writing in a German journal) suggested that observation during an eclipse could detect the deflection of light passing near the Sun. A single copy of that paper, passed through neutral Holland, reached the British astronomer Arthur S. Eddington.

The technique involves taking a photograph six months earlier of the stars in front of which the eclipsed Sun will appear, and measuring the position of all the stars accurately. Then the same stars are photographed during the eclipse. This is when the starlight has to travel by the Sun through warped spacetime, and thus its path is no longer a straight line. As seen from Earth, the stars closest to the Sun will be "out of place"—slightly away from their regular positions.

The next suitable eclipse was on May 29, 1919. The British organized two expeditions to observe it, one on the island of Principe, off the coast of West Africa, and the other in Sobral, in northern Brazil. Despite some problems with the weather, both expeditions obtained successful photographs. The stars seen near the Sun were indeed displaced, and to the accuracy of the measurements, which was about 20 percent, the shifts were consistent with the predictions of relativity. It was a triumph that made Einstein a world celebrity.

The measurements made in 1919 were good enough to distinguish between Newton's theory of gravity, which predicts no deflection of starlight, and Einstein's theory of gravity, which does predict a deflection. Nevertheless, the accuracy of the measurements was still so low that they could not be considered a convincing demonstration of the theory's absolute correctness. Far higher accuracy has been obtained recently at radio wavelengths.

Radio waves traveling by the Sun should be deflected as well, and because the Sun is less bright in radio waves, such measurements can be carried out more easily. At first, as discussed in Chapter 5, radio astronomers could not pinpoint the exact direction of a radio wave source as accurately as could those working with visible light. But now, the technique of interferometry and the introduction of widely spaced arrays of radio antennas has allowed very

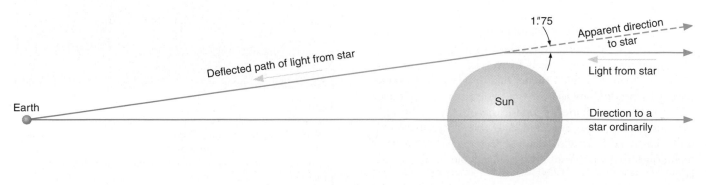

Figure 23.8
Starlight passing near the Sun is deflected slightly, so that it no longer travels in a straight line.

precise radio position measurements. As the Sun moves through the sky, measurements of the shifts in radio positions have allowed astronomers to confirm the predictions made by Einstein to within 1 percent.

Time in General Relativity

A Prediction

General relativity theory makes various predictions about the behavior of space and time. One of these predictions, put in everyday terms, is that *the stronger the gravity, the slower the pace of time.* Such a statement goes very much counter to our intuitive sense of time as a flow we all share. Time has always seemed the most democratic of concepts: all of us, regardless of wealth or status, appear to move together from the cradle to the grave in the great current of time.

But Einstein would argue that it only seems this way to us because all humans so far have lived and died in the gravitational environment of the Earth. We have had no chance to test the idea that the pace of time might depend on the strength of gravity because we have not experienced radically different gravities. And, the differences in the flow of time are extremely small until truly large masses are involved. Nevertheless, Einstein's prediction can now be tested, both on Earth and in space.

The Tests of Time

In an ingenious experiment in 1959, Robert Pound and Glenn Rebka used the most accurate atomic clock known to compare time measurements on the ground floor and the top floor of the physics building at Harvard University. For a clock, the experimenters used the frequency (the number of cycles per second) of gamma rays emitted by radioactive cobalt. Einstein's theory predicts that such a cobalt clock on the ground floor, being a bit closer to the Earth's center of gravity, should run very slightly slower than the same clock on the top floor. This is precisely what the experimenters observed. Later, atomic clocks were taken up in high-flying aircraft and even on one of the Gemini space flights. In each case, the clocks farther from the Earth ran a bit faster.

The effect is more pronounced if the gravity involved is the Sun's and not the Earth's. If stronger gravity slows the pace of time, then it will take longer for a light or radio wave that passes very near the edge of the Sun to reach the Earth than we would expect on the basis of Newton's law of gravity. (It also takes longer because spacetime is curved in the vicinity of the Sun.) The smaller the distance between the ray of light and the edge of the Sun at closest approach, the longer the delay in the arrival time.

In November 1976, when the two Viking spacecraft (see Chapter 9) were operating on the surface of Mars, Mars, as seen from Earth, went behind the Sun (Figure 23.9). Scientists had preprogrammed Viking to send a radio wave toward the Earth that would go extremely close to the outer regions of the Sun. According to general rela-

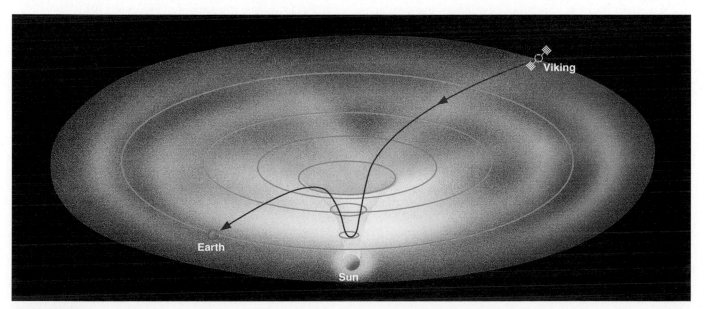

Figure 23.9
Radio signals from the Viking lander on Mars are delayed when they pass near the Sun, where spacetime is curved relatively strongly. In this picture, spacetime is pictured as a two-dimensional rubber sheet.

tivity, there would be a delay because the radio wave would be passing through a region where time ran more slowly. The experiment was able to confirm Einstein's theory to within 0.1 percent.

Gravitational Redshift

What does it mean to say that time runs more slowly? When light enters a region of strong gravity and time slows down, the light experiences a change in its frequency and wavelength; we call this a *gravitational redshift*.

To understand what happens in this situation, let's recall that a wave of light is a repeating phenomenon—crest follows trough follows crest with great regularity. In this sense, each light wave is a little clock, keeping time with its wave cycle. If stronger gravity slows down the pace of time, then the rate at which crest follows crest must slow down—the waves become less *frequent*. This means that there must be more space between the crests; in other words, the wavelength of the wave must increase. This kind of increase (when caused by the motion of the source) is just what we called a *redshift* in Chapter 4. Here, it is not motion but gravity that causes the redshift.

When Einstein's theory was first proposed, it was not possible to measure the gravitational redshift for the Earth or the Sun because the effects were too small for our instruments. But the much stronger gravity of white dwarf stars made them better candidates for observing such a redshift. For the average white dwarf, we do not know how to disentangle the redshift due to its gravity from the redshift caused by its motion through the Galaxy. But if a white dwarf is a member of a binary star system, the motions can be clarified. In such systems astronomers have successfully measured gravitational redshifts, starting in the 1920s.

Far-higher accuracy has been attained recently in the near-Earth environment with space-age technology. In the mid-1970s a hydrogen *maser*, a device (akin to a laser) that produces a microwave radio signal at a particular wavelength, was carried by a rocket to an altitude of 10,000 km. The rocket-borne maser was used to detect the radiation from a similar maser on the ground. The radiation showed a redshift due to the Earth's gravity that confirmed the relativity predictions to within a few parts in 10,000.

So far, then, each prediction of general relativity that can be tested has been confirmed within the accuracy of the experiments. Today, general relativity is accepted as our best description of gravity and is used by astronomers and physicists to understand the behavior of active galaxies, the beginning of the universe, and the subject with which we began the chapter—the death of truly massive stars.

Black Holes

Let's now apply what we have learned about gravity and spacetime curvature to the collapsing core in a very massive star. We saw that if the core's mass is greater than about 3 M_{Sun}, our theoretical understanding forces us to conclude that nothing can stop the core from collapsing forever. We will examine this situation from two perspectives—first from a pre-Einstein point of view, and then with the aid of general relativity.

Classical Collapse

Let's begin with a thought experiment. We want to know what speeds are required to escape from the gravitational pull of different objects. A rocket must be launched from the surface of the Earth at a very high speed if it is to escape the pull of the Earth's gravity. In fact, any object—rocket, ball, astronomy book—that is thrown into the air with a velocity less than 11 km/s will soon fall back to the Earth's surface. Only those objects launched with a velocity greater than this *escape velocity* can get away from the Earth.

The escape velocity from the surface of the Sun is higher yet—618 km/s. Now imagine that we begin to compress the Sun, forcing it to shrink in diameter. Recall that the pull of gravity depends on both the mass that is pulling you and your distance from the center of gravity of that mass. If the Sun is compressed, its *mass* will remain the same, but the *distance* between a point on the Sun's surface and the center will get smaller and smaller. Thus the pull of gravity for an object on the shrinking surface will get stronger and stronger (Figure 23.10).

When the shrinking Sun reaches the diameter of a neutron star (less than 100 km), the velocity required to escape its gravitational pull will be about half the speed of light. Suppose we continue to compress the Sun to a smaller and smaller diameter. (We saw this can't happen to our Sun in the real world because of electron degeneracy, but it will happen to the supermassive star core.) Ultimately, as the Sun shrinks the escape velocity will exceed the speed of light. If the speed you need to get away is faster than the fastest possible speed in the universe, then nothing, not even light, is able to escape. An object with such large escape velocity emits no light, and anything falling into it can never return.

In modern terminology, we call an object from which light cannot escape a **black hole,** a name suggested by the American scientist John Wheeler (Figure 23.11) in 1969. The idea that such objects might exist is, however, not a new one. Cambridge professor and amateur astronomer John Michell wrote a paper in 1783 about the possibility that stars with escape velocities exceeding that of light might exist. And in 1796 the French mathemati-

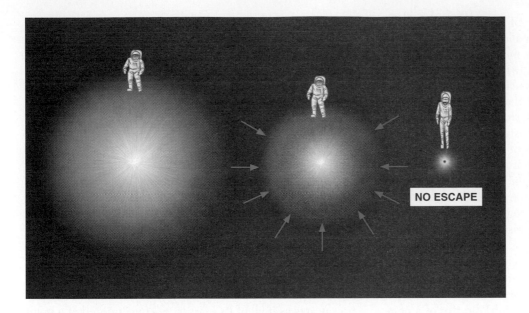

Figure 23.10
At left, an (imaginary) astronaut floats near the surface of a massive star-core about to collapse. As the same mass falls into a smaller sphere, the gravity at the surface goes up, making it harder to escape. Eventually the mass collapses into so small a sphere that the escape velocity exceeds the speed of light and nothing can get away. Note that the size of the astronaut has been exaggerated so you can see him. In the last picture the astronaut is also stretched and squeezed by the strong gravity.

cian Pierre Simon, Marquis de Laplace, made similar calculations using Newton's theory of gravity; he called the resulting objects "dark bodies."

While these early calculations provided a hint that something strange should be expected if very massive objects collapse under their own gravity, we really need general relativity theory to give an adequate description of what happens in such a situation.

Collapse with Relativity

General relativity tells us that gravity is really a curvature in spacetime. As gravity increases (as in the collapsing Sun of our thought experiment) the curvature gets larger and larger. Eventually, if the Sun could shrink down to a diameter of about 6 km, the curvature would become so great that light, trying to go in a straight line, would actually describe a circle and come back upon itself. Anything trying to escape the collapsing star would be unable to do so, because all paths "straight out" would curve back in.

Keep in mind that gravity is not pulling on the light. Gravity has curved spacetime, and light (like the trained ant of our earlier example), "doing its best" to go in a straight line, is now confronted with a world in which straight lines have become circular paths. The collapsing star is a *black hole,* in this view, because the very concept of "out" has no geometrical meaning. The star has become trapped in its own little pocket of spacetime, from which there is no escape.

The star's geometry cuts off communication with the rest of the universe at precisely the moment when,

in our earlier picture, the escape velocity becomes equal to the speed of light. The size of the star at this moment defines a surface that we call the **event horizon.** It's a wonderfully descriptive name: just as objects that sink below our horizon cannot be seen on Earth, so any event inside the event horizon can no longer affect the rest of the universe.

Imagine a future spacecraft foolish enough to sit close to the surface of a massive star that collapses in the way we have been describing. Perhaps the captain is asleep at the gravity meter, and before they can say Albert Einstein, they have collapsed with the star inside the event horizon. Frantically, they send an escape pod straight outward. But paths outward twist around to become paths inward, and the pod quickly returns. In desperation now, they send a radio message to their loved ones, bidding good-bye. But radio waves, like light, must travel through spacetime, and spacetime allows nothing to get out. Their woeful message returns to them unheard. Events inside the event horizon can *never again* affect events outside it.

Figure 23.11
John Wheeler, a brilliant physicist who did much pioneering work in general relativity theory, coined the term *black hole* in 1969. (Photo courtesy Roy Bishop)

The existence and behavior of an event horizon was first worked out by astronomer and mathematician Karl Schwarzschild (Figure 23.12). A member of the German army in World War I, he died in 1916 of an illness he contracted while doing artillery shell calculations on the Russian front. His paper on the theory of event horizons was among the last things he finished as he was dying; it was the first exact solution to Einstein's equations of general relativity. We call the radius of the event horizon the *Schwarzschild radius* in his memory.

The event horizon is the boundary of the black hole; calculations show that it does not get smaller once the star has collapsed inside it. It is the region that separates the things trapped inside it from the rest of the universe. Anything coming from the outside is also trapped once it comes inside the event horizon. The horizon's size turns out to depend only on the mass inside it. For a black hole of 1 M_{Sun}, the Schwarzschild radius is about 3 km; thus the entire black hole is about one-third the size of a neutron star of that same mass. Feed the black hole some mass, and the horizon will grow—but not very much. Doubling the mass will make a black hole 6 km in radius, still very tiny on the cosmic scale.

The event horizons of more-massive black holes have greater radii. For example, if a globular cluster of 100,000 stars could collapse to a black hole, it would be 300,000 km in radius, a little less than half the radius of the Sun. If the entire Galaxy could collapse to a black hole, it would

be only about 10^{12} km in radius—about a tenth of a LY. Smaller masses have correspondingly smaller horizons: for the Earth to become a black hole, it would have to be compressed to a radius of only 1 cm—about the size of a grape. A typical asteroid, if crushed to a small enough size to be a black hole, would have the dimensions of an atomic nucleus!

A Black Hole Myth

Much of the modern folklore about black holes is misleading. One idea you may have heard is that black holes are monsters that go about sucking things up with their gravity. Actually, it is only close to a black hole that the strange effects we have been discussing come into play. The gravitational attraction far away from a black hole is the same as that of the star that collapsed to form it.

Remember that the gravity of any star some distance away acts as if all of its mass were concentrated at a point in the center, which we call the center of gravity. For real stars, we merely *imagine* that all the mass is concentrated there; for black holes, we will see that all the mass *really is* concentrated at a point in the center. So if you are a star or distant planet, orbiting around a star that becomes a black hole, your orbit may not be significantly affected by the collapse of the star (although it may be affected by any mass loss that precedes the collapse). If, on the other hand, you venture close to the event horizon, it would be very hard for you to resist the pull of the warped spacetime near the black hole.

If another star or a spaceship were to pass one or two solar radii from a black hole, Newton's laws would be adequate to describe what would happen to it. Only very near the surface of a black hole is its gravitation so strong that Newton's laws break down. A massive star coming into our neighborhood would be far, far safer to us as a black hole than it would have been earlier in its life as a brilliant, hot star.

A Trip into a Black Hole

The fact that they cannot *see* inside black holes has not kept scientists from trying to calculate what they are like. One of the first things these calculations showed was that the formation of a black hole obliterates nearly all information about the star that collapsed to form it. Physicists like to say "black holes have no hair," meaning that nothing sticks out of a black hole to give us clues about what kind of star produced it. The only information a black hole can reveal about itself is its mass, its spin (rotation), and whether it has any electrical charge.

What happens to the collapsing star-core that made the black hole? Our best calculations predict that the material will continue to collapse under its own weight, forming an infinitely *squozen* point—a place of zero volume and infinite density—to which we give the name

Figure 23.12
Karl Schwarzschild was the first to demonstrate mathematically that a black hole is possible. (Yerkes Observatory)

(text cont. on page 472)

Gravity and Time Machines

Time machines are one of the favorite devices of science fiction. Such a device would allow you to move through time at a different pace or in a different direction from everyone else. General relativity suggests that it is possible, in theory, to construct a time machine using gravity that could take you into the future.

Let's imagine a place where gravity is terribly strong, such as a black hole. General relativity predicts that the stronger the gravity, the slower the pace of time. So imagine a future astronaut, with a fast spaceship, who volunteers to go on a mission to such a high-gravity environment. The astronaut leaves in the year 2200, just after graduating from college at age 22. She takes (let's say) exactly ten years to get to the black hole. Once there, she orbits some distance from it (taking care not to get pulled in).

She is now in a high-gravity realm where time passes much more slowly than it does on the Earth. This isn't just an effect on the mechanism of her clocks—*time itself* is running slowly. That means that every way she has of measuring time will give the same slowed-down reading when compared to time passing on Earth. Her heart will beat more slowly, her hair will grow more slowly, her antique wristwatch will tick more slowly, etc. She is not aware of this slowing down, since all her readings of time, whether made by her own bodily functions or with mechanical equipment, are measuring the same—slower—time. Meanwhile, back on Earth, time passes as it always does.

Our astronaut now emerges from the region of the black hole, her mission of exploration finished, and returns to Earth. Before leaving she carefully notes that (according to her timepieces) she spent about two weeks around the black hole. She then takes exactly ten years to return to Earth. Her calculations tell her that since she was 22 when she left the Earth, she will be 42 plus two weeks when she returns. So the year on Earth, she figures, should be 2242, and her

classmates should now be approaching their midlife crises.

But our astronaut should have paid more attention in her astronomy class! Because time slowed down near the black hole, much less time passed for her than for the people on Earth. While her clocks measured two weeks spent near the black hole, more than 2000 weeks (depending on how close she got) could well have passed on Earth. That's equal to 40 years, meaning her classmates will be senior citizens in their 80s when she (a mere 42-year-old) returns. On Earth it will be not 2242, but 2282—and she will say that she has arrived in the future.

Is this scenario real? Well, it has practical flaws—we don't think there are any black holes close enough for us to reach in ten years, and we don't think any spaceship or human can survive around a black hole. But the key point, about the slowing down of time, is a natural consequence of Einstein's general theory of relativity, and we saw that its predictions have been confirmed by experiment after experiment.

Science-fiction writers who pay attention to developments in astronomy and physics have seized upon this idea for wonderful plot devices. In Larry Niven's novel *World Out of Time* (1976, Ballantine Books), the protagonist travels around an extremely massive black hole and arrives back on Earth some 3 million years in the future. And in Fred Pohl's *Gateway* (1977, Ballantine Books), the "hero" has to push his girlfriend into a black hole so he can escape its strong gravity. Not only can they never speak or touch again, he must then deal with the additional guilt of knowing that, now that he is far from the black hole, his time is passing much more quickly than hers. To him she seems almost suspended in time, and (if he had a good enough telescope) he could see her there for the rest of his life, looking back at him in disbelief that he could have done so horrible a thing. (Indeed, the "hero" spends much of the book talking to a robot psychiatrist named Albert about how guilty he feels.)

Gateway by F. Pohl
(Cover copyright Ballantine Books)

singularity. At the singularity, spacetime ceases to exist. The laws of physics as we know them break down. We do not yet have the physical understanding or the mathematical tools to describe the singularity itself. From the outside, however, the entire structure of a basic black hole (one that is not rotating) can be described as a singularity surrounded by an event horizon. In comparison with humans, black holes are really very simple objects!

Scientists have also calculated what would happen if a daring (perhaps we should say suicidal) astronaut were to fall into a black hole. Let's take up an observing position a long, safe distance away from the event horizon and watch this astronaut fall toward it. At first he falls away from us, moving ever faster, just as though he were approaching any massive star. However, as he nears the event horizon of the black hole, things change. As we saw, the strong gravitational field around the black hole will make his clocks run more slowly, when seen from our outside perspective.

If, as he approaches the event horizon, he sends out a signal once per second according to his clock, we will see the spacing between signals grow longer and longer until it becomes infinitely long when he reaches the event horizon. (If the infalling astronaut uses a blue light to send his signals every second, we will also see the light get redder and redder until its wavelength is nearly infinite.) As the spacing between clock ticks approaches infinity, it will appear to us that the astronaut is slowly coming to a stop, frozen in time at the event horizon.

In the same way, all matter falling into a black hole will also appear to an outside observer to stop at the event horizon, frozen in place and taking an infinite time to fall through it. For this reason, black holes are sometimes called *frozen stars.* But don't think that matter falling into a black hole will therefore be easily visible at the event horizon. The tremendous redshift will make it very difficult to observe any radiation from the "frozen" victims of the black hole.

This, however, is only how we, located far away from the black hole, see things. To the astronaut, his time goes at its normal rate and he falls right on through the event horizon into the black hole. (Remember, this horizon is not a physical barrier, but only a region in space where the curvature of spacetime makes escape impossible.)

You may have trouble with the idea that you and the astronaut have such very different ideas about what has happened. This is the reason Einstein's ideas about space and time are called theories of *relativity;* what each observer measures about the world depends on (is relative to) his frame of reference. The observer in strong gravity measures time and space differently from the one sitting in weaker gravity. When Einstein proposed these ideas, many scientists also had difficulty with the idea that two such very different views of the same event could be correct, each in its own "world," and tried to find a mistake in the calculations. There were no mistakes: we and the astronaut really would see the same fall into a black hole very differently.

For the astronaut, there is no turning back. Once inside the event horizon, the astronaut, any signals from his radio transmitter, and any cries of regret are doomed to remain hidden forever from the universe outside. He will, however, not have a long time (from his perspective) to feel sorry for himself as he approaches the black hole. Suppose he is falling feet first. The force of gravity exerted by the singularity on his feet will be slightly greater than on his head, and he will be stretched slightly. Since the singularity is a point, the left side of his body will be pulled slightly toward the right, and the right side slightly toward the left, bringing each side closer to the singularity. The astronaut will therefore be slightly squeezed in one direction and stretched in the other.

The Earth exerts similar *tidal forces* on an astronaut performing a spacewalk. In the case of the Earth, the tidal forces are so small that they pose no threat to the health and safety of the astronaut. Not so in the case of a black hole. Sooner or later, as the astronaut approaches the black hole, the tidal forces will become so great that the astronaut will be ripped apart. His legs will be ripped from his torso, his ankles from his legs, his toes from his feet, and so forth. Soon it will only be the individual atoms from his shredded body that will continue their inexorable fall into the singularity. (Jumping into a black hole is definitely a once-in-a-lifetime experience!)

One Way to Make a Black Hole

Theory tells us what a black hole must be like. But can black holes really exist? For example, we know of nothing in nature that could crush the Earth or the Sun into a small enough sphere to become a black hole. But, as we have seen, the collapse of a massive star-core may be exactly the kind of situation that could produce a black hole.

The critical question, and one that theory has not yet been able to answer, is this: Are the cores of some stars actually so massive that they exceed the upper limit on the mass of a neutron star (about three times the mass of the Sun)? If so, then nothing that we are presently aware of can halt their collapse, and they will form black holes.

If theory cannot yet tell us what conditions produce a black hole rather than a neutron star, then do observations provide any evidence that black holes actually do exist? At first, you might think it pretty hopeless to look for something many light years away that is on the order of 10 km across and completely black! But it *is* possible to detect the presence of a black hole by the unspeakable things it does to a companion star.

Evidence for Black Holes

As stars collapse into black holes, they leave behind their gravitational influence. We've already seen that a companion star some distance away may be unaffected by the collapse (although any accompanying supernova explosion may take its toll on the companion). If a member of a double-star system becomes a black hole, then we may be able to detect it by studying the motion and other properties of its companion.

Requirements for a Black Hole

So here is a prescription for finding a black hole: First look for a star whose motion (determined from the Doppler shift of its spectral lines) shows it to be a member of a binary star system. If both stars are visible, neither can be a black hole, so focus your attention just on those systems where only one star of the pair is visible even with our most sensitive telescopes.

Being invisible is not enough, however, because a relatively faint star might be unseen next to the glare of a brilliant companion, or in a shroud of dust. And even if the star really is invisible, it could be a neutron star, which does not emit any light. Therefore, we must also have evidence that the unseen star has a mass too high to be a neutron star, and that it is a collapsed object—one of extremely small size.

We can use Kepler's law (see Chapter 2) and our knowledge of the visible star to measure the mass of the invisible member of the pair. If the mass is greater than about 3 M_{Sun}, then we are likely seeing (or, more precisely, not seeing) a black hole—as long as we can make sure the object really is a collapsed star.

If matter falls toward a compact object of high gravity, the material is accelerated to high speed. Near the event horizon of a black hole, matter is moving at velocities that approach the speed of light. As the atoms rush chaotically toward the event horizon, they rub against each other; internal friction can heat them to temperatures of 100 million K or more. Such hot matter emits radiation in the form of flickering x rays. The last part of our prescription, then, is to look for a source of x rays associated with the binary system. Since x rays do not penetrate the Earth's atmosphere, such sources must be found using x-ray telescopes in space.

The infalling gas that produces the x-ray emission comes from the black hole's companion star. As we saw in Chapter 22, stars in close binary systems can exchange mass, especially as one of the members expands into a red giant. Suppose that one star in a double-star system has evolved to a black hole, and that the second star begins to expand in size. If the two stars are not too far apart, the outer layers of the expanding star may reach the point where the black hole exerts more gravitational force on them than do the inner layers of the red giant to which the atmosphere belongs. The outer atmosphere has then passed through the point of no return between the stars, and falls toward the black hole.

The mutual revolution of the giant-star and the black hole causes the material falling toward the black hole to spiral around it rather than flowing directly into it. The in-falling gas whirls around the black hole in a pancake of matter called an *accretion disk*. It is in the inner part of this disk that matter is revolving about the black hole so fast that internal friction heats it up to x-ray-emitting temperatures.

Another way to form an accretion disk in a binary star system is to have a powerful stellar wind come from the black hole's companion. Such winds are a characteristic of several stages in a star's life. Some of the ejected gas in the wind will then flow close enough to the black hole to be captured by it into the disk (Figure 23.13).

We should point out that, as often happens, these measurements are not quite as simple as they are described in introductory textbooks. In real life, Kepler's law only allows us to calculate the combined mass of the two stars in the binary system. We must learn more about the visible star of the pair and its history to disentangle its mass from the unseen companion. Also, the calculations are affected by how the orbit of the two stars is tilted toward the Earth, something we rarely know how to measure. And neutron stars can also have accretion disks that produce x rays, so astronomers must study the properties of these x rays carefully when trying to determine what kind of object is at the center of the disk. Nevertheless, the evidence for black holes in such systems is quite strong.

Candidate Objects

Because the x rays are such an important clue that we are dealing with a black hole that is having some material from its companion for lunch, the search for black holes had to await the launch of sophisticated x-ray telescopes into space. These instruments must have the resolution to distinguish the location of x-ray sources and allow us to match them to the positions of binary star systems. Such a case, we think, is the binary system that was our first black hole candidate—Cygnus X-1.

The visible star in Cygnus X-1 is spectral type B. Measurements of the Doppler shifts of the B star's spectral lines show that it has an unseen companion. The x rays flickering from it strongly suggest that it is a small collapsed object—but is it a neutron star or a black hole? (Remember, neutron stars are likely to be much more common, since they come from lower-mass stars, which are more numerous.) To find out, astronomers have tried to determine the mass of the invisible member of the pair: it appears to be nearly ten times that of the Sun. The companion is therefore too massive to be either a white dwarf or a neutron star.

There are several other binary systems that meet all the conditions to be prime black hole candidates. Table 23.1 lists the characteristics of some of our best cases.

Feeding a Black Hole

We saw that when an isolated star, even if it is in a binary star system, becomes a black hole, there is usually a limit to how much it can grow. After all, material must ap-

(text cont. on page 475)

Figure 23.13
Mass lost from a giant star through a stellar wind streams toward a black hole and swirls around it before finally falling in. In the inner portions of the accretion disk, the matter is revolving so fast that internal friction heats it to very high temperatures and x rays are emitted.

TABLE 23.1
Some Black Hole Candidates in Binary Star Systems

Name or Catalog Designation	Companion Star Spectral Type	Orbital Period (days)	Black Hole Mass Estimates (M_{Sun})
Cygnus X-1	B supergiant	5.6	6–15
LMC X-3	B main sequence	1.7	4–11
A0620-00 (V616 Mon)*	K main sequence	7.8	4–9
GS2023+338 (V404 Cyg)	K main sequence	6.5	more than 6
GS2000+25 (QZ Vul)	K main sequence	0.35	5–14
GS1124-683 (Nova Mus 1991)	K main sequence	0.43	4–6
GRO J1655-40 (Nova Sco 1994)	F main sequence	2.4	4–5
H1705-250 (Nova Oph 1977)	K main sequence	0.52	more than 4

* As you can tell, there is no standard way of naming these candidates. The chain of numbers are the location of the source in right ascension and declination (the longitude and latitude system on the sky); the letter(s) preceding them designate the satellite that discovered the candidate—A for Ariel, G for Ginga, etc. The notations in parentheses are those used by astronomers who study binary star systems or novae.

proach close to the event horizon before the gravity is any different from that of the star that became the black hole. Out in the suburban regions of the Milky Way Galaxy where we live (see Chapter 24), stars and star systems are much too far apart for other stars to provide "food" to a hungry black hole.

But, as we will see, the central regions of galaxies are quite different from the outer parts. Here stars and raw material can be quite crowded together, and they can interact much more frequently with each other. Therefore, black holes in the centers of galaxies may have a much better opportunity to find mass close enough to their event horizons to pull it in. Black holes are not particular about what they "eat": they are happy to consume other stars, asteroids, gas, dust, and even other black holes. (If two black holes merge, you just get a black hole with more mass and a larger event horizon.)

As a result, black holes in crowded regions can grow, eventually swallowing thousands or even millions of times the mass of the Sun. Observations with the Hubble Space Telescope have recently shown dramatic evidence for the existence of a black hole containing about three billion M_{Sun} in a galaxy called M87 (see opening figure for this chapter). The feeding frenzy of such *supermassive black holes* may be responsible for some of the most energetic phenomena in the universe (see Chapters 24 and 26).

More and more evidence is accumulating that some stars end their lives as black holes. Furthermore, the observational tests of Einstein's general theory of relativity have convinced even the most skeptical scientists that his picture of warped or curved spacetime is indeed our best description of the effects of gravity. The ideas discussed in this chapter can seem strange and overwhelming, especially the first time you read them. But you have to admit that they make the universe even more interesting and bizarre than you might have thought before you took this class.

Summary

23.1 The **equivalence principle** is the foundation of **general relativity.** According to this principle, there is no way that anyone or any experiment in a sealed environment can distinguish between free-fall and the absence of gravity.

23.2 By considering the consequences of this principle, Einstein concluded that we live in a curved **spacetime.** The distribution of matter determines the curvature of spacetime; other objects (and even light) entering a region of spacetime must follow its curvature.

23.3 At low speeds and in weak gravitational fields, the predictions of general relativity agree with the predictions of Newton's theory of gravity. However, general relativity predicts, while Newtonian theory does not, that starlight (or radio waves) will be deflected when they pass near the Sun, and that the position where Mercury is at perihelion should change by 43 arcsec per century even if there are no other planets in the solar system to perturb its orbit. These predictions have been verified by experiment.

23.4 General relativity predicts that the stronger the gravity, the more slowly time must run. Experiments on Earth and with spacecraft have confirmed this prediction with remarkable accuracy.

23.5 Theory suggests that stars with stellar cores more massive than three times the mass of the Sun at the time they exhaust their nuclear fuel will collapse to become **black holes.** The surface surrounding a black hole, where the escape velocity equals the speed of light, is called the **event horizon,** and the radius of the surface is called the Schwarzschild radius. Nothing, not even light, can escape through the event horizon from the black hole. At its center each black hole has a **singularity,** a point of infinite density and zero volume. Matter falling into a black hole appears, as viewed by an outside observer, to freeze in position at the event horizon. However, if we were riding on the infalling matter, we would pass through the event horizon. In the process, tidal forces exerted by the singularity would tear our bodies apart.

23.6 Black holes that collapse from stars can only be detected in binary star systems having (a) one star of the pair that is invisible; (b) x-ray emission characteristic of an accretion disk around a compact object; and (c) an indication from the orbit and characteristics of the visible star that the mass of its invisible companion is greater than 3 M_{Sun}. Several systems with these characteristics have been found. More-massive black holes are also possible.

1. How does the principle of equivalence lead us to suspect that spacetime might be curved?

2. If general relativity offers the best description of what happens in the presence of gravity, why do physicists still make use of Newton's equations in describing gravitational forces on Earth (when building a bridge, for example)?

3. Einstein's general theory of relativity made predictions about the outcome of several experiments that had not yet been carried out at the time the theory was first published. Describe three experiments that verified the predictions of the theory.

4. If a black hole emits no radiation, why do astronomers and physicists today believe that the theory of black holes is correct?

5. What characteristics must a binary star have to be a good candidate for a black hole? Why is each of these characteristics important?

6. Suppose the Earth were to fall into a black hole. What would happen to it?

7. A student becomes so excited by the whole idea of black holes, he decides to jump into one.
  a. What is the trip like for him?
  b. What is it like for the rest of the class, watching from afar?

Thought Questions

8. Imagine that you have built a large room around the boy and girl in Figure 23.2, and that this room is falling at exactly the same rate as they are. Galileo showed that, if there is no air friction, light and heavy objects that are dropping due to gravity will fall at the same rate. Suppose this were not true, and that instead heavy objects fall faster; also suppose that the boy in Figure 23.2 is twice as massive as the girl. What would happen? Would this violate the equivalence principle?

9. A monkey hanging from a tree branch sees a hunter aiming a rifle directly at him. The monkey then sees a flash, telling him that the rifle has been fired. Reacting quickly, he lets go of the branch and drops so the bullet can pass harmlessly over his head. Does this act save the monkey's life? Why?

10. During the 1970s, the United States had in orbit a small space station called Skylab. Some of the astronauts exercised by running around the inside wall of their cylindrical vehicle.

How could they stay against the wall while running, rather than float aimlessly inside the Skylab? What physical principles were involved?

11. Why would we not expect to detect x rays from a disk of matter about an ordinary star?

12. Look elsewhere in this book for the necessary data, and indicate what the final stage of evolution—white dwarf, neutron star, or black hole—will be for each of the following kinds of stars.
  a. Spectral type-O main-sequence star
  b. B main-sequence star
  c. A main-sequence star
  d. G main-sequence star
  e. M main-sequence star

13. Which is likely to be more common in our Galaxy—white dwarfs or black holes? Why?

Problems

14. The formula for the radius of the event horizon is $R = 2GM/c^2$, where c is the speed of light. Show that this equation is equivalent to saying that the circumference of the event horizon is equal to 18.5 km times the mass of the black hole in units of the mass of the Sun.

15. What would be the circumference of a black hole with the mass of Jupiter?

16. Suppose that the Earth collapsed to the size of a golf ball.

  a. What would be the revolution period of the Moon around it, at a distance of 400,000 km?
  b. Of a spacecraft orbiting at a distance of 6000 km?
  c. Of a miniature spacecraft orbiting at a distance of 0.1 m?
  d. Calculate the orbital speed of this mini-spaceship, and compare it with the speed of light.

17. If the Sun could suddenly collapse to a black hole, how would the period of the Earth's revolution about it differ from what it is now?

Suggestions for Further Reading

Chaisson, E. *Relatively Speaking.* 1988, W. W. Norton. An astronomer introduces relativity and its astronomical applications.

Greenstein, G. *Frozen Star.* 1984, Freundlich Books. Wonderful, lucid introduction to the death of stars, with a good section on black holes.

Kaufmann, W. *Cosmic Frontiers of General Relativity.* 1977, Little, Brown. Good nontechnical primer on general relativity theory and its applications in astronomical contexts.

McClintock, J. "Do Black Holes Exist?" in *Sky & Telescope,* Jan. 1988, p. 28.

Overbye, D. "God's Turnstile: The Work of John Wheeler and Stephen Hawking" in *Mercury,* Jul./Aug. 1991, p. 98.

Parker, B. "Where Have All the Black Holes Gone?" in *Astronomy,* Oct. 1994, p. 36. On candidates within our Galaxy.

Thorne, K. *Black Holes and Time Warps.* 1994, W. W. Norton. The definitive introduction to black holes.

Wheeler, J. *A Journey into Gravity and Spacetime.* 1990, Scientific American Library/W. H. Freeman. A brilliant introduction by one of the foremost scientists of our time.

Will, C. *Was Einstein Right?—Putting General Relativity to the Test.* 1986, Basic Books. Superb guide to the many tests.

Zirker, J. "Testing Einstein's General Relativity During Eclipses" in *Mercury,* Jul./Aug. 1985, p. 98.

Using REDSHIFT ™

None of the companions to the black hole candidates in Table 23.1 are visible in *RedShift*.

1. If stars with initial masses greater than 40 solar masses may eventually become black holes, use Table 21.1 and *RedShift* to identify which main-sequence stars might become black holes.

2. If *RedShift* can display up to 250,000 stars, estimate the percentage of stars that end their lives as black holes.

Infrared radiation can penetrate the dust that lies between us and the center of the Milky Way. This infrared image shows the central 150 LY of our Galaxy. The bright region in the middle is the nucleus of the Galaxy, which harbors large amounts of gas, dense groups of stars, and perhaps a massive black hole. The image is a color-coded composite of views of the same scene at three infrared wavelengths: blue is 1.2 μm, green is 1.65 μm, and red is 2.2μm. (Ian Gatley, Michael Merrill, and Richard Joyce/National Optical Astronomy Observatories)

CHAPTER 24

The Milky Way Galaxy

Thinking Ahead

In the midst of a dark forest, a campfire illuminates the night. The light of the fire shows us the nearby trees and orients us to our immediate surroundings. But beyond the trees that catch the firelight lies a much larger darkness, where other trees and much else we don't know about can be found. In the same way, is it possible that as we examine our Galaxy of stars by the familiar light they cast, we are missing a much larger region whose nature we don't fully understand?

Across the height of heaven there runs a road Clear when the night is bare, the Milky Way, Famed for its sheen of white . . .

Ovid, *Metamorphoses* (lines 169–175), quoted by E. C. Krupp in *Beyond the Blue Horizon* (1991)

Having completed our survey of the life stories of the stars, we now turn our attention to the way they are grouped in the universe. We find stars arranged in vast islands scattered in the dark oceans of space—islands we call *galaxies.* Let us begin our examination of these great star systems by studying the one in which we find ourselves—the Milky Way Galaxy.

One of the most striking features in a truly dark sky is a band of faint white light that stretches from one horizon to the other (Figure 24.1). Because of its appearance, this band of light is called the Milky Way. In the Northern Hemisphere the band is brightest in the region of the constellation Cygnus, and is best viewed in the summer. In the Southern Hemisphere the Milky Way is even brighter—so bright, in fact, that native people in South America gave names to various portions of it just as northern astronomers gave constellation names to conspicuous groups of stars. Unfortunately, the faint light of the Milky Way is swamped by the artificial lighting in our urban areas (see "Making

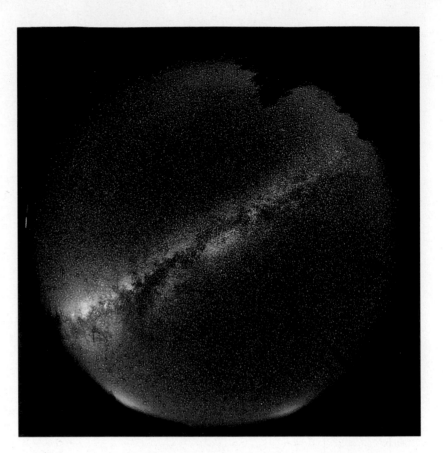

Figure 24.1
This full-sky view of the Milky Way, made with a special lens that can show a 360° panorama, was taken from the summit of Mount Graham, a 3200-m-high mountain in southeastern Arizona. Note the dark rift through the center of the Milky Way. Around the horizon in a counterclockwise direction, beginning from the bottom, you can see the lights of Willcox, Tucson, and Phoenix, as well as the silhouettes of test towers and fir trees. The brightest point-like object in the lower left of the picture is the planet Jupiter. (Roger Angel, Steward Observatory/University of Arizona)

Connections" box), so many city dwellers (perhaps you among them) have never seen the Milky Way.

Although the Milky Way was more easily visible to them on a cloudless night, the ancients did not really have a good idea what it was. The name comes from legends that compare its faint white splash of light to a stream of spilled milk. But folktales differ from culture to culture: one East African tribe thought of it as the smoke of ancient campfires; several Native American stories tell of a path across the sky traversed by sacred animals; and in Siberia it was known as the seam in the tent of the sky.

In 1610 Galileo made the first telescopic survey of the Milky Way and discovered that it is composed of a multitude of individual stars. From far away, the unaided eye sees their combined light as a splash of white. But why is the splash in the form of a strip or arc across the sky? All the stars we see with the unaided eye (and not just the arc of the Milky Way) are part of an enormous wheel or frisbee-shaped system of stars. The stars close to us lie all around us in every direction. But as we look at the more distant parts of the system, the frisbee shape becomes more apparent. The frisbee is not oriented in the same direction as the plane of our solar system or the celestial equator, so we see the arc of the Milky Way flung diagonally across the sky.

We call this great stellar system, which we now know includes hundreds of billions of stars and much else, the **Milky Way Galaxy** or, more simply, just the **Galaxy.** The process of understanding the extent of the Galaxy, and our place in it, took many centuries.

24.1

The Architecture of the Galaxy

In 1785 William Herschel showed that the stellar system to which the Sun belongs has the shape of a wheel or disk (Figure 24.2). He used a telescope to count the numbers of stars that he saw in various directions. To understand how he did this, imagine that you are a member of the band standing in formation during halftime at a football game. If you count the band members that you see in different directions and get about the same number each time, you can conclude that the band has arranged itself in a circular pattern with you at the center. Since you see no band members above you or underground, you know that the circle made by the band is much flatter than it is wide.

Herschel saw much the same thing when he counted stars in different directions in the Galaxy. He found that most of the stars lay in a narrow band, the Milky Way, cir-

Figure 24.2
William Herschel (1738–1822), a German musician who emigrated to England and took up astronomy in his spare time, was the discoverer of infrared radiation and the planet Uranus. He built several large telescopes and made measurements of the Sun's place in the Galaxy, the Sun's motion through space, and the comparative brightnesses of stars. (NASA)

ter of the solar system, but at least our Sun appeared to be in the center of the distribution of stars.

We now know that Herschel was right about the shape of our system, but wrong about where the Sun lies within that disk. As we saw in Chapter 19, we live in a dusty Galaxy. Because interstellar dust absorbs the light from stars, Herschel could see only those stars within about 6000 LY of the Sun. This is a very small section of the much larger disk of stars that makes up the Galaxy: the full disk is actually about 100,000 LY across.

In the same way, if we stand on the streets of a large urban area on a smoggy day, we can see only a small section of the city; the dust and pollution in the air block the light from more distant neighborhoods. Herschel, confined to using visible light for his observations, had no way of penetrating the dust between the stars and discerning that the Sun is actually located fairly far from the center of the Galaxy.

Globular Clusters and the Center of the Galaxy

Until early in this century, astronomers generally accepted Herschel's conclusions. The discovery of the Galaxy's true size and our actual location came about largely through the efforts of Harlow Shapley (see "Voyagers in Astronomy" box). To understand what he did, let us return to the analogy of a smoggy day in a big city. Suppose that, despite your inability to see through the smog, you nevertheless need to get some sense of the extent of the city and where you are currently located.

Being a keen observer, you notice that the smog is actually confined to a relatively thin layer right at ground level (this is not true of real smog, but humor us for a minute). If you look upwards instead of through the smog, your view is somewhat clearer. As it happens, the local baseball team has just won the World Series, and all over

cling the sky, and that the numbers of stars were about the same in any direction around this band. He therefore concluded that the Sun must be near the center of a disk of stars (Figure 24.3). There was some philosophical comfort in this view: the Earth may not have been in the cen-

Sun

Figure 24.3
A copy of Herschel's diagram, showing a cross section of the Milky Way system based on counting stars in various directions. The large circle shows the location of the Sun.

Harlow Shapley: Mapmaker to the Stars

Born in 1885 on a farm in Missouri, Harlow Shapley at first dropped out of school with only the equivalent of a fifth-grade education. He studied at home and at age 16 got a job as a newspaper reporter covering crime stories. Frustrated by the lack of opportunities for someone who had not finished high school, Shapley went back and completed a six-year high school program in only two years, graduating as class valedictorian.

In 1907, at age 22, he went to the University of Missouri, intent on studying journalism, but found that the School of Journalism would not open for a year. Leafing through the college catalog (or so he told the story later), he chanced to see "Astronomy" among the subjects beginning with "A." Recalling his boyhood interest in the stars, he decided to study astronomy for the next year (and the rest, as the saying goes, was history).

Upon graduation Shapley received a fellowship for graduate study at Princeton and began to work with the brilliant Henry Norris Russell (see "Voyagers in Astronomy" box in Chapter 17). For his PhD thesis, he made major contributions to the methods of analyzing the behavior of eclipsing binary stars. He was also able to show that cepheid variable stars are not binary systems, as some people thought at the time, but individual stars that pulsate with striking regularity.

Impressed with Shapley's work, George Ellery Hale offered him a position at the Mount Wilson Observatory, where the young man took advantage of the clear mountain air

Harlow Shapley (1885–1972)
(Harvard College Observatory Archives)

and the 60-in. reflector to do his pioneering study of variable stars in globular clusters.

Shapley subsequently accepted the directorship of the Harvard College Observatory, and over the next 30 years he and his collaborators made contributions to many fields of astronomy, including the study of neighboring galaxies, the discovery of dwarf galaxies, a survey of the distribution of galaxies in the universe, and much more. He wrote a series of nontechnical books and articles and became known as one of the most effective popularizers of astronomy. Shapley enjoyed giving lectures around the country, including at many smaller colleges where students and faculty rarely got to interact with scientists of his caliber.

During World War II Shapley helped rescue many scientists and their families from Eastern Europe; later he helped found UNESCO, the United Nations Educational, Scientific, and Cultural Organization. He wrote a little pamphlet called *Science from Shipboard* for men and women in the armed services who had to spend many weeks onboard transport ships to Europe. And during the difficult period of the 1950s when congressional committees began their "witch hunts" for communist sympathizers (including such liberal leaders as Shapley), he spoke out forcefully and fearlessly in defense of the freedom of thought. A man of many interests, he was fascinated by the behavior of ants, and wrote scientific papers about them as well as about galaxies.

By the time he died in 1972, Shapley was acknowledged as one of the pivotal figures of modern astronomy, a "20th century Copernicus" who had mapped the Milky Way and showed us our place in the Galaxy.

---

the city civic leaders have released giant balloons bearing the team name as part of the celebration. Because these balloons rise above the smog layer, they can be spotted much farther away. If the balloons have been released democratically all around the city, you can make a catalog of their locations and get a pretty good sense of the city's size and your location.

What Shapley realized is that the Galaxy has provided us with just such signposts above (and below) its dusty disk. They are the globular clusters (see Chapter 21)—collections of some 100,000 or more stars that can be seen (with telescopes) at much larger distances than stars within the Galaxy's disk.

Most globular clusters contain at least a few RR Lyrae variable stars that—along with the Cepheids—are excellent distance indicators (see Section 18.3). By comparing

the known intrinsic luminosity of these stars to how bright they appeared, he could derive their distance. (Recall that it is distance that makes the stars look dimmer than they would be "up close," and that we know the brightness fades as the distance squared.) Knowing the distance to any star in a cluster then tells us the distance to the cluster itself.

In 1917 Shapley used the distances and directions of 93 globular clusters to map out their positions in space (Figure 24.4). He found that the clusters are distributed in a spherical-shaped volume, which has its center not at the Sun but at a distant point along the Milky Way in the direction of Sagittarius. Shapley then made the bold assumption that the point on which the system of globular clusters is centered is also the center of the entire Galaxy. Today this assumption has been verified by many pieces of

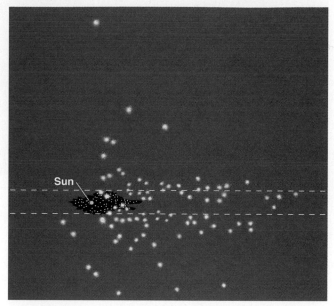

Figure 24.4
A copy of a diagram by Shapley, showing a map of the distribution of globular clusters and the positions of the Sun and the center of the Galaxy. The black area shows Herschel's old diagram (Figure 24.3), centered on the Sun, approximately to scale.

Overview of the Galaxy

With modern instruments, astronomers can now penetrate the "smog" of the Milky Way. Just as you can listen to your favorite radio station whether or not it is smoggy in the city, astronomers can "tune in" on radio and infrared radiation from distant parts of the Galaxy. Measurements at these wavelengths (as well as observations of other galaxies like ours) have fleshed out Shapley's first picture of the Milky Way. We now know it is a thin, circular, rotating disk of luminous matter (Figure 24.5) distributed across a region about 100,000 LY in diameter, and about 1000 LY thick. (Given how thin the disk is, perhaps a French crepe is a more appropriate analogy than a frisbee.)

Bright stars, as well as the dust and gas from which stars form, are closely confined to the galactic disk. This disk, however, is embedded in a spherical **halo** of faint stars that extends to a distance of at least 50,000 LY from the galactic center. The globular clusters also trace out a large spherical halo around the galactic center.

It is not an easy task to pin down the characteristics of this halo. When we see a faint star above or below the plane of the Galaxy with a powerful telescope, it is rarely possible to determine its distance. Thus we do not know if it is an intrinsically faint star that is nearby, or a brighter star that is far away. Luckily, there are variable stars that are distance indicators in the halo: individual RR Lyrae stars have been found in significant numbers as far away as 50,000 LY above and below the galactic plane. This shows that the halo has an overall diameter of at least 100,000 LY.

Close in to the galactic center (within about 3000 LY), the stars are no longer confined to the disk, but form a **nuclear bulge** consisting of old stars. When we use ordinary visible radiation, we can only glimpse the stars in the bulge in directions where there happens to be rela-

evidence, including the discovery that the Galaxy rotates around that same point, which is about 26,000 LY from the Sun.

Shapley's work showed once and for all that our star has no special place in the Galaxy. We are in a nondescript region of the Milky Way, one of at least 100 billion stars that circle the distant center of our Galaxy. Whatever importance humanity finds in the vastness of the cosmos is much more likely to come from our own actions than the location we have been assigned by nature.

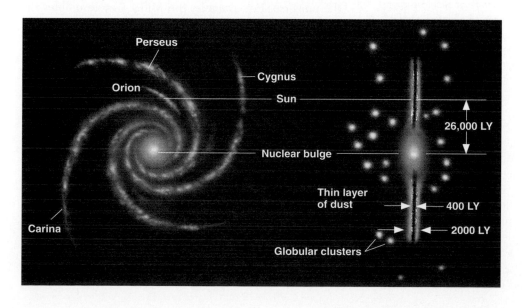

Figure 24.5
A schematic representation of the Galaxy. The Sun is located on the inside edge of the short Orion spur.

Figure 24.7
This 1990 image presents a view of the inner part of the Milky Way Galaxy obtained by an experiment aboard the Cosmic Background Explorer Satellite (COBE). Taken in the near-infrared part of the spectrum, the image permits us to see, for the first time, the bulge of old stars that surround the center of our Galaxy. Redder colors correspond to regions with more dust, and this dust forms a thin disk of material that shows the plane of the Milky Way. The same dust, building up into a kind of interstellar smog, prevents us from seeing the inner parts of our Galaxy with visible light. (NASA).

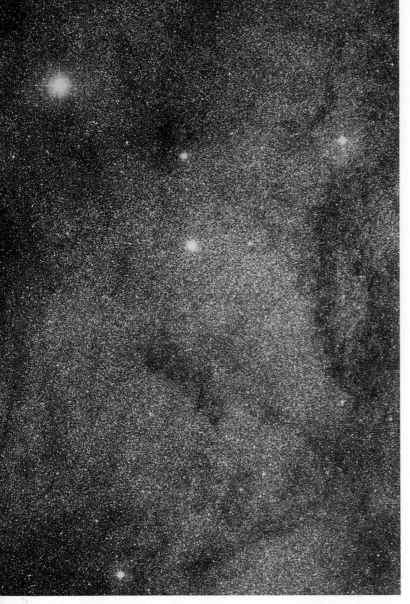

Figure 24.6
Although the nucleus of our Galaxy is hidden by dust at visible wavelengths, a few places where the obscuration is low allow us to see into the cloud of old stars concentrated toward the center and forming the nuclear bulge. This is a picture of such a region, called Baade's window after the astronomer who first pointed it out. The density of stars in this photograph peaks at a distance of about 26,000 LY. The brightest star in the picture is Gamma Sagittarii, a deep yellow star in the foreground. At a distance of about 100 LY, it can be seen with the unaided eye. (© Roe/AAT Board 1987)

tively little interstellar dust (Figure 24.6). The first picture that actually succeeded in showing the bulge as a whole was taken at infrared wavelengths (Figure 24.7).

The fact that much of the bulge is obscured by dust makes its shape difficult to determine. For a long time astronomers assumed it was spherical. However, infrared images and other data suggest that the bulge may be elongated—shaped like a very fat cigar. We know many other galaxies that also have bar-shaped concentrations of stars in their central regions; for that reason they are called barred spirals.

The stars and raw material in the disk of the Galaxy are not distributed smoothly. As the diagram in Figure 24.5 shows, the young stars in the disk are concentrated in a series of spiral arms. We will explore these features in more detail shortly. For now we want to note that establishing this overall picture of the Galaxy from our suburban, dust-shrouded vantage point has been one of the great achievements of 20th-century astronomy (and one that took decades of effort by astronomers working with a wide range of telescopes).

One thing that helped enormously was the discovery that our Galaxy is not unique in its characteristics. There are many other such spiral-shaped aggregations of stars in the universe. For example, the Milky Way resembles the Andromeda galaxy, which at a distance of about 2.3 million LY is one of our nearest neighbors (see Figure 24.11 later in the chapter). Just as you can get a much better picture of yourself if someone else takes the photo from some distance away, pictures (and other data) from nearby galaxies that resemble ours have been vital for understanding the properties of the Milky Way.

Interstellar Matter in the Galaxy

A major step in understanding the detailed structure of the Galaxy came with the discovery of 21-cm radio waves from cold hydrogen (see Section 19.2). Hydrogen is abundant throughout the Galaxy's disk, and radiation at this wavelength easily comes through dust. Furthermore, as we will see, different parts of the Galaxy move around the center at different rates. This means that various hydro-

gen clouds have different Doppler shifts as detected from Earth. Astronomers can use 21-cm radiation to sort out the different structures within the Galaxy, and also to get a much better sense of the full extent of its disk.

Radio observations of the 21-cm line show that the Galaxy's cold atomic hydrogen is confined to an extremely flat layer that is only about 400 LY thick (Figure 24.8). In the plane of the Galaxy, this hydrogen extends well beyond the Sun to a distance of about 80,000 LY from the galactic center. As discussed in Chapter 19, the Milky Way has plenty of raw material for making new stars in its disk, and it is in the disk that we see young, recently formed stars.

Dust, too, is confined to the disk, with the highest concentrations occurring in the spiral arms. The thickness of the dust layer is also about 400 LY. But it seems more concentrated toward the inner parts of the disk; very little dust is found outside the Sun's orbit around the galactic center.

The most-massive molecular clouds (discussed in Chapters 19 and 20) are found in the spiral arms of the Galaxy. In many cases, individual clouds have gathered into large complexes containing a dozen or more discrete clumps. Since the large molecular clouds and cloud complexes are the sites of star formation, most young stars are also found in spiral arms. Remember that the hot, brilliant stars are the easiest to see from a great distance. But these stars don't live very long, and are found during their brief lives quite close to the reservoirs of raw material that gave them birth. When we look at photographs of other spiral galaxies (Figure 24.9), it is these brilliant stars that we see clearly outlining the spiral arms.

Figure 24.8
A map showing the distribution of cold hydrogen in the Galaxy. Prepared from observations at 21 cm, the map is color-coded to mark different amounts of the gas: black and dark blue show regions with the least hydrogen, while red and white indicate the largest concentrations. Most of the hydrogen is in a thin disk that shows the plane of the Galaxy. The wisps and filaments of hydrogen above and below the disk are nearby structures, which we see all around us. (C. Jones and W. Forman/CfA, from data by many investigators)

Spiral Structure of the Galaxy

The Arms of the Milky Way

Using the techniques just discussed, astronomers are beginning to pin down the precise layout of the Milky Way. It appears that the Galaxy has four major **spiral arms,** with some smaller spurs (shown in Figure 24.5). The Sun is near the inner edge of a short arm or spur called the *Orion arm,* which is about 15,000 LY long and contains such conspicuous features as the Cygnus Rift (the great dark nebula in the summer Milky Way) and the Orion Nebula.

More distant, and therefore less conspicuous, are the *Sagittarius–Carina* and *Perseus arms,* located, respectively, about 6000 LY inside and outside the Sun's position relative to the galactic center. Both of these arms, and the more distant *Cygnus arm,* are about 80,000 LY long. The fourth arm is unnamed and difficult to detect because it runs along the other side of the Galaxy. Radiation from it is mingled with radiation from the central regions of the Galaxy.

Formation of Spiral Structure

At the Sun's distance from its center, the Galaxy does not rotate like a solid wheel. Individual stars obey Kepler's third law, just as planets in our solar system do. Remember that Pluto takes longer than the Earth to complete one full circuit around the Sun. In just the same way, stars in larger orbits in the Galaxy trail behind those in smaller ones. This effect is called *differential galactic rotation.*

Differential rotation explains why so much of the interstellar material in our Galaxy is concentrated into elongated features that resemble spiral arms. No matter what the original distribution of the material might be, the dif-

ferential rotation of the Galaxy can stretch it out into spiral features. Figure 24.10 shows the development of spiral arms from two irregular blobs of interstellar matter, as the portions of the blobs closest to the galactic center move fastest, while those farther out trail behind.

But this picture of spiral arms presents astronomers with an immediate problem. If that's all there were to the story, differential rotation would—over the 15-billion-year history of the Galaxy—have wound the Galaxy's arms tighter and tighter until all semblance of structure had disappeared. Somehow, the graceful spiral arms must maintain themselves over time, even as the material in them turns and turns around the center of the Milky Way.

We can think of the spiral arms as regions where matter is more densely concentrated. The arms are defined by the presence of a concentration, but individual stars and gas clouds may move into and out of the arms. To see what we mean, consider this analogy. Imagine that you are driving on a highway, and that cars with very nervous student drivers are moving unusually slowly in two of the lanes ahead of you. Traffic behind these cars will be forced to slow down, and the density of cars around them will increase. (So will the noise level, as people honk their horns and make rude remarks.)

Over time, individual cars like yours will manage to get past the slowly moving cars, but others will take their place in the traffic jam. Viewed from high above the highway, there will be a clump of cars moving along the freeway at the same speed as the two slowly moving cars. The clump will be moving more slowly than the traffic both in front of and behind the jam. The cars stuck in the clump will be different as time passes; each driver will only be caught in the traffic jam for a while.

If the same highway has other pairs of slow drivers farther back, there will be other clumps of traffic through which anyone on that road will have to move. In the same way, the "traffic" of stars and interstellar matter in the Galaxy slows down and clumps up when it meets the

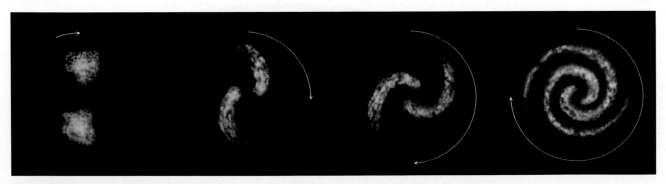

Figure 24.10
How spiral arms might form from irregular clouds of interstellar material stretched out by the different rotation rates throughout the Galaxy. The regions farthest from the galactic center rotate most slowly.

denser regions of the spiral arms. The concentrations of higher density rotate more slowly than does the actual material in the Galaxy, so that any group of stars, gas, and dust eventually passes through the spiral arms.

Astronomers have made models of how stars and gas clouds should move in response to the gravitational force exerted by matter in the Galaxy. The calculations show that objects moving around the Galaxy should indeed slow down slightly in the regions of the spiral arms, and linger there longer than elsewhere in their orbits. Thus a coiled wave of higher density builds up where the spiral arms are.

This theory for how spiral arms form is called the **spiral density wave** model. Spiral density waves have been observed directly in the rings of Saturn, another region containing a disk with many objects circling a common center. Astronomers are still debating how such density wave patterns begin, and how they can continue. But the model explains many of the observed properties of the Milky Way's disk (and those of other spiral galaxies).

As gas and dust clouds approach the inner boundaries of an arm and encounter the higher density of slowly moving matter, they collide with it. It is just where the shock of the collision occurs that the compression required for star formation is most likely to take place. This explains why the spiral arms are sites of vigorous star formation, and the home of the youngest stars. In some other galaxies, whose spiral arms we can view face-on, we see young stars, along with the densest dust clouds, occupying the inner boundaries of spiral arms—just as this theory predicts.

Not all the properties of each galaxy can be explained by the spiral density model. Even our galaxy has shorter spurs or segments between spiral arms that are not so easy to explain. Other spiral galaxies have even more chaotic or complicated patterns. Theorists have found other ways, unrelated to spiral density waves, to produce elongated structures that look like spiral arms. For example, great chain reactions of star formation may move progressively through molecular clouds and produce extended regions of young stars that mimic spiral arms (see Section 20.1). It is likely that more than one mechanism contributes to building up the specific design of galaxies such as the Milky Way.

24.3
Stellar Populations in the Galaxy

Two Kinds of Stars

The stars in our Galaxy can be roughly divided into two different groups, or populations, a discovery first made by Walter Baade of the Mount Wilson Observatory in southern California during World War II. As a German national, Baade was not allowed to do war research as many other scientists were, so he was able to make regular use of the Mount Wilson telescopes. Aided by the darker skies resulting from the wartime blackout of Los Angeles, he distinguished two distinct populations of stars in the nearby Andromeda galaxy (Figure 24.11).

Baade was impressed by the similarity of the stars in the Andromeda galaxy's nuclear bulge to those in our own Galaxy's globular clusters and the halo, and by the differences between all of these stars and those found in the spiral arms near the Sun. On this basis, he called all the stars in the halo, nuclear bulge, and globular clusters **population II,** and the bright blue stars in the spiral arms **population I.**

We now know that the populations differ not only in their locations in the Galaxy, but also in other characteristics such as chemical composition and age. Population I stars, found only in the disk, are especially concentrated in the spiral arms, and follow nearly circular orbits around the galactic center. Examples are bright supergiants, main-sequence stars of high luminosity (spectral classes O and B), and members of young open star clusters. Interstellar matter and molecular clouds are found in the same places as population I stars.

Population II stars show no correlation with the location of the spiral arms. These objects are found throughout the disk of the Galaxy, with the greatest concentration toward the nuclear bulge. Many follow elliptical orbits that carry them high above the galactic disk into the halo. Examples include planetary nebulae and RR Lyrae variables. The stars in globular clusters, found almost entirely in the halo and central bulge of the Galaxy, are also classified as population II.

Today we know much more about stellar evolution than did astronomers in the 1940s, and we can explain what causes the differences in the two stellar populations. One important difference has to do with the *ages* of the stars. Population I includes all the stars formed more recently: while some are as much as 10 billion years old, others are still forming today. The Sun, which is about 5 billion years old, is a good example of a population I star. Population II consists entirely of old stars that formed early in the history of the Galaxy: typical ages are 13 billion years and higher.

Clues from Chemical Composition

Measurements of stellar spectra show that a key difference between population I and population II stars is their chemical composition. Nearly all stars appear to be composed mostly of hydrogen and helium, but their abundances of the heavier elements differ. In the Sun and other population I stars, the heavy elements (those heavier than hydrogen and helium) account for about 1 to 4 percent of the total stellar mass. Population II stars in the outer galactic halo and in globular clusters have much lower abundances of the heavy elements—often

Figure 24.11
The spiral galaxy in Andromeda (M31) looks very much like our own Galaxy. Note the bulge of older, yellowish stars in the center, the bluer and younger stars in the outer regions, and the dust in the disk that blocks some of the light from the bulge. (Photo by Tony Hallas) *Inset:* Walter Baade (1893–1960). (Caltech Archives)

less than one-tenth or even one-hundredth the concentrations found in the Sun.

As discussed in earlier chapters, heavy elements are created in stars. They return to the Galaxy's reserves of raw material when stars die, only to be recycled into new generations of stars. Thus, as time goes on, stars are born with larger and larger supplies of heavier elements. The abundances of the elements demonstrate that the two stellar populations must have been born at different times. Population II stars seem to represent the early generations of the Galaxy. We might call them the senior citizens of the Milky Way; like many human seniors, they lived during simpler times (at least as far as the abundances of the elements are concerned).

The Real World

With rare exceptions, we should never trust any theory that divides the world into just two categories. Since Baade's pioneering work, we have learned that the notion that all stars can be characterized as either old and poor in heavy elements, or younger and rich in heavy elements, is an oversimplification. The idea of two populations helps organize our initial thoughts about the Galaxy, but we now know it cannot explain everything we observe.

For example, the stars in the nuclear bulge of the Galaxy are all fairly old. Their average age is in the range of 11 to 14 billion years, and none is younger than about 5 billion years old. Yet the abundance of heavy elements in these stars is about that of the Sun. Astronomers think that star formation in the crowded nuclear bulge occurred very rapidly just after the Milky Way Galaxy formed, so the stars there were soon enriched with heavy elements expelled in supernova explosions from the first generations of massive stars. Thus even stars that formed in the bulge 11 to 14 billion years ago started with a good supply of heavier elements.

Exactly the opposite situation occurs in the Small Magellanic Cloud, a small galaxy that orbits our own. Even the youngest stars in this galaxy are deficient in heavy elements, presumably because star formation has occurred so slowly that there have been, so far, relatively few supernova explosions. (Smaller galaxies also have more trouble holding onto the gas produced in supernova explosions long enough to recycle it; low-mass galaxies exert only a modest gravitational force, and the high-speed gas ejected by supernovae can easily escape from them.) The elements with which a star is endowed depend not only on when it forms in the history of its galaxy, but also on how many stars in its part of the galaxy have already completed their lives by the time the star is ready to form.

The Mass of the Galaxy

The Sun, like all the other stars in the Galaxy, orbits the center of the Milky Way. Our star's orbit is nearly circular and lies in the Galaxy's disk. The speed of the Sun in its orbit is about 220 km/s, which means it takes us approximately 225 million years to go once around. We call the period of the Sun's revolution about the nucleus the *galactic year*; it is a long time compared to human time scales. During the entire lifetime of the Earth, only about 20 galactic years have passed. And we have gone a mere fraction of the way around the Galaxy in all the time that humans have gazed into the sky.

Kepler Helps Weigh the Galaxy

We can use this information to estimate the mass of the Galaxy inside the Sun's orbit (just as we could "weigh" the Sun by monitoring the orbit of a planet around it—see Chapter 2). Let's assume that the Sun's orbit is circular and that the Galaxy is roughly spherical, both of which are fairly accurate assumptions. Long ago, Newton showed that if you have matter distributed in the shape of a sphere, then it's simple to calculate the pull of gravity on some object just outside that sphere: you can assume gravity acts as if all the matter were concentrated at a point in the center of the sphere. So, for our purposes we have a point in the center of the Galaxy, containing all the mass that lies inward of the Sun's position, and we have the Sun orbiting that point about 26,000 LY away.

This is the sort of situation to which Kepler's third law (as modified by Newton) can be directly applied. To use the formula in Section 2.3, we must put the distance in astronomical units and the period in years. There are 6.3×10^4 AU in 1 LY, so 26,000 LY becomes 1.6×10^9 AU. If it takes about 2.25×10^8 years for the Sun to orbit the center of the Milky Way, Kepler's law says:

$$M_{Galaxy} + M_{Sun} = a^3/P^2 = (1.6 \times 10^9)^3/(2.25 \times 10^8)^2$$
$$= \text{approximately } 10^{11} \, M_{Sun}$$

Note that Kepler's law gives you the sum of the masses, but since the mass of the Sun is completely trivial compared to the mass of the Galaxy, then for all practical purposes the result is the mass of the Milky Way. More sophisticated calculations based on complicated models give a similar result.

This estimate tells us only how much mass is contained in the volume inside the Sun's orbit. It is a good estimate for the *total* mass of the Galaxy if and only if there is hardly any mass outside the Sun's orbit. For many years astronomers thought this assumption was reasonable. The number of bright stars and the amount of *luminous matter* (meaning any material from which we can detect elec-

tromagnetic radiation) both drop off dramatically at distances of more than about 30,000 LY from the galactic center.

A Galaxy of Mostly Invisible Matter

In science, what seem to be reasonable assumptions can later turn out to be wrong (which is why we continue to do observations and experiments every chance we get). Observations now show that there is a lot more to the Milky Way than meets the eye (or the telescope)! While there is relatively little luminous matter beyond 30,000 LY, there turns out to be a lot of *invisible* material at large distances from the galactic center.

We can understand how astronomers reached this strange conclusion by remembering that, according to Kepler's third law, objects orbiting at large distances from a massive object move more slowly than do objects closer to that central mass. We have already seen an example of this idea in the case of the solar system. The outer planets move more slowly in their orbits than do those close to the Sun.

There are a few objects, including some gas clouds and globular clusters, that lie well outside the luminous boundary of the Milky Way. If most of the mass of our Galaxy were concentrated within the luminous region, then these very distant objects should travel around their galactic orbits at lower speeds than, for example, that of the Sun.

It turns out, however, that the few objects seen at large distances from the luminous boundary of the Milky Way Galaxy are *not* moving more slowly than the Sun. There are some globular clusters and RR Lyrae stars between 30,000 and 150,000 LY from the center of the Galaxy, and their orbital velocities are even greater than the Sun's (Figure 24.12).

What do these higher speeds mean? Kepler's third law tells us how fast objects must orbit a source of gravity if they are neither to fall in (because they move too slowly), or to escape (because they move too fast). If the Galaxy only had the mass calculated in the previous section, the high-speed outer objects should long ago have escaped the grip of the Milky Way. The fact that they have not done so means that our Galaxy must have more gravity than can be supplied by the luminous matter—in fact, a lot more gravity. The increasing speed of these outer objects tells us that the source of this gravity must extend outward from the center far beyond the Sun's orbit.

If the gravity were supplied by stars or something else that gives off radiation, we should have spotted this outer material long ago. We are therefore forced to the reluctant conclusion that this matter is *invisible* and has, except for its gravitational pull, gone entirely undetected!

Studies of the motions of the most remote globular clusters show that the total mass of the Galaxy out to a ra-

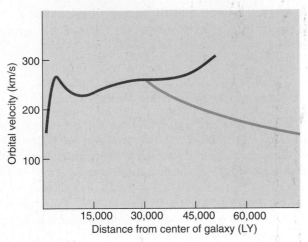

Figure 24.12
The orbital speed of carbon monoxide (CO) gas at different distances from the center of the Milky Way Galaxy is shown in red. The blue curve shows what the rotation curve would look like if all of the matter in the Galaxy were located inside a radius of 30,000 LY. Instead of going down, the speed of gas clouds farther out goes up, indicating a great deal of mass beyond the Sun's orbit.

dius of 150,000 LY is about 10^{12} M_{Sun}, ten times greater than the amount of matter inside the solar orbit. Theoretical arguments suggest that this **dark matter** (as astronomers have come to call it) is distributed in an enormous sphere around the Galaxy. Some astronomers call this the Galaxy's **corona,** a name taken from the very faint but extensive outer atmosphere that surrounds the Sun. But what could this corona be made of?

Let's look at a list of suspects taken from our study of astronomy so far. Since this matter is invisible, it clearly cannot be in the form of ordinary stars. And it cannot be gas in any form (remembering that there has to be a *lot* of it). If it were neutral hydrogen, its 21-cm radiation would have been detected. If it were ionized hydrogen, it should be hot enough to emit visible radiation. If a lot of hydrogen atoms out there had combined into hydrogen molecules, these should produce dark features in the ultraviolet spectra of objects lying beyond the Galaxy: such features have not been seen. Nor can the corona consist of interstellar dust, since dust in the required quantities would block the light from distant galaxies.

The dark matter cannot be black holes of stellar mass or old neutron stars, since the accretion of interstellar matter onto such objects would produce more x rays than are observed. Also, recall that the formation of black holes and neutron stars is preceded by supernova explosions, which scatter heavy elements into space to be incorporated into subsequent generations of stars. If the corona consisted of an enormous number of black holes and neutron stars, then the young stars we observe in our Galaxy

today would contain much larger abundances of heavy elements than they actually do.

The possibilities that remain to account for the dark matter in the Galaxy are low-mass objects such as planets or brown dwarfs, white dwarfs formed from an early generation of stars that have now cooled and ceased to shine, black holes with masses a million times the mass of the Sun, or exotic subatomic particles of a type not yet detected on Earth. Very sophisticated (and difficult) experiments are now under way to look for evidence in favor of one or another of these possibilities. Recent measurements suggest that white dwarfs make up at least some of the dark matter (see Chapter 27).

One important thing to note is that the problem of dark matter is not by any means confined to the Milky Way. Observations show that dark matter must also be present in other galaxies (whose outer regions also orbit too fast "for their own good") and, as we will see, even in great clusters of galaxies, whose members move around under the influence of far more gravity than can be accounted for by luminous matter alone.

Stop a moment and consider how astounding this conclusion is. Perhaps as much as 90 percent of the mass in our Galaxy (and many other galaxies) is invisible, and we do not even know what it is made of! The stars and raw material we *can* observe may be merely the tip of the cosmic iceberg (to use an old cliche); underlying it all may be other matter, perhaps familiar, perhaps startlingly new. Understanding the nature of this dark matter is one of the great challenges of astronomy today, and we will return to this problem in later chapters.

24.5

The Nucleus of the Galaxy

Another complex and puzzling region of the Galaxy is its center, which lies in the direction of the constellation Sagittarius. As discussed, we cannot see the nucleus in visible light because of absorption by the interstellar dust that lies between us and the galactic center (Figure 24.13). Light from the central region of the Galaxy is dimmed by a factor of a trillion (10^{12}). Infrared and radio radiation, however, whose wavelengths are long compared with the sizes of the interstellar grains, flows around the dust particles and can reach us (see the opening figure for this chapter). In fact, the very bright radio source in the nucleus, known as Sagittarius A, was the first cosmic radio source discovered.

Astronomers have used observations at all these wavelengths to try to determine what lies at the center of the Milky Way. It is a crowded, complicated region, full of gas, dust, stars, and stellar corpses; its exact geography is very difficult to sort out. Perhaps the question of most interest to astronomers is whether or not a massive black hole resides there. A great deal of effort has gone into trying to answer this question, and as we will see next, the jury is still out.

The Central Few Light Years

Let's take an imaginary journey to the heart of our Galaxy, starting from a distance of a few dozen light years and going as close to the center as observations permit. The first structure we encounter is a clumpy and rather irregular donut consisting of clouds of dust and gas, with much of the gas cool enough to form molecules. These clouds form a thick ring with an outer diameter of at least 25 LY and an inner diameter of about 10 LY. The individual molecular clouds typically have sizes in the range of 0.5 to 1.5 LY (Figure 24.14). The ring is rotating about the galactic center, but the motions of the individual clumps are turbulent, and the clouds sometimes collide with one another.

We can examine the gas in the ring with spectrographs, and use the Doppler effect to derive its orbital speed. Just as the Sun holds the planets in their orbits, so too there must be some mass inside the ring that keeps the fast-moving molecular clouds from flying off into space. We can use Kepler's third law as modified by Newton to estimate just how large this mass is. Given the observed rotational velocity of the ring, we calculate that the

Figure 24.13
This wide-angle picture covers over 50° of the sky in the direction of the center of the Milky Way. The center itself cannot be seen because of the vast quantities of dust in that direction. (Anglo-Australian Telescope Board)

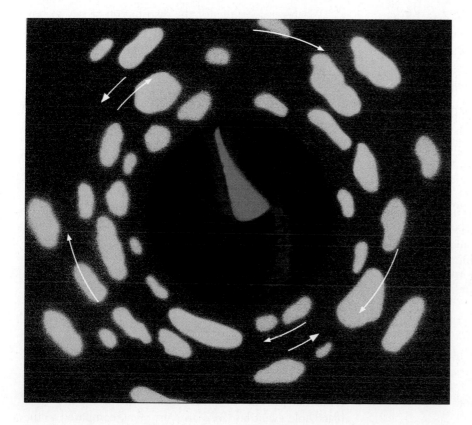

Figure 24.14
Schematic drawing of the central 20 LY of the Galaxy. Streamers of ionized gas are falling into the galactic center. Surrounding the central regions is a ring of dust, and beyond that lie individual clouds of gas. The reddish-brown region in this drawing corresponds to the region of radio emission shown in Figure 24.15.

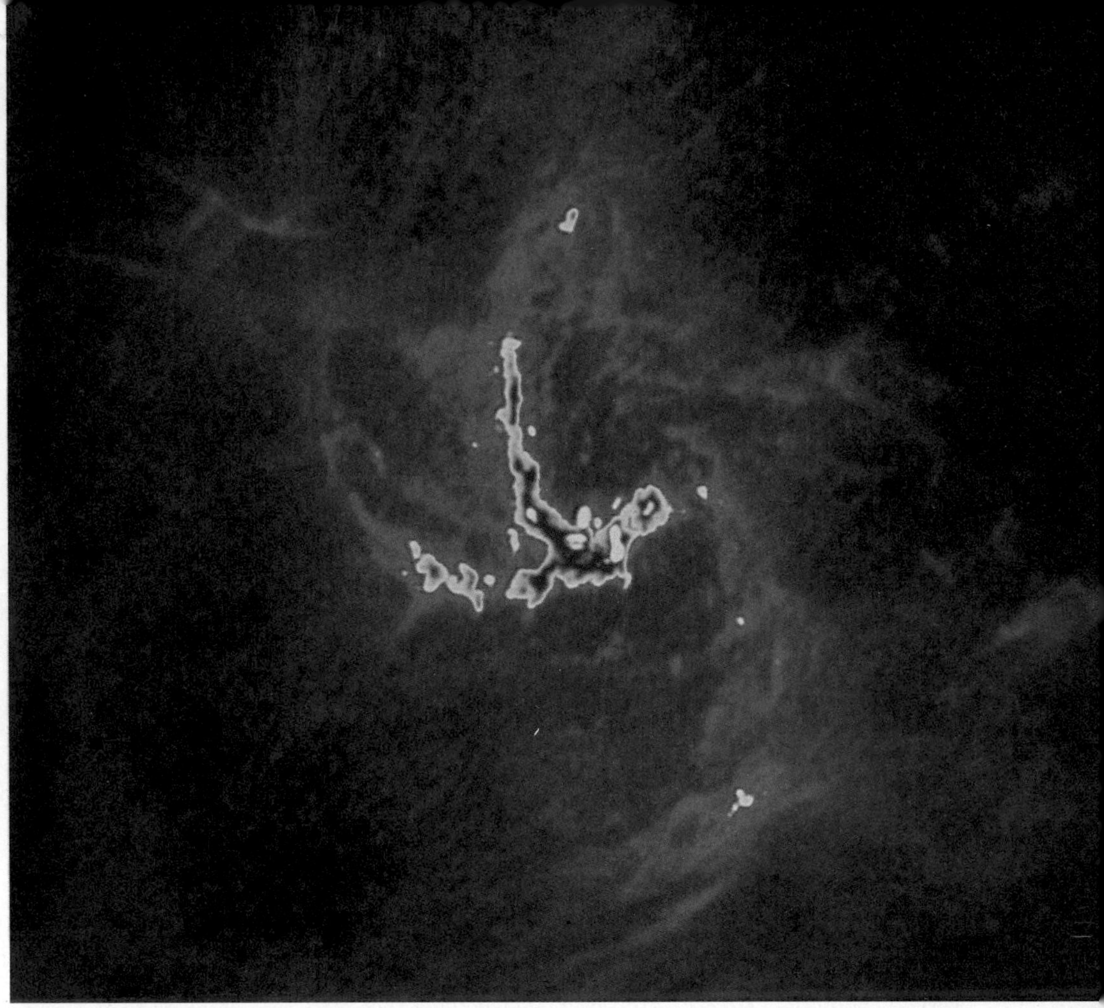

Figure 24.15
This radio image of a region approximately 10 LY across at the center of our Galaxy shows emission from hot gas. The colors indicate the amount of gas, ranging from the least (blue) to the most (red). The most likely interpretation of the image is that we see gas that is falling into the central region from the ring around it. The bright red region in the middle is the location of the compact radio source at the center of the Galaxy. (K. Lo et al., National Radio Astronomy Observatory/AUI)

total mass required is 2 to 5 million times the mass of the Sun. This is an impressive amount, but remember we are dealing with the crowded center of the Galaxy, so we should probably expect to find that it contains a lot of material. Let's move inside the ring to see whether we can figure out what and where this mass is.

To our surprise, just inside the ring we find hardly any interstellar matter at all. Measurements show that the density of gas and dust is 10 to 100 times higher in the ring than in the region just inside it. That is, the ring surrounds a cavity containing very little dust or cold gas. There are some streamers of hot, ionized gas (Figure 24.15), but the total mass of the ionized gas in the central few light years of the Galaxy is only about 100 M_{Sun}. This is not very much at all, considering how many stars we find in the inner region. Some remarkable event must have cleared out this cavity within the last 100,000 years; one possibility is an explosion of some kind at the center of the Galaxy, but we really don't know.

We do know that whatever cleared out the cavity must have happened recently (on a cosmic time scale), because material falling inward would soon blur the sharp edge that now separates the ring from its nearly empty interior. Collisions between the clouds also tend to eradicate individual clumps and produce a smooth ring of material, which is not what we observe.

Whatever accounts for the distribution of raw material in this area, the amount of gas we can measure in the cavity does not come close to accounting for the mass the ring motions require to be in the center of the Galaxy. What else is inside the ring? Infrared observations show a huge number of stars (perhaps as many as a million) packed into a region a few light years in diameter (Figure 24.16), where the average distance between stars is so small that light takes weeks, not years, to travel between neighboring stars. If there are planets around any of them, their nights must be bright with the brilliance of the star-studded sky.

But even the mass of all these stars is not enough to add up to the 2 to 5 million solar masses that we are looking for. To discover what else is inside the ring, we must continue our journey to the very core of this inner region (and think a bit about how we could detect a black hole if one is, in fact, there).

The Central Energy Source

As we saw in Chapter 23, showing the existence of a black hole is a challenge because the hole itself emits no radiation. We must examine the effect of the black hole on material outside the event horizon. One problem is that black hole event horizons are very small compared to the

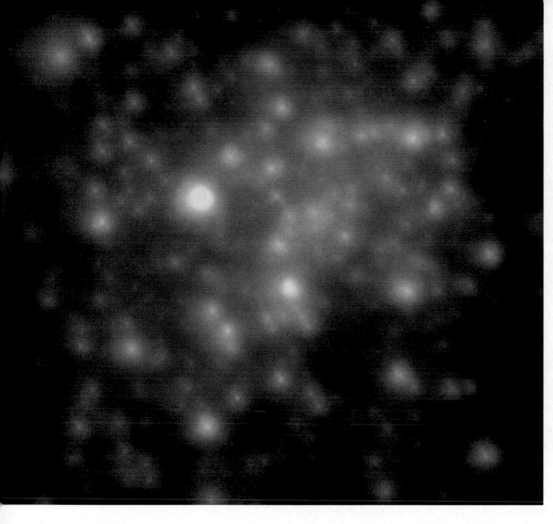

Figure 24.16
This marvelous new infrared image shows a region at the center of the Galaxy that is about 6 LY across and is filled with closely packed stars. All the stars seen on this image are red giants; the differences in color are caused by the different amounts of reddening of their light by dust between us and the center. However, not all the reddish objects are stars. Some are regions of ionized hydrogen (HII), inside which there must be some hot stars whose energetic light does the ionizing. None of these objects can be seen with visible light since the dust completely obscures our view at visible wavelengths. (Ian Gatley and Michael Merrill/National Optical Astronomy Observatories)

kinds of measurements we can make from our vantage point 26,000 LY away. A black hole with a mass of about 3 million M_{Sun} would have a radius similar to that of our own Sun.

However, such a black hole might be surrounded by a whirling *accretion disk* of material not much bigger than the orbits of the planets in our own solar system. In theory, we could again measure the speed of the orbiting material to see how much mass would have to be squeezed inside to account for the motions. Unfortunately, we cannot see fine enough detail with existing telescopes to measure the velocities of gas and stars that are only a few tens of astronomical units from the center of the Galaxy.

Indirect evidence, however, suggests that the mass may be very highly concentrated at the core of the Milky Way. An intense radio source, called Sagittarius A°, is at or very near what we believe to be the exact center of the Galaxy. Measurements with the VLA radio telescope show that the diameter of this radio source is no larger than the diameter of Jupiter's orbit (10 AU), and it may even be somewhat smaller—close to the size we are looking for. The radio radiation would come from the disk around the black hole, from material being heated as it spirals inward toward the event horizon.

Another argument in favor of thinking that Sagittarius A° has a high mass is that it is stationary. Ordinary stars in the galactic center change their positions by measurable amounts because of the gravitational tugs and pulls of other nearby stars, but Sagittarius A° does not. The most likely explanation is that Sagittarius A° is so much more massive than ordinary stars that it is not measurably moved by their gravitational influence.

Suppose there is a black hole of several million solar masses in the center of the Galaxy. Where did its mass come from? To start such a massive black hole, all you need is one very massive star early in the history of the Galaxy, going quickly through its life and dying by collapsing within its event horizon. Then gas, dust, and other stars that collide with this black hole could slowly be swallowed. Over billions of years, this could increase the mass of the original black hole enormously.

At the present time, matter is falling into the galactic center at the rate of about 1 M_{Sun} per 1000 years. If matter had been falling in at the same rate for about 5 billion years, then it would have easily been possible to accumulate the matter needed to form a black hole with a mass of several million solar masses. And it isn't merely gas and dust that could serve as a "meal" for such a black hole. The density of stars near the galactic center is such that we would expect a star to pass near the black hole and be swallowed by it every few thousand years.

So scientists find a great deal of circumstantial evi-

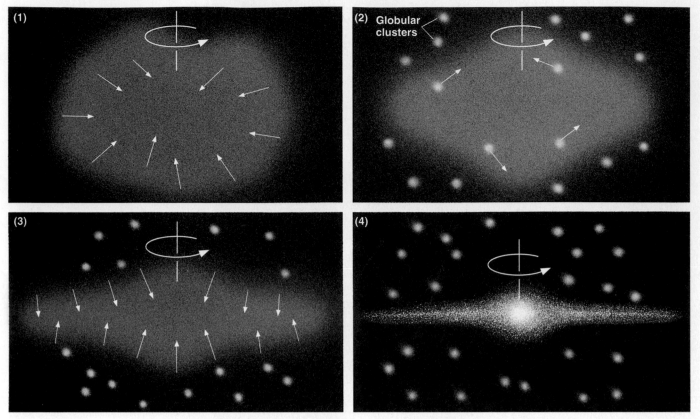

Figure 24.17
The Galaxy probably formed from an isolated, rotating cloud of gas that collapsed due to gravity. Halo stars and globular clusters either formed prior to the collapse or were formed elsewhere and attracted by gravity to the Galaxy early in its history. Stars in the disk formed late, and the gas from which they were made was contaminated with heavy elements produced in early generations of stars.

dence to support the idea that there is a black hole in the center of the Galaxy. But this may not be the only explanation. Over the next few years, astronomers will turn a host of new observational tools—gamma-ray, x-ray, infrared, and radio telescopes—toward the galactic center in an attempt to prove or disprove the existence of a black hole at the heart of the Milky Way. Whatever the final answer turns out to be, the possibility of such a supermassive black hole is not restricted to the center of our own Galaxy. As we will see in later chapters, there is also evidence that such black holes exist in other galaxies, and in the mysterious powerhouses called quasars as well.

24.6

The Formation of the Galaxy

Once astronomers understood what our Galaxy is like, they began to develop models for its formation and evolution. The first models were quite simple in that they assumed the Galaxy developed in isolation, uninfluenced by interactions with other galaxies. The flattened disk shape of the Galaxy suggests that it formed through a process similar to the one that created the Sun and solar system (see Chapter 13). Building on this idea, early models assumed that the Galaxy, like the solar system, formed from a single rotating cloud.

The Protogalactic Cloud

Since the oldest stars—those in the halo and in globular clusters—are distributed in a sphere centered on the nucleus of the Galaxy, it makes sense to assume that such a *protogalactic cloud* was roughly spherical in shape. The oldest stars in the halo have ages of about 15 billion years, and so we estimate that the formation of the Galaxy began about that long ago. Then, just as in the case of the solar system, the protogalactic cloud collapsed and formed a thin rotating disk. Any stars born before the cloud collapsed did not participate in the collapse, but continue to orbit in the halo to the present day (Figure 24.17).

Gravitational forces caused the gas in the thin disk to fragment into clouds or clumps with masses like those of star clusters. These individual clouds then fragmented further to form individual stars. Since the oldest stars in

Light Pollution and the Milky Way

Bright lights have robbed most city dwellers of the pleasure of looking at stars. As we mentioned, many people in urban areas have never even seen the Milky Way. When light from street lamps and advertising signs shines into the sky, where it has no practical value (and wastes energy), we call it "light pollution." The amount of light that now flows upward has brightened the night sky so much that it is difficult or impossible to see the constellations, watch for meteor showers, or identify the occasional passing comet.

Artificial lights are also a problem for many major observatories, making it difficult for astronomers to detect faint stars and galaxies against the bright sky. More and more, astronomers must move their instruments to remote islands or mountains to escape the persistent glow of civilization.

But such lights, you may say, are the price we must pay for living in large cities. Aren't lights essential for safety and security? Fortunately, there is a solution that provides high-quality lighting while minimizing the amount of light pollution. The key first of all is to direct the light to where it is most useful. This means that outdoor lighting should be shielded so that light is emitted only downward to streets, sidewalks, and sports fields. Engineers have designed lighting fixtures that shield street lights in such a way that no light is projected upward. Some reflected light will brighten the sky no matter what we do, but with proper shielding and top reflectors, we could in theory direct much more of our light to ground level.

For astronomers, the concern is not just the amount of light shining upwards into the sky, but also the specific wavelengths in which that light is emitted. Much of modern astronomy depends on spectroscopy, where we must be able to detect light at the specific wavelengths corresponding to various elements. We therefore need a source of artificial light that will interfere with as few wavelengths as possible.

The accompanying diagram shows the spectra of three types of light sources. Two pollute at many wavelengths. But one, the low-pressure sodium lamp, emits light at only a few discrete wavelengths, while the rest of the spectrum is free of contaminating light. This is the kind of light that astronomers prefer cities to use. Fortunately, low-pressure sodium lights are extremely energy-efficient and thus are the most cost-effective form of street lighting. They are also the best light for visual acuity, since they emit most of their energy in the yellow part of the spectrum where the eye is most sensitive.

Check the lights in your own city. What type are they? (Your astronomy instructor may be able to lend you an inexpensive *diffraction grating* so you can observe the spectra of some street lights and compare them to the accompanying spectra.) Are the lights shielded? Notice the city lights next time you are in a plane at night. Do you see the lights themselves? If so, then some of the light is being emitted directly upward, where it does no good. Or do you see only the light reflected from the pavement? In this case, the lighting is properly shielded. If you live in or near a large city, check the sky. What part of it is contaminated with artificial light? In what directions is it brightest? What is the faintest star you can see? Now drive away from the city and notice how the sky changes. How far away do you have to go before the sky appears truly dark and you can see the Milky Way?

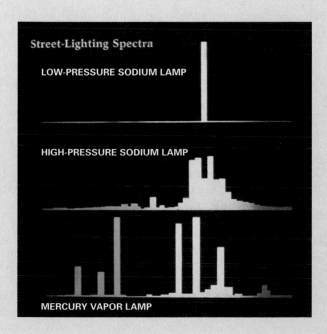

Comparison of the spectra of three forms of street lighting. Mercury vapor lights are the most common, but much of their energy is emitted in the red and blue regions of the spectrum, at wavelengths to which the eye is not very sensitive. In contrast, low-pressure sodium lamps emit all of their energy in a very narrow wavelength region where the eye is highly sensitive. This type of lighting leaves most of the spectrum free of radiation that would contaminate observations of faint stars and galaxies.

the disk are nearly as old as the stars in the halo, the collapse must have been rapid (astronomically speaking), requiring perhaps no more than a few hundred million years.

Recently, astronomers have begun to find hints that the evolution of the Galaxy has not been quite as peaceful as this traditional model suggests. Globular clusters differ in age by as much as 3 billion years; this means they could not all have formed during the few hundred million years before the interstellar matter collapsed to form the galactic disk, as the traditional model predicts. Studies also show that some globular clusters orbit the center of the Galaxy in a direction opposite that of the majority of the globulars and of the galactic disk. These "backward" globular clusters appear to be 2 to 3 billion years younger than globular clusters that rotate in the normal direction, and they tend to lie farther from the center of the Galaxy.

Collision Victims

Still more recently, in the direction of Sagittarius, astronomers have discovered a small new galaxy that is a satellite of our own Galaxy—just as the Moon is a satellite of the Earth. The Sagittarius dwarf galaxy is only 50,000 LY away from the nucleus of the Milky Way and is the closest galaxy known. It is very elongated, and its shape indicates that it is being torn apart by the Galaxy's gravitational force—just as Comet Shoemaker-Levy was torn apart when it passed too close to Jupiter (see Chapter 12).

Within another 100 million years, its stars will be captured and dispersed through the halo of the Milky Way.

The Milky Way may have more collisions in store. The Large and Small Magellanic Clouds (see Figure 14 in the Prologue), two other nearby satellite galaxies, are spiraling ever closer to our Galaxy and, according to calculations, will merge with it in the distant future. The Milky Way Galaxy has eight other small satellite galaxies, and at least some of these dwarfs appear to have been split off from the Magellanic Clouds when they passed close to the Milky Way.

All of this evidence suggests that the evolution of the Galaxy has been influenced by interactions and collisions with other galaxies. Astronomers now believe that the Milky Way formed in two stages. The first stage proceeded as shown in Figure 24.17. About 2 billion years after the formation of the majority of globular clusters, which rotate in the forward direction, the Galaxy acquired additional stars and globular clusters from one or more satellite galaxies that were captured when they ventured too close. (In the same way, you might recall, some of the satellites in our own solar system have retrograde orbits and are thought to be captured asteroids or comets.)

It appears likely that over the next few billion years, the Galaxy will gather still more stars from neighboring galaxies. We are coming to realize that environmental influences play an important role in determining the properties of our own Galaxy. In future chapters we will see that collisions and mergers are a major factor in the evolution of many other galaxies as well.

Summary

24.1 The Sun is located in the outskirts of the **Milky Way Galaxy.** The Galaxy consists of a disk containing dust, gas, and young stars; a **nuclear bulge** containing old stars; and a spherical **halo** containing very old stars, globular clusters, and RR Lyrae variables. Analysis by Shapley of the distribution of globular clusters gave the first indication that the Sun is not located at the center of the Galaxy. Radio observations at 21 cm show that cold hydrogen is confined to a flat disk with a thickness near the Sun of only 400 LY. Dust is found in the same locations; very little exists outside the Sun's orbit. The most-massive molecular clouds, where star formation is active, are concentrated in the spiral arms.

24.2 The Galaxy has four main **spiral arms** and several short spurs; the Sun is located on one of these spurs. Measurements show that the Galaxy does not rotate as a solid body, but instead its stars follow Kepler's laws; those closer to the galactic center complete their orbits more quickly. The **spiral density wave** theory is one way to ac-

count for the spiral arms. Calculations show that the gravitational forces within the Galaxy cause stars and gas clouds to slow down in the vicinity of the spiral arms, thereby leading to higher densities of material. When molecular clouds attempt to pass through these regions of higher density, star formation is triggered.

24.3 We can roughly divide the stars in the Galaxy into two categories. Old stars with few heavier elements are referred to as **population II** stars and are found in the halo, in globular clusters, and in the nuclear bulge. **Population I** stars, containing more heavier elements, are found in the disk and are especially concentrated in the spiral arms. The Sun is a member of population I. Population I stars formed after previous generations of stars produced heavy elements and ejected them into the interstellar medium.

24.4 The mass of the Galaxy can be determined by measuring the orbital velocities of stars or interstellar matter.

The Sun revolves completely around the galactic center in about 225 million years. The total mass of the Galaxy is about 10^{12} M_{Sun}, and about 90 percent of this mass consists of **dark matter** that emits no electromagnetic radiation and can be detected only because of the gravitational force it exerts on visible stars and interstellar matter. This dark matter is located mostly in the Galaxy's **corona;** its nature is not understood at present.

24.5 The central region of the Galaxy contains a ring of molecular clouds surrounding a cavity that is surprisingly empty of gas and dust, but that contains a very crowded cluster of stars. Measurements of the velocities of stars and gas show that the region within 2 LY of the galactic center contains a mass roughly 3 million times that of the Sun. A massive black hole can explain these observations, but it is not yet certain that a black hole is the only possibility.

24.6 The Galaxy formed about 15 billion years ago. Models suggest that the stars in the halo and globular clusters formed first, while the Galaxy was spherical. The gas, somewhat enriched in heavy elements by the first generation of stars, then collapsed from a spherical distribution to a disk-shaped distribution. Stars are still forming today from the gas and dust that remain in the disk. Star formation occurs most rapidly in the spiral arms, where the density of interstellar matter is highest. There is evidence that the Galaxy captured additional stars and globular clusters from small galaxies that ventured too close to the Milky Way.

Review Questions

1. Explain why we see the Milky Way as a faint band of light stretching across the sky.

2. Explain where (and why) in a spiral galaxy you would expect to find globular clusters, molecular clouds, and atomic hydrogen.

3. Describe several characteristics that distinguish population I from population II stars.

4. Briefly describe the three parts of our Galaxy—the disk, the halo, and the corona.

5. Describe the evidence indicating that a black hole may be at the center of our Galaxy.

6. Explain why the abundances of heavy elements in stars correlate with their positions in the Galaxy.

Thought Questions

7. Suppose the Milky Way were a band of light extending only halfway around the sky (that is, in a semicircle). What, then, would you conclude about the Sun's location in the Galaxy? Give your reasoning.

8. The globular clusters revolve around the Galaxy in highly elliptical orbits. Where would you expect the clusters to spend most of their time? (Think of Kepler's laws.) At any given time, would you expect most globular clusters to be moving at high or low speeds with respect to the center of the Galaxy? Why?

9. Consider the following five kinds of objects: (1) open cluster, (2) giant molecular cloud, (3) globular cluster, (4) group of O and B stars, and (5) planetary nebulae.
   a. Which occur only in spiral arms?
   b. Which occur only in the parts of the Galaxy *other than* the spiral arms?
   c. Which are thought to be very young?
   d. Which are thought to be very old?
   e. Which have the hottest stars?

10. The dwarf galaxy in Sagittarius is the one closest to the Milky Way, yet it was discovered only in 1994. Can you think of a reason it was not discovered earlier? (*Hint:* Think about what else is in its constellation.)

11. Why does star formation occur primarily in the disk of the Galaxy?

12. Where in the Galaxy would you expect to find Type II supernovae, which are the explosions of massive stars that go through their lives very quickly? Where would you expect to find Type I supernovae, which involve the explosions of white dwarfs?

13. You are captured by space aliens who take you inside a complex cloud of gas and dust. To escape, you need to make a map of the cloud. Luckily, the aliens have a complete astronomical observatory with equipment for measuring all the bands of the electromagnetic spectrum. Use what you have learned in this chapter to explain what kinds of maps you would make of the cloud to plot your most effective escape route.

14. Suppose that stars evolved without losing mass—that once matter was incorporated into a star, it remained there forever. How would the appearance of the Galaxy be different from what it is now? Would there be population I and population II stars? What other differences would there be?

15. Use the data in Appendix 12 to identify the galaxies that are satellites of the Milky Way. How far away are they on average? How do their sizes compare with that of our own Galaxy?

16. Suppose the average mass of a star in the Galaxy were one-third of a solar mass. Use the value for the mass of the Galaxy given in the text to find how many stars the system contains.

17. Assume that the Sun orbits the center of the Galaxy at a speed of 220 km/s and a distance of 26,000 LY from the center.
   a. Calculate the circumference of the Sun's orbit, assuming it to be approximately circular.
   b. Calculate the Sun's period, the "galactic year."
   c. Use Newton's formulation of Kepler's third law to calculate the mass of the Galaxy inside the orbit of the Sun.
   d. It is estimated that the total mass of the Galaxy is ten times that within the Sun's orbit. If this mass were to collapse inside the orbit of the Sun, what would be the length of the galactic year?

   e. In this case, what would be the Sun's new orbital speed?

18. Construct a rotation curve for the solar system by using the orbital velocities of the planets. How does this curve differ from the rotation curve for the Galaxy? What does it tell you about where most of the mass in the solar system is concentrated?

19. Calculate the rotation speed of the gas ring at the galactic center if the mass inside the ring is 2×10^6 M_{Sun} and the radius of the ring is 25 LY.

20. Calculate how much mass would fall into a black hole at the center of the Galaxy in a billion years at the current rate of infall of 1 M_{Sun} per 1000 years.

Suggestions for Further Reading

Bok, B. "Harlow Shapley and the Discovery of the Center of Our Galaxy" in Neyman, J., ed. *The Heritage of Copernicus.* 1974, MIT Press.

Croswell, K. *Alchemy of the Heavens.* 1995, Doubleday/Anchor. A popular-level review of current Milky Way research.

Dame, T. "The Molecular Milky Way" in *Sky & Telescope,* July 1988, p. 22.

Davis, J. *Journey to the Center of the Galaxy.* 1991, Contemporary Books. A journalist reviews our understanding of the Galaxy.

Ferris, T. *Coming of Age in the Milky Way.* 1988, Morrow. Historical survey.

Henbest, N. and Couper, H. *The Guide to the Galaxy.* 1994, Cambridge U. Press. Illustrated history of the growth of our understanding, plus a tour of the Milky Way.

Jayawardhana, R. "Destination: Galactic Center" in *Sky & Telescope,* June 1995, p. 26.

Mateo, M. "Searching for Dark Matter" in *Sky & Telescope,* Jan. 1994, p. 20.

Townes, C. and Genzel, R. "What Is Happening at the Center of Our Galaxy" in *Scientific American,* Apr. 1990, p. 46.

van den Bergh, S. and Hesser, J. "How the Milky Way Formed" in *Scientific American,* Jan. 1993, p. 72.

Verschuur, G. "In the Beginning" in *Astronomy,* Oct. 1993, p. 40. On globular clusters and the Galaxy.

Verschuur, G. "Journey into the Galaxy" in *Astronomy,* Jan. 1993, p. 32. Excellent tour.

1. Choose the *Mercator Projection* with a zoom factor of 0.2. Compare the distributions of stars with spectral types O and B with that for types K and M.

Which plot shows the Milky Way and why?

Can you still see the Milky Way if you display all stars?

Select all the types of nebulae (stars and all other deep-sky objects off). Do they show the outline of the Milky Way?

Which object most clearly shows the direction of the Galaxy's center?

2. Display only galaxies. Why is there a band where galaxies do not appear?

This is part of the deepest image ever taken of the universe of galaxies. Two hundred seventy-six exposures, taken with the Hubble Space Telescope over 150 consecutive orbits, were combined to make an image that shows fainter galaxies than have ever been glimpsed before, including some that are a record-breaking 4 billion times fainter than the human eye can see. Hundreds of galaxies are visible in this section, which is about one-fourth of the full Hubble image. Some are so far away that we are seeing them as they were only a billion years or so after the universe began its expansion. The region photographed, a tiny patch of sky just above the Big Dipper, was selected because it is far from the disk of the Milky Way and contains very few nearby galaxies. (R. Williams, the Hubble Deep Field Team, & NASA)

CHAPTER **25**

Galaxies

Thinking Ahead

In this chapter we reach vistas so great they completely stagger the human imagination. Galaxies dot the sky in every direction, as far as our telescopes permit us to peer. How can we measure the distances to galaxies so far away that their light takes millions and billions of years to reach us?

Our voyage now leaves the confines of the Milky Way as we begin our exploration of the realm of the other galaxies. Looking outward, we are awestruck at the vast number of "island universes"—each containing billions of stars. There are so many galaxies that a modern telescope allows us to see millions and millions of them just in the bowl of the Big Dipper. Like tourists from a small town, amazed by the extent and number of the great cities they are visiting, we are just coming to realize how much exists beyond the borders of our home.

We begin our exploration of galaxies with a guide to their properties, which is the thrust of this chapter. In the chapters that follow we will look more carefully at how such galaxies change with time, and at the role environment plays in their development.

The very idea that other galaxies exist was controversial for a long time. As late as 1920, many astronomers, including Harlow Shapley, thought the Milky Way encompassed all there was in the universe. The proof in 1924 that our Galaxy is not unique was one of the great advances in our understanding of our place in the universe.

There is nothing like
astronomy to pull the stuff
out of man.
His stupid dreams and red-
rooster importance: let him
count the star-swirls.

From the poem *Star-Swirls* by
Robinson Jeffers (1924)

The Great Nebula Debate

Growing up at a time when the Hubble Space Telescope orbits above our heads, and giant telescopes are springing up on the great mountaintops of the world, you may be surprised to learn that we were not sure about other galaxies for such a long time. But bear in mind that today's giant telescopes and electronic detectors are recent additions to the astronomer's toolbox (see Chapter 5).

With the telescopes available in earlier centuries, galaxies looked like small fuzzy patches and were difficult to distinguish from the star clusters and gas and dust clouds that are part of our own Galaxy. All objects that were not sharp points of light were given the same name, "nebulae," the Latin word for clouds. Because even their precise shapes were often hard to make out, and no techniques had yet been devised for measuring their distances, the nebulae were the center of much debate and discussion among astronomers.

As early as the 18th century, the philosopher Immanuel Kant (1724–1804) suggested that some of the nebulae might be distant systems of stars (other Milky Ways). By 1908 nearly 15,000 nebulae had been cataloged and described. Some had been correctly identified as star clusters and others (such as the Orion Nebula) as gaseous nebulae. The nature of most of them, however, remained unexplained, particularly the ones that looked small and indistinct. (For more on how such nebulae are named, by the way, see the "Astronomy Basics" box in Chapter 19.)

If these nebulae were nearby, with distances comparable to those of observable stars, they were most likely clouds of gas within our Galaxy. If, on the other hand, they were remote, far beyond the edge of the Galaxy, they could be other systems containing billions of stars. To determine which idea was correct, astronomers had to find a way of measuring the distances to at least some of the nebulae. And for that, larger telescopes were necessary. When the 100-in. (2.5-m) telescope on Mount Wilson in southern California went into operation, astronomers finally had the tool to settle the controversy.

Working with the 100-in. telescope, Edwin Hubble (see "Voyagers in Astronomy" box) was able to resolve individual stars in several of the brighter spiral-shaped nebulae (including M31, the great spiral in Andromeda already mentioned in Chapter 24). Among these stars he discovered some faint variables that—when he analyzed their light curves—turned out to be cepheids. Here were reliable indicators that Hubble could use to measure the distances to the nebulae in the way that Henrietta Leavitt had pioneered (see Chapter 18). He estimated that the Andromeda galaxy was about 900,000 LY away from us. (Today we know it is actually slightly more than twice as distant as Hubble's estimate, but his conclusion about its true nature remains unchanged.)

No one in human history had ever measured a distance so great. The Milky Way could not possibly extend

Figure 25.1
The nearby spiral galaxy M83. This galaxy is about 10 million LY away and has a diameter of 30,000 LY. (© Anglo-Australian Telescope Board, 1977)

to cover so large a volume. The spiral nebulae had to be separate galaxies (Figure 25.1). When Hubble's paper on the distances to nebulae was read before a meeting of the American Astronomical Society on the first day of 1925, the entire room erupted in a standing ovation. A new era had begun in the study of the universe, and a new field—extragalactic astronomy—had just been born.

Types of Galaxies

Having established the existence of other galaxies, Hubble and others then began to observe them more closely—noting their shapes, their contents, and as many other properties as they could measure. This was a daunting task in the 1920s, when a single photograph or spectrum of a galaxy could take a full night of tireless observing. In recent decades, larger telescopes and electronic detectors have made this task less difficult, although the most distant galaxies (those that show us the universe in its earliest phases) still require enormous effort. (See this chapter's opening image, which required 10 days of observation with the Hubble Space Telescope.)

The first step in trying to understand a new type of object is often simply to describe its appearance. As it

Edwin Hubble: Expanding the Universe

The son of a Missouri insurance agent, Edwin Hubble graduated from high school at age 16. He excelled in sports, winning letters in track and basketball at the University of Chicago, where he studied both science and languages. Both his father and grandfather wanted him to study law, however, and he gave in to family pressure. He received a prestigious Rhodes scholarship to Oxford University in England, where he studied law with only middling enthusiasm. Returning to the United States, he spent a year teaching high school physics and Spanish, as well as coaching basketball, while trying to determine his life's direction.

The pull of astronomy eventually proved too strong to resist, and so he went back to the University of Chicago for graduate work. Just as he was about to get his degree and accept an offer to work at the soon-to-be-completed 100-in. telescope, the United States entered World War I and Hubble enlisted as an officer. Although the war ended by the time he had arrived in Europe, he received more officer's training abroad and enjoyed a brief time of further astronomical study at Cambridge before being sent home.

In 1919, at age 30, he joined the staff at Mount Wilson and began working with the world's largest telescope. Ripened by experience, energetic, disciplined, and a skillful observer, Hubble soon established some of the most important ideas in modern astronomy.

Edwin Hubble (1889–1953)
(Photo by J. Stokley/A.S.P. Archives)

He showed that other galaxies existed, classified them on the basis of their shapes, found a pattern to their motion (and thus put the notion of an expanding universe on a firm observational footing), and began a lifelong program to study the distribution of galaxies in the universe. Although a few others had glimpsed pieces of the puzzle, it was Hubble who put it all together and showed that an understanding of the large-scale structure of the universe was feasible.

His work brought Hubble much renown and many medals, awards, and honorary degrees. As he became better known (he was the first astronomer to appear on the cover of *Time* magazine), he and his wife enjoyed and cultivated friendships with movie stars and writers in southern California. Hubble was instrumental (if you'll pardon the pun) in the planning and building of the 200-in. telescope on Mount Palomar, and had begun to use it for studying galaxies when he was felled by a stroke in 1953. When astronomers built a space telescope that would allow them to extend Hubble's work to distances he could only dream about, it seemed natural to name it in his honor.

---

turns out, most of the bright nearby galaxies come in one of two shapes—they either have spiral arms, like our own Galaxy, or they appear to be elliptical (something like the Goodyear blimp). Many faint galaxies, on the other hand, have a more irregular shape.

Spiral Galaxies

Our own Galaxy and M31 (see Figure 24.11), which is believed to be much like it, are typical large **spiral galaxies.** They consist of a nucleus, a halo, a disk, and spiral arms. Interstellar material is usually spread throughout the disks of spiral galaxies. Bright emission nebulae and hot young stars are present, especially in the spiral arms (Figure 25.2), showing that new star formation is still going on.

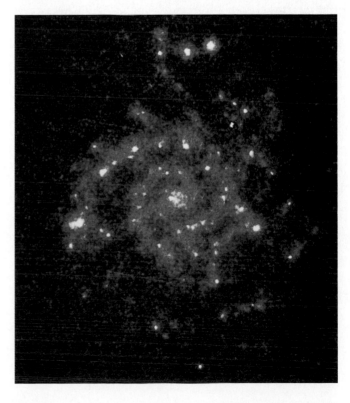

Figure 25.2
This ultraviolet image of the spiral galaxy M74 was taken in 1990 with NASA's Ultraviolet Imaging Telescope aboard the Space Shuttle. The colors have been added to show different intensities of ultraviolet light. The bright splotches outlining the spiral arms are regions of recent star formation where a cluster of hot stars is putting out a great deal of UV radiation. (NASA)

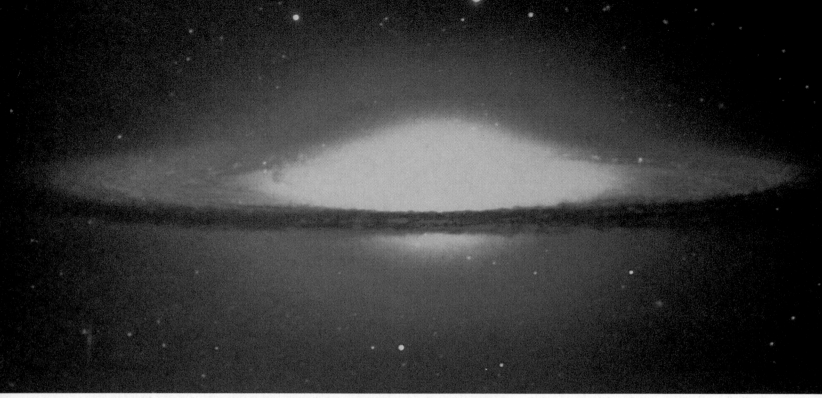

Figure 25.3
The Sombrero galaxy, M104. More than 24 million LY away, this galaxy is surrounded by a prominent dust ring almost 50,000 LY across. The central spheroidal region has the reddish appearance characteristic of an older population of stars. (European Southern Observatory)

Figure 25.4
NGC 1365, a barred spiral galaxy about 55 million LY away. This is a large galaxy, more luminous than either the Milky Way or M31. (European Southern Observatory)

The disks are often dusty, which is especially noticeable in those systems that we view almost edge-on (Figure 25.3). In galaxies we see face-on, the bright stars and emission nebulae make the arms of spirals stand out like those of a Fourth-of-July pinwheel (Figure 25.1). Open star clusters can be seen in the arms of nearer spirals, and globular clusters are often visible in their halos. Spiral galaxies contain a mixture of young and old stars, just as the Milky Way does. All spirals rotate, and the direction of their spin is such that the arms appear to trail, like the coattails of a brisk runner.

Perhaps a third or more of spiral galaxies have conspicuous bars of stars running through their nuclei. The spiral arms of such a system usually begin from the ends of the bar, rather than winding out directly from the nucleus. Such galaxies are called barred spirals (Figure 25.4). Some astronomers believe that almost all spirals, including the Milky Way, contain at least a weak bar.

In both normal and barred spirals, we observe a range of different shapes. At one extreme the nuclear bulge is large and luminous, the arms are faint and tightly coiled, and bright emission nebulae and supergiant stars are inconspicuous. At the other extreme the nuclear bulge is small—almost absent—and the arms are loosely wound. In these latter galaxies, luminous stars and emis-

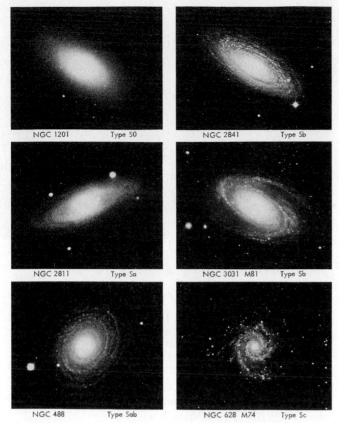

NGC 1201   Type S0   NGC 2841   Type Sb

NGC 2811   Type Sa   NGC 3031 M81   Type Sb

NGC 488   Type Sab   NGC 628 M74   Type Sc

Figure 25.5

Types of spiral galaxies, going from S0, which are mostly bulge and very little disk, to Sc, which are mostly disk and very little bulge. (Palomar Observatory, Caltech)

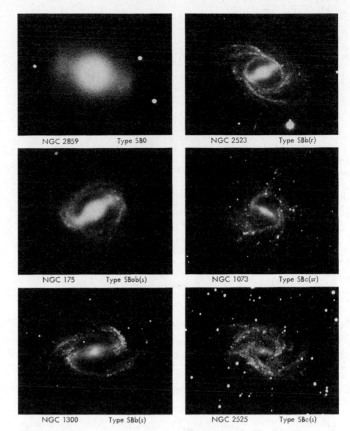

NGC 2859   Type SB0   NGC 2523   Type SBb(r)

NGC 175   Type SBab(s)   NGC 1073   Type SBc(sr)

NGC 1300   Type SBb(s)   NGC 2525   Type SBc(s)

Figure 25.6

Types of barred spirals, paralleling the types of regular spirals. We suspect that most spirals may have some sort of bar-like structure in them, but it is much more pronounced in galaxies such as the ones shown here. (Palomar Observatory, Caltech)

sion nebulae are very prominent. Our Galaxy and M31 are both intermediate between the two extremes. Photographs of spiral galaxies, illustrating the different types, are shown in Figures 25.5 and 25.6.

The luminous parts of spiral galaxies range in diameter from about 20,000 to more than 100,000 LY, and the atomic hydrogen in the disks often extends to far greater diameters. There may also be considerable dark matter in them, in a much larger invisible corona, just as there is in the Milky Way. From the limited observational data available, their masses are estimated to range from 10^9 to 10^{12} M_{Sun}. The total luminosities of most spirals fall in the range of 10^8 to 10^{11} L_{Sun}. Our Galaxy and M31 are relatively large and massive, as spirals go.

Elliptical Galaxies

Elliptical galaxies consist almost entirely of old stars, and have shapes that are spheres or ellipsoids (somewhat squashed spheres). They contain no trace of spiral arms. Their light is dominated by older reddish stars (the population II stars discussed in Chapter 24), and in this respect ellipticals resemble the nuclear bulge and halo components of spiral galaxies. In the larger nearby ellipticals, many globular clusters can be identified (Figure 25.7).

Figure 25.7

This image, prepared with special photographic masking techniques, shows a swarm of globular clusters surrounding the giant elliptical galaxy M87. A jet of material is shooting out of the galaxy's core, which is not visible here. M87 is an active galaxy, with strong radio and x-ray emissions; such galaxies are discussed in Chapter 26. (Photo by David Malin; © Anglo-Australian Telescope Board)

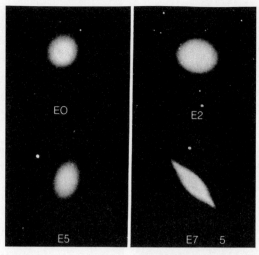

Figure 25.8
Types of elliptical galaxies. (Palomar Observatory, Caltech)

Figure 25.9
Leo I, a dwarf elliptical galaxy. (R. Schild, Center for Astrophysics)

Dust and emission nebulae are not conspicuous in elliptical galaxies, but many do contain a small amount of interstellar matter.

Elliptical galaxies show various degrees of flattening, ranging from systems that are approximately spherical to those that approach the flatness of spirals (Figure 25.8). The rare *giant ellipticals* (for example, M87 in Figure 25.7) reach luminosities of 10^{11} L_{Sun}. The mass in a giant elliptical can be as large as 10^{13} M_{Sun}. The diameters of these large galaxies extend over at least several hundred thousand light years and are considerably larger than the largest spirals. Although individual stars orbit around the center of an elliptical galaxy, the orbits are not all in the same direction as they are in spirals. Therefore, ellipticals don't rotate in a systematic way, and it is hard to estimate how much dark matter they contain.

This category of galaxies ranges all the way from the giants, just described, to *dwarf ellipticals,* which, astronomers have recently come to realize, may be the most common kind of galaxy. They escaped our notice for a long time because they are very faint and difficult to discern. An example of a dwarf elliptical is the Leo I system shown in Figure 25.9. There are so few bright stars in this galaxy that even its central regions are resolved. However, the total number of stars (most of which are too faint to show in our photo) is probably at least several million. The luminosity of this typical dwarf is approximately 10^6 L_{Sun}, about equal to that of the brightest globular clusters.

Intermediate between the giant and dwarf elliptical galaxies are systems such as M32 and NGC 205, two companions of M31. They can be seen in the photograph of M31 in Figure 24.11.

Irregular Galaxies

Hubble classified all star systems that did not have the regular shapes associated with the categories just de-

scribed into the catchall bin of **irregular galaxies,** and we continue to use his term. These galaxies often appear chaotic, and many are undergoing relatively intense star formation activity; they contain both population I and population II stars. Nearby irregular galaxies, for which we can make good measurements, have lower masses and luminosities than typical spirals.

The two best-known irregular galaxies are the Large and Small Magellanic Clouds (Figures 25.10 and 25.11), which are among our nearest extragalactic neighbors. Their name reflects the fact that Ferdinand Magellan and his crew, making their round-the-world journey, were the first European travelers to notice them. Although not visible from the United States and Europe, these two systems are prominent from the Southern Hemisphere, where they look like wispy clouds detached from the Milky Way. They are only about one-tenth as distant as the Andromeda spiral. The Large Cloud contains the 30 Doradus complex (also known as the Tarantula Nebula), one of the largest and most luminous groupings of supergiant stars and associated gas known in any galaxy.

The Small Magellanic Cloud is greatly elongated and considerably less massive than the Large Cloud. The length of this narrow wisp of material is six times greater than its width, and the galaxy points directly toward our Galaxy like an arrow, stretching from about 150,000 LY out to 250,000 LY. The Small Cloud was probably the victim of a near-collision with the Large Cloud some 200 million years ago. It is now being pulled apart by the gravity of the Milky Way. Similar interactions may explain the strange shapes and active star-forming regions of other irregulars.

Do Galaxy Types Evolve?

Encouraged by the success of the H–R diagram for stars (see Chapter 17), astronomers studying galaxies hoped to find some sort of comparable scheme, where differences

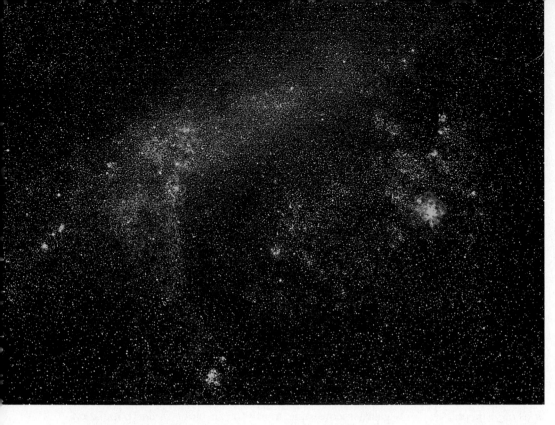

Figure 25.10
The Large Magellanic Cloud, a satellite of our own Galaxy, is visible to the naked eye from the Southern Hemisphere. The large red nebula (called the Tarantula) is the site of active star formation and contains many young supergiant stars. (National Optical Astronomy Observatories)

in appearance could be tied to different evolutionary stages. Wouldn't it be nice if every elliptical galaxy evolved into a spiral, for example, just as every main-sequence star evolves into a red giant? Several simple ideas of this kind were tried, some by Hubble himself, but none has stood the test of time (and observation).

While galaxies do change in appearance over billions of years, an isolated galaxy will not change from a spiral to an elliptical or vice versa. As spirals consume their gas,

star formation stops, and the spiral arms gradually become less conspicuous. Over long periods, spirals therefore begin to look more like the galaxies at the top left of Figures 25.5 and 25.6 (which astronomers refer to as S0 types). We have also come to realize that collisions and mergers of smaller galaxies, including spirals, in the centers of dense galaxy clusters can build up the massive ellipticals such as M87. We will discuss the long-term evolution of galaxies in more detail in Chapter 27.

25.3

Properties of Galaxies

Masses of Galaxies

The technique for deriving the masses of galaxies is basically the same as that used to estimate the mass of the Sun, the stars, and our own Galaxy: we apply Kepler's third law as modified by Newton. When we measure how fast objects in the outer regions of the galaxy are orbiting the center, we can calculate how much mass is inside that orbit. For spiral galaxies, astronomers can measure the rotation speed by obtaining spectra of either stars or gas in the galaxy and looking for wavelength shifts produced by the Doppler effect.

Such observations of M31, for example, show it to have a mass (within the main visible part of the galaxy, out to a distance of 100,000 LY from the center) of about 4×10^{11} M_{Sun}, which is about the same as the mass of our own Galaxy. (The total mass of M31 is higher than 4×10^{11} M_{Sun} because we have not included the material that lies more than 100,000 LY from the center.) Like our own

Figure 25.11
The Small Magellanic Cloud. This dwarf irregular galaxy is another satellite of the Milky Way. (National Optical Astronomy Observatories)

TABLE 25.1
Characteristics of the Different Types of Galaxies

Characteristic	Spirals	Ellipticals	Irregulars
Mass (M_{Sun})	10^9 to 10^{12}	10^5 to 10^{13}	10^8 to 10^{11}
Diameter (thousands of LY)	15 to 150	3 to 600	3 to 30
Luminosity (L_{Sun})	10^8 to 10^{11}	10^6 to 10^{11}	10^7 to 2×10^9
Populations of stars	Old and young	Old	Old and young
Interstellar matter	Gas and dust	Almost no dust; little gas	Much gas; some have little dust, some much dust
Mass-to-light ratio in the visible part	2 to 10	10 to 20	1 to 10
Mass-to-light ratio for total galaxy	100	100	?

Galaxy, Andromeda appears to have a large amount of dark matter beyond its luminous boundary.

Elliptical galaxies do not rotate; for them we must use a slightly different technique to measure mass. Their stars are still moving in orbit around the galactic center, but not in the organized way that characterizes spirals. Since elliptical galaxies contain stars that are billions of years old, we can assume that the galaxies themselves are not flying apart. Therefore, if we can measure the various speeds with which the stars are moving in their orbits around the center of the galaxy, we can calculate how much mass the galaxy must contain in order to hold the stars within it.

In practice, the spectrum of a galaxy is a composite of the spectra of its many stars, whose different motions produce different Doppler shifts (some red, some blue). The result is that the lines we observe from the entire galaxy contain the combination of many Doppler shifts; they look much wider in a spectrum than would the same lines in a hypothetical galaxy in which the stars had no orbital motion. Astronomers call this phenomenon *line broadening*. The amount by which each line broadens indicates the range of speeds at which the stars are moving with respect to the center of the galaxy. The range of speeds depends, in turn, on the force of gravity that holds the stars within the galaxies. With information about the speeds, it is possible to calculate the mass of the elliptical galaxy.

Table 25.1 summarizes the range of masses (and other properties) of the various types of galaxies. The most-massive galaxies are the giant ellipticals, but the lowest-mass galaxies are ellipticals as well. On average, irregular galaxies have less mass than spirals.

Mass-to-Light Ratio

A useful way of characterizing a galaxy is by stating the ratio of its mass, in units of the Sun's mass, to its light output, in units of the Sun's luminosity. For the Sun, given this definition, the **mass-to-light ratio** is 1. Galaxies, however, are not just composed of stars that are like the Sun. The overwhelming majority of stars are less lumi-

nous than the Sun, and usually these stars contribute most to the mass of a system, without accounting for very much light. Since the average star has more mass than light, a galaxy's mass-to-light ratio is generally greater than 1. Galaxies in which star formation is still occurring tend to have mass-to-light ratios in the range of 1 to 10, while in galaxies consisting mostly of an older stellar population, such as ellipticals, the ratio is 10 to 20.

But these figures refer only to the inner, conspicuous parts of galaxies (Figure 25.12). In Chapter 24 we discussed the evidence for invisible matter in the outer regions of our own Galaxy, extending much farther from the galactic center than do the bright stars and gas. Recent measurements of the rotations of the outer parts of nearby galaxies, such as the Andromeda spiral, suggest that they, too, have extended distributions of dark matter around the visible disk of stars and dust. This largely invisible matter adds to the mass of the galaxy while contributing nothing to its luminosity, thus increasing the mass-to-light ratio. If dark invisible matter is present in a galaxy, its mass-to-light ratio can be as high as 100. The two different mass-to-light ratios measured for various types of galaxies are given in Table 25.1.

These measurements of other galaxies support the conclusion already reached from studies of the rotation of our own Galaxy—namely, that as much as 90 percent of all the material in the universe cannot at present be observed directly in any part of the electromagnetic spectrum. An understanding of the properties and distribution of this invisible matter is crucial to our understanding of galaxies. Through the gravitational force that it exerts, dark matter probably plays a dominant role in their formation and early evolution. As we will see in Chapter 28, it may also determine the ultimate fate of the universe.

There is an interesting parallel here between our time and the time during which Edwin Hubble was receiving his training in astronomy. By 1920 many scientists were aware that astronomy stood on the brink of important breakthroughs if only the nature and behavior of the nebulae could be settled with better observations. In the

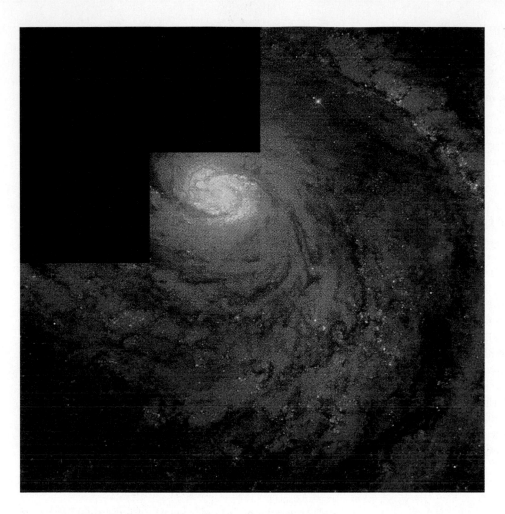

Figure 25.12
The Hubble Space Telescope took this dramatic image of part of the spiral galaxy M100, about 50 million LY away. With the Hubble, astronomers can make much better measurements of the galaxy's mass, structure, and distance. What cannot be seen on such images is any dark matter this galaxy contains. (J. Trauger and NASA)

same way, many astronomers today feel we may be closing in on a far more sophisticated understanding of the large-scale structure of the universe, if only we can learn more about the nature and properties of dark matter. If you follow astronomy in the newspapers or news magazines (as we hope you will), you should be hearing more about dark matter in the years to come.

The Extragalactic Distance Scale

To determine many of the key properties of a galaxy, such as its luminosity or its size, we must first know how far away it is. If we know the distance to a galaxy, we can convert how bright the galaxy appears to us in the sky into its true luminosity, because we know the precise way light is dimmed by distance. (The same galaxy ten times farther away, for example, would look a hundred times dimmer.) But the measurement of galaxy distances is one of the most difficult problems in modern astronomy: all the galaxies are far away, and most are so distant we cannot even make out individual stars in them.

For decades after Hubble's initial work, the techniques used to measure galaxy distances were relatively inaccurate, and different astronomers derived distances that differ by as much as a factor of two. Imagine if the distance between your home or dorm and your astronomy class were this uncertain; it would be difficult to make sure you got to class on time. In the past few years, however, astronomers have devised new techniques for measuring distances to galaxies; all of them give the same answer to within an accuracy of about 10 percent. As we will see, this means we may finally be able to make reliable estimates of the scale of the universe.

Variable Stars

Before we could measure distances to other galaxies, we first had to establish the scale of cosmic distances using objects in our own Galaxy. We described the chain of these distance methods in Chapter 18 (and we recommend that you review that chapter if it has been a while since you've read it). We saw that astronomers were especially delighted to learn how to measure distances using certain kinds of variable stars, including the bright cepheids. Such stars could be seen in star clusters within the Milky Way and even, as Hubble found, in nearby galaxies.

After the variables had been used to make distance measurements for a few decades, Walter Baade (whose work was discussed in Chapter 24) showed that there were actually two kinds of cepheids, and that astronomers had been unwittingly mixing them up. As a result, in the early 1950s all the distances to the galaxies had to be increased by about a factor of two. We mention this because we want you to bear in mind, as you read on, that science is always a progress report. Our first tentative steps in such difficult investigations are always subject to future revision as our techniques become more sophisticated.

The amount of work involved in finding cepheids and measuring their periods can be enormous. Hubble, for example, obtained 350 long-exposure photographs of the nearby spiral M31 over a period of 18 years, and identified only 40 cepheids. Even though cepheids are fairly luminous stars, they can be detected in only about 30 of the nearest galaxies with the world's largest ground-based telescopes. As mentioned in Chapter 18, one of the main projects being carried out with the Hubble Space Telescope is the measurement of cepheids in more-distant galaxies (out to at least 65 million LY), to improve the accuracy of the extragalactic distance scale (Figure 18.7).

Nevertheless, we can only use cepheids to measure distances within a small pocket of the universe of galaxies. After all, to use this method we must be able to resolve single stars and follow their subtle variations. Beyond a certain distance, even our finest space telescopes cannot help us do this. We needed to find other ways to measure the distances of galaxies.

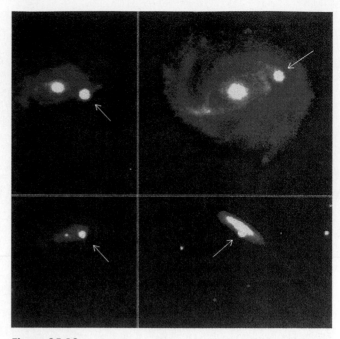

Figure 25.13
Four images of Type I supernovae in different galaxies, taken with the 1.2-m telescope at the Whipple Observatory. If Type I supernovae are all the same maximum luminosity, then those in more-distant galaxies will appear dimmer and can be used to estimate distances. (A. Riess et al./Harvard-Smithsonian Center for Astrophysics)

Standard Bulbs

We also discussed in Chapter 18 the great frustration that astronomers felt when they realized that the stars in general were not *standard bulbs*. If every light bulb in a huge auditorium is a standard 100-W bulb, then bulbs that look brighter to us must be closer, while those that look dimmer must be further away. If every star were a standard luminosity (or wattage), then we could similarly "read off" their distances based on how bright they appear to us. Alas, as we have seen, neither stars nor galaxies come in one standard-issue luminosity. Nonetheless, astronomers have been searching for objects out there that do act in some way like a standard bulb—that have the same intrinsic (built-in) brightness wherever they are.

A number of suggestions have been made for what sorts of objects might be standard bulbs, including the brightest supergiant stars, planetary nebulae (which give off a lot of ultraviolet radiation), and the average globular cluster in a galaxy. Particularly useful are Type I supernovae—those that involve the explosion of a white dwarf in a binary system (see Section 22.4). Observations show that supernovae of this type all reach nearly the same luminosity (about 10^{10} L_{Sun}) at maximum light. With such tremendous luminosities, these supernovae can be detected out to distances of at least 2 billion LY (Figure 25.13).

At very large distances we can use the total light emitted by an entire galaxy as a standard bulb. This technique, however, will not work for a single isolated galaxy because galaxies span an enormous range in intrinsic luminosity (see Table 25.1), and we cannot tell from appearance alone what the luminosity of any one galaxy really is. (For example, ellipticals with different luminosities often look similar when seen in isolation.) We can only tell which galaxies are highly luminous and which are not if we see a collection of them, with various brightnesses, side by side. Luckily, galaxies (as we will see in Chapter 27) are social creatures, and come in groups and clusters. Thus we can use the apparent brightness of the brightest elliptical members (say, the average of the brightest five) in a large cluster to estimate the distance to the cluster.

There is much debate among astronomers about how good these various standard bulbs are. It's as if, to continue our analogy, standard 100-W bulbs came from a slipshod manufacturer who made some of them 90 W and others 110 W. In this case you could only use the brightness of any given bulb to estimate its brightness plus or minus 10 percent. The universe is not in the business of making its objects into exact standard bulbs either, so there is still some variation in the luminosities of the best of these distance indicators. We must use such methods with a sense of their fallibility and calibrate them carefully every chance we get.

TABLE 25.2
Some Methods for Estimating Distances to Galaxies

Method	How Reliable	Galaxy Type	Approximate Distance Range
(SB = standard bulb)			(millions of LY)
Cepheid variables	Very	Spirals, irregulars	0–65
Brightest stars (SB)	Moderate	Spirals, irregulars	0–150
Planetary nebulae (SB)	Very	All	0–70
Globular clusters (SB)	Moderate	All	0–100
Surface brightness fluctuations	Very	Ellipticals	0–100
Type I supernovae (SB)	Very	All	0–2,000
Tully–Fisher method (21-cm line widths)	Very	Spirals, irregulars	0–300
Brightest galaxy in cluster (SB)	Very	Ellipticals in clusters	70–13,000
Redshifts (Hubble law)	Very	All	300–13,000

New Techniques

Some powerful new techniques for measuring distances have recently been developed. All start with observations of nearby galaxies whose distances have been measured through the use of cepheids and standard bulbs. The characteristics of these nearby galaxies are carefully calibrated and used to measure distances to galaxies so far away that individual stars and clusters can no longer be detected. Let's examine two of these techniques briefly.

The first method makes use of an interesting relationship noticed in the late 1970s by Brent Tully of the University of Hawaii and Richard Fisher of the National Radio Astronomy Observatory. They found that the luminosity of a spiral galaxy is related to its rotational velocity. The more mass a galaxy has, the faster the objects in its outer regions must orbit. A more massive galaxy has more stars in it, and is thus more luminous (ignoring dark matter for a moment). Using our terminology from the previous section, we can say that the mass-to-light ratios for various spiral galaxies are pretty similar.

Tully and Fisher used the 21-cm radiation from cold hydrogen to determine how rapidly material in a spiral galaxy was orbiting around its center. Since 21-cm radiation comes in a nice narrow line, the amount of broadening tells us the range of orbital velocities. The broader the line, the faster material is orbiting in the galaxy, and the more massive and luminous it turns out to be.

It is somewhat surprising that this technique works, since much of the mass associated with galaxies is dark matter, which does not contribute at all to the luminosity. There is also no obvious reason that the mass-to-light ratio should be similar for all spiral galaxies. Nevertheless, observations show that measuring the rotational velocity of a galaxy provides an accurate estimate of its intrinsic luminosity. Once we know how luminous the galaxy really is, we can compare the luminosity to the apparent brightness and use the difference to calculate its distance.

Another new technique involves measuring fluctuations in the apparent surface brightness of elliptical galaxies. This technique has been explored by John Tonry at MIT. Elliptical galaxies contain mostly very old stars and very little gas or dust. A perfectly sharp picture of an elliptical galaxy would look much like that of a globular cluster, with many individual stars appearing as discrete points of light. Even with the blurring caused by the Earth's atmosphere, the image of an elliptical galaxy does not appear perfectly smooth. Rather, it is mottled or bumpy due to the naturally lumpy distribution of light emitted by the individual stars belonging to that galaxy.

The amount of bumpiness depends on the distance to the galaxy. In a nearby galaxy we can see individual stars or clusters of stars, and the image has many bumps of varying brightness. In a very distant galaxy the stars cannot be resolved at all, and the image is smooth.

The distance to an elliptical galaxy can therefore be estimated by measuring the degree of bumpiness in the light distribution. This technique will not work for spiral galaxies, whose disks contain large, random amounts of dust that can also cause fluctuations in brightness distribution. But the method seems to give good results for elliptical galaxies, which are the very ones that cannot be effectively examined with the Tully–Fisher technique.

Table 25.2 lists the type of galaxy for which each of the techniques described here is useful, the range of distances over which the technique can be applied, and the reliability of the distance estimates derived with each technique.

25.5

The Expanding Universe

We now come to one of the most important discoveries ever made in astronomy—the pattern that underlies the motion of all galaxies. Before we describe how the discovery was made, we should point out that the first steps in the study of galaxies came at a time when the techniques of spectroscopy were also making great strides. Astronomers using larger telescopes could record the spectrum of a faint star or galaxy on more sensitive photographic plates, guiding their telescopes so they remained pointed to the same object for many hours. The resulting spectra of galaxies contained a wealth of information about the composition, mass, and motion of these great star systems.

Slipher's Moving Observations

Curiously, the discovery of the galaxies' motions began with the search for Martians and other solar systems. In 1894 the controversial (and wealthy) astronomer Percival Lowell established an observatory in Flagstaff, Arizona, to study the planets and to search for life in the universe (see Chapter 9). Lowell thought that the spiral nebulae might be solar systems in the process of formation—like the solar nebula described in Chapter 13. He therefore asked one of the observatory's young astronomers, Vesto M. Slipher (Figure 25.14), to photograph the spectra of some of the spiral nebulae to see if their spectral lines might show chemical compositions like those expected for newly forming planets.

The Lowell Observatory's major instrument was a 24-in. refracting telescope, which was not at all well suited to observations of faint spiral nebulae. With the technology available in those days, it took 20 to 40 hours to expose a good spectrum (in which Doppler shifts could reveal a galaxy's motion). This often meant continuing to expose the same photograph over several nights. Beginning in 1912, and making heroic efforts over a period of about 20

years, Slipher managed to photograph the spectra of more than 40 nebulae.

To his amazement, the Doppler shifts revealed huge speeds for the spiral nebulae. His first announcement came in 1914, years before Hubble showed that these objects were distant galaxies. A few spirals, such as M31, now known to be our close neighbors, turned out to be approaching us. But he found the overwhelming majority to be receding at speeds as high as 1800 km/s. Other observers soon confirmed his findings and made measurements of the motions of other spiral nebulae, all showing astounding **redshifts** (Doppler shifts in which the wavelength of the light gets longer) and thus rapid motion away from us.

The Hubble Law

The profound implications of Slipher's work became apparent only during the 1920s, when Hubble found ways of estimating the distances of the spiral nebulae. Hubble carried out the key observations in collaboration with a remarkable man, Milton Humason (Figure 25.15), who dropped out of school in the 8th grade and began his astronomical career by driving a mule train up the trail on Mount Wilson to the observatory. In those early days, supplies had to be brought up that way; even astronomers hiked up to the mountaintop for their turns at the telescope. Humason became interested in the work of the astronomers and, after marrying the daughter of the observatory's electrician, took a job as janitor there. After a time he became a night assistant, helping the astronomers run the telescope and take data. Eventually he made such a mark that he became a full astronomer at the observatory.

By the late 1920s Humason was collaborating with Hubble by photographing the spectra of faint galaxies

Figure 25.14
Vesto M. Slipher, 1875–1969. (Lowell Observatory)

Figure 25.15
Milton Humason worked with Hubble to establish the expansion of the universe. (Caltech Archives)

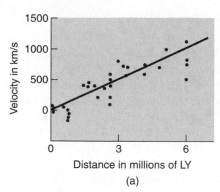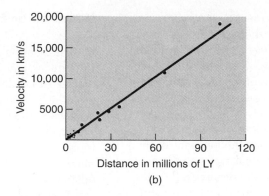

(a)   (b)

Figure 25.16

(a) Hubble's original velocity–distance relation, adapted from his 1929 paper in the *Proceedings of the National Academy of Sciences*. (b) Hubble and Humason's velocity–distance relation, adapted from their 1931 paper in *The Astrophysical Journal*. The red dots at the lower left are the points in the diagram in the 1929 paper. Comparison of the two graphs shows how rapidly the determination of galactic distances and redshifts progressed in the two years between these publications.

with the 100-in. telescope. (By then there was no question that the spiral nebulae were in fact galaxies.) When Hubble laid his own distance estimates next to Slipher's and Humason's measurements of the speed with which the galaxies were moving away, he found something stunning: there was a relationship between distance and velocity for galaxies. *The more distant the galaxy, the faster it was receding from us.*

In 1931 Hubble and Humason jointly published a classic paper in *The Astrophysical Journal*, the foremost technical publication read by U.S. astronomers (Figure 25.16). They compared distances and velocities of remote galaxies moving away from us at speeds as high as 20,000 km/s, and were able to show that the recession velocities of galaxies are directly proportional to their distances from us. We now know that this relationship holds for nearly every galaxy (except a few of the nearest ones) whose independent distance we have been able to measure! (The few galaxies approaching us turn out to be part of the Milky Way's own group, which has its own internal motions.)

Written as a formula, the relationship between velocity and distance is

$$v = H \times d$$

where v is the recession speed, d is the distance, and H is a number called the **Hubble constant.** This equation is now known as the **Hubble law.**

ASTRONOMY BASICS

Constants of Proportionality

Mathematical relationships such as the Hubble law are pretty common in life. To take a simple example, suppose you get hired by your college or university to call rich alumni and ask

for donations. You are paid $1.50 for each call; the more calls you squeeze in between studying astronomy and other courses, the more money you take home. We can set up a formula that connects p, your pay, and n, the number of calls:

$$p = A \times n$$

where A is the Alumni constant, with a value of 1.50 dollars. If you make 20 calls, you will earn 1.50 dollars times 20, or $30.

Suppose your boss forgets to tell you what you will get paid for each call. You can calculate the Alumni constant that governs your pay by keeping track of how many calls you make and noting your gross pay each week. If you make 100 calls the first week and are paid $150, you can deduce that the constant is 1.5 (in units of dollars per call). Hubble, of course, had no higher agent to tell him what his constant would be—he had to calculate its value from the measurements of distance and velocity.

Astronomers express the value of Hubble's constant in units that relate to how they measure speed and velocity for galaxies. In this book, we will use km/s per million LY as that unit. For many years, estimates of the value of the Hubble constant have been in the range of 15 to 30 km/s per million LY. The most recent work appears to be converging on a value near 25 km/s per million LY. If H is 25 km/s per million LY, a galaxy moves away from us at a speed of 25 km/s for every million light years of its distance. As an example, a galaxy 100 million LY away is moving away from us at a speed of 2500 km/s.

If every galaxy in the universe for which we can make distance measurements obeys this relationship, it cannot be a coincidence. (A few astronomers actually glimpsed this relationship in their data before Hubble's work, but dismissed it as a coincidence.) The Hubble law tells us something fundamental about the universe. Since all but the nearest galaxies appear to be in motion *away* from us, with the most distant ones moving the fastest, we must be living in an *expanding universe*. We will further explore the implications of this idea shortly, as well as in Chapter 27; for now we will just say that Hubble's observation underlies all our theories about the origin and evolution of the universe.

As we will see, Einstein's equations of general relativity (see Chapter 23), when solved for the behavior of the entire universe, implied that all the galaxies should show just this sort of expansion. But Einstein dismissed the idea as too disturbing, and introduced a kind of fudge factor into his equations to keep the universe at rest. Later he called his failure to predict what Hubble would soon discover "the greatest blunder of my life."

Hubble's Law and Distances

The regularity expressed in the Hubble law has a built-in bonus: it gives us a new way to determine the distances to remote galaxies. First we must reliably establish Hubble's constant by measuring *both* the distance and the velocity of many galaxies in many directions, to be sure the Hubble law is truly a universal property of galaxies. Then, a measurement of the speed with which a galaxy is moving away from us will tell us (by using the Hubble law) what its distance is. For example, if we find a galaxy moving away at 25,000 km/s, the Hubble law tells us it must be at a distance of a billion LY.

This is a very important technique, because, as we have seen, our best methods for determining galaxy distances don't take us much beyond 600 million LY (and they have many uncertainties). But to use the Hubble law as a distance indicator, all we need to do is get a spectrum of a galaxy and measure the Doppler shift, something astronomers are now very good at.

With large telescopes and modern spectrographs, such spectra can be taken of extremely faint galaxies. As we will see in the chapters to come, astronomers can now measure galaxies with redshifts that imply a velocity away from us over 90 percent the speed of light. At such high velocities, lines in the ultraviolet region of the spectrum—which normally cannot be observed from the ground—are shifted to yellow and red wavelengths and can be photographed as ordinary light.

Models for an Expanding Universe

At first, hearing about Hubble's law and being a fan of Copernicus and Shapley, you might be shocked. Are all the galaxies really moving away *from us?* Is there, after all, something special about our position in the universe? Worry not; the fact that galaxies obey the Hubble law only shows that the universe is expanding uniformly. A uniformly expanding universe is one that is expanding at the same rate everywhere. In such a universe, we and all other observers within it, no matter where they are located, *must* observe a proportionality between the velocities and distances of remote galaxies.

To see why, first imagine a ruler made of flexible rubber, with the usual lines marked off at each centimeter. Now suppose someone with strong arms grabs each end of the ruler and slowly stretches it so that, say, it doubles in length in 1 min (Figure 25.17). Consider an intelligent ant sitting on the mark at 2 cm—a point that is not at either end or in the middle of the ruler. He measures how fast other ants, sitting at the 4-, 7-, and 12-cm marks, move away from him as the ruler stretches.

The ant at 4 cm, originally 2 cm away from our ant, has doubled its distance in 1 min; it therefore moved away at a speed of 2 cm/min. The one at the 7-cm mark, which was originally 5 cm away from our ant, is now 10 cm away; it thus had to move at 5 cm/min. And the one that started at the 12-cm mark, which was 10 cm away from the ant doing the counting, is now 20 cm away, meaning it must have raced away at a speed of 10 cm/min. Ants at different distances move away at different speeds, and their speeds are proportional to their distances (just as the Hubble law states for galaxies). Yet all the ruler was doing was stretching uniformly.

Now let's repeat the analysis, but put the intelligent ant on some other mark—say, on 7 or 12. We discover that, as long as the ruler stretches uniformly, this ant also finds every other ant moving away at a speed proportional to its distance. In other words, the kind of relationship expressed by the Hubble law can be explained by a uniform stretching of the "world" of the ants. And *all* the ants on our simple diagram will see *the other ants* moving away from them as the ruler stretches.

Figure 25.17
Stretching a ruler. See text for explanation.

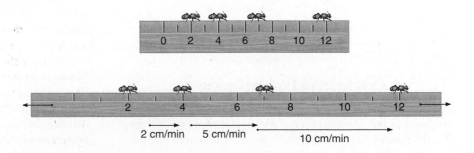

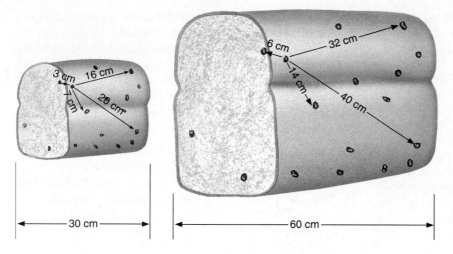

Figure 25.18
Expanding raisin bread.
See text for explanation.

For a three-dimensional analogy, let's look at the loaf of raisin bread in Figure 25.18. The chef has put too much yeast in the dough, and when she sets the bread out to rise, it doubles in size during the next hour, causing all the raisins to move farther apart. We again pick a representative raisin (that is not at the edge or the center of the loaf) and show the distances from it to several others in the figure (before and after the loaf expands).

Measure the increases in distance and calculate the speeds for yourself. Since each distance doubles during the hour, each raisin moves away from our selected raisin at a speed proportional to its distance. The same is true no matter which raisin you start with. Each raisin (if it could talk) would say, "Look! All the other raisins are expanding away from me. And their speeds are proportional to their distances."

Our two analogies are useful for clarifying our thinking, but you must not take them literally. On both the ruler and the raisin bread, there are points that are at the end or edge. You can use these to pinpoint the middle of the ruler and the loaf. While our models of the universe have some resemblance to the properties of the ruler and the loaf, we will see that they do not have a center or an edge.

What is useful to notice about both the ants and the raisins is that they themselves did not "cause" their motion. It isn't as if the raisins decided to take a trip away from each other, and then put on roller skates to get away. No, in both our analogies it was the stretching of the medium (the ruler or the bread) that moved the ants or the raisins farther apart. In the same way, we will see that the galaxies don't have rocket motors propelling them away from each other. Instead, they are passive participants in the expansion of space. As space stretches, the galaxies are carried farther and farther apart.

The expansion of the universe, by the way, does not imply that the individual galaxies and clusters of galaxies themselves are expanding. The raisins in our analogy do not grow in size as the loaf expands. Similarly, gravity holds galaxies and clusters together, and they merely get farther away from each other—without themselves changing in size—as the universe expands.

In the final chapter of our book, we will explore some of the implications of this expansion for the past history and ultimate fate of the universe. But first we turn to some of the most intriguing and energetic objects in the expanding universe.

Summary

25.1 Faint star clusters, clouds of glowing gas, dust clouds reflecting starlight, and galaxies all appeared as faint patches of light (or nebulae) in the telescopes available at the beginning of the 20th century. It was only when Hubble measured the distance to the Andromeda galaxy using cepheid variables in 1924 that the existence of other galaxies similar to the Milky Way in size and content was established.

25.2. The majority of bright galaxies are either **spirals** or **ellipticals.** Spiral galaxies contain both old and young stars, as well as interstellar matter, and have typical masses in the range of 10^9 to 10^{12} M_{Sun}. Our own Galaxy is a large spiral. Ellipticals are spheroidal or elliptical systems that consist almost entirely of old stars, with very little interstellar matter. Elliptical galaxies range in size from giants, more massive than any spiral, down to dwarfs, with masses of only about 10^6 M_{Sun}. A small percentage of galaxies with more chaotic shapes are classified as **irregulars.**

25.3 The masses of spiral galaxies are determined from measurements of their rates of rotation. The masses of elliptical galaxies are estimated from analyses of the motions of the stars within them. Galaxies can be characterized by their **mass-to-light ratios.** The luminous parts of galaxies with active star formation typically have mass-to-light ratios in the range of 1 to 10; the luminous parts of elliptical galaxies, which contain only old stars, typically have mass-to-light ratios of 10 to 20. The mass-to-light ratios of whole galaxies, including their outer regions, are as high as 100, indicating the presence of a great deal of dark matter.

25.4 Astronomers determine the distances to galaxies using a variety of methods, including the period–luminosity relationship for cepheid variables; objects such as Type I supernovae, which appear to be standard bulbs; and the Tully–Fisher method that connects the line-broadening of 21-cm radiation to the luminosity of spiral galaxies. Each method has limitations in terms of its precision, the kinds of galaxies it can be used with, and the range of distances over which it can be applied.

25.5 The universe is expanding. Observations show that the spectral lines of distant galaxies are **redshifted,** and that their recession velocities are proportional to their distances from us, a relationship known as the **Hubble law.** The rate of recession, called the **Hubble constant,** is approximately 25 km/s per million LY. We are not at the center of this expansion; an observer in any other galaxy would see the same expansion that we do.

Review Questions

1. Describe the main distinguishing features of spiral, elliptical, and irregular galaxies.

2. Why did it take so long for the existence of other galaxies to be established? What finally convinced astronomers that they do exist?

3. Explain what the mass-to-light ratio is and why it is smaller in regions of star formation in spiral galaxies than in the central regions of elliptical galaxies.

4. If we now realize dwarf ellipticals are the most common type of galaxy, why did they escape our notice for so long?

5. Describe the best ways to measure the distance to:
  a. a nearby spiral galaxy
  b. a nearby elliptical galaxy
  c. a distant isolated elliptical galaxy
  d. a distant isolated spiral galaxy
  e. a distant elliptical galaxy that is a member of a galaxy cluster

6. Why is the Hubble law considered one of the most important discoveries in the history of astronomy?

7. What does it mean to say that the universe is expanding? For example, is your astronomy classroom expanding?

Thought Questions

8. In 1920, Harlow Shapley and another astronomer, Heber Curtis, held a debate. Although it began as a discussion of one issue (the size of our Galaxy), it eventually came to involve the larger question of the existence of other galaxies. Use the resources in your library to look up information about this debate. Summarize the points that each participant made, and compare to what we know today. (*Hint:* You can start by looking up the works by Smith in "Suggestions for Further Reading.")

9. Where might the gas and dust (if any) in an elliptical galaxy come from?

10. Why can we not determine distances to galaxies by the same method used to measure the parallaxes of stars?

11. Starting with the determination of the size of the Earth, outline the steps necessary to obtain the distance to a remote cluster of galaxies.

12. Suppose that the Milky Way Galaxy were truly isolated, and that no other galaxies existed within 100 million LY. Suppose that galaxies were observed in larger numbers at distances greater than 100 million LY. Why would it be more difficult to determine accurate distances to those galaxies than if there were also galaxies relatively close-by?

13. Suppose you were Hubble and Humason, working on the distances and Doppler shifts of the galaxies. What sorts of things would you have to do to convince yourself (and others) that the relationship you were seeing between the two quantities was a real feature of the behavior of the universe? (For example, would data from two galaxies be enough to demonstrate the Hubble law?)

14. Measurements of the Andromeda galaxy show that it is rotating at a speed of 230 km/s, at a distance of about 100,000 LY or 6×10^9 AU from the center. Use Kepler's third law to calculate the mass of Andromeda. Is the true mass of the galaxy likely to be larger or smaller than this estimate? Why?

15. Calculate the mass-to-light ratio for a globular cluster with a luminosity of $10^6 L_{Sun}$ and 10^5 stars. (Assume that the average mass of a star in such a cluster is 1 M_{Sun}.) Do the same for a superluminous star of 100 M_{Sun} having the same luminosity of $10^6 L_{Sun}$.

16. Plot the velocity–distance relation for the raisins in the bread analogy from the numbers given in Figure 25.18.

17. Repeat Problem 16, but use some other raisin for a reference and measure the distances with a ruler. Is your new plot the same as the last one?

18. Suppose a supernova explosion occurred in a galaxy at a distance of 10^8 LY. If we are only now detecting it, how long ago did the supernova actually occur? According to the Hubble law, what is the velocity with which this galaxy is moving away from us?

19. A cluster of galaxies is observed to have a radial velocity of 60,000 km/s. Find the distance to the cluster.

Suggestions for Further Reading

Christianson, G. *Edwin Hubble: Mariner of the Nebulae.* 1995, Farrar, Straus, Giroux. The definitive biography.

Dressler, A. *Voyage to the Great Attractor.* 1994, Knopf. Outstanding book by a noted astronomer on how modern extragalactic astronomy is done.

Eicher, D. "Candles to Light the Night" in *Astronomy,* Sep. 1994, p. 33. Introduction to standard bulbs and the cosmic distance scale.

Ferris, T. *The Red Limit.* 1983, Morrow. Well-written account of how we discovered the large-scale properties of the universe.

Freedman, W. "The Expansion Rate and Size of the Universe" in *Scientific American,* Nov. 1992, p. 76.

Hartley, K. "Elliptical Galaxies Forged by Collision" in *Astronomy,* May 1989, p. 42.

Hodge, P. "The Extragalactic Distance Scale: Agreement at Last?" in *Sky & Telescope,* Oct. 1993, p. 16.

Kinney, A. "Fourteen Billion Years Young" in *Mercury,* Mar./Apr. 1996, p. 29. On recent work with the Hubble Space Telescope.

Lake, G. "Understanding the Hubble Sequence [of Galaxies]" in *Sky & Telescope,* May 1992, p. 515.

Osterbrock, D. "Edwin Hubble and the Expanding Universe" in *Scientific American,* July 1993, p. 84.

Parker, B. "The Discovery of the Expanding Universe" in *Sky & Telescope,* Sep. 1986, p. 227.

Smith, R. "The Great Debate Revisited" in *Sky & Telescope,* Jan. 1983, p. 28.

Smith, R. *The Expanding Universe: Astronomy's Great Debate.* 1982, Cambridge U. Press.

Trefil, J. "Galaxies" in *Smithsonian,* Jan. 1989, p. 36. Nice long review article.

Using **REDSHIFT**™

1. Turn on only the stars and galaxies. Set the *Stellar Limiting Magnitude* to 6 and the *Deep-Sky Limiting Magnitude* to 8 to see which galaxies you can view tonight through binoculars or a small telescope.

2. Display only galaxies with a limiting magnitude of 20; find a region with galaxies, center on it, and then set the *Zoom Level* to 15. Count the number of ellipticals, spirals, and irregulars.

Which type appears most common?

According to the chapter text, which is the most common type of galaxy? Can you explain any difference?

3. Set the *Limiting Magnitude* to select the 20 brightest galaxies. Which type is most common? Is it the same type as in Exercise 2? Explain any difference.

In this wonderful example of gravitational lensing, a cluster of galaxies (the yellow objects) is distorting the light of more-distant blue galaxies. The result is a complicated series of blue arcs that are—in a sense—ghost images of objects that would otherwise look like blue points of light.
(Tony Tyson/AT&T Bell Laboratories)

CHAPTER 26

Quasars and Active Galaxies

Thinking Ahead

Suppose that at the center of a galaxy a massive black hole develops with access to a great deal of "fuel" (material relatively close to its event horizon). What would such a galaxy look like from far away?

Quasar, quasar, burning

bright

In the forests of the night

What immortal hand or eye

Can frame thy fearful

luminosity.

With profound apologies
to William Blake.

During the first half of the 20th century, astronomers viewed the universe of galaxies as a mostly peaceful place. They assumed that galaxies had started billions of years ago and then evolved slowly as the populations of stars within them formed, aged, and died. That placid picture has completely changed in the last few decades of the 20th century.

Today astronomers can see that the universe is often shaped by violent events, including cataclysmic explosions of supernovae, collisions of whole galaxies, and the tremendous outpouring of energy as matter interacts in the environment surrounding massive black holes. The key event that led to our changed view of the nature of the universe was the discovery of the bizarre objects we now call *quasars*.

Quasars

The Sun does not radiate much energy at radio wavelengths. Astronomers expected that other stars would also be quiet in the radio region of the spectrum. They were, therefore, quite surprised when in 1960 two sources of radio waves were identified with what appeared to be faint blue stars (Figure 26.1). Naturally, astronomers rushed to take spectra of these radio "stars," but the spectra only deepened the mystery: they had emission lines that could not be identified with any known substance.

Huge Redshifts

Bear in mind that by the 1960s, astronomers had a century of experience in identifying elements and compounds in the spectra of stars. Elaborate tables had been published showing the lines that each element would produce under a wide range of conditions (see Chapter 16). A star with unidentifiable lines in the ordinary visible light spectrum was either an insult in the face of a hundred years of progress or, as turned out to be the case, a hint of something completely new.

The breakthrough came in 1963. At Caltech's Palomar Observatory, Maarten Schmidt (Figure 26.2) was puzzling over the spectrum of one of the radio stars, named 3C 273 because it was the 273rd entry in the third Cambridge catalog of radio sources. Schmidt recognized that the emission lines in its spectrum had the same spacing between them as the Balmer lines of hydrogen, but were shifted far to the red of the wavelengths at which the Balmer lines are normally located. Indeed, these lines were at such long wavelengths that if the redshifts were attributed to the Doppler effect, 3C 273 was receding from us at a speed of 45,000 km/s, or about 15 percent the speed of light!

After Schmidt's proposed identification of the lines in 3C 273, the puzzling emission lines in other star-like radio sources were re-examined to see if they, too, might be

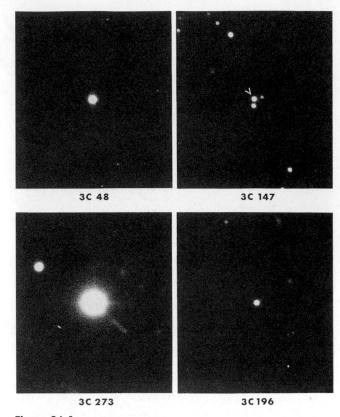

Figure 26.1
Quasi-stellar radio sources photographed with the 5-m Hale telescope. (Palomar Observatory, Caltech)

well-known lines with large redshifts. Such proved to be the case, but the other objects were found to be receding from us at even greater speeds. Since stars don't show Doppler shifts this large, no one had thought of considering high redshifts to be the cause of the strange spectra.

Their astounding speeds show that radio stars cannot possibly be nearby stars. Any true star moving at more than a few hundred kilometers per second would be able to overcome the gravitational force of the Galaxy and completely escape from it. (Eventually, astronomers also realized that there was more to some of these "stars" than just a point of light; see Figure 26.3.)

As we will see, radio stars only look like stars because they are compact and very far away. Since they superficially resemble stars, but do not share their properties, they were given the name *quasi-stellar radio sources.* Later, astronomers discovered objects with large redshifts that appear star-like but have no radio emission. Today, all these objects are referred to as *quasi-stellar objects (QSOs)* or, as they are more popularly known, **quasars.**

Figure 26.2
Maarten Schmidt (left), who solved the puzzle of the quasar spectra in 1963, shares a joke in this 1987 photo with Allan Sandage, who took the first spectrum of a quasar. Sandage has also been instrumental in measuring the value of the Hubble constant. Schmidt remains active in observing the properties and evolution of quasars. (Andrew Fraknoi)

(The name was soon appropriated by a manufacturer of home electronics.)

Today we know that about 90 percent of all known quasars are not radio sources. We're glad some do give off radio waves, however, because without that telltale oddity, it might have taken a lot longer to discover just how interesting these faint objects really are.

Thousands of QSOs have now been discovered, and spectra are available for a representative sample. All the spectra show large to very large redshifts (and none show blueshifts). The largest measured to date (March 1996) corresponds to a shift in wavelength of $\Delta\lambda/\lambda = 4.9$, where $\Delta\lambda$ is the difference between the wavelength that a spectral line would have if the emitting source were stationary, and the wavelength that it actually has. The stationary or laboratory wavelength is indicated by the Greek letter λ.

In the record-holding quasar, the first Lyman series line of hydrogen, with a laboratory wavelength of 121.5 nm in the ultraviolet portion of the spectrum, is shifted all the way through the visible region to 700 nm! At such high redshifts, the simple formula for converting a Doppler shift to speed (see Section 4.6) must be modified to take into account the effects of the theory of relativity.

Figure 26.3

A modern image of 3C 273 shows a brilliant (overexposed) point of light in the center, a fainter "host galaxy" around that point, and a wiggly jet that we now know measures some 150,000 LY in length. The four spikes coming from the central point are artifacts caused by the telescope mirror supports, but the jet is real. This image was taken with the High-Resolution Camera at the Canada-France-Hawaii Telescope on Mauna Kea. (R. McClure and J. Hutchings/DAO/National Research Council, Canada)

If we apply the relativistic form of the Doppler shift formula, we find that a redshift of 4.9 corresponds to a velocity of more than 94 percent the speed of light.

Explaining the Redshifts

What can these quasars be? At first, many theories were suggested to explain their properties. Could the large redshifts be caused by something else besides motion? Strong gravity causes a redshift (as we saw in Chapter 23), but such strong gravity would leave other evidence in the spectrum, and that evidence is simply not there. Could the quasars be high-speed projectiles shot out from galaxies by some unknown process? That might explain their high speeds, but then some galaxy somewhere would surely be shooting out a projectile that happened to come *toward us* and show a blueshift. Yet only redshifts have been observed.

Eventually, astronomers settled on the most straightforward explanation. If the quasars are like galaxies, and thus participating in the expansion of the universe, their redshifts make perfect sense. According to the Hubble law, their high redshifts simply mean that the quasars are far away from us. Could the QSOs be distant galaxies?

If they are, they certainly don't show the properties of normal galaxies. Galaxies contain stars, so the spectra of normal galaxies have absorption lines, as do the spectra of stars. Quasar spectra are dominated by emission lines. When they do have absorption lines, these frequently have redshifts that are less than the emission lines show. And we run into another difficulty when we try to account for the energy these objects emit.

The Luminosity Puzzle

If the quasars are as far away as their redshifts imply, then they must be incredibly luminous to be visible to us at all—far brighter than any normal galaxy could be. Even in visible light alone, most are far more energetic than the brightest elliptical galaxies. But quasars also emit energy at x-ray and ultraviolet wavelengths, and some are radio sources as well. When all their radiation is added together, some QSOs have total luminosities as large as 10^{14} L_{Sun}, or 10 to 100 times the brightness of the brighter elliptical galaxies.

Finding a mechanism to produce this much energy would be difficult under any circumstances. But there is an additional problem. When astronomers began looking for other quasars and then monitoring them carefully, they found that some vary in luminosity on time scales of months, weeks, or even, in some cases, days. This variation is irregular and can change the brightness of a quasar by a few tens of percent in both its light and radio output.

Think about what such a change in luminosity means. A quasar at its dimmest is still more brilliant than any normal galaxy. Now imagine that the brightness increases by 30 percent in a few weeks. Whatever mechanism is re-

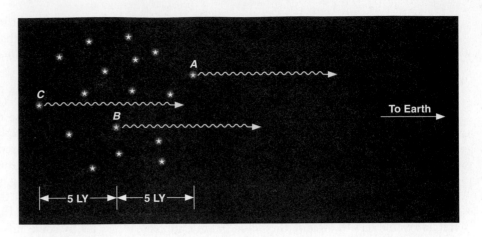

Figure 26.4
A diagram showing why light variations from a large region in space appear to last for an extended period of time as viewed from Earth. Suppose all the stars in this cluster, which is 10 LY across, brighten simultaneously and instantaneously. From Earth, star *A* will appear to brighten five years before star *B*, which in turn will appear to brighten five years earlier than star *C*. It will take ten years for an Earth observer to get the full effect of the brightening.

sponsible must be able to release new energy at rates that stagger our imaginations. The most dramatic changes in quasar brightness are equivalent to the energy released by 100,000 billion Suns. To produce this much energy we would have to convert the total mass of about ten Earths into energy every minute!

Moreover, because the fluctuations occur in such short times, the part of a QSO that is varying must be smaller than the distance light travels in the time it takes the variation to occur—typically a few months. To see why this must be so, let's consider a cluster of stars 10 LY in diameter (Figure 26.4) at a very large distance from the Earth (the Earth is off to the right in the figure). Suppose every star in this cluster somehow brightens simultaneously and remains bright. When the light from this event arrives at the Earth, we would first see the brighter light from stars on the near side; five years later we would see increased light from stars at the center. Ten years would pass before we detected more light from stars on the far side.

Even though all stars in the cluster brightened at the same time, the fact that the cluster is 10 LY wide means that ten years must elapse before the light from every part of the cluster reaches us. From Earth we would see the cluster get brighter and brighter, as light from more and more stars began to reach us, and not until ten years after the brightening began would we see the cluster reach maximum brightness. In other words, if an extended object suddenly flares up, it will seem to brighten over a period equal to the time it takes light to travel across the object from its far side.

We can apply this idea to brightness changes in quasars to estimate their diameters. Because QSOs typically vary (get brighter and dimmer) over periods of a few months, the region where the energy is generated can be no larger than a few light months. If it were larger, it would take longer than a few months for the light from the far side to reach us.

How large is a region of a few light months? Pluto, usually the outermost planet in our solar system, is about 5.5 light hours from us. The Oort comet cloud may extend out to about 10 light months, while the nearest star is 4 LY away. Clearly a region a few light months across is insignificantly small compared to a galaxy. And some quasars vary even more rapidly, which means their energy is generated in an even smaller region. Whatever mechanism powers the quasars must be able to generate more energy than that produced by an entire galaxy in a volume of space that, in some cases, is no larger than our solar system. When quasars were first discovered, no one could think of a way to explain so much energy in so small a space.

Solving the Quasar Paradox

In the 1960s, then, astronomers faced a dilemma. If the quasars were at the distances given by the Hubble law, some new mechanism was needed to explain their huge energy output. There *was* a way to keep the quasars from being such disturbingly strong cosmic powerhouses. If their redshifts were not caused by the expansion of the universe, then they would not be a measure of distance. The quasars could then be at any distance at all. If they were much closer to us, their energies would be correspondingly smaller and much easier to explain. But in that case, we had no explanation for what *did* cause their enormous redshifts. Either way, the quasars seemed to mean we were going to learn something new about the universe.

One way to decide between these two interesting alternatives was to get an independent estimate of how far away the quasars are. An enormous amount of effort, by astronomers all over the world, was directed to this task for about two decades. Some tried to show that there were quasars and galaxies that were physically connected but had *different* redshifts. Since the redshift of the *galaxy* is clearly caused by its expansion with the universe, this would show that at least some part of the quasar's redshift must have another cause. But none of the cases of connection was particularly convincing; most astronomers thought it was likely that the two objects just appeared to lie near each other on the dome of the sky. There seemed no reason to assume they were also at similar distances

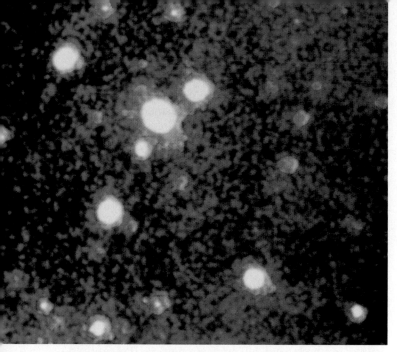

Figure 26.5
Quasar 3C 275.1, the first to be found at the center of a cluster of galaxies, appears as the brightest object near the center of this image. The quasar nucleus is surrounded by a gas cloud that is elliptical in shape. Its redshift indicates that this quasar and its cluster are both 7 billion LY away. (National Optical Astronomy Observatories)

from us. Another approach to the problem bore considerably more fruit.

As we will discuss in more detail in the next chapter, galaxies throughout the universe are found in groups and clusters. We know these groups are physically associated because they are not only close together in the sky, but their members show the same redshift. If quasars behave like galaxies (that is, if their redshifts are evidence of participation in the expansion of the universe), then we might find some quasars in distant clusters of galaxies. It is not enough for a quasar to be seen in the same area of the sky as a cluster; after all, that could be a coincidence, like seeing a bird and an airplane together, only to realize that the plane is really much farther away. To show that the quasar belongs to the cluster, the galaxies in the cluster and the quasar must share the same redshift.

This was not an easy idea to test. Quasars are typically far away; we see them because they are so luminous. Any ordinary galaxies with the same redshift (and thus the same large distance) as a quasar would be fainter and extremely difficult to detect. After much effort, astronomers have identified a number of faint galaxy clusters that contain quasars with the same redshifts as their other members (Figure 26.5). When only one such cluster-with-a-quasar was known, it was possible to suggest that the redshift similarity was a mere coincidence. Perhaps the quasar's redshift was caused by some other mechanism, and it just happened to lie in the same direction as the cluster.

Such coincidences do occasionally happen, but as gamblers trying to outsmart the house in Las Vegas can attest, they do not happen very often. As more and more clusters were found with quasars inside possessing the same redshift as all the cluster members, it became clear that the redshifts of quasars are caused by the same effect that causes the redshifts of ordinary galaxies—the expanding universe.

Other observations also support the hypothesis that quasars are located at the distances indicated by their redshifts. One key discovery was that many relatively nearby quasars are not true point sources but rather are embedded in a fainter, fuzzy-looking patch of light (Figure 26.6). The colors of this fuzz are like those of spiral galaxies. In a few cases, spectra have been obtained; they indicate that the light of the fuzz is derived from stars, clearly demonstrating that quasars are located in galaxies.

This still leaves us with the mystery of how the quasars can produce so much energy in such a tiny region. Clues to the answer come from the study of peculiar galaxies that provide a link between ordinary galaxies and the quasars.

Figure 26.6
The Hubble Space Telescope reveals the much fainter "host" galaxy around a quasar known only by its position in the sky (QSO 1229+204). The host galaxy is involved in a collision with a dwarf galaxy, which may be the reason the quasar at the center is bright. (J. Hutchings and NASA)

Figure 26.7
The Seyfert galaxy NGC 1566, at a distance of about 50 million LY, appears on this photograph to be a normal spiral. However, it has a very luminous nucleus with many of the characteristics of a quasar, though much less energetic. The active region at the center of NGC 1566 has been found to vary on a time scale of less than a month, indicating that it is extremely compact. Spectra show that hot gas near the tiny nucleus is moving at an abnormally high velocity, suggesting that it may be in orbit around a massive black hole. (Anglo-Australian Telescope Board)

26.2

Active Galaxies

When quasars were first discovered, they appeared to be much more luminous than any galaxies. And they are indeed more luminous than *normal* galaxies, but we have now found a class of galaxies that fills in the luminosity gap. These peculiar galaxies share many of the properties of the quasars, although to a less-spectacular degree. Members of this class are referred to as **active galaxies.** Since abnormal amounts of energy are produced in their centers, they are said to have **active galactic nuclei.** In effect, active galaxies have mini-quasars embedded in their nuclei.

Seyfert Galaxies

A good example of an active galaxy type is a group called *Seyfert galaxies,* which are spirals with point-like energetic nuclei (Figure 26.7). They are named after the astronomer who discovered the first examples. Like quasars, Seyferts have strong, broad emission lines in their spectra. Since stars typically produce absorption lines in the spectrum of a galaxy, emission lines indicate that there are hot gas clouds near the Seyfert nuclei. The width of the lines indicates that the gas is moving at speeds up to thousands of kilometers per second.

One of the most impressive images taken with the Hubble Space Telescope shows the nuclear region of the Seyfert galaxy NGC 1068 (Figure 26.8). The image, freed from the blurring of the Earth's atmosphere, shows the complex structure, including clouds of hot gas as small as 10 LY across, in the very center of the galaxy.

Some Seyferts, again like quasars, are radio or x-ray sources or both, and all emit strongly in the infrared. Some show brightness variations over a period of a few months, so, as we concluded for quasars, the region from which the radiation comes can be no more than a few light months across. The visible-light luminosities of Seyferts are about normal for spiral galaxies, but when their infrared emission is taken into account, their total luminosities are found to be about 100 times normal.

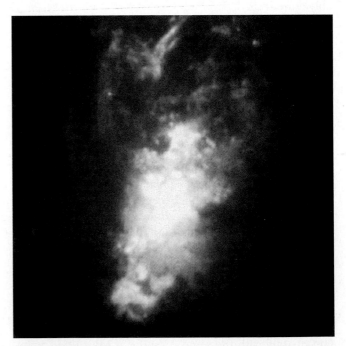

Figure 26.8
The central region of the Seyfert galaxy NGC 1068, about 150 LY across, can be seen in this image taken with the Hubble Space Telescope. Clouds of ionized gas as small as 10 LY in diameter are visible at the energetic center of this galaxy, which is about 60 million LY away in the constellation of Virgo.
(D. Macchetto et al./NASA)

Studies of many such objects show that Seyfert and other peculiar galaxies tend to be more luminous than normal galaxies, but less luminous than quasars. Their bright but point-like nuclei indicate that enormous amounts of energy are being emitted from a small region at their centers. The crucial point about Seyferts and other active galaxies is that a significant fraction of their power comes from a source other than individual stars.

Active Elliptical Galaxies

Spirals are not alone in boasting peculiar activity in their centers. Some elliptical galaxies also have powerful central energy sources. The first clue came from radio observations. Many giant elliptical galaxies that appear comparatively normal when seen with visible light turn out to be powerful emitters of radio energy; hence we call them *radio galaxies*. The galaxy M87, discussed in Chapter 25, is a fine example. Figure 25.7 shows it looking deceptively

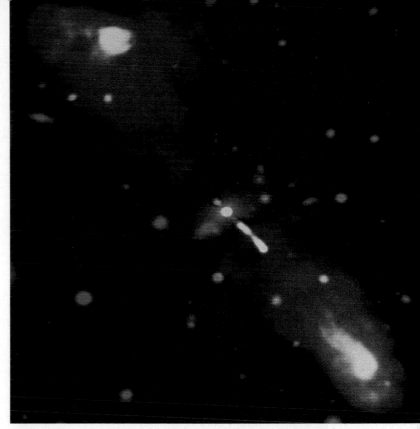

Figure 26.10
In this view of the elliptical radio galaxy 3C 219, a visible-light image (in blue) has been combined with a radio image (in red). The lobes of radio emission extend far beyond the visible parts of the galaxy; a short jet connects the lobes and the galaxy. (D. Clark et al., National Radio Astronomy Observatory)

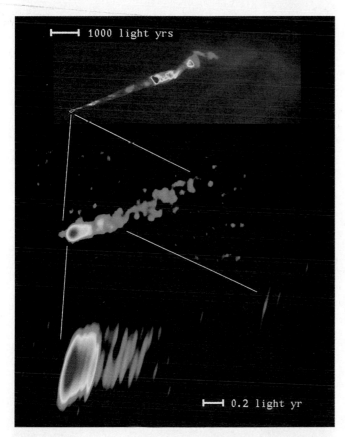

Figure 26.9
Three images of M87 made at radio wavelengths. The top view, made with the Very Large Array, shows the radio source at the galaxy's center (the red spot at left) and the full extent of the impressive jet. The center and bottom images show what can be done with the Very Long Baseline Array of radio telescopes, spanning the continent. They show the inner part of the jet with unprecedented detail; the smallest features we can discern are only light weeks in size. (J. Biretta, W. Junor, National Radio Astronomy Observatory)

peaceful; a radio image (Figure 26.9), however, allows us to "look" deeper inside the galaxy. It reveals an intense source of radio emission in the center, and a complicated jet of material extending from it for more than 6000 LY. (The jet is also visible in the opening image for Chapter 23.) Recent measurements show that knots of material in this jet are moving outward at speeds approaching two-thirds the speed of light. We will have more to say about what's happening at the center of M87 in the next section.

In some radio galaxies, most of the radio emission comes from small regions near their centers (reminding us of quasars). In others, we not only find bright sources in the nucleus, but also larger extended regions of radio emission surrounding the galaxy. In about three-quarters of the radio galaxies, the radio source is double, with most of the radiation coming from extended regions on opposite sides of each galaxy (Figure 26.10). Typically, the two emitting regions (called radio lobes) are far larger than the galaxy itself, and their centers are a few hundred thousand light years away from it. Often, observations also reveal two narrow jets of radio radiation pointing away from the galaxy toward the large extended sources. These jets are similar to the one in Figure 26.9, but can be more than a million LY long (Figure 26.11).

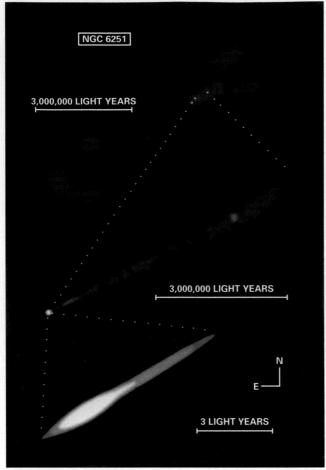

Figure 26.11
Several radio images of galaxy NGC 6251, whose unusually long jet is over 300,000 LY in length. The upper image shows two lobes of radio emission on either side of the galaxy, the middle image shows the long jet, and the bottom image shows one of the intense knots of emission in the jet. (National Radio Astronomy Observatory)

It appears that in all these objects hot ionized gases are "shot out" along the jets into the radio lobes by an extremely intense energy source in the nucleus of the galaxy. Some mechanism in the center is able to direct the flow of energy into narrowly focused beams that spread out and interact with colder neutral gas as they get beyond the confines of the galaxy. Interestingly, with modern instruments similar jets can now be seen in more than half of all quasars (Figures 26.3 and 26.12).

A Variety of Active Galactic Nuclei

Our examples barely scratch the surface of the many different types of active galaxies astronomers have uncovered in the last few decades. Although at first the variety in their appearance seemed bewildering, today we are coming to understand that the differences are often due to the angles at which we see the jets or lobes. (The dis-

tances of the galaxies and the environments in which their activity takes place also play a role.)

Remember that galaxies can be oriented in every possible way to our line of sight from Earth. Sometimes we look right down into the jet, as into the barrel of a gun; in these cases we see a brilliant spot. The jets of other galaxies shoot out to the side from our perspective, giving us the best view. Many times we see the central source, the jets, or the lobes at an angle, making our interpretations more difficult. And, of course, the farther away the object is, the more challenging is the task of observing it. Some of the quasars that looked like dim points of light with the technology of the 1960s now show jets coming from them, or the fuzzy hint of a surrounding galaxy.

Based on many such observations and decades of constructing theoretical models, most astronomers now view all of these objects as part of the same basic phenomenon. Some powerful source of energy at the centers

Figure 26.12
A radio image of the quasar 1007+417 taken with the VLA. The quasar is located at the bright point near the top of the picture; a jet leads to a lobe of radio emission at the bottom. (National Radio Astronomy Observatory)

of galaxies seems to be ultimately responsible for all the different forms of activity. In this view, the quasars are more-distant and extreme examples of this general behavior; often all we can see is the outrageously bright activity source in the center. (That quasars are extreme cases is born out by the fact that we have found only one quasar for every 10 million normal spiral galaxies.)

As we saw in Chapter 24, even our own Galaxy has a compact source of energy at its center. While the energy emitted by this compact source is tiny relative to that generated by a quasar, the center of the Milky Way may represent the low-energy extreme of the range of activity in compact galactic nuclei. Let us now turn to the key remaining question: what is the mechanism that can explain all this activity?

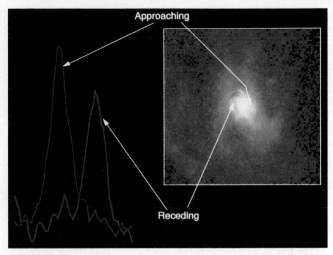

Figure 26.13

A disk of gas was discovered at the center of M87 with the Hubble Space Telescope. Observations made on opposite sides of the disk show that one side is approaching us (the spectral lines are blueshifted by the Doppler effect) while the other is receding (lines redshifted), a clear indication that the disk is rotating. The rotation speed is about 550 km/s or 1.2 million mi/h. Such a high rotation speed is clear evidence that there is a very massive black hole at the center of M87. (H. Ford et al. and NASA)

26.3

The Power Behind the Quasars

Once they realized that all these active objects have a compact central source of enormous energy in common, astronomers quickly began to come up with models to explain the energy source. These included collisions of stars in the crowded cores of galaxies; supermassive stars undergoing violent episodes in their regular life cycles; extraordinarily powerful supernovae; and other suggestions, some of which lay on the very frontiers of science.

Over the past few years, however, compelling evidence has been obtained to support the idea that quasars and other types of active galaxies derive their energy output from a massive black hole at the center. As we will discuss shortly, black holes are capable of producing large amounts of energy as long as material is falling toward them. (We saw this in Chapter 23 when we looked at binary star systems with black holes that were strong sources of x-ray emission.)

The ability of a black hole to produce energy depends on its mass and the amount of material that comes close enough to be "processed" by it. For a black hole with a billion Suns' worth of mass inside (10^9 M_{Sun}), even relatively modest amounts of additional material—only about 10 M_{Sun} per year—falling into the black hole would produce as much energy as a thousand normal galaxies. This could account for the total energy of a quasar.

Observational Evidence

Some of the strongest observational evidence for the existence of massive black holes in the centers of galaxies has been obtained with the Hubble Space Telescope. The opening figure for Chapter 23 shows that the stars at the center of the giant elliptical galaxy M87 become densely concentrated toward the center, forming a bright sharp core. The central density of stars in M87 is at least 300 times greater than expected for a normal giant elliptical galaxy, and over 1000 times denser than the distribution of stars in the neighborhood of our own Sun. Detailed analysis of the distribution of light in M87 shows that a central black hole with a mass of close to 2.5 billion M_{Sun} would cause just this kind of concentration of stars.

With the Hubble, astronomers were able to identify and study an inner disk of hot (10,000 K) gas swirling around the center of M87 (Figure 26.13). It was surprising to find such hot gas in an elliptical galaxy, but the discovery was extremely useful for pinning down the existence of the black hole. Astronomers were able to measure the Doppler shift of spectral lines emitted by this gas, and to find its speed of rotation. They then used the speed to derive the amount of mass inside the disk—applying Kepler's third law. These measurements confirm that there is indeed a mass of about 2.5 billion M_{Sun} concentrated in a tiny region at the very center of M87. So much mass in a small volume of space must be a black hole.

We really don't know how such an enormous black hole was created. As discussed in earlier chapters, one possibility is that it formed from the merger of small black holes created by the explosion of massive stars when M87 was a young galaxy. However it got started, the black hole would have continued to grow by feeding on gas and stars that passed too close to it. As we will see in Chapter 27, whole galaxies sometimes collide and merge. Material from one or more galaxies swallowed by M87 may have fallen toward the black hole and added to its mass. (Such mergers may also explain why M87 is such a large elliptical galaxy.)

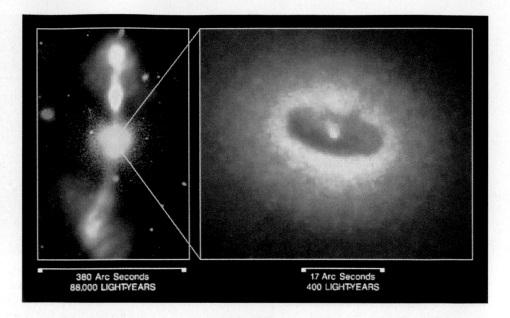

Figure 26.14
The picture on the left shows the active galaxy NGC 4261, located in the Virgo cluster. This elliptical galaxy—the white circular region in the center—is shown the way it looks in visible light, while the jets are seen at radio wavelengths. A Hubble Space Telescope image of the central portion of the galaxy is shown on the right. It contains a ring of dust and gas about 400 LY in diameter, surrounding what is probably a supermassive black hole. Note that the jets emerge from the galaxy in a direction perpendicular to the plane of the ring, consistent with the models shown in Figure 26.15. (W. Jaffe, H. Ford, and NASA)

The Hubble Space Telescope recently found similar evidence for a 2-billion-M_{Sun} black hole at the center of the galaxy NGC 3115 from the motions of the stars in its central region. A 1.2-billion-M_{Sun} black hole is required to explain observations of the elliptical galaxy NGC 4261 (Figure 26.14). Indeed, it is becoming clearer and clearer that many galaxies, including our neighbor the Andromeda galaxy, harbor black holes in their centers.

Energy Production Around a Black Hole

But how does a black hole become one of the most powerful sources of energy in the universe? As we saw in Chapter 23, a black hole itself can radiate no energy. The energy comes from material very close to the black hole, but not inside its event horizon. The black hole attracts matter—stars, dust, and gas—orbiting in the dense nuclear regions of the galaxy. This matter spirals in toward the spinning black hole and forms an accretion disk of material around it. As the material spirals ever closer to the black hole, it accelerates and becomes compressed, heating up to temperatures of millions of degrees. Such hot matter can radiate prodigious amounts of energy as it falls in toward the black hole.

To convince yourself that falling into a region with strong gravity can release a great deal of energy, imagine dropping your astronomy textbook out the window of the ground floor of the library. It will land with a thud, and maybe give a surprised pigeon a nasty bump, but the energy released by its fall will not be very great. Now take the same book up to the 15th floor of a tall building and drop it from there. For anyone below, astronomy could suddenly become a deadly subject; when the book hits, it does so with a great deal of energy. Dropping things from far away into the much stronger gravity of a black hole is much more effective. Just as the falling book can heat up

the air, shake the ground, or produce sound energy that can be heard some distance away, so the energy of material falling toward a black hole can be converted to significant amounts of electromagnetic radiation.

The exact way in which the energy of infalling material is converted to radiation near a black hole is far more complicated than our simple example. To cite just one factor, the material does not go straight in, but whirls around in a chaotic accretion disk. In fact, to understand what happens in the "rough-and-tumble" region around a massive black hole, astronomers and physicists must resort to computer simulations (and they require supercomputers, fast machines capable of awesome numbers of calculations per second). The details of these models are beyond the scope of our book, but they look very promising.

Connecting the Model and the Observations

A number of the phenomena that we observe can be explained naturally in terms of this model. First, reasonable amounts of material falling into a very massive black hole *can* produce the vast quantities of energy that are emitted by quasars and active galactic nuclei. Detailed calculations show that about 10 percent of the mass of matter falling into a black hole can be converted to energy. This is a very efficient process: remember that during the entire course of the evolution of a star like the Sun, less than 1 percent of its mass is converted to energy by nuclear fusion.

The event horizon of a black hole is very small; recall from Chapter 23 that a black hole with a mass of a billion M_{Sun} would have a radius of approximately 3 billion km, about the distance between the Sun and the planet Uranus. The emission produced by infalling matter comes from a small volume of space closely surrounding the

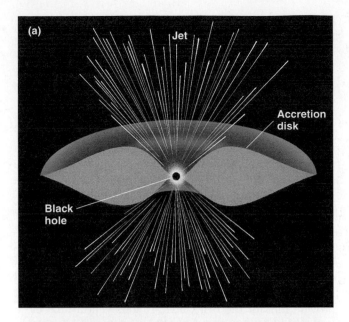

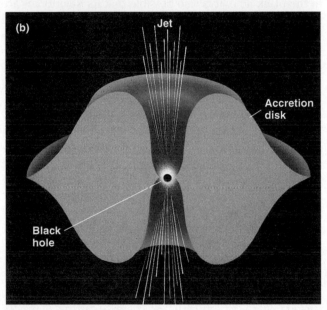

Figure 26.15
Schematic drawings of two accretion disks around large black holes. (a) A thin accretion disk. (b) A "fat" disk—the type needed to account for channeling outflow of hot material into narrow jets oriented perpendicular to the disk.

black hole. The small size of the radiating region is exactly what we need to explain the fact that quasars vary on time scales of weeks to months.

We also discussed that quasar spectra are dominated by strong emission lines. Such lines form when a hot glowing gas is present, which is just what we would expect to see around such an energetic black hole. Our models suggest that the broad emission lines are formed in relatively dense clouds within about half a light year of the black hole. These clouds are located farther out than the accretion disk and may provide additional fuel for the black hole in the future.

Radio Jets

So far our model seems to effectively explain the central energy source in quasars and active galaxies. But, as we have seen, quasars and other active galaxies also have long jets that glow with radio waves, light, and sometimes even x rays, and that extend far beyond the limits of the parent galaxy. Can we find a way for our black hole and its accretion disk to produce these jets as well?

A wide range of observations have traced the jets to within 3 to 30 LY of the parent quasar or galactic nucleus. While the black hole and accretion disk are typically smaller than 1 LY, we nevertheless presume that if the jets come this close, they probably originated in the vicinity of the black hole.

Why are energetic electrons and other particles ejected into jets, and often into two oppositely directed jets, rather than in all directions? Again, we must use theoretical models and supercomputer simulations of what happens when a lot of material whirls inward in a crowded black hole accretion disk. There is no agreement on exactly how jets form, but it is clear that any material escaping from the neighborhood of the black hole has an easier time doing so perpendicular to the disk.

In some ways the inner regions of black hole accretion disks resemble a baby that is just learning to eat by herself. As much food can sometimes wind up being spit out in various directions as going into the baby's mouth. In the same way, some of the material whirling inward toward a black hole finds itself under tremendous pressure, and orbiting with tremendous speed. Under such conditions, our simulations show that a significant amount of material can be flung outward—not back along the disk, where more material is crowding in, but above and below the disk. If the disk is thick (as it tends to be when a lot of material falls in quickly), it can channel the outrushing material into narrow beams (Figure 26.15).

Figure 26.16 is an artist's summary of some of the ideas we have been discussing. At the center of an active galaxy, a small bright spot marks the region of maximum radiation (and the greatest thickness for the accretion disk). On its perimeter the disk gets thinner. Two jets are seen emerging perpendicular to the disk. Not shown, of course, is the rest of the galaxy, which can sometimes obscure the inner regions and allow us only glimpses of the hungry monster within.

Evolution of Quasars

There is one other important fact about quasars that our black hole model can explain. Recall that when first introducing quasars, we mentioned that they generally tend to be far away—and are, in fact, the most distant objects we can detect in the universe. This result is very interesting, because when we see extremely distant objects, we are seeing them as they were long ago. Radiation from a quasar 8 billion LY away is telling us what that quasar was

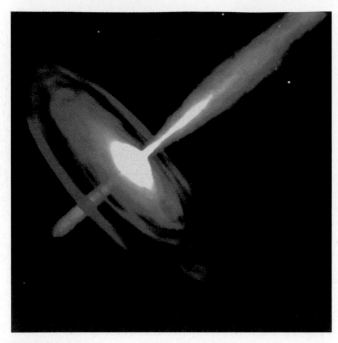

Figure 26.16
An artist's conception of a black hole accretion disk and perpendicular jets. (Painting by Dana Berry, STScl)

like eight billion years ago, much closer to the time the galaxy formed. If there are more quasars far away, there must have been more quasars long ago.

Recently completed counts of quasars at different redshifts show us how dramatic this trend really is (Figure 26.17). The number of quasars was the greatest at the time when the universe was only 20 percent of its present age.

Our model says that quasars are black holes with enough fuel to make a brilliant accretion disk right around

them. Why were there more quasars long ago (far away) than there are today (nearby)? Perhaps in the days of its youth, a black hole could find abundant fuel. But it may be that later in its life, much of the available fuel has been used up, and the hungry black hole has very little left with which to light up the galaxy's central regions.

In other words, if matter in the accretion disk is continually being depleted by falling into the black hole or being blown out from the galaxy in the form of jets, then a quasar can continue to radiate only as long as there is gas available to replenish the accretion disk. Where does this matter come from originally and how might it be replenished?

One possibility for the original fuel source is the very dense star clusters that form near the centers of galaxies. As these stars lose some of their mass during the normal course of stellar evolution—by means of stellar winds and supernova explosions—their combined material could provide fuel for the black hole. If stars venture too close to the black hole, it is also possible that its tidal forces are strong enough to tear the stars apart. Any raw material (gas and dust) near the center of the galaxy would also be a good source of fuel, and we would expect to find more gas in the central regions early in a galaxy's life than later on. (Recall that both ellipticals and the central bulges of spirals today have very little raw material left.)

As hinted above, an alternate source of fuel may come from collisions of galaxies. If two galaxies collide and merge, then gas and dust from one may come close enough to the black hole in the other to be devoured by it and so provide the necessary fuel. Astronomers are just beginning to understand how common collisions of galaxies are (see Chapter 27). But they present a possible way that a quasar or active galaxy might be restarted long after all the local fuel is used up.

Figure 26.17
A graph showing the number of quasars at earlier and earlier times (that is, farther and farther away). An age of 0 corresponds to the beginning of the universe; an age of 1 corresponds to the present time. Note that quasars were most abundant when the universe was about 20 percent of its current age.

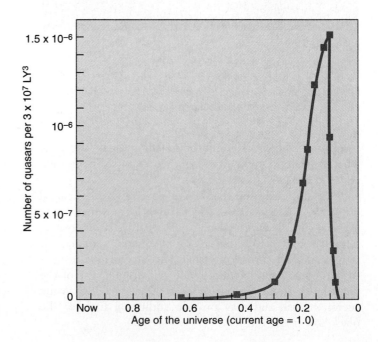

If this scenario is correct, we should generally see more quasars long ago (far away)—as we in fact do. And when we see a nearby quasar or active galaxy, we should check to see if it has been involved in any cosmic traffic accidents recently. And, indeed, many of the relatively nearby, still-active quasars appear to be embedded in galaxies that have experienced collisions with other galaxies. Gas and dust from these other galaxies have apparently been swept up by a dormant black hole and so provided the new source of fuel required to rekindle it.

This model also suggests that we should be able to detect low-level quasar-like activity in some nearby galaxies. That is, black holes might exist in the nuclei of nearby galaxies, but they would be producing only low levels of activity because they have already accreted nearly all of the matter available in their vicinity.

Emission lines like those seen in quasars and Seyferts but of much lower intensity are indeed seen in the nuclei of many otherwise normal nearby galaxies. Some astronomers speculate that a black hole may lurk in the center of every galaxy with a significantly crowded center, but that most of these black holes are now dark and quiet—mere shadows of their former selves. Whether or not this speculation is true remains to be seen.

26.4

Gravitational Lenses

Whatever the quasars are, their brilliance and large distance make them ideal probes of the far reaches of the universe. Like a distant searchlight that illuminates the hazards of the sea on a dark night, the radiation from distant quasars passes through or by whatever lies between them and our telescopes.

For example, while the spectra of quasars are dominated by emission lines, some quasars show absorption lines in their spectra as well. However, these absorption lines typically have smaller redshifts than the emission lines that characterize the quasar. Lines like this arise in clouds of gas within or between galaxies. These clouds are closer to us than the quasar, but difficult to make out in other ways. By studying them carefully, astronomers can learn about the distribution of raw material in the universe at different epochs.

But there is another—and very intriguing—way that the light from quasars can be affected on its way to us. Recall from Chapter 23 that the general theory of relativity predicts that light will be deflected in the vicinity of a strong gravitational field. This can happen when rays of light from a distant quasar pass near a compact massive galaxy on its way to Earth and are bent in a variety of ways. Calculations show that, like a reflection in a funhouse mirror, such bending can produce multiple or twisted images. These bizarre effects (called **gravitational lensing**) have now actually been observed.

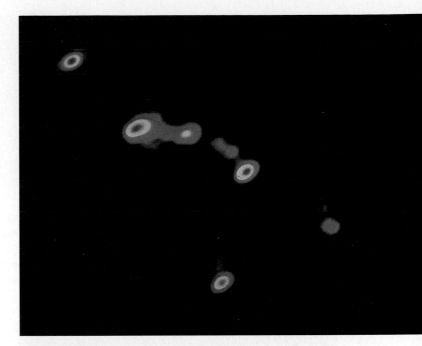

Figure 26.18
A radio image of the double quasar QSO 0957+561. A massive galaxy acting as a gravitational lens forms multiple images of a quasar lying far beyond it. The two images of the quasar are the bright point-like objects just above and below the center of the picture. The weak blue area just above the lower quasar image is the galaxy that acts as the gravitational lens. Several other radio sources are seen above and to the left of the double quasar. (National Radio Astronomy Observatory)

The First Gravitational Lens

In 1979 astronomers Dennis Walsh, Robert Carswell, and Ray Weymann of the University of Arizona noticed that a pair of quasars, very close to each other in the sky and known by the single name QSO 0957+561 (the numbers give their coordinates in the sky), were remarkably similar in appearance and spectra (Figure 26.18). Both had the same redshift. The astronomers suggested that the two quasars might actually be only one, and that the two images were produced by an intervening object—an effect first predicted by Einstein himself.

We now know that there is a dim galaxy lying in the same direction as one of the quasar images. In fact, the galaxy turns out to be a member of a galaxy cluster that is much closer to us than is the quasar. The geometry and estimated mass of the galaxy are just right to produce gravitational lensing. Figure 26.19 is a schematic diagram showing this effect.

Gravitational lenses can produce not only double images but also multiple images, arcs, or rings (Figure 26.20). A remarkable example is shown in this chapter's opening figure, in which we see the images of a blue background galaxy distorted into gracefully curving blue arcs by the foreground (yellow) cluster of galaxies.

(text cont. on page 533)

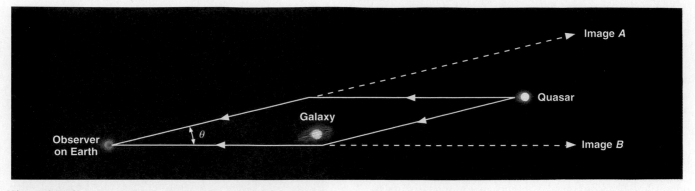

Figure 26.19

An explanation of how a gravitational lens could make two images, as in Figure 26.18. Two light rays from a distant quasar are shown being bent while passing a foreground galaxy; they then arrive together at Earth. Although the two beams of light contain the same information, they now appear to come from two different points on the sky. This sketch is oversimplified and not to scale, but it gives a rough idea of the lensing phenomenon.

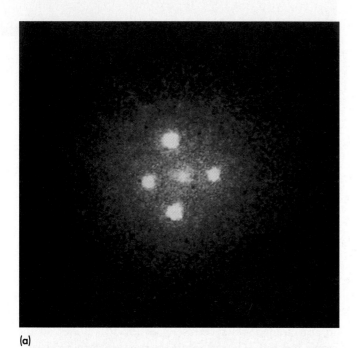

(a)

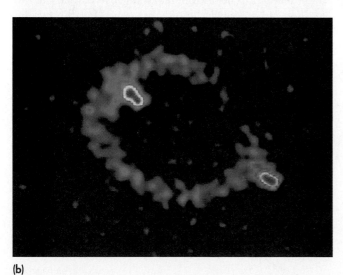

(b)

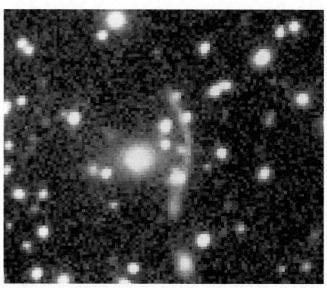

(c)

Figure 26.20

A gallery of gravitational lenses, with multiple or distorted images of distant sources. (a) The Hubble Space Telescope reveals four separate images of the same distant quasar, a configuration called an Einstein cross. The quasar is approximately 8 billion LY away. A galaxy (the slightly fuzzy object in the center) lying at a distance of 400 million LY serves as the gravitational lens. (NASA) (b) If the alignment is just right, a gravitational lens can turn a point source into what is affectionately called an Einstein ring. In this radio map, made with the Very Large Array, a dense cluster of galaxies creates two point images and a ring image of a distant quasar. (J. Hewitt, E. Turner, and NRAO) (c) The blue arc is a distant galaxy smeared out by the gravitational lensing effect of the foreground cluster of galaxies. (NOAO)

Quasars and the Attitudes of Astronomers

The discovery of quasars in the early 1960s was the first in a series of surprises astronomers had in store. Within another decade they would find neutron stars (in the form of pulsars), the first hints of black holes (in binary x-ray sources), and even the radio echo of the Big Bang itself. Many new possibilities seemed just around the horizon.

As Maarten Schmidt reminisced in 1988, "This had, I believe, a profound impact on the conduct of those practicing astronomy. Before the 1960s, there was much authoritarianism in the field. New ideas expressed at meetings would be instantly judged by senior astronomers and rejected if too far out." We saw a good example of this in the trouble Chandrasekhar had in finding acceptance for his ideas about the death of stars with cores greater than 1.4 M_{Sun}.

"The discoveries of the 1960s," Schmidt continued, "were an embarrassment, in the sense that they were totally unexpected and could not be evaluated immediately. In reaction to these developments, an attitude has evolved where even outlandish ideas in astronomy are taken seriously. Given our lack of solid knowledge in extragalactic astronomy, this is probably to be preferred over authoritarianism."[1]

This is not to say that astronomers (being human) don't continue to have prejudices and preferences. Those who had hoped to show that the redshifts of quasars were not connected with their distances (which was definitely a minority opinion), often felt excluded from meetings or from access to telescopes in the 1960s and 1970s. It's not so clear that they actually *were* excluded, as much as that they felt the very difficult pressure of knowing that most of their colleagues strongly disagreed with them. As it turned out, the evidence—which must ultimately decide all scientific questions—was not on their side either.

But today, as better instruments bring solutions to some problems and starkly illuminate our ignorance about others, the entire field of astronomy seems more open to discussing unusual ideas. Of course, before any hypotheses become accepted, they must be tested—again and again—against the evidence that nature itself reveals. Still, the many strange proposals published about what dark matter might be (we'll mention just a few in the next chapter) attest to the new openness that Schmidt described.

[1] From M. Schmidt: "The Discovery of Quasars" in *Modern Cosmology in Retrospect,* ed. B. Bertotti et al. (1990, Cambridge U. Press).

Gravitational Lenses and the Search for Dark Matter

Visible galaxies are not the only possible gravitational lenses. Concentrations of dark matter can also reveal their presence by producing this effect. Figure 26.21 shows a foreground cluster of yellowish galaxies; many blue arcs or small portions of arcs can also be seen in this image. Each arclet is the image of a very distant galaxy whose apparent shape was distorted when the light from it passed through the foreground cluster of yellowish galaxies.

These arclets can be used to map the total mass in the foreground cluster—including dark matter. The resulting maps show that the foreground galaxy cluster contains ten times more matter than is actually seen in visible light. If the dark matter were not there, we would see many fewer distorted images of the background blue galaxies. Astronomers hope to use such lens images to get a better sense of how much dark matter is contained in galaxies and clusters, but already the observations offer strong support for the idea that most of the mass in the universe is in the form of dark matter.

Figure 26.21

This image shows a cluster of galaxies (yellowish objects) that acts as a gravitational lens; it distorts the image of a very distant blue galaxy lying behind it, producing the blue arcs. From the shape and location of these arcs, astronomers can estimate how much dark matter is contained within the foreground cluster. (Tony Tyson/AT&T Bell Labs)

Summary

26.1 The first **quasars** discovered looked like stars but had strong radio emission. The quasar spectra obtained so far show redshifts ranging from 15 to nearly 95 percent of the speed of light. The evidence shows that quasars are at the great distances implied by these redshifts, and that they therefore have 10 to 100 times the luminosity of the brighter normal galaxies. Their variations show that this tremendous energy output is generated in a small volume—in some cases, in a region no larger than our own solar system. Some quasars are members of small groups or clusters with the same redshift as the quasar. Other quasars have fuzz around them that has the spectrum of a normal galaxy, which shows that quasars are located inside galaxies.

26.2 Most astronomers now view quasars as the most extreme example of a class of peculiar galaxies that generate large amounts of energy in a small **active galactic nucleus.** Seyferts are an example of such **active galaxies.** Even some giant elliptical galaxies are strong radio sources. In many cases, there are jets of radio emission extending to large radio-emitting regions located on either side of the galaxy or quasar. Active galaxies have luminosities intermediate between those of normal galaxies and quasars.

26.3 Both active galactic nuclei and quasars derive their energy from material falling toward, and forming a hot accretion disk around, a massive black hole. Quasars were much more common billions of years ago than they are now, and astronomers speculate that they mark an early stage in the formation of galaxies, when fuel for the accretion disk and jets was more available. Quasar activity can apparently be retriggered by collisions between galaxies, which provide a new source of fuel to feed the black hole.

26.4 Quasars can be used as probes of distant reaches of the universe. For example, observations of quasars give evidence that the **gravitational lensing** effects predicted by general relativity actually do occur. Gravitational lenses can produce various types of quasar images, including double or multiple images and even arcs and rings. Analysis of lensed quasars supports the idea that most of the mass in the universe is in the form of dark matter.

Review Questions

1. Describe some differences between quasars and normal galaxies.

2. Describe the arguments supporting the idea that at least some quasars are at the distances indicated by their redshifts.

3. In what ways are active galaxies like quasars and different from normal galaxies?

4. Describe the process by which the action of a black hole can explain the energy radiated by quasars.

5. What is a gravitational lens? Show with a diagram how it is possible to see two quasars in the sky when in reality there is only one.

6. Describe the observations that convinced astronomers that M87 is an active galaxy.

Thought Questions

7. Suppose you observe a star-like object in the sky. How can you determine whether it is actually a star or a quasar?

8. Why don't any of the methods for establishing distances to galaxies, described in Chapter 25, work for quasars?

9. One of the early hypotheses to explain the high redshifts of quasars was that these objects had been ejected at very high velocities from galaxies. This idea was rejected, because no quasars with large blueshifts have been found. Explain why we would expect to see quasars with both blueshifted and redshifted lines if they were ejected from nearby galaxies.

10. If we see a double image of a quasar produced by a gravitational lens, and can obtain a spectrum of the galaxy that is acting as the gravitational lens, we can then put limits on the distance to the quasar. Explain how.

11. A friend of yours who has watched many *Star Trek* episodes says, "I thought that black holes pulled everything into them. Why then do astronomers think that black holes can explain the great *outpouring* of energy from quasars?" How would you respond?

Problems

12. Rapid variability in quasars indicates that the region in which the energy is generated must be small. Show why this is true. Specifically, suppose that the region in which the energy is generated is a transparent sphere 1 LY in diameter. Suppose that in 1 s this region brightens by a factor of 10 and remains bright for two years, after which it returns to its original luminosity. Draw its light curve (a graph of its brightness over time) as viewed from Earth.

13. A quasar has a redshift of $\Delta\lambda/\lambda = 4.1$. What is the observed wavelength of its Lyman α line of hydrogen, which has a laboratory or rest wavelength of 121.6 nm? Would this line be observable with a ground-based telescope in a quasar with zero redshift? Would it be observable from the ground in a quasar with a redshift of $\Delta\lambda/\lambda = 4.1$? Explain your answer.

Suggestions for Further Reading

Burns, J. "Chasing the Monster's Tail: New Views of Cosmic Jets" in *Astronomy,* Aug. 1990, p. 28.

Croswell, K. "Have Astronomers Solved the Quasar Enigma?" in *Astronomy,* Feb. 1993, p. 29.

Finkbeiner, A. "Active Galactic Nuclei: Sorting Out the Mess" in *Sky & Telescope,* Aug. 1992, p. 138. Good introduction by a science writer.

Osmer, P. "Quasars as Probes of the Distant and Early Universe" in *Scientific American,* Feb. 1982, p. 126.

Preston, R. "Beacons in Time: Maarten Schmidt and the Discovery of Quasars" in *Mercury,* Jan./Feb. 1988, p. 2.

Preston, R. *First Light.* 1987, Atlantic Monthly Books. On doing astronomy with the 200-in. telescope; contains a number of sections on quasar research.

Rees, M. "Black Holes in Galactic Centers" in *Scientific American,* Nov. 1990, p. 56.

Schild, R. "Gravity Is My Telescope" in *Sky & Telescope,* Apr. 1991, p. 375.

Smith, H. "Quasars and Active Galaxies" in *The Universe,* B. Preiss and A. Fraknoi, eds. 1987, Bantam.

Turner, E. "Gravitational Lenses" in *Scientific American,* July 1988, p. 54.

Weedman, D. "Quasars: A Progress Report" in *Mercury,* Jan./Feb. 1988, p. 12.

Wright, A. and Wright, H. *At the Edge of the Universe.* 1989, Ellis Horwood/John Wiley. Introductory book on extragalactic astronomy.

Using **REDSHIFT** ™

1. Turn off the planets and stars, and set the deep-sky filter to show only quasars. Set the *Limiting Magnitude* to 20.

How many quasars do you see?

Is there a relation between quasar brightness and its redshift? (The redshift is given after "emission line" in the information window displayed for the object.)

2. Display only spiral galaxies and set the *Limiting Magnitude* to 20.

Estimate how many galaxies are displayed.

If quasars are related to spiral galaxies, how do you reconcile the number of spirals with the number of quasars counted in Exercise 1?

The inner parts of galaxies NGC 4038 and 4039, which are in the process of colliding. We can see many young star clusters, surrounded by reddish regions of ionized hydrogen, that formed as a result of the collision. The nuclei of the two galaxies can be seen as two diffuse yellowish "blobs" of light. Note the massive, dusty cloud of interstellar matter between the two galaxies. Such collisions alter the shape of each galaxy and result in bursts of star formation as the raw material in the galaxies interacts. (Photo by David Malin; copyright Anglo-Australian Telescope Board)
The inset is a negative image of the same galaxies with a larger field of view, showing the long streamers or tails of stars and gas pulled out of the galaxies by the interaction. (F. Schweizer, Carnegie Institution of Washington)

CHAPTER 27

The Organization of the Universe

Thinking Ahead

From the vantage point of our own island of stars, we look out and try to fathom how the other islands (galaxies) are organized in the dark ocean of space. How can we be sure that the islands we can see are representative of the whole universe?

"But I want to make sure of our whereabouts and whenabouts," said Van, "it is a philosophical need."

Vladimir Nabokov in Ada (1969)

Now that we are familiar with the different types of galaxies, we are ready to examine how they are organized in space and time. In this chapter we expand our perspective to the largest scales of all: we explore the distribution of galaxies over millions and billions of light years, and use that information to examine the way they change over the lifetime of the universe.

Only in the past decade have astronomers begun to make real progress in both these areas. Their work has been slow for several reasons. To understand the way galaxies are distributed in space, we must make a three-dimensional map of the positions of many galaxies. While it is easy to see where a galaxy lies on the dome of the sky, the third dimension—its distance away from us—is much more difficult to measure. (See the opening figure in the Prologue for a wonderful image of galaxies that lie at many different distances in the same part of the sky.) We must take a spectrum of each galaxy, collect enough light to measure a redshift, and then use the Hubble law to convert it to a distance. Only recently have we been able to do this for a sufficient number of remote galaxies to get a meaningful sample of the universe.

When it came to figuring out the evolution of galaxies, there were equally formidable difficulties. First, galaxies are made up of stars, and it was only after the evolution of individual stars was understood that astronomers could begin to explore how whole systems of stars change with time. Thirty years ago, any attempt to describe the evolution of galaxies would have been pointless: we simply did not know enough about the life histories of stars.

A second difficulty in the study of galaxies is the fact that they are very, very faint. Until recently it was extremely hard, even with large telescopes, to see individual stars in most galaxies, or to determine the shapes of the most remote galaxies. So measuring how they change with time became possible only when our instruments could show us the details of faint galaxies with greater resolution. Today, large telescopes on the ground and instruments like the Hubble in space are finally making such a task possible.

We do, however, find one built-in advantage in studying galactic evolution. As we have seen, the universe itself is a kind of time machine that permits us to observe remote galaxies as they were long ago. For the closest galaxies, the time the light takes to reach us is on the order of a few hundred thousand to a few million years. Typically not much changes over times that short. But when we observe a galaxy that is a billion LY distant, we are seeing it as it was when the light left it a billion years ago. By observing more and more distant objects, we look ever farther back toward a time when both galaxies and the universe were young (Figure 27.1).

Figure 27.1

An illustration of astronomical time travel. This true-color, long-exposure image, made during 48 orbits of the Earth with the Hubble Space Telescope, shows a small area in the direction of the constellation of Hercules. We can see galaxies an estimated 3 to 8 billion LY away, including a series of faint blue galaxies that were much more common in that earlier time than they are today. While the nature of these blue galaxies is still controversial, they appear blue because they are undergoing active star formation and making hot, bright blue stars. (R. Windhorst et al. and NASA)

27.1

The Distribution of Galaxies in Space

Celestial objects rarely journey through space alone. The Earth is one of nine planets orbiting the Sun. More than half of all stars are found in double or triple star systems, and many are members of star clusters. All of the stars and clusters in the Milky Way Galaxy are kept close together by the force of gravity, orbiting a common center. Astronomers wondered whether this cosmic togetherness persists on still-larger scales. Are galaxies found in groups as well, or do they all exist in isolation? And how, in general, are galaxies distributed in space? Are there as many in one direction of the sky as in any other, for example?

Edwin Hubble found answers to some of these questions only a few years after he first showed that other galaxies exist. As he examined galaxies all over the sky, Hubble made two discoveries that are crucial for studies of the evolution of the universe.

The Cosmological Principle

Hubble made his observations with what were then the world's largest telescopes—the 100-in. and 60-in. reflectors on Mount Wilson. Although these telescopes can probe to great depths, they can do so only in small fields of view. To photograph the entire sky with the 100-in. telescope would take thousands of years. So instead, Hubble sampled the sky in many regions, much as Herschel did with his star gauging (see Section 24.1). In the 1930s Hubble photographed 1283 sample areas, and on each print he carefully counted the numbers of galaxy images.

The first discovery Hubble made was that the number of galaxies visible in each area of the sky is about the same. (Strictly speaking this is true only if the light from distant galaxies is not absorbed by dust in our own Galaxy, but Hubble made corrections for this absorption.) He also found that the numbers of galaxies increase with faintness, as we would expect if the density of galaxies is about the same at all distances from us.

To understand what we mean, imagine you are taking snapshots in a crowded stadium during a sold-out concert. The people sitting near you look big, so only a few of them will fit into a photo. But if you focus on the people sitting in seats way on the other side of the stadium, they look so small that many more will fit into your picture. If all parts of the stadium have the same seat arrangements, then as you look farther and farther away, your photo will get more and more crowded with people. In the same way, as Hubble looked at fainter and fainter galaxies, he saw more and more of them.

Hubble's findings are enormously important, for they indicate that the universe is both **isotropic** and **homogeneous**—it looks the same in all directions, and a large volume of space at any given redshift is much like any other volume at that redshift. If that's so, it doesn't matter what section of it we observe (as long as it's a sizable portion)—any section will look the same as any other.

In other words, his results suggest not only that the universe is about the same everywhere, apart from changes with time, but also that the part we can see around us, aside from small-scale local differences, is representative of the whole. The idea that the universe is the same everywhere is called the **cosmological principle** and is the starting assumption for nearly all theories that describe the entire universe (see Chapter 28).

Without having established the cosmological principle, we could make no progress at all in studying the universe. Suppose our own local neighborhood were unusual in some way. Then we could no more understand what the universe is like than if we were marooned on a warm south seas island without outside communication and were trying to understand the geography of the Earth. From our limited vantage point, we could not know that some parts of the planet are covered with snow and ice, or that large continents exist with a much greater variety of terrain than that found on our island.

Hubble merely counted the numbers of galaxies in various directions without knowing how far away most of them were. But in the past decade astronomers have measured the velocities and distances of thousands of galaxies, and so built up a meaningful picture of the large-scale structure of the universe. In the rest of this section we describe what we know about the distribution of galaxies, beginning with those nearby.

The Local Group

The region of the universe for which we have the most detailed information is, as you would expect, our own local neighborhood. It turns out that the Milky Way Galaxy is a member of a small group of galaxies called, not too imaginatively, the **Local Group.** It is spread over about 3 million LY and contains at least 30 members. There are 3 large spiral galaxies (our own, the Andromeda galaxy, and M33), at least 12 dwarf irregulars, 2 intermediate ellipticals, and 14 known dwarf ellipticals (see Appendix 12).

New members of the Local Group are still being discovered. We mentioned in Chapter 24 that a dwarf galaxy only about 80,000 LY from the Earth and about 50,000 LY from the center of the Galaxy was recently found in the constellation of Sagittarius. (You may remember that this dwarf is actually venturing too close to the much larger Milky Way and will eventually be consumed by it.) Such newly discovered neighbors are typically very faint and often lie in directions where their light is dimmed by the dust in the disk of the Milky Way.

Figure 27.2 is a rough sketch of how the brighter members of the Local Group are distributed. The average of the motions of all the galaxies in the Group indicates that its total mass is about 5×10^{12} M$_{Sun}$. A substantial amount of this mass is in the form of dark matter.

Neighboring Groups and Clusters

Surveys of more-distant galaxies reveal that the clustering of our nearest neighbors into the Local Group is not exceptional. We find other galaxy groupings at every distance scale we have examined. To start locally, Figure 27.3 shows a galaxy called Dwingeloo 1 (don't you just love astronomical names!). Discovered in 1994, it is part of a small group about 10 million LY away.

However, small groups like ours are hard to notice once we turn our sight to much larger distances. Astronomers therefore focus their attention on more substantial groups called **galaxy clusters.** Such clusters are described as poor or rich depending on how many galaxies they contain. Note that as we examine clusters that are very far away, it gets more difficult to see any dwarf galaxies, and so these are typically not even counted when determining the richness of a cluster. Instead we must (by necessity) use the brighter galaxies in each cluster. (Some of this will now change as instruments such as the Keck and Hubble telescopes show us much fainter galaxies than we have been able to detect in the past.)

The nearest moderately rich galaxy cluster is called the Virgo cluster, after the constellation in which it is seen. It is about 50 million LY away and contains thousands of members, of which a few are shown in Figure 27.4. The giant elliptical (and very active) galaxy M87, which you came to know and love in Chapter 26, belongs to the Virgo cluster. Although M87 is not shown in Figure 27.4, two other giant ellipticals in the cluster are.

A good example of a cluster that is much larger than the Virgo complex is the Coma cluster, with a diameter of at least 10 million LY and thousands of observable galaxies (Figure 27.5). Lying about 250 to 300 million LY away, this cluster is centered on two giant ellipticals whose luminosities equal about 400 billion Suns. Dwarf galaxies are too faint to be seen at this distance, but if they are present, as we expect they are, then Coma likely contains tens of thousands of galaxies. The total mass of this cluster is about 4×10^{15} M$_{Sun}$ (enough mass to make 4 million billion stars like the Sun!).

Let's pause here for a moment of perspective. We are now discussing numbers by which even astronomers sometimes feel overwhelmed. The Coma cluster may have 10, 20, or 30 thousand galaxies, and each galaxy has billions and billions of stars. If you were traveling at the speed of light, it would still take you more than 10 million

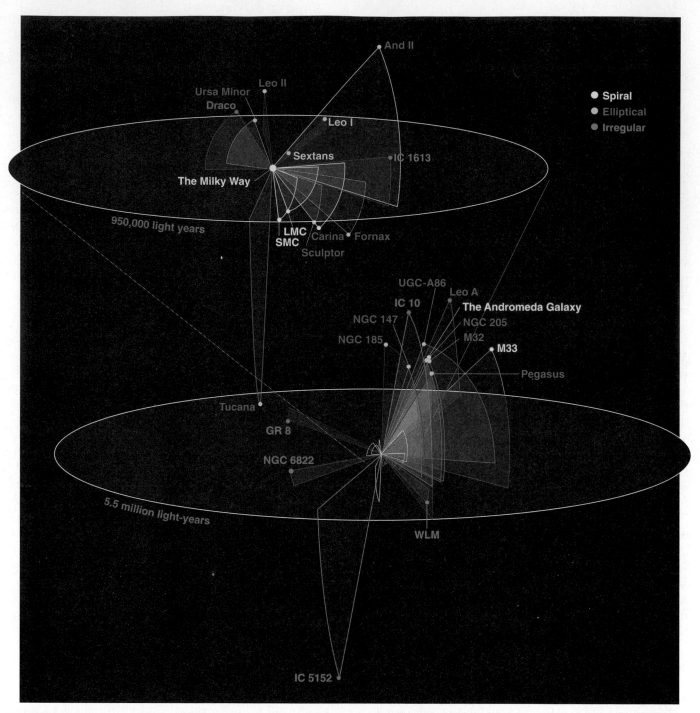

Figure 27.2

A three-dimensional view of some members of the Local Group of galaxies, with our Milky Way at the center. The exploded view at the top shows the region closest to the Milky Way. The three largest galaxies among the three dozen or so members of the Local Group are all spirals; most of the others are dwarf galaxies. (From an original drawing by Thomas L. Hunt, *Astronomy* Magazine; from data by Paul Hodge)

years (longer than the history of the human species) to cross this giant swarm of galaxies. And if you lived on a planet on the outskirts of one of these galaxies, many other members of the cluster would be close enough to be noteworthy sights in your nighttime sky.

Really rich clusters such as Coma are usually spherical in shape, with a high concentration of galaxies near the center. Giant elliptical galaxies are present in these central regions, but few if any spiral galaxies are found there. The spirals that do exist generally occur on the outskirts of clusters. We might say that ellipticals are highly "social": they are often found in groups, and very much enjoy "hanging out" with other ellipticals in crowded situations.

(*text continued on page 542*)

Figure 27.3
This member of a neighboring group of galaxies, called Dwingeloo 1 (after the location of the Dutch telescope that found it), was discovered in 1994. It is about 10 million LY away. The spiral galaxy is in the middle of the picture. The objects with colored spikes are bright stars in our own Galaxy that are overexposed. (S. Hughes and S. Maddox, Isaac Newton Telescope, Royal Greenwich Observatory)

Figure 27.4
The central region of the Virgo cluster of galaxies, the nearest rich cluster (50 million LY away). With its hundreds of bright galaxies, Virgo is the dominant feature of the Local Supercluster of galaxies. In this picture you can see two giant elliptical galaxies, M84 and M86, plus a number of spirals. (© Anglo-Australian Telescope Board)

Figure 27.5
The central part of the rich Coma cluster of galaxies, named because it can be seen in the constellation of Coma Berenices. The two dominant galaxies in this image are huge elliptical galaxies, and they have the yellowish color characteristic of old stars. (The bright blue object is a star in our own Galaxy.) More than 10,000 galaxies belong to this cluster. Astronomers see almost no spirals like the Milky Way in the central regions of such rich clusters. (National Optical Astronomy Observatories)

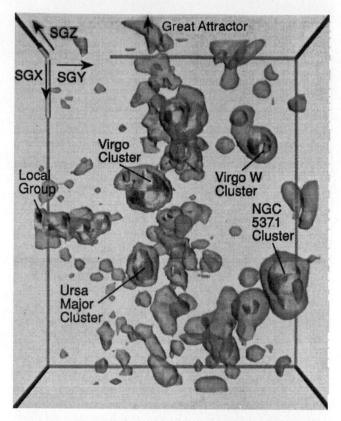

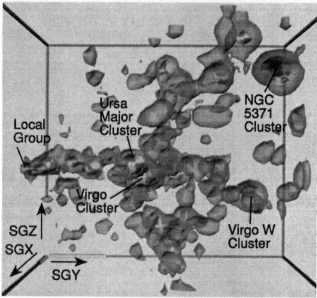

Figure 27.6
The distribution of galaxies in the Local Supercluster, a volume of space over 100 million LY across. Using a computer, astronomer Brent Tully has mapped the concentration of galaxies in an arbitrary box of space from two different perspectives. The Local Group is at the left wall in each case, and the Virgo cluster is in the center. The green areas show where the galaxies are found; the regions colored red and yellow have the highest density of galaxies. Galaxies are found in clumps and small groups, while much of space contains no galaxies at all. (Courtesy of Brent Tully, University of Hawaii)

Spirals, on the other hand, are more "shy": they are more likely to be found in poor clusters or on the edges of rich clusters. We will return to why these differences exist later in the chapter.

Superclusters and Voids

After astronomers discovered clusters of galaxies, they naturally wondered whether there were still-larger structures in the universe. Do the clusters of galaxies also gather together? It turns out they do, forming very large-scale structures that we call **superclusters.** And among the superclusters we have begun to glimpse great **voids,** emptier zones where few if any galaxies can be found.

The best-studied of the superclusters is the *Local Supercluster.* So-named because we are part of it, there is really nothing very "local" about it. With a diameter of at least 100 million LY and a mass of about 10^{15} M_{Sun}, it contains our Local Group of galaxies as well as the Virgo cluster, which lies near its center. The Milky Way lies on the outskirts of the Local Supercluster (Figure 27.6).

Perhaps the most surprising fact revealed by studies of the Local Supercluster is that space is mostly empty. Most of the galaxies are concentrated into clusters, and these clusters occupy only about 5 percent of the total volume of space contained within the boundaries of the Local Supercluster. We find similar results for other superclusters as well.

Slices of the Universe

Once we move outside the Local Supercluster, our view must take in enormous volumes of space. Surveying these distant regions is one of the most difficult and time-consuming projects in astronomy, although it is also one of our most important steps toward understanding the large-scale structure of the universe. How can we make such a task manageable?

The situation is similar to that faced by the first explorers in a huge, uncharted territory on Earth. Since there is only one band of explorers and an enormous amount of land, they have to make choices about where to go first. One strategy might be to strike out in a straight line in order to get a sense of the terrain. They might, for example, cross some mostly empty prairies and then hit a dense forest of trees. As they make their way through the forest, they learn how thick it is in the direction they are traveling, but not its width to their left or right. Then a river crosses their path; as they wade across, they can measure its width but learn nothing about its length. Still, as they go on in their straight line, they begin to get some sense of what the landscape is like and can make at least part of a map. Other explorers, striking out in other directions, will someday help fill in the remaining parts of that map.

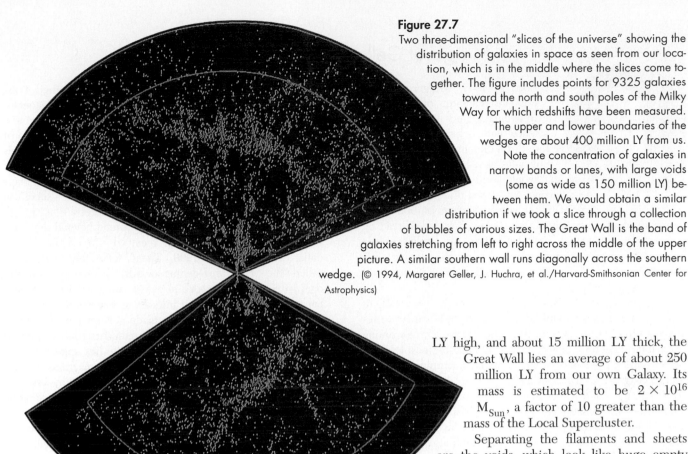

Figure 27.7
Two three-dimensional "slices of the universe" showing the distribution of galaxies in space as seen from our location, which is in the middle where the slices come together. The figure includes points for 9325 galaxies toward the north and south poles of the Milky Way for which redshifts have been measured. The upper and lower boundaries of the wedges are about 400 million LY from us. Note the concentration of galaxies in narrow bands or lanes, with large voids (some as wide as 150 million LY) between them. We would obtain a similar distribution if we took a slice through a collection of bubbles of various sizes. The Great Wall is the band of galaxies stretching from left to right across the middle of the upper picture. A similar southern wall runs diagonally across the southern wedge. (© 1994, Margaret Geller, J. Huchra, et al./Harvard-Smithsonian Center for Astrophysics)

Astronomers must make the same sort of choices. We cannot explore in every direction: there are far too many galaxies and far too few telescopes (and graduate students) to do the job. But we can pick a single direction or a small slice of the sky and start mapping the galaxies. Margaret Geller (see "Voyagers in Astronomy" box), John Huchra, and their students at the Harvard-Smithsonian Center for Astrophysics have examined such slices for a decade and a half; a computer plot of their results is shown in Figure 27.7. (In this figure, the radial velocities of the galaxies are plotted, but remember that according to the Hubble law, distance is proportional to radial velocity.) Other groups around the world are undertaking similar probes in other directions, and tens of thousands of galaxies have been mapped so far. Although this is still an insignificantly small section of what is out there, the first surveys have already yielded major surprises.

Galaxies are not arranged uniformly throughout the universe, but are found in huge filamentary superclusters that look like great arcs of inkblots splattered across a page. A dramatic example of such a filamentary structure, nicknamed the "Great Wall," is shown in Figure 27.7. A sheet of galaxies at least 500 million LY long, 200 million LY high, and about 15 million LY thick, the Great Wall lies an average of about 250 million LY from our own Galaxy. Its mass is estimated to be 2×10^{16} M_{Sun}, a factor of 10 greater than the mass of the Local Supercluster.

Separating the filaments and sheets are the voids, which look like huge empty bubbles walled in by the great arcs of galaxies. The existence of these voids came as a surprise when galaxy surveys first began showing them in the late 1970s and early 1980s. They have typical diameters of 150 million LY, with the clusters of galaxies concentrated along their walls. The whole arrangement of filaments and voids reminds us of a sink full of soap bubbles, the inside of a honeycomb, or a hunk of Swiss cheese with very large holes. If you take a good slice or cross-section through any of these, you will see something that looks like Figure 27.7.

Before the voids were discovered, most astronomers would probably have predicted that the regions between giant clusters of galaxies are filled with many small groups of galaxies, or even with isolated individual galaxies. Careful searches within these voids have so far confirmed that few galaxies of any kind are found there. Apparently, 90 percent of the galaxies occupy less than 10 percent of the volume of space.

We have said that the universe is expanding, but what exactly in Figure 27.7 is expanding? The galaxies and clusters of galaxies are held together by gravity and do not expand as the universe does. However, the voids do grow larger and the filaments move farther apart.

In this discussion we have not considered the presence of dark matter, which may make up a much larger part of many galaxies than the matter we can detect. Nor have we considered whether the voids might actually con-

Margaret Geller: Cosmic Surveyor

Born in 1947, Margaret Geller was the daughter of a chemist who encouraged her interest in science and helped her visualize the three-dimensional strucure of molecules as a child. (It was a skill that would come in very handy for visualizing the three-dimensional structure of the universe.) She remembers being bored in elementary school, but was encouraged to read on her own by her parents. Her recollections also include subtle messages from teachers that mathematics (her strong early interest) was not a field for girls, but she did not allow herself to be deterred.

Geller obtained a B.A. in physics from the University of California at Berkeley and became the second woman to receive a PhD in physics from Princeton. There, while working with James Peebles, one of the world's leading cosmologists (see Chapter 28), she became interested in problems relating to the large-scale structure of the universe. In 1980 she accepted a research position at the Harvard-Smithsonian Center for Astrophysics, one of the most dynamic institutions for astronomy research in the country. She saw that to make progress in understanding how galaxies and clusters are organized, a far more intensive series of surveys was required. Although it would not bear fruit for many years, Geller and her collaborators began the long, arduous task of mapping the galaxies.

Her team was fortunate to be given access to a telescope that could be dedicated to their project, the 60-in. reflector on Mount

Dr. Margaret Geller. (CfA)

Hopkins, near Tucson, Arizona, where they and their assistants continue to take spectra to determine galaxy distances. To get a slice of the universe, they point their telescope at a determined position in the sky, and then let the rotation of the Earth bring new galaxies into their field of view. In this way they have measured the positions and redshifts of over 15,000 galaxies, and made a wide range of interesting maps to display their data. Their surveys now include "slices" in both the Northern and Southern Hemispheres.

As news of her important work spread beyond the community of astronomers, Geller received a MacArthur Foundation Fellowship in 1990. These fellowships, popularly called "the MacArthur genius awards," are designed to recognize truly creative work in a wide range of fields. Geller continues to have a strong interest in visualization, and has (with filmmaker Boyd Estus) made several award-winning videos explaining her work to nonscientists (one is entitled *So Many Galaxies, So Little Time!*). She has appeared on a variety of national news and documentary programs, including the *MacNeil/Lehrer Newshour, The Astronomers,* and *The Infinite Voyage.* Energetic and outspoken, she has given talks on her work to many audiences around the country, and works hard to find ways to explain the significance of her pioneering surveys to the public.

> It's exciting to discover something that nobody's seen before. [To be] one of the first three people to ever see that slice of the universe . . . [was] sort of being like Columbus. . . . Nobody expected such a striking pattern!   —Margaret Geller

---

tain dark matter. Many astronomers who make models of galaxy formation feel that dark matter plays a key role in determining the large-scale structure we observe. We will return to the role of dark matter in Section 27.4.

Implications of the Structure

Knowledge of exactly *how* galaxies are distributed may provide clues as to *why* they are distributed that way. The existence of such large filaments of galaxies and voids is a puzzle, because we have evidence (to be discussed in Chapter 28) that the universe was extremely smooth a few hundred thousand years after forming. The challenge for the theoretician is to understand how such a featureless universe changed into the complex and lumpy one that we see today.

For example, based on data such as that plotted in Figure 27.7, which shows that galaxies are distributed on the walls of bubble-like structures, some astronomers argued that matter used to be inside the bubbles and then was somehow cleaned out and pushed toward the walls. But how? One possibility is that there were giant explosions during the first few billion years of the universe that swept the gas into bubble-like shells, and that galaxies subsequently formed in those shells. Unfortunately, despite considerable ingenuity, astronomers have yet to devise a way to produce sufficiently energetic explosions to account for the largest voids.

Even if explosions do not seem to explain why the voids exist, this discussion should serve to show how measurements of galaxy distribution in space can help generate ideas about what kinds of events must have occurred when superclusters, clusters, and galaxies were just beginning to form. We will discuss more theories about the earliest stages of the universe in Chapter 28.

If all this structure exists, you may want to ask whether the universe can really be described as homogeneous and isotropic. The answer is probably yes, provided we consider regions of the universe large enough to include a number of superclusters and voids. This is similar to making a meaningful survey of small towns and wide-open spaces in the American midwest. If you live in a town and look only at the houses nearest to you, your perspective will be biased. Similarly, if you live on an isolated farm and look only at the nearest few acres, you will miss all the things that happen exclusively in urban areas. A representative survey must catch enough people or territory to include towns, villages, farms, and open country.

27.2

The Evolution of Galaxies: The Observations

Measuring the positions and distances of galaxies is not the only way to learn more about their origins. We can also examine their spectra, colors, and shapes at different periods of cosmic history to see how galaxies have evolved.

Spectra, Colors, and Shapes

The spectrum of a galaxy provides a great deal of information. We have already seen how the redshift can be used to estimate its distance. Studies of a galaxy's rotation can be used to calculate its mass. Detailed spectral line analysis can determine what types of stars inhabit a galaxy, what their compositions are, and whether a galaxy contains large amounts of interstellar matter.

Unfortunately, many galaxies are so faint that collecting enough light to produce a measurable spectrum is impossible. Astronomers thus have to use a much rougher guide to estimate what kinds of stars inhabit the faintest galaxies—their colors (Figure 27.8). To understand how this works, remember that hot luminous blue stars have lifetimes of only a few million years. If we see a very blue galaxy, we know that it must have many hot luminous blue stars, and that star formation must have occurred in the past few million years. A yellow or red galaxy, on the other hand, contains mostly old stars, formed billions of years before the light that we now see was emitted.

Another important clue to the nature of a galaxy is its shape. As we have seen, spiral galaxies contain young stars and large amounts of interstellar matter, while elliptical galaxies have mostly old stars and very little interstellar matter. Elliptical galaxies appear to have turned most of

Figure 27.8
Elliptical galaxy NGC 1199 and companions. NGC 1199 is the large elliptical galaxy at bottom left. It is accompanied by a face-on barred spiral galaxy (NGC 1189) seen toward the upper left; another spiral galaxy (NGC 1190) of spindly appearance is found near the right edge. We can see the yellow-orange color characteristic of older population II stars in the elliptical galaxy, and in the central bar and bulge of the barred spiral. Younger population I stars produce the bluish color of the barred spiral's disk. This picture shows how colors can be used to determine the ages of stars found in distant galaxies. (N. A. Sharp/NOAO)

their interstellar matter into stars many billions of years ago, while star formation has continued until the present day in spiral galaxies. If we can count the number of galaxies of each type that exist during each epoch of the universe, it will help us understand how the pace of star formation changes with time.

Unfortunately, very distant galaxies appear small on the sky. Often we cannot tell whether a distant galaxy is a spiral or an elliptical. One research program being carried out with the Hubble Space Telescope involves obtaining images of distant galaxies, unblurred by the Earth's atmosphere, in order to determine whether the same types of galaxies visible nearby in the present-day universe existed billions of years ago (see Figure 27.1 and the opening figure for Chapter 25).

The Ages and Composition of Galaxies

One of the most important conclusions from our observations of galaxies is that nearly all of them are very old. For example, our own Galaxy contains globular clusters with stars that are at least 13 billion years old, and some may be even older than that. The Milky Way must be at least as old as the oldest stars in it.

As we will discuss in Chapter 28, astronomers have traced the expansion of the universe backwards in time and discovered that the universe itself is not significantly older than the globular-cluster stars. Thus it appears that the globular-cluster stars in the Milky Way must have formed during the first billion years or so after the expansion began.

The light from the most-distant elliptical galaxies for which we have some composition information left them when the universe was only about half its present age. Yet they resemble nearby elliptical galaxies in their luminosity and colors, implying they have the same sorts of stars in them—even though the nearby galaxies are about twice as old. The similarity of ellipticals that span half the age of the universe suggests that star formation in this type of galaxy has either been absent or very slow for the last several billion years. Star formation in elliptical galaxies probably began less than a billion years or so after the universe began, and new stars continued to form for at most a few billion years.

The most-distant galaxy observed to date is so far away that the light we see now left it when the universe was about one-tenth its present age—or only about 1.5 billion years after the beginning. Yet the spectrum of this galaxy contains lines of heavy elements, including carbon, silicon, aluminum, and sulfur. These elements were not present when the universe began. This means that when the light from this galaxy was emitted, an entire generation of stars had already been born, lived out their lives, and died—spewing out the new elements made in the interiors of stars.

We can look even farther back in time by studying quasars (see Chapter 26); a few have been observed at even larger distances than the most-distant galaxies we can see. Remember that we have good evidence that quasars are found in the centers of galaxies. The gas in these distant quasars also contains some heavier elements such as carbon, nitrogen, and oxygen. Therefore, we again conclude that at least one generation of stars in the galaxy surrounding the quasar must already have completed its evolution before the light that we now see was emitted. Given the distance to quasars, this means that some stars must have formed when the universe was *even less* than a billion years old. Stars may already have been going through their life cycles before galaxies were fully developed. Any theory of galaxy formation must therefore include a way for star formation to begin close to the time that galaxies *start* their lives.

Star Formation in Different Galaxies

As we have discussed, the evolution of spiral and elliptical galaxies seems to differ in a fundamental way. In spirals, star formation is a continuous process that is still occurring today. In elliptical galaxies, even the youngest stars are older than the Sun. Since ellipticals contain very little dust or gas, star formation cannot take place in the present era.

Where did the gas and dust in ellipticals go? Much of it must have been consumed very rapidly during the formation of the first generations of stars. But star formation alone would not be efficient enough to consume all of the gas and dust originally present in elliptical galaxies. In any case, as stars evolve, they lose mass either via stellar winds, by forming planetary nebulae, or by exploding. Through all of these processes they replenish the interstellar material in their galaxy.

We must therefore conclude an efficient mechanism exists for removing gas and dust from elliptical galaxies. A clue to one such mechanism came from observations of galaxy clusters made with x-ray telescopes in orbit. In recent years, astronomers have discovered that galaxy clusters, particularly rich ones such as the Coma cluster, are usually sources of x rays (Figure 27.9). The x rays are produced by hot gas, with temperatures between 10 and 100 million degrees, located between the galaxies. This hot gas typically makes up a significant fraction of the total cluster mass. Where does it come from?

We noted earlier that spiral galaxies are absent from the central regions of rich galaxy clusters. The presence of

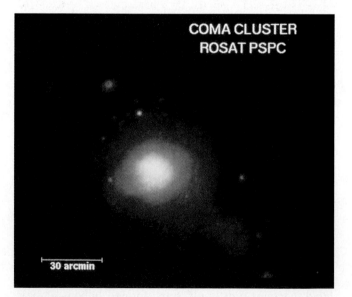

Figure 27.9

An x-ray image of the Coma cluster, taken with the Rosat x-ray satellite. You can see a large x-ray "cloud" of emission from hot gas in the cluster. Notice a smaller cloud near the bottom right, which appears to be a smaller cluster on a collision course with Coma. (Max Planck Institute for Extraterrestrial Physics)

hot gas and the absence of spirals are related. Lines in the x-ray spectrum indicate that the hot gas is not just hydrogen and helium, the composition we would expect if the gas were *primordial*—if it were the raw material from which the cluster galaxies originally formed. Instead, the hot gas contains heavy elements in abundances similar to that of the Sun. This means that the gas must have at some point passed through the interiors of stars, where heavy elements are built up. Since stars are found *in* galaxies and not between them, we conclude that the hot gas must have been processed inside the galaxies at one time.

As we saw in Chapter 22, massive stars not only make a variety of heavier elements during their lives, but they then have the courtesy to explode and recycle them into the gas among the stars. So over the billions of years, gas inside galaxies naturally builds up increasing concentrations of heavier elements. But how did this gas then get outside its parent galaxies in rich clusters such as Coma?

Galaxies in rich (and crowded) clusters are much more likely to collide with each other than in thin regions such as our Local Group. We look at collisions and mergers of galaxies in the next section, but for now we merely note that one effect of such collisions is to strip gas from both galaxies, allowing it to escape into the space around them. (See the "Astronomy Basics" box for more on why collisions among galaxies should not surprise you.)

In addition, the galaxies are in motion around the center of the cluster (just as stars and star clusters in our Galaxy are in motion around our center). As the galaxies move through the gas that is already out among them—traveling at speeds up to thousands of kilometers per second—the motion sweeps gas out of them. Figure 27.10 shows a beautiful example of a galaxy with windswept jets reminiscent of a long-haired bicyclist riding fast on a windy afternoon. Supernova explosions are also effective in driving gas out of galaxies.

Such removal of interstellar matter from galaxies not only increases the amount of gas between the galaxies, but *decreases* the amount of gas within each galaxy. As time passes, galaxies have less and less material with which to make new stars. Collisions in rich clusters also tend to destroy spiral structure. Ultimately, only rounded and featureless galaxies containing mostly old stars are left, which is just what we see in the centers of rich clusters. The swept-out gas remains hot because the galaxies continue to move through it at speeds up to thousands of kilometers per second, stirring and heating it.

In poor clusters, both the chances of collision and the speed with which galaxies move are significantly less. As a result, there is much less stripping of gas going on, and so more spirals can be seen throughout the cluster. Here again we meet the idea introduced in Chapter 25—that a galaxy's appearance depends not just on how it was born, but on the interactions with other galaxies in its environment.

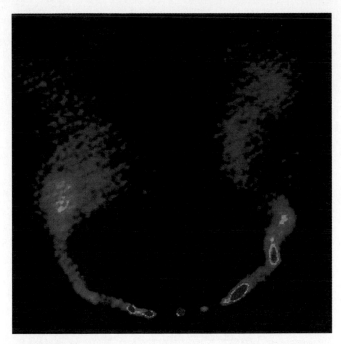

Figure 27.10
A radio image of galaxy NGC 1265, located in the Perseus cluster, which moves at about 2000 km/s through the hot gas that fills the cluster. Like a boat leaving a wake in a river, the jets (which are about 60,000 LY long) are swept back by the galaxy's motion through the gas. (National Radio Astronomy Observatory)

As a result, in the present era relatively few spirals are found near the centers of rich clusters. Conversely, isolated galaxies found in regions outside of clusters or groups of galaxies, where the density of material is low, are mostly spirals. Spiral galaxies, such as the Milky Way and the Andromeda galaxy, have managed to retain their gas and dust because they lie isolated in regions of space where the density of intergalactic gas is too low to sweep them clean.

ASTRONOMY BASICS

Why Galaxies Collide and Stars Rarely Do

Throughout this book we have emphasized the large distances between objects in space. You might therefore have been surprised to hear about frequent collisions between galaxies in a rich cluster. Yet (except for the very cores of galaxies) we have not worried at all about *stars* colliding. Let's see why there is a difference.

The reason is that stars are pathetically small compared to the distances between them. Let's use our Sun as an example. The Sun is about 1.4 million km wide, but is separated from the closest other star by about 4 LY, or about 38 trillion km. In other words, the Sun is 27 million of its own diameters from its nearest neighbor. This is typical of stars that are not in the nuclear bulge of a galaxy, or inside star clusters. Let's contrast this with the separation of galaxies.

The visible disk of the Milky Way is about 100,000 LY in diameter. We have three satellite galaxies that are just one or two Milky Way diameters away from us (and we have already seen that these will eventually collide with us). The closest major spiral is M31, about 2.4 million LY away. Therefore our nearest large galaxy neighbor is only 24 of our Galaxy's diameters from us. (And that does not even take into consideration that both galaxies probably have a much larger corona of dark matter!)

Galaxies in rich clusters are even closer together than the members of our poor Local Group. Thus the chances of galaxies colliding are far greater than the chances of stars in the disk of a galaxy colliding. And we should note that the difference between the separation of galaxies and stars also means that when galaxies do collide, their stars can often pass right by each other like ships passing unknowingly in the night.

Colliding Galaxies: Mergers and Cannibalism

It is becoming increasingly clear that collisions between galaxies play an important role in determining how galaxies evolve. During the past two decades, astronomers have begun to see mounting evidence of such collisions all around the universe. We have already discussed some of the effects of collisions. Let's take a closer look at what happens when galaxies collide.

Stars in colliding galaxies are not much affected. Since the stars are very far apart, a direct collision of two stars is very unlikely, although their orbits are altered. Interstellar matter, however, is much more affected by galaxy interactions. Interstellar gas clouds are large and likely to experience direct impacts with other clouds. These violent collisions compress the gas in the clouds, and the increased density can lead to star formation (Figure 27.11).

In some interacting galaxies, star formation is so intense that all the available gas is exhausted in only a few

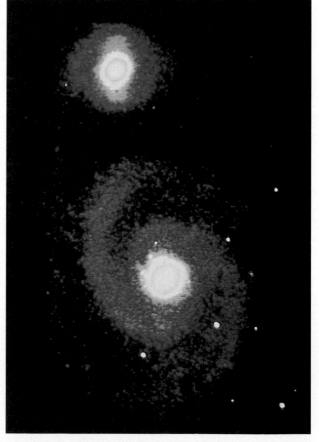

Figure 27.11
Visible light (left) and false-color infrared (right) images of M51, the Whirlpool galaxy. The outlying arm, which reaches to the companion galaxy, is much brighter in visible light than in the infrared. This difference in color suggests that most of the stars in this arm are hot young stars, which are blue in color, and that the formation of these stars was stimulated by the interaction of the two galaxies. (NOAO)

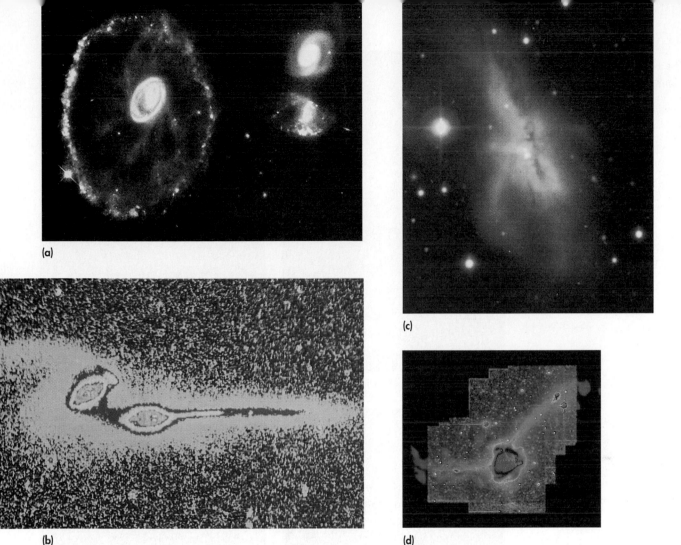

Figure 27.12

A gallery of interacting galaxies: (a) The Cartwheel galaxy, seen in this Hubble Space Telescope image, is the result of a head-on collision. The spiral galaxy at left collided with another galaxy (probably one of the two seen at right), which produced a ring of vigorous star formation (seen in blue). The ring is about 150,000 LY across and contains several billion new stars. (K. Borne and NASA) (b) NGC 4676 A and B are nicknamed "The Mice." This is a visible-light image that has been computer processed and color coded to show subtle details in the levels of light. You can see the long narrow tails of stars caused by the interactions of the two spirals. (NOAO) (c) NGC 6240 shows two galaxy nuclei quite close together in the center, with tails of material indicating that two spiral galaxies must have been involved in a collision. Observations with the IRAS satellite have shown that this galaxy puts out a tremendous amount of energy in the infrared. This is consistent with the idea that vigorous star formation in the center is warming vast quantities of dust (which hides much of this activity from our view in the visible region of the spectrum). (W. Keel and ESO) (d) This image of NGC 7252 combines visible light and radio views of the same pair of colliding galaxies. The red regions at the center of the collision are where we see the brightest visible light. The green areas are much fainter visible-light "tails" from stars and gas thrown out by the interaction. The blue color shows radio emission at 21 cm, the wavelength of hydrogen gas. Note that the hydrogen is found with the green tails and not in the red zone, confirming the idea that such collisions tend to strip gas from the galaxies involved. (J. Hibbard, J. van Gorkom, L. Schweizer, and National Radio Astronomy Observatory)

million years; the burst of star formation is clearly only a temporary phenomenon. Such bursts are very rare in isolated galaxies, but astronomers see them frequently in galaxies that are undergoing collisions. (A gallery of interesting colliding galaxies is shown in Figure 27.12.)

As you can see in these images, when galaxies interact, their shapes can change significantly. Great rings, huge tendrils of stars and gas, and other complex structures can form. Many of the larger irregular galaxies we observe may owe their chaotic shapes to past interactions.

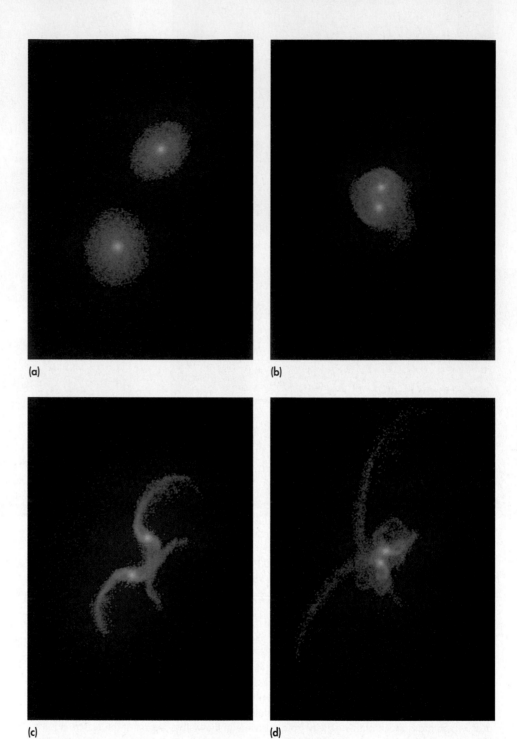

Figure 27.13
Given the right initial conditions, a computer simulation of the collision of two galaxies can produce a structure that strongly resembles NGC 4038/39, the pair of interacting galaxies shown in the opening figure of this chapter. The sequence shows the galaxies at (a) 60 million years, (b) 185 million years, (c) 310 million years, and (d) 435 million years after the interaction began. (Computer images courtesy of Josh Barnes, U. of Hawaii)

The details of galaxy collisions are complex and best simulated on a large computer (Figure 27.13). We lack much in our understanding of them, but it is nevertheless clear that collisions play a larger role in the development of galaxies than we thought even a decade ago.

If the collision is slow, the colliding galaxies may coalesce to form a single galaxy. When two galaxies of equal size are involved, we call such an interaction a **merger** (the term applied in the business world to two equal companies that join forces). But, small galaxies can also be swallowed by larger ones—**galactic cannibalism.**

The very large elliptical galaxies found in the centers of rich galaxy clusters probably form by cannibalizing a variety of smaller galaxies in their clusters. These "monster" galaxies frequently possess more than one nucleus, and have probably acquired their unusually high luminosities by swallowing nearby galaxies. The multiple nuclei are the remnants of their victims (Figure 27.14). Slow collisions and mergers can transform spiral galaxies into elliptical galaxies by stripping out their gas and causing great bursts of star formation that turns whatever gas remains into stars.

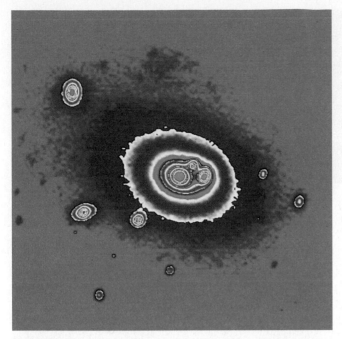

Figure 27.14
False-color image of light intensity in the multiple-nucleus galaxy NGC 6166 in the constellation of Hercules. This giant elliptical has apparently consumed several companion galaxies and lies at the center of a galaxy cluster. (R. Schild)

The Blue Galaxy Mystery

The galaxies we see today appear to differ from the types that were most common several billion years ago. Rich clusters at distances of about 5 billion years contain many more blue galaxies than do nearby rich clusters. These blue galaxies are mainly spirals (see Figure 27.1), and we have now observed them in a sufficient number of clusters to know that they were common when the universe was younger. Since a blue galaxy must contain young stars, this difference in color indicates that there were more spiral galaxies actively forming stars then than now—another indication that the rate of star formation has on average declined during the past 5 billion years or so.

Two processes have been suggested for producing a high rate of star formation in rich clusters at earlier times. First, as we saw earlier, collisions of galaxies can compress the gas within them and stimulate the formation of stars. Another agent could be the hot gas that exists in these clusters. A galaxy, as it moves on its orbit through the cluster, may run into a thicker clump of this high-temperature gas. In the ensuing collision, the cold molecular clouds in the galaxy may be compressed, again accelerating the rate of star formation.

Why have most of these luminous blue spirals vanished from the universe over the past 5 billion years? One possibility is that mergers have reduced the number of galaxies, thereby reducing the likelihood of collisions and the enhanced rates of star formation that they cause. Another is that the rate of star formation has slowed dramati-

cally as interstellar matter has been used up, and the spirals are simply fainter in the current era.

The Evolution of Galaxies: The Theories

Having examined the observational evidence—much of it still fragmentary and inconclusive—we are now ready to take a look at some of the models that astronomers have proposed for how galaxies and groups of galaxies formed during the early history of the universe. Two possibilities have been explored in recent years.

Top-Down or Bottom-Up

Top-down theories assume that supercluster-sized concentrations of matter formed first and then fragmented to form galaxies. *Bottom-up theories* hypothesize that small structures formed first and then merged to build larger ones. If this bottom-up picture is correct, individual galaxies formed first and then gradually assembled to build clusters and then superclusters.

Both top-down and bottom-up theories assume that initially the universe was not absolutely smooth, but contained small fluctuations in density—regions where, by chance, more material had accumulated. (Some of this material, as we will see, may have been dark matter.) As the universe expanded, the regions of higher density acquired additional mass because they exerted a slightly larger than average gravitational force on surrounding material.

As in the case of star formation, the fate of these higher-density regions depended on the balance between pressure and gravity. Initially, each individual region expanded because the universe was expanding. However, if the inward pull of gravity exceeded the outward pressure, the individual region would ultimately have stopped expanding. It would have then begun to collapse to form a cluster of stars, a galaxy, a cluster of galaxies, or even a supercluster of galaxies.

Unfortunately, calculating what might have happened in the early universe is very difficult, primarily because we do not know what the dark matter is made of (see Section 27.4) and the most likely size of the first high-density regions to collapse is not clear. There are two possibilities that seem especially promising. In one, the typical initial condensations are very large and contain total masses equal to 10^{15} M_{Sun}, which is the mass of a supercluster. In the other, we start with clumps containing only 10^6 M_{Sun}, about the mass of a large globular cluster. Calculations indicate that condensations between these two sizes are not as likely to form.

Superclusters First

Top-down theories examine the consequences if very large-scale density fluctuations were the first to collapse.

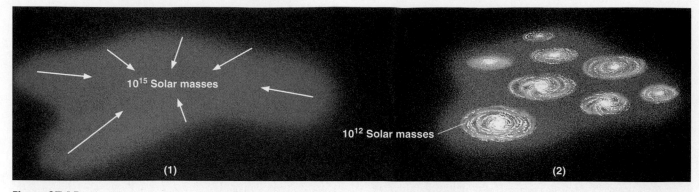

Figure 27.15
Schematic diagram showing how galaxies might have formed if large-scale supercluster structures formed first and then fragmented to form galaxies.

Calculations show that when gas clouds with masses of about 10^{15} M_{Sun} collapse, they form an irregular, pancake-shaped blob. Within the pancake many individual regions of high density are then formed, and these too will collapse under their own gravity. Stable structures can form only in regions with masses of less than 10^{12} M_{Sun} and diameters less than 300,000 LY—just the size of the galaxies we have now (Figure 27.15). Larger fragments are extremely diffuse and will be destroyed in collisions with other fragments before stars can form within them. Galaxy-sized fragments, however, can persist, and eventually the material within them can condense into stars and star clusters.

This top-down model has the advantage of being able to explain why galaxies are no larger than they are observed to be—but it has one fatal flaw. The original pancake takes a long time to collapse and fragment into still-smaller galaxy-sized structures. This model predicts that galaxies should still be forming in the present era. While astronomers have found a few nearby galaxies that may be young, it is clear that most of them formed billions of years ago. Another problem is that no way has yet been found to use top-down models to account for the existence of the large numbers of galaxies and quasars that were already present when the universe was only 10 percent of its present age.

Star Clusters First

The basic assumption of the bottom-up models is that the first higher-density regions to begin collapsing had masses of approximately a million M_{Sun}, about the same as a large globular cluster. Their collapse began when the universe was no more than 2 percent of its present age. As time passed, regions containing ever-larger mass, possibly as large as the mass of a giant elliptical, began to collapse as well. Galaxies could thus be formed either from the collapse of single galaxy-sized clouds or through the merger of several smaller structures.

Clusters of galaxies formed as individual galaxies congregated, drawn together by their mutual gravitational attraction (Figure 27.16). First a few

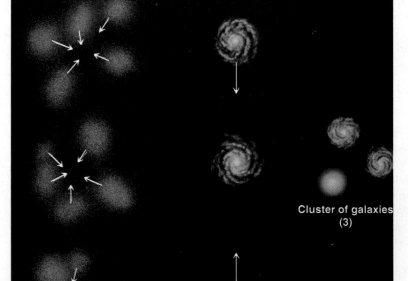

Figure 27.16
Schematic diagram showing how galaxies might have formed if small clouds formed first and then congregated to form galaxies and then clusters of galaxies.

galaxies came together to form groups, much like our own Local Group. Then the groups began combining to form clusters and eventually superclusters.

This model explains in a natural way why there are more small galaxies than large ones. The collapsing clouds are initially composed of gas, and they collide and merge to form galaxies. The more collisions and mergers that occur, the larger the galaxy that finally emerges. Since a given cloud is likely to experience only a few collisions and mergers, very large galaxies should be rare—just as 2-car accidents are more common than 50-car pileups. A detailed theory of galaxy building through this process, however, has not yet been worked out.

Giant elliptical galaxies are round, and occur in the regions of highest density. They were likely formed through the collision and merger of many smaller fragments. A collision of two systems of gas and stars tends to stir up the orbits of the individual stars within each system, and also to strip away from the system's outer regions any matter that is not held by gravity. The result of many collisions is thus to build round systems lacking extended disks.

According to this theory, spiral galaxies are formed in relatively isolated regions. A single cloud of gas collapses to a disk in which stars are then formed. A spiral might, through collisions with smaller systems, acquire some additional stars that populate its halo and nuclear bulge. Like the stars in ellipticals, which are also built through mergers, these stars are distributed in a spherical fashion. As the isolated cloud collapses to form a spiral, it may leave behind some fragments that become dwarf galaxies. Many spirals, including the Milky Way Galaxy, are surrounded by a swarm of small galaxies. As we have seen, the Milky Way is still capturing these small galaxies and adding them to its halo.

The bottom-up model predicts very few young galaxies in the present era, and indeed young nearby galaxies appear to be very rare. This model also predicts that clusters and superclusters should still be in the process of gathering together. Observations do in fact suggest that galaxy clusters are still accumulating their members.

Overall, the evidence appears to favor the idea that low-mass regions formed first, and that galaxies and clusters were formed as these regions interacted and merged under the influence of gravity. However, galaxy formation theory is still in its infancy, and cannot fully explain the individual differences we see in many galaxies and clusters. We are far from having a theory that can account for *all* of the observations, as we will see in Chapter 28.

27.4

A Universe of (Mostly) Dark Matter?

So far this chapter has focused almost entirely on matter that radiates electromagnetic energy. But, as we have pointed out in several earlier chapters, it is now clear that galaxies contain large amounts of *dark matter* as well. There may be much more dark matter, in fact, than matter we have detected—which means it could be foolish to ignore the effect of this material in our theories about the structure of the universe. (As many a ship captain in the polar seas found out too late, the part of the iceberg visible above the ocean's surface was not necessarily the only part he needed to pay attention to.) The dark matter could be extremely important in determining the evolution of galaxies and of the universe as a whole.

The idea that much of the universe is filled with dark matter may seem like a bizarre concept, but there is one historical example of dark matter that we have already described in this book. In the mid-19th century, measurements showed that the planet Uranus did not follow exactly the orbit predicted by adding up the gravitational forces of all of the known objects in the solar system. Its orbital deviations were attributed to the gravitational effects of an (at the time) invisible planet. Calculations showed where that planet had to be, and Neptune was discovered just about in the predicted location.

In the same way, astronomers are now trying to determine the location and amount of dark matter in galaxies by measuring its gravitational effects on objects we can see. And, by measuring the way that galaxies move in clusters, scientists are discovering that dark matter may play an important role in galaxy evolution as well. It appears that dark matter makes up at least 90 percent of all the matter in the universe. The following topics describe the search and evidence for the dark matter, and offer some speculations about what it might be made of.

Dark Matter in the Local Neighborhood

The first place we might look for dark matter is in our own solar system. Astronomers have examined the orbits of the known planets, and of spacecraft as they journey to the outer planets and beyond. No deviations have been found from the orbits predicted on the basis of the objects already discovered in our solar system, and so no evidence exists for large amounts of nearby dark matter.

Astronomers have also looked for evidence of dark matter in the region of the Milky Way Galaxy that lies within a few hundred light years of the Sun. In this vicinity most of the stars are restricted to a thin disk. It is possible to calculate how much mass the disk must contain in order to keep the stars from wandering far above or below it. The total matter that must be in the disk is less than twice the amount of luminous matter. This means that no more than half the mass near the Sun is dark.

Dark Matter Around Galaxies

In contrast with our local neighborhood, there is (as we saw in Chapter 24) evidence suggesting that 90 percent of the mass in the entire Galaxy may be in the form of a

Figure 27.17
The edge-on spiral galaxy NGC 5746 and a graph of the velocity with which it is rotating at each point. As is true of the Milky Way, its rotation does not decrease with distance from the center, as we would expect. This indicates the presence of a corona of dark matter, under whose influence the outer regions go around faster than the observed matter alone could explain.
(William Keel and ASP)

corona of dark matter. In other words, there could be nine times more dark matter than visible matter. The stars in the outer region are revolving very rapidly around the center of the Galaxy—so rapidly that the mass contained in all the stars and all the interstellar matter in the Galaxy cannot exert enough gravitational force to keep those distant stars in their orbits. The same result is found for other spiral galaxies as well (Figure 27.17).

Mathematical analyses of the rotation of spiral galaxies suggest that the dark matter is found in a large corona surrounding the luminous parts of each galaxy. The radius of these coronae may be as large as 300,000 LY.

Dark Matter in Clusters of Galaxies

Galaxies in clusters orbit around the cluster's center of mass. It is not possible for us to follow a galaxy around its entire orbit. For example, it takes 10 billion years or more for the Andromeda and Milky Way galaxies to complete a single orbit around each other. It is possible, however, to measure the velocities with which galaxies in a cluster are moving, and then estimate what the total mass in the cluster must be to keep the individual galaxies from flying off into space. The observations indicate that the total amount of dark matter in clusters probably exceeds that contained within the galaxies themselves, indicating that dark matter exists between galaxies as well as inside them.

Dark Matter in Superclusters

The universe is expanding, but the expansion is not perfectly uniform. Some galaxies are moving away from us at a slightly faster than average rate. Others are moving away at a slower than average rate. Suppose, for example, that a galaxy lies outside but relatively close to a rich cluster of galaxies. The gravitational force of the cluster will tug on that neighboring galaxy and slow down the rate at which it moves away from the cluster due to the expansion of the universe.

As a specific example, consider the Local Group of galaxies, lying on the outskirts of the Virgo supercluster. The mass concentrated at the center of the Virgo cluster exerts a gravitational force on the Local Group. As a result, the Local Group is moving away from the center of the Virgo cluster at a velocity a few hundred kilometers per second slower than the Hubble law predicts. By measuring deviations from a smooth expansion, astronomers can estimate the total amount of mass contained in large clusters.

Astronomers have now measured accurate distances and velocities for thousands of galaxies within about 150 million LY of the Milky Way (Figure 27.18). Superim-

Figure 27.18
A drawing showing the concentrations of matter within about 150 million LY of the Milky Way, shown at the center of the drawing. Our Galaxy is located in a region where the density is intermediate between that of the Great Attractor and a void. The mass concentration that corresponds to the Great Attractor is shown on the right-hand side of the drawing. All of the galaxies, including those in the Perseus–Pisces region, are flowing toward the Great Attractor at a velocity of about 425 km/s. You can think of it as a giant river in the sky, containing thousands of galaxies.

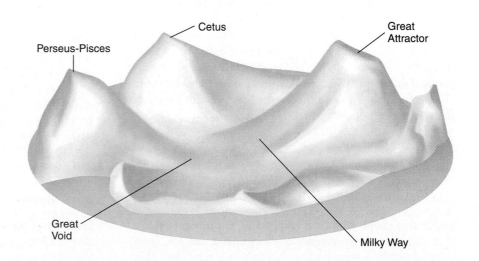

TABLE 27.1	
Mass-to-Light Ratios	
Type of Object	Mass-to-Light Ratio
Sun	1
Matter in vicinity of Sun	2
Total mass in Milky Way	10
Small groups of galaxies	50–150
Rich clusters of galaxies	250–300

posed on the more local motions we find a new and surprising trend. These galaxies tend to be flowing toward an enormous concentration of mass, dubbed the *Great Attractor*. The mass of the Great Attractor is estimated to be 3×10^{16} M$_{Sun}$, equivalent to tens of thousands of galaxies. This mass is much larger than the amount of luminous matter seen in this direction, and so most of the matter in the Great Attractor must be dark.

Mass-to-Light Ratio

Section 25.3 described the use of the mass-to-light ratio to characterize the matter in galaxies or clusters of galaxies. For systems containing mostly old stars, the mass-to-light ratio is typically 10 to 20, where mass and light are measured in units of the Sun's mass and luminosity. Mass-to-light ratios of 100 or more are a signal that a substantial amount of dark matter is present. Table 27.1 summarizes the results of measurements of mass-to-light ratios for various classes of objects. Very large mass-to-light ratios are found for all systems of galaxy size and larger, and this result indicates that dark matter is present in all of these types of objects. Dark matter apparently makes up most of the total mass of the universe. If observations of the motions of galaxies toward the Great Attractor prove to be correct, then as much as 99 percent of the mass in the universe may be dark.

The Search for Dark Matter

Suppose most of the universe really is made up of dark matter. How do we go about detecting it? The technique depends on its composition. For now, let's suppose that dark matter is made up of normal particles—protons and neutrons and electrons. We already know that these particles are not assembled into stars that shine, or we would see them. Neither can dark matter be in the form of dust and gas, or, again, we could detect it (see the discussion in Section 24.4). The protons and neutrons could be in the form of black holes, brown dwarfs, or white dwarfs. The latter two types of objects do emit some radiation but

have such low luminosities that they cannot be seen at distances greater than a few thousand light years. While we cannot *see* dark matter, astronomers can detect invisible assemblages of protons and neutrons because they act as gravitational lenses (see Chapter 26) and distort the radiation emitted by objects that lie behind them.

Two experiments have been designed to look for gravitational lensing by dark matter. The first involves the study of light from distant galaxies. As we saw in Chapter 26, foreground galaxies can act as a gravitational lens and distort the images of more-distant galaxies. These distorted images (such as Figure 26.21) can then be used to determine how much mass—including nonluminous dark matter—is present in the foreground cluster. The resulting maps prove that the foreground clusters of galaxies contain ten times more matter than is actually seen in visible light.

A second technique has been devised to look for black holes, brown dwarfs, and white dwarfs in the halo of our own Galaxy. These objects have been whimsically dubbed MACHOs (massive compact halo objects). If an invisible MACHO passes directly between a distant star and the Earth, it acts as a gravitational lens, focusing the light from the distant star and causing it to appear to brighten over a time interval of several days before returning to its normal brightness.

Research teams making observations of millions of stars in the Large Magellanic Cloud have recently reported several examples of the type of brightening expected if MACHOs are present in the halo of the Milky Way. From the number of such events, they can estimate how many MACHOs there are. It appears that MACHOs contribute no more than 50 percent of the dark matter in the Milky Way Galaxy.

Astronomers are eager to determine what the dark matter is, and this first result is a bit disappointing. Even if it is confirmed, we still don't know what at least 50 percent of the dark matter is made of. In the next chapter we will explore some possibilities. As we will see, it may turn out that most of the dark matter is composed of a new type of particle not yet detected on Earth.

If most of the universe is made of dark matter, we must consider what effect this material has on the clustering and evolution of galaxies. For example, while we do not presently detect anything inside the great voids, it is possible that they have significant amounts of dark matter within them. Dark matter may have played the role of "seed cores" in the formation of galaxies. And if many galaxies have large coronae made of dark matter, this will clearly influence how they will interact with one another and what shapes and types of galaxies their collisions will create. Astronomers armed with various theories are working hard to produce models of galaxy structure and evolution that take dark matter into account in just the right way. Unfortunately, it is too early to tell just what that "right way" is.

Summary

27.1 Counts of galaxies in various directions establish that the universe on the large scale is **homogeneous** and **isotropic** (the same everywhere and in all directions apart from evolutionary changes with time). The sameness of the universe everywhere is referred to as the **cosmological principle.** Galaxies tend to group together to form clusters. The Milky Way Galaxy is a member of the **Local Group,** which contains at least 30 member galaxies. Rich clusters (such as Virgo and Coma) contain thousands or tens of thousands of galaxies. **Galaxy clusters** often group together with other clusters to form large-scale structures called **superclusters,** which can extend over distances of several hundred million light years. Clusters and superclusters fill only a small fraction of space. Most of space consists of **voids** between superclusters, with nearly all galaxies confined to less than 10 percent of the total volume. Rich clusters of galaxies often contain hot (10^7 to 10^8 K) x-ray-emitting gas that has been stripped from member galaxies.

27.2 Observations place important constraints on models of galaxy formation. Galaxies were formed when the universe was no more than 1 or 2 billion years old. Star formation in spirals was much more active 5 billion years ago than it is today. There were probably more galaxies several billion years ago than there are today, with the number being reduced by collisions and mergers. Elliptical galaxies tend to be found in the centers of dense galaxy clusters, while spiral galaxies tend to be relatively isolated.

Galaxies in clusters are close enough together that collisions are likely. These collisions may trigger star formation through compression of interstellar clouds, cause the **merger** of galaxies of comparable size, or result in **galactic cannibalism,** in which a small galaxy is swallowed by a much larger one.

27.3 The challenge for galaxy formation theories is to show how an initially smooth distribution of matter can develop the structures—galaxies and galaxy clusters—that we see today. Calculations show that the first condensations of matter are likely to have contained either the mass of a galaxy supercluster or of a globular cluster. Observations seem to favor the initial condensation of globular-cluster-sized masses, which then congregate to form galaxies and clusters of galaxies.

27.4 The visible matter in the universe does not exert a large enough gravitational force to hold stars in their orbits within galaxies, or to hold galaxies in their orbits around other galaxies. There is at least 5 to 10 times, and perhaps as much as 100 times, more dark matter than luminous matter. Astronomers do not yet know whether the dark matter is made of ordinary matter—protons and neutrons, for example—or of some totally new type of particle not yet detected on Earth. Observations of gravitational lensing effects on distant objects have been used to detect some dark matter in the outer region of our Galaxy.

Review Questions

1. Explain what we mean when we call the universe homogeneous and isotropic. Would you say that the distribution of elephants on the Earth is homogeneous and isotropic? Why?

2. Describe the organization of galaxies into groupings, from the Local Group to superclusters.

3. Suppose you see a group of faint galaxies in a small part of the sky. How would you go about determining their distances?

4. What is the evidence that galaxies formed 1 or 2 billion years after the universe began?

5. Describe two possible ways in which galaxies might form. Which seems more likely? Why?

6. What is the evidence that a large fraction of the matter in the universe is invisible?

Thought Questions

7. Suppose you are standing in the center of a large, densely populated city that is exactly circular, surrounded by a ring of suburbs with lower-density population, surrounded in turn by a ring of farmland. Would you say the population distribution is isotropic? Homogeneous?

8. Use the data in Appendix 12 to determine which is more common in the Local Group—large luminous galaxies, or small faint galaxies. Which is more common—spirals or ellipticals?

9. Based on data in Appendix 12, would you describe the Milky Way Galaxy as a typical member of the Group? Why or why not?

10. Describe how you might use the color of a galaxy to determine something about what kinds of stars it contains.

11. Suppose a galaxy formed stars for a few million years and then stopped. What would be the most-massive stars on the main sequence after 500 million years? After 10 billion years? How would the color of the galaxy change over this time span? (Refer to Table 21.1.)

12. Suppose that the Milky Way Galaxy were truly isolated, with no other galaxies within 100 million LY. Then suppose that galaxies were observed in large numbers at distances greater than 100 million LY. Why would it be more difficult to determine accurate distances to those galaxies than if there were also galaxies relatively close by?

13. Given the ideas presented here about how galaxies form, would you expect to find a giant elliptical galaxy in the Local Group? Why or why not? *Is* there a giant elliptical in the Local Group?

14. Can an elliptical galaxy evolve into a spiral? Explain your answer.

15. Why do we know less about the formation of galaxies than about the formation of stars?

16. Suppose you developed a theory to account for the evolution of New York City. Would it most closely resemble a bottom-up or a top-down theory as we have applied those terms to galactic evolution?

17. Margaret Geller and John Huchra have been making maps by observing a slice of the universe and seeing where the galaxies lie within that slice. If the universe is isotropic and homogeneous, why do they need more than one slice? Suppose they now want to make each slice extend farther into the universe. What do they need to do?

Problems

18. Suppose that on one survey you count galaxies to a certain limiting faintness. On a second survey you count galaxies to a limit that is four times fainter.

    a. To how much greater distance does your second survey probe?

    b. How much greater is the volume of space you are reaching in your second survey?

    c. If galaxies are distributed homogeneously, how many times more of them would you expect to count on your second survey?

19. Calculate the mass-to-light ratios for the stars listed in Table 17.1. Can stars alone explain a mass-to-light ratio of 100, which is measured for some elliptical galaxies?

20. Show just how dominant the dark matter halo of our own Galaxy is by making a scale drawing. Use data given in the text for the diameter of the luminous disk and the sphere of dark matter that surrounds it.

Suggestions for Further Reading

Barnes, J. et al. "Colliding Galaxies" in *Scientific American,* Aug. 1991, p. 40.

Dressler, A. *Voyage to the Great Attractor.* 1994, A. Knopf. A noted astronomer describes how we find large-scale structure.

Geller, M. and Huchra, J. "Mapping the Universe" in *Sky & Telescope,* Aug. 1991, p. 134.

Hodge, P. "Our New Improved Cluster of Galaxies" in *Astronomy,* Feb. 1994, p. 26. On the Local Group.

Keel, W. "The Real Astrophysical Zoo: Colliding Galaxies" in *Mercury,* Mar./Apr. 1993, p. 44.

Lake, G. "Cosmology of the Local Group" in *Sky & Telescope,* Dec. 1992, p. 613.

Lemonick, M. *The Light at the Edge of the Universe.* 1993, Villard/Random House. A journalist's tour of extragalactic astronomy.

MacRobert, A. "Mastering the Virgo Cluster" in *Sky & Telescope,* May 1994, p. 42. How to observe the galaxies through small telescopes.

Parker, B. *Colliding Galaxies.* 1990, Plenum Press. Good nontechnical introduction.

Schramm, D. "Dark Matter and the Origin of Cosmic Structure" in *Sky & Telescope,* Oct. 1994, p. 28.

Using REDSHIFT ™

1. Spot the brightest galaxies in the Local Cluster by turning on only the stars and galaxies and setting the *Limiting Magnitude* to 6.5. All of these galaxies are visible with the unaided eye, but can also be viewed through binoculars or a small telescope.

2. Pick out the Virgo and Coma clusters by setting the *Limiting Magnitude* for galaxies to 14. You need at least a small telescope to see these galaxies.

From the map, what appears to be the most common type of galaxy in the Virgo cluster?

What would you expect the most common type to be? Can you explain any difference?

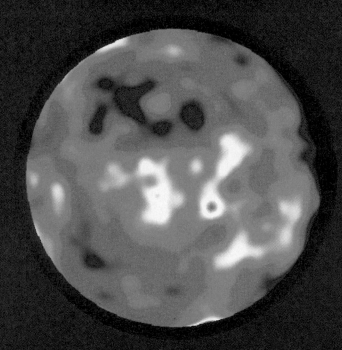

North Galactic Hemisphere

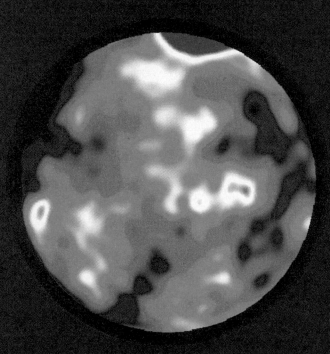

South Galactic Hemisphere

This false-color map shows tiny variations in the temperature of the cosmic microwave background, the radiation left over from the very hot, early phases of the universe. It is constructed from four years of observations with the Cosmic Background Explorer (COBE) satellite in Earth orbit. This radiation has been redshifted by the expansion of the universe until today it is mostly in the infrared and radio regions of the spectrum. The blue and red spots on the map correspond to regions of greater and lesser density in the early universe. It is out of such tiny fluctuations of density that all the structure we now observe in the universe grew. (C. Bennett, NASA)

$-100 \, \mu K$ $+100 \, \mu K$

CHAPTER 28

The Big Bang

Thinking Ahead

As we look farther and farther out in the universe, we look farther and farther back in time. What lies at the beginning of time? What is the earliest information about the universe that our instruments can detect?

We are now ready to complete our voyage through the universe by asking the largest questions that astronomers can ask. How did the entire universe come into being, how has it changed since the beginning, and what will its ultimate fate be? In past centuries such questions were considered the realm of religion and philosophy. The methods of science could not be applied to them because science depends on experiments and observations to decide among possible models or theories. Only recently have we been able to carry out the sorts of observations and experiments that can help us understand the past and future of the cosmos (Figure 28.1).

The branch of astronomy devoted to the study of the universe as a whole is called **cosmology.** An enormous field, it has been the subject of more popular books than any other astronomical subject. In this brief chapter we can give you only an overview of the highlights of current cosmological thinking. But at the end you will find a list of our favorite nontechnical books on the subject; we urge you to consult them if our discussion whets your appetite for more.

Let us begin by reviewing, in Table 28.1, some of the observational discoveries about the universe as a whole that have already been covered in this book. For example, as discussed in Chapter 25, we know from observations of redshifts and the Hubble law that the universe is expanding. Thus we do not need to consider

Making a model of the universe is like trying to pitch a tent on a moonless night in a howling Arctic wind. The tent is theory. The wind is experiment. Progress is made whenever a tent peg proves sturdy enough to hold.

Timothy Ferris in "Minds and Matter," a brief article on cosmology in *The New Yorker*, May 15, 1995

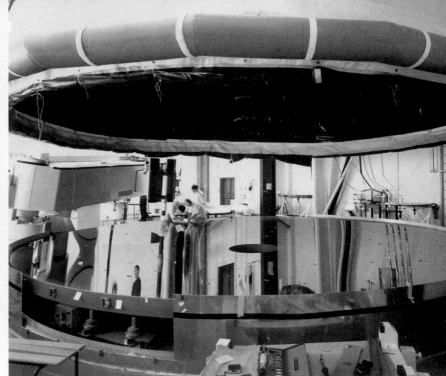

Figure 28.1
Two of the instruments that will be used to observe the most-distant objects in the universe to answer questions in cosmology: (Left) The Hubble Space Telescope, seen in the Shuttle during the December 1994 repair mission. (NASA) (Right) One of the 8.2-m-wide mirrors for the European Southern Observatory's Very Large Telescope being tested in a French optical workshop. It is the first of four such giant mirrors that will be coupled together on a mountaintop in Chile toward the end of this century to make the world's largest visible-light telescope. (ESO)

TABLE 28.1
Some Observed Characteristics of the Universe

1. All galaxies show a redshift proportional to distance, implying that the universe is expanding. (Chapter 25)
2. The distribution of galaxies on the largest scale is isotropic and homogeneous (the cosmological principle). (Chapter 27)
3. The contents of the universe evolve with time: hydrogen and helium are changed into heavier elements inside stars. (Chapters 15, 21, 22)
4. Gravity warps (curves) the fabric of spacetime. (Chapter 23)

---

any model that does not account in a natural way for this expansion. In the discussion that follows, we build on this fundamental observation (and others) to construct the best model we can currently make of the cosmos.

28.1

The Age of the Expanding Universe

With hindsight, it is surprising that scientists in the 1920s and 1930s were so shocked to discover that the universe is expanding. In fact, our theories of gravity demand that the universe *must* be either expanding or contracting. To show what we mean, let's begin with a universe of finite size—say a giant ball of a thousand galaxies. All these galaxies attract each other because of their gravity. If they were initially stationary, they would inevitably begin to move closer together and eventually collide. They could avoid this collapse only if for some reason they happened to be moving away from each other at high speeds. In just the same way, a rocket launched at high enough speed can avoid falling back to Earth.

The problem of what happens in an infinite universe is harder to solve, but Einstein used his theory of general relativity to show that even infinite universes cannot be static. Since astronomers at that time did not yet know the universe was expanding, Einstein changed his equations with the introduction of an arbitrary new term called the **cosmological constant.** It represented a hypothetical force of repulsion that could balance gravitational attraction on the largest scales and permit galaxies to remain at fixed distances from one another. When Hubble and his co-workers reported that the universe was expanding, Einstein realized—to his chagrin—that the constant was not needed.

The Hubble Time

If we had a movie of the expanding universe, and ran the film backward, what would we see? The galaxies, instead of moving apart, would move *together*—getting closer and closer all the time. Eventually, we would find that *all* matter was once concentrated in an infinitesimally small volume. Astronomers identify this time with the *beginning of the universe*. The explosion of that concentrated universe at the beginning of time is called the **big bang** (not a bad term, since you can't have a bigger bang than one that creates the entire universe!). But when did this occur?

We can make a reasonable estimate of the time since the expansion began. To see how astronomers do this, let's begin with an analogy. Suppose your astronomy class decides to have a party (a kind of "big bang") at someone's home to celebrate the end of the semester. Unfortunately, everyone is celebrating with so much enthusiasm that the neighbors call the police, who arrive and send everyone away at the same moment. You get home at 2 A.M., still somewhat bitter about the way the party ended, and realize you forgot to look at your watch to see what time the police got there. But you use a map to measure that the distance between the party and your house is 40 km. And you also remember that you drove the whole trip at a steady speed of 80 km/h (since you were worried about the police cars following you). Therefore the trip must have taken:

$$\text{Time} = \frac{\text{Distance}}{\text{Velocity}} = \frac{40\ \text{km}}{80\ \text{km/h}} = 0.5\ \text{h}$$

So the party must have broken up at 1:30 A.M.

There were no humans around to look at their watches when the universe began, but we can use the same technique to estimate when the galaxies began moving away from each other. (Remember that, in reality, it is space that is expanding, not the galaxies that are somehow traveling on their own!) If we can measure how far apart the galaxies have gotten, and how fast they are moving, we can figure out how long a trip it's been.

Let's call the age of the universe measured in this way T_0. The time it has taken a galaxy to move a distance, d, away from the Milky Way (remember that at the beginning the galaxies were all together in a very tiny volume) is (as in our example)

$$T_0 = d/v,$$

where v is the velocity of the galaxy. Since individual galaxies have their own local motions, we want to make measurements not for just one galaxy, but for a good sample of them. If we can measure the speed with which many different galaxies are moving away, and also the distances between them, we can establish how long ago the expansion began.

Making such measurements should sound very familiar. This is just what Hubble and many astronomers after him needed to do in order to establish the Hubble law and the Hubble constant. We learned in Chapter 25 that a

Figure 28.2
A key target for measuring distances to galaxies (and thus both the Hubble constant and the age of the universe) with the Hubble Space Telescope is this cluster of galaxies in the southern constellation of Fornax. You can see a number of bright elliptical galaxies on this wide-angle photograph, and a beautiful barred spiral in the lower right. (© Anglo-Australian Telescope Board)

galaxy's distance and its velocity in the expanding universe are related by

$$v = H \cdot d$$

where H is the Hubble constant. Combining these two expressions gives us

$$T_0 = d/v = d/(H \cdot d) = 1/H$$

We see, then, that the work of calculating this time was already done for us when astronomers measured the Hubble constant. The age of the universe estimated in this way turns out to be just the *reciprocal of the Hubble constant* (that is, $1/H$). This age estimate is sometimes called the *Hubble time*. For a Hubble constant of 25 km/s per million LY, the Hubble time is about 13 billion years.

Remember that to determine the Hubble constant (and thus our rough gauge of the age of the universe), astronomers have to be able to measure the velocities and distances of many galaxies. We can measure velocity using Doppler shift of the lines in a galaxy's spectrum, but—as we saw in Chapter 25—distances are much more difficult to obtain (Figure 28.2). Controversy still exists over the accuracy of our various methods. If distance estimates were in error by a factor of two, the Hubble constant would be wrong by a factor of two, and the estimate of the age of the universe would also be wrong by a factor of two. Thus the age of the universe and the extragalactic distance scale are inextricably connected.

Over the past 20 years, estimates of the Hubble constant have ranged from about 15 to 35 km/s per million LY. Since the Hubble constant is the reciprocal of the age, the bigger the constant, the faster the universe expands, and hence the younger it must be. The Hubble time implied by these values ranges from 20 billion years on the old end to about 10 billion years on the young end. In just the past five years, several new techniques for estimating distances have all converged on a Hubble constant of about 25 km/s per millon LY, which makes the Hubble time about 13 billion years. (In units used by professional astronomers and often quoted in the press, 25 km/s per million LY corresponds to 80 km/s per million parsecs; the current most likely value of the Hubble constant appears to be 70–75 km/s per million parsecs or 21–23 km/s per million LY, which we have rounded to 25 km/s per million LY for this text.)

The Role of Deceleration

The Hubble time turns out to be the maximum possible age of the universe, but its actual age may be less, because in figuring out the Hubble time we have assumed that the universe has always been expanding at the same rate. Continuing with our analogy, this is equivalent to assuming that you traveled home from the party at a constant rate, when in fact this may not have been the case. At first, angry about having to leave, you may have driven fast, but then as you calmed down—and thought about police cars on the highway—you may have begun to slow down until you were driving at a more socially acceptable speed (such as 80 km/h). In this case, given that you were driving faster at the beginning, the trip home would have taken less than a half-hour.

In the same way, in calculating the Hubble time, we have assumed that H has been constant throughout all of time. This may not be a good assumption. Matter creates gravity, whereby all objects pull on all other objects. This mutual attraction will slow the expansion as time goes on, which means that in the past, the rate of expansion must have been faster than it is today. How much faster depends on the importance of gravity in slowing the expansion. We say the universe has been *decelerating* since the beginning.

If the universe were nearly empty, the role of gravity would be minor. Then the deceleration would be close to zero and the universe would have been expanding at a constant rate. But in a universe with any significant density of matter, the deceleration means that the expansion is slower now than it used to be. In this case, the age of the universe is smaller than our estimate assuming a constant expansion rate. If we use the current rate of expansion to estimate how long it took the galaxies to reach their current separations, we will overestimate the age of the universe—just as we may have overestimated the time it took for you to get home from the party.

We will see next that the age for a universe that is decelerating just enough to eventually stop its expansion is two-thirds of the maximum age. Therefore, astronomers believe that the range of likely ages for a Hubble constant of 25 km/s per million LY is between 9 and 13 billion years.

Comparing Ages

How else can we estimate the age of the universe besides measuring the rate at which it expands? One way is to find the oldest objects whose ages we can measure. After all, the universe has to be at least as old as the oldest objects in it. In our Galaxy and others, the oldest stars are found in the globular clusters (Figure 28.3), which can be dated using the methods described in Section 21.3.

Our best models indicate that some globular clusters are at least 15 billion years old. If it took a billion years or so after the expansion began for the first stars to form, which theorists say is likely, then globular cluster ages suggest that the universe is 16 billion years old. This result is inconsistent with the age estimated from the rate of expansion for a Hubble constant of 25!

Astronomers are working hard to determine whether there is a real inconsistency in the different approaches to estimating the age of the universe. Many things can go

Figure 28.3
Globular clusters, such as 47 Tucanae, are among the oldest objects in our Galaxy and can be used to estimate its age. This image was taken with the 1-m Schmidt telescope at La Silla in Chile. (European Southern Observatory)

wrong in arriving at the numbers we have cited. The observational determinations of the Hubble constant still have some uncertainties. If the constant were 20 km/s, for example, the range of ages measured from the expansion would be 10.5 to 16 billion years.

The models of stellar evolution that we use to estimate the ages of the oldest stars might be inaccurate. In that case, the globular clusters may not be as old as we currently think. And it may turn out that our models of the universe are too simple—that some of our assumptions about its various properties are not quite right. (For example, Einstein's cosmological constant may not be equal to zero after all.) Given the uncertainties, we have in many places in this book referred to the age of the universe in round numbers as 10 to 15 billion years.

One of the major challenges of modern astrophysics is to reconcile the various estimates of the age of the universe. Good agreement would support our current models of its birth and evolution. Proof of disagreement could lead to a revolution in our understanding comparable to what occurred when Hubble discovered the redshift–distance relation. It is much too early to declare a major crisis in astronomy (as some popular accounts of the controversy have described it). On the other hand, astronomers would sleep better if the different ages could be brought closer together.

Playing with Dimensions

To appreciate what happens when the gravity of all the matter in the universe affects the curvature of spacetime for the whole universe, we must first discuss the concept of *dimensions*. A dimension can be defined as an independent option to move. For example, if you are standing at the foot of a wall with a ladder on it, you have three independent options to move (Figure 28.4). You can go to the left or the right; you can step forward or backward; and you can climb up or down. Each of these motions is independent, in the sense that you can do one without doing any of the others. In most life situations, however, you combine the three: for example, you might climb a grand curving staircase by moving up, forward, and to the right simultaneously.

We live in a world of three dimensions and take them pretty much for granted. With minimal effort we can

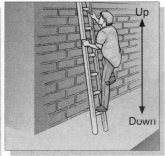

The 3 dimensions (options to move)

28.2

The Geometry of Spacetime

We now turn to another of the basic observed features of the universe, which we must include as a prime ingredient in our modeling recipe. As we saw in Chapter 23, the effect of gravity is to bend or curve the fabric of space and time. We observed that this effect is difficult to measure on Earth, but it becomes overwhelming as we approach the event horizon of a black hole.

Einstein's equations of general relativity describe the relationship between gravity and spacetime for any system, including the entire universe. The predictions of general relativity are wonderfully mind-boggling. We describe them briefly in this section, but warn you that we are unable to do them full justice.

Figure 28.4

In a world of three dimensions, there are three independent options to move: forward–backward, left–right, and up–down.

move in any of them, and everything from our bodies to the most distant galaxies extends in all three of the dimensions we have discussed. What general relativity says—translated into everyday language—is that gravity can curve or warp the universe in a *fourth* dimension. (This is *not* time, but a fourth *spatial* dimension.)

Such a dimension is impossible to picture in your mind (although there are now very nice computer programs that can show three-dimensional slices of four-dimensional solids—much as an architect's drawings show two-dimensional slices through a three-dimensional house). Since no one (including your authors) can think in four dimensions, let's try to understand the properties of such a universe by reducing our dimensions by one. We will examine one possible two-dimensional universe that is curved in the third dimension.

A Balloon Analogy

Imagine that the "universe" of interest to us is the outer skin of a large spherical balloon. We can represent the galaxies by pasting grains of sand onto this skin. If we are restricted to moving on the *surface* of the balloon—and not inside or above it—then we are dealing with a two-dimensional world. (Check this by counting the number of options to move: you can move forward–backward or left–right, but you cannot move up–down and remain in the world of the skin.) But the two-dimensional universe of the balloon's surface is curved in the third dimension—it is wrapped around the space inside the balloon.

Let's explore the properties of our two-dimensional balloon-skin world. We imagine that we ourselves are tiny two-dimensional specks that live in this world. If we travel around, we can determine some of its characteristics (Figure 28.5):

- If you keep going in a straight line (say always forward), you eventually come back to where you started. The world of the balloon skin has no *edge*.
- Since you can go around in any direction, there is no point in this world that could be considered a *center*. All points on the skin could equally well call themselves the center; no point is any more special than any other.
- If someone starts blowing up the balloon, the grains of sand pasted to the skin move away from each other. Each piece of the balloon skin stretches equally. All the grains move away from each other, but as they look out at other grains, they see something interesting. The more balloon there is between two sand grains, the faster they appear to move away from each other. This is the Hubble law! (We saw the same thing with our raisin bread in Chapter 25.)
- As the balloon is blown up, we find its surface gets bigger and bigger. Where does this new surface come from? It is being stretched out of the existing balloon skin. And the motion of the dust grains is caused by this stretching (not by anything the grains themselves are doing).

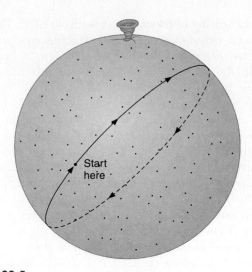

Figure 28.5
A "universe" that consists of only the surface of a balloon (with grains of sand pasted to the skin) has neither center nor edge. If you move straight ahead in one direction, you eventually return to where you started. If you blow up such a balloon, all the grains of sand move away from all the other grains.

If we were one of the two-dimensional specks living on the balloon skin, we would find our world mighty strange! A world with no center, no edge, and straight lines that come back upon themselves—all of its properties would go against our two-dimensional intuition. Yet, as a three-dimensional astronomy student, you probably don't find the situation all that odd. After all, the surface of the Earth is just this kind of "world." Our planet's *surface* has no center and no edge, and if you keep going straight ahead on it, you come back to where you started.

A speck might also ask where in the world of the balloon surface the expansion began. The only proper answer is that the world of the balloon skin began expanding everywhere at once. There is no grain of sand, no spot of balloon skin that was uniquely the site of the first expansion. As air filled the balloon, all points on its skin moved away from all other points at the same time.

The difference between you and a two-dimensional speck is that you can picture what is happening in one more dimension than the specks can. The surface of the balloon (or the Earth) has the "strange properties" we have outlined because it is curved in another (third) dimension—a dimension that the poor speck simply cannot imagine.

The kind of curved "universe" we have been describing in our analogy is called a **closed universe** because as you keep going, it closes in on itself until you eventually get back to where you started. The event horizon of a black hole encloses this sort of a "pocket universe." We hasten to point out that this is not the only kind of curvature that can exist. In an **open universe** the opposite occurs: as you move outward, more space than you expect opens up in front of you. An open universe curves away from where you started in all directions. It is much more difficult to picture, but it too has neither center nor edge.

The Curvature of Space

Now let's apply to the real universe what we learned in our analogy. Suppose the gravity of all the matter in the universe curves space in some (to us unimaginable) fourth dimension. Then we—like the specks on the balloon skin—might discover some intuitively odd things. For example, there would be no center or edge to the universe. Every galaxy would have an equal right to call itself the center or to demand that no galaxy be considered the center.

The expansion of the universe means that space stretches, with more space emerging from the existing space. The galaxies move apart because space is stretching and carrying them apart. The more distant galaxies, which have more space between us and them, stretch away the fastest, giving us the Hubble law.

And if an enterprising travel guide wanted to start tours to the place where the expansion began, this effort would be doomed to failure. The expansion of the universe began *everywhere at once* throughout the universe we can see.

Just as a two-dimensional speck could not fathom the idea that there might be a third dimension in which its world is bent, so we three-dimensional creatures resist the idea that there might be a fourth dimension of which we cannot become aware through our senses. But remember that in science, it is not whether a model or theory is likeable or simple that counts; rather, the ultimate judge must be whether experiments support the theory.

What evidence do we have that gravity can warp or bend three-dimensional space into a fourth dimension? We have already discussed a number of experiments confirming this view in Chapters 23 and 26, including the deflection of light and the advance in the perihelion of Mercury, as well as the existence of gravitational lenses in the realm of the galaxies. Today, scientists have learned to accept both the curvature of space and their inability to picture it.

The Redshift Re-Examined

If it is space that is stretching, and not the galaxies that are moving through space, why do the galaxies show redshifts in their spectra? When you were young and naive—three chapters ago—it was fine to discuss the redshifts of distant galaxies as if they resulted from their motion away from us. But now that you are an older and wiser student of cosmology, this view will simply not do!

A more accurate view of the redshifts of galaxies is that the waves are stretched by the stretching of the space they travel through. Think about the light from a remote galaxy. As it moves away from its source, the light has to travel through space. If space is stretching during all the time the light is traveling, the light waves will be stretched as well. As we saw in Chapter 4, a redshift is a stretching of waves—the wavelength of each wave increases. (Continuing with our balloon analogy, if we paint thin wavy

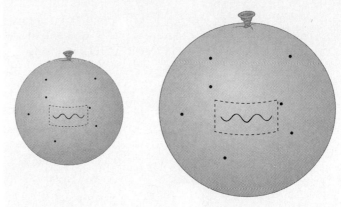

Figure 28.6
As the surface of a balloon expands, a wave on its surface stretches. For light waves, the increase in wavelength would be seen as a redshift.

lines on our expanding balloon's skin to represent light waves, the crests of those waves will get farther and farther apart as the balloon is stretched; see Figure 28.6.) Light from more distant galaxies is stretched more than light from closer ones—and thus shows a greater redshift.

Models of the Universe

Having examined a few of the key ideas in cosmology in more detail, we are now ready to look at some of the specific models astronomers have made to describe the large-scale behavior of the universe. These models begin with the facts outlined in Table 28.1. Then they make predictions about how the universe has evolved so far and what will happen to it in the future.

The Expanding Universe (with Deceleration)

Every model of the universe must include the expansion we observe. In addition, the cosmological principle shows that the universe is homogeneous. As a result, the expansion rate must be uniform (the same everywhere apart from any possible deceleration over time), causing the universe to undergo a uniform change in *scale* over time. By scale we mean, for example, the distance between two clusters of galaxies. It is customary to represent the scale by a scale factor R; if R doubles, then the distance between the clusters has doubled. Since the universe is expanding at the same rate everywhere, the change in R tells us how much it has expanded (or contracted) at any given time. For a static universe, R would be constant. In an expanding universe, R increases with time.

The simplest scenario of an expanding universe would be one in which R increases with time uniformly. But, as we have seen, the effect of gravity makes this case

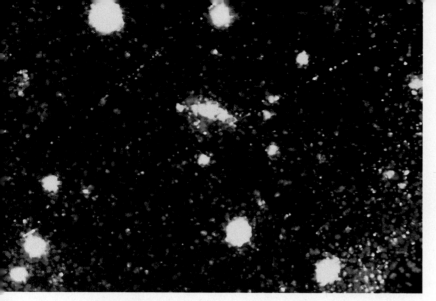

Figure 28.7
This image of one of the most-distant known galaxies, 4C 41.17, combines two views taken with the fully operational Keck telescope in the infrared with one visible-light view taken with the 4-m telescope on Kitt Peak. The galaxy has a redshift of 3.8, which means that the light we see in the infrared (colored red and green on the image) was emitted as visible light. The light we see as visible light (blue) was actually emitted as ultraviolet. The shape of the galaxy (the blob in the center) suggests it is interacting with a companion that is not visible on the image. The only way we have of estimating the distance of remote galaxies such as this is the Hubble law. This means the actual distance to 4C 41.17 cannot be determined until we know the deceleration of the universe.

unlikely. Gravity decelerates the expanding universe. We can make different models depending on the amount of gravity and thus the amount of deceleration. In some models—as we will see—the universe expands forever. In others it stops expanding and starts to contract. If we could measure the precise amount by which the universe is decelerating, we could select the correct model and collect our Nobel Prize.

Unfortunately, it is very difficult to estimate the rate of deceleration. One way might be to measure the distances and speeds of very distant galaxies. We are seeing these as they were long ago, and so we could see how much faster they were moving when the universe was young. However, think about how we actually measure the distance to very remote groups of galaxies (Figure 28.7). We use the Hubble law, which allows for only one rate of expansion, not for a rate that changes with time. So when we use this method to estimate how far the galaxies are from us, we are *assuming* that we live in a uniformly expanding universe.

To check on how much the universe is slowing down, we need an *independent* way of measuring the distance. For example, if a certain type of galaxy were a standard bulb (see Chapter 27), we could use its apparent brightness to tell us how far away each galaxy is. Alas, not only are galaxies not standard bulbs at any *given* time in his-

tory, but we also know that their brightness can change over time. As we have seen, galaxies evolve, experience collisions and mergers, and sometimes undergo bursts of star formation that make them unusually bright for several million years. Although we can't use whole galaxies as standard bulbs, we saw that Type I supernovae may play such a role. They can be seen out to distances of a few billion light years, and astronomers are now trying to estimate the deceleration by finding such supernovae in distant galaxies.

The Cosmic Tug-of-War

Since the deceleration is produced by gravity, another approach to estimating its rate is to measure how much material the universe contains. Here is where the cosmological principle really comes in handy. Since the universe is the same all over, we only need to measure how much material exists in a (large) representative sample of it. (Such sampling is how pollsters can describe how we feel about political issues without asking each and every person in the country.) What astronomers look at is the *average density* of the universe.

By average density we mean the mass of matter (including the equivalent mass of energy)[1] that would be contained in each unit of volume (say, 1 cm^3) if all the stars, galaxies, and other objects were taken apart, atom by atom, and if all those particles, along with the light and other energy, were distributed throughout all of space with absolute uniformity. If the average density is low, the universe will not decelerate very much and can expand forever. High average density, on the other hand, means that a lot of gravity is pulling the galaxies together, and so the expansion will eventually stop.

In a sense, we can say that there is a "tug-of-war" going on between the expansion of the universe, which pushes everything apart, and gravity, which pulls everything together. One of the great questions of modern astronomy is who will win this tug-of-war; the answer can tell us what the ultimate fate of the universe will be.

Since we know how fast the galaxies are moving away from each other, we can calculate the **critical density** for the universe—the mass per unit volume that will be just enough to slow the expansion to zero at some time infinitely far in the future. If the actual density is higher than this critical density, then the expansion will ultimately reverse and the universe will begin to contract. If the actual density is lower, then the universe will expand forever.

These various possibilities are illustrated in Figure 28.8. Time increases to the right, and the scale, R, increases upward in the figure. Today, marked "present" along the time axis, R is increasing in each model. We are

---

[1] By equivalent mass we mean that which would result if the energy were turned into mass using Einstein's formula, $E = mc^2$.

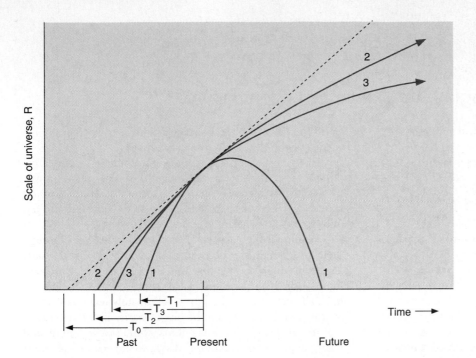

Figure 28.8
A plot of R, the scale of the universe, against time for various cosmological models. Curve 1 represents a closed universe (where the density is greater than the critical value), curve 2 represents an open universe, and curve 3 is a universe with a critical density. The dashed line is for an empty universe, one in which the expansion is not slowed by gravity. Note that the age of the universe equals the Hubble time only for this (unlikely) possibility. In all the other cases, the universe is less old than the Hubble time.

still "early" in the history of the universe, so the galaxies are expanding no matter which model is right. (The same situation holds for a baseball thrown high into the air. While it may eventually fall back down, at the beginning of the throw it moves upward.) The straight dashed line corresponds to the empty universe with no deceleration; it intercepts the time axis at a time, T_0 (the Hubble time), in the past. The other curves represent models with varying amounts of deceleration, starting from the Big Bang at shorter times in the past. Let's take a look at the future according to the different models.

Facing the Closed Future

One of the things we learn from general relativity is that the future of the expansion and the geometry of space-time are intimately related. The differently curved universes we explored briefly in Section 28.2 correspond to different futures. For example, let's take curve 1 in Figure 28.8. In this case the actual density of the universe is higher than the critical density. This universe will stop expanding some time in the future and begin contracting. Eventually the scale drops to zero, which means that space will have shrunk down to infinitely small size. The noted physicist John A. Wheeler (who gave black holes their name) calls this the "big crunch," because matter, energy, space, and time would all be crushed out of existence.

This scenario corresponds to the *closed* geometry discussed earlier, and is therefore called the closed universe. It is closed in two senses: at any time, space curves so that if you could keep going, you would eventually come back to where you started; and at the end of time, space closes down on itself. (See "Making Connections" box for more on what happens toward the end.)

It is tempting to speculate that another big bang might follow the "crunch," giving rise to a new expansion phase, and then another contraction—perhaps oscillating between successive big bangs indefinitely in the past and future. Such speculation is sometimes referred to as the *oscillating theory* of the universe, but it is not really a theory because we know of no mechanism that can produce another big bang. General relativity predicts, instead, that at the crunch the universe would collapse into a universal black hole. The oscillating model is more a philosophical idea than a scientific one; scientific hypotheses, after all, must be tested with experiments. There is no way to test whether another cycle of the universe could arise from the big crunch because nothing—not matter, not energy, not space, not time—could survive the big crunch.

Facing the Open Future

If the density of the universe is less than the critical density, (curve 2 in Figure 28.8), gravity is never important enough to stop the expansion, and so the universe expands forever. This corresponds to the (more difficult to imagine) open geometry discussed earlier, and is thus called an open universe. Such a universe is infinite and always has even more room in it than naive three-dimensional observers would expect. In this case, time and space begin with the Big Bang, but they have no end; the universe simply continues expanding, always a bit more slowly as time goes on. Groups of galaxies eventually get so far apart that it would be difficult for observers in any of them to see the others (see "Making Connections" box).

At the critical density (curve 3) the universe can just barely expand forever. The critical-density universe has an age of exactly two-thirds T_0, where T_0 is the age of the

What Might It Be Like in the Distant Future?

Some say the world will end in fire,
Some say in ice.
From what I've tasted of desire
I hold with those who favor fire.

— From the poem "Fire and Ice"
by Robert Frost (1923)

Given the destructive power of impacting asteroids, expanding red giants, and nearby supernovae, our species may not be around in the remote future. Nevertheless, you might enjoy speculating about what it would be like to live in a much, much older universe.

The far future in a *closed* universe would be exciting but not very healthy for living things. As space contracted (R would get smaller and smaller), the galaxies would see each other's light blueshifted and not redshifted; that is, the compression of space would compress the waves. But shorter waves have more energy. Thus the temperature of the background radiation (see Section 28.4) would increase with time.

As the universe got smaller, it would become easier to circumnavigate. Rays of light could eventually make it around the universe and come back to where they started. In theory, you could see a star die in one direction, and then see the light of its birth come around from the other direction. "Ghost galaxies"—images of galaxies whose light had gone completely around the universe—would become visible, adding to the number of galaxies we observed.

Ultimately, as space shrank and the temperature of the background radiation increased, the temperature of space would become higher than that of any planets. Heat would then flow from space to a planet, instead of the other way around (which is what we now take for granted). As a planet heated up, any life-forms on its surface would be "cooked" by the heat of space. As the radiation of space got even hotter, heat would flow from space into the stars, breaking them apart. Their end could be mourned only briefly, however, be-cause soon all matter and energy would be crushed out of existence by the closing down of spacetime.

The *open* universe scenario is unsettling in a completely different way. In this case the universe would expand forever (R would just increase) and the clusters of galaxies would spread ever farther apart with time. As the eons passed, the universe would get thinner, colder, and darker.

Within each galaxy, stars would continue to go through their lives, eventually becoming white dwarfs, neutron stars, and black holes. Low-mass stars might take a long time to finish their evolution, but in this model we would literally have all the time in the world. Ultimately, even the white dwarfs would cool down to be black dwarfs, any neutron stars that revealed themselves as pulsars would stop spinning, and black holes with accretion disks would complete their "meals." The final states of stars would be dark and difficult to observe.

Therefore the light that now reveals galaxies to us would eventually go out. Even if a small pocket of raw material were left in one unsung corner of a galaxy, ready to be turned into a fresh cluster of stars, we would only have to wait until the time that their evolution was also done. And time is one thing this model of the universe has plenty of. There would surely come a time when all the stars were out, galaxies were as dark as space, and no source of heat remained to help living things survive. Then the lifeless galaxies would just continue to move apart in their lightless realm.

If these views of the future seem discouraging (from a human perspective), you might take heart in the knowledge that science is always a progress report. The most-advanced ideas about the universe from a hundred years ago now strike us as rather primitive. It may well be that our best models of today will in a hundred or a thousand years also seem rather childish, and that there are other factors determining the ultimate fate of the universe of which we are still completely unaware.

empty universe. Interestingly, this model (called a **flat universe**) has zero curvature, and resembles the three-dimensional universe you probably expected before you took this class. Universes that will expand forever have ages between two-thirds T_0 and T_0. Universes that will someday begin to contract have ages less than two-thirds T_0. The various models we have described are summarized in Table 28.2.

Who's Winning the Tug-of-War?

What kind of universe do we live in? Is its density larger or smaller than the critical density? The critical density depends on H_0. If the Hubble constant is 25 km/s per million LY, the critical density is about 10^{-29} g/cm^3. Since we cannot find a way at present to measure the deceleration directly, astronomers have been focusing their efforts on finding the actual density of the universe.

There are several observational tests by which we can try to determine the average density of matter in space. One way is to count all the galaxies out to a given distance and use estimates of their masses, including dark matter, to calculate the average density. Such estimates indicate a density of about 1 to 2×10^{-30} g/cm^3 (10 to 20 percent of critical) and suggest that the universe is open. However,

TABLE 28.2
Some Models of the Universe

Kind of Universe (Model)	Age (billions of years)*	Average Density (g/cm³)	Ultimate Fate of Universe
Closed	Less than 9	More than 10^{-29}	Stop and contract
Flat	9	About 10^{-29}	Barely expand forever
Open	9–13	Less than 10^{-29}	Expand forever

* The numbers here assume a Hubble constant of 25 km/s per million LY. If the Hubble constant is smaller than this, the ages increase correspondingly.

we may have underestimated the amount of dark matter in galaxy clusters, and there may also be a lot of dark matter between galaxy clusters (including in the great voids) where we cannot detect it. Therefore, we cannot really be certain that we live in an open universe based on this measurement.

The ages of stars also suggest that we live in an open universe. The best estimate of H_0 is 25 km/s per million LY, which corresponds to a maximum age of 13 billion years. If we live in a critical-density universe, the actual age would be only two-thirds of 13 billion years, which is 9 billion years. It seems highly unlikely that our models of how stars evolve, which give ages of 15 billion years for the oldest stars, are bad enough to give stellar ages that are wrong by a factor of nearly two. As you can see in Table 28.2, an open universe provides the oldest possible age.

All such conclusions, however, depend on the models of the universe we choose to test. There are more-complex cosmological models—such as those where Einstein's cosmological constant (the repulsion force) is not zero—in which the density comes closer to the critical value. In this introductory text we cannot explore these alternative models; suffice it to say that astronomers are now trying to design observational tests to determine whether more-complex models are required.

TABLE 28.3
Lookback Times for Different Redshifts

Redshift	Percent of Current Age of Universe When the Light Was Emitted*
0	100% (now)
0.5	55%
1.0	35%
2.0	19%
3.0	12%
4.0	9%
4.5	8%
4.9	7%
Infinite	0%

* Assumes a universe with a density equal to the critical density.

Lookback Time

In Chapter 25 we discussed how we can use the Hubble law to measure the distance to a galaxy. The Hubble law applies quite simply to galaxies that are not moving too fast (that is, are not too far away). Once we get to large distances, we are looking so far into the past that we must take the deceleration of the universe into account. Since we do not know how much the universe is decelerating, we must *assume* one of the models of the universe to be able to convert large redshifts into distances.

This is why astronomers squirm when reporters and students ask them *exactly* how far away some newly discovered distant quasar or galaxy is. We really can't give an answer without first explaining the model of the universe we are assuming in calculating it (by which time a reporter or student is long gone!).

Once we assume a model, we can use it to calculate the *lookback time* for an object—the measure of how long ago the light we see was emitted from it. As an example, Table 28.3 lists the lookback times as fractions of the current age of the universe (assuming a model in which the universe has critical density). The numbers are not so important, but notice that as we find objects with higher and higher redshifts, we are looking back to very small fractions of the age of the universe.

We have already learned some important things by observing objects at large redshifts. For example, we saw that quasars were most abundant when the universe was roughly 20 percent of its current age (Figure 26.17). At a still-earlier time, quasars were exceedingly rare. Perhaps galaxies had not yet formed, or had formed only recently and not yet had time to produce massive black holes. Such observations provide clear evidence that our universe is evolving with time.

28.4

The Beginning of the Universe

As we look farther and farther back, we see the galaxies and quasars thin out, and we approach the era when matter had still not settled down into the structures we observe today. What were things like when the universe was young, and space had not yet stretched very significantly? In other words, what was it like just after the Big Bang?

The History of the Idea

It is one thing to say the universe had a beginning (as the equations of general relativity imply), and quite another to describe that beginning. The Belgian priest and cosmologist Georges Lemaître (1894–1966) was probably the first to propose a specific model for the Big Bang itself (Figure 28.9). He envisioned all the matter of the universe starting in one great bulk he called the *primeval atom,* which then broke into tremendous numbers of pieces. Each of these pieces continued to fragment further until they became the present atoms of the universe, created in a vast nuclear fission. In a popular account of his theory, Lemaître wrote, "The evolution of the world could be compared to a display of fireworks just ended—some few red wisps, ashes and smoke. Standing on a well-cooled cinder we see the slow fading of the suns and we try to recall the vanished brilliance of the origin of the worlds."

Physicists today know much more about nuclear physics than was known in the 1920s, and have shown that the primeval fission model cannot be correct. Yet Lemaître's vision was in some respects quite prophetic. We still believe that everything was together at the beginning; it was just not in the form of matter as we now know it.

In the 1940s the American physicist George Gamow (Figure 28.10) suggested a universe with a different kind of beginning—involving nuclear fusion instead of fission.

Figure 28.9
Abbé Georges Lemaître (1894–1966), Belgian cosmologist, studied theology at Mechelen, and mathematics and physics at the University of Leuven. It was there that he began to explore the expansion of the universe, and postulated its explosive beginning. He actually predicted the Hubble law two years before its verification, and he was the first to seriously consider the physical processes by which the universe began. (Yerkes Observatory)

Figure 28.10
This montage of images shows George Gamow emerging like a genie from a bottle of YLEM, a Greek term for the original substance from which the world formed. Gamow revived the term to describe the material of the hot Big Bang. Flanking him are Robert Herman (left) and Ralph Alpher, with whom he collaborated in working out the physics of the Big Bang. The modern composer Karlheinz Stockhausen was inspired by Gamow's ideas to write a piece of music called *Ylem,* in which the players actually move away from the stage as they perform, simulating the expansion of the universe. (Courtesy of Ralph Alpher)

He worked out the details with Ralph Alpher, and they published the results in 1948. (Gamow, who had a quirky sense of humor, decided at the last minute to add the name of physicist Hans Bethe to their paper, so that the coauthors would be Alpher, Bethe, and Gamow, a pun on the first three letters of the Greek alphabet: alpha, beta, and gamma.) Gamow's universe started with fundamental particles that built up the heavy elements by fusion in the Big Bang.

His ideas were close to our modern view, except we now know that only the three lightest elements—hydrogen, helium, and a small amount of lithium—were formed in appreciable abundances at the beginning. The heavier elements formed later in stars. Since the 1940s many astronomers and physicists have worked on a detailed theory of what happened in the early stages of the universe. The result of their efforts is now called the *standard model* of the Big Bang.

Standard Model

Three basic ideas hold the key to tracing the changes that occurred during the first few minutes after the universe began. The first is that the universe cools as it expands, much as gas cools when sprayed from an aerosol can. Figure 28.11 shows how the temperature changes with the passage of time. In the first fraction of a second, the uni-

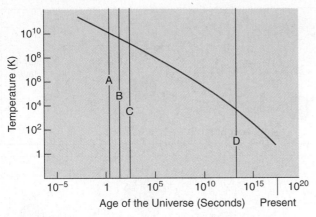

Figure 28.11
This graph shows how the temperature of the universe varies with time as predicted by the standard model of the Big Bang. Note that both the temperature (vertical axis) and the time in seconds (horizontal axis) change over vast scales on this compressed diagram. The vertical line labeled A designates approximately the time at which neutrinos stop interacting with matter. B denotes the time when the temperature becomes too cool for new positrons and electrons to form from radiation. Helium synthesis occurs at time C, and the universe becomes transparent to radiation at time D. The line showing the drop in temperature ends at the point in time labeled "the present."

verse was unimaginably hot. By the time 0.01 s had elapsed, the temperature had dropped to 100 billion (10^{11}) K. After about 3 min, it had fallen to about 1 billion (10^9) K, still some 70 times hotter than the interior of the Sun. After a few hundred thousand years, the temperature was down to a mere 3000 K, and the universe has continued to cool since that time.

All of these temperatures but the last are derived from theoretical calculations, since (obviously) no one was there to measure them directly. As we will see, however, we have actually detected the feeble glow of radiation emitted at a time when the universe was a few hundred thousand years old. Indeed, the fact that we have done so is one of the strongest arguments in favor of the Big Bang model.

The second step in understanding the evolution of the universe is to realize that at very early times it was so hot that it contained mostly radiation (and not the matter that we find dominating today). The photons—the packets of pure electromagnetic energy described in Chapter 4—that filled the universe could collide and produce material particles. That is, under the conditions just after the Big Bang, energy could turn into matter (and matter could turn into energy). We can calculate how much mass is produced from a given amount of energy by using Einstein's formula $E = mc^2$ (see Chapter 15).

The idea that energy can turn into matter is a new one for many students since it is not part of our everyday experience. (That's because when we compare the universe today to what it was like right after the Big Bang, we live in cold, hard times! The photons in the universe today

typically have far less energy than the amount required to make new matter.) In Chapter 15 we briefly mentioned that when subatomic particles of matter and *antimatter* collide, they turn into pure energy. But is it really true that energy can turn into matter (as our theories predict)? This process has been observed in particle accelerators around the world. If we have enough energy, under the right circumstances, new particles of matter (and antimatter) are indeed created.

Our third key point is that the hotter the universe (see Figure 28.11), the more energetic the photons available to make matter (and antimatter). To take a specific example, at a temperature of 6 billion (6×10^9) K, the collision of two typical photons can create an electron and its antimatter counterpart, a positron. The much more massive proton can be created only in an environment with a temperature in excess of 10^{14} K.

The Evolution of the Early Universe

Keeping these three ideas in mind, we can trace the evolution of the universe from the time it was about 0.01 s old and had a temperature of about 100 billion K. Why not begin at the very beginning? There are as yet no theories that allow us to penetrate to a time before about 10^{-43} s. When the universe was that young, its density was so high that the theory of general relativity is not adequate to describe it; at present, we have no theory that can deal with such extreme conditions.

Scientists, by the way, have been somewhat more successful in describing the universe when it was older than 10^{-43} s but still less than about 0.01 s old. During that time the universe was filled with energy and strongly interacting subatomic particles. Although the theory of these particles is difficult to deal with, very recently theoretical physicists have begun to speculate about what things may have been like during this very, very early time. We will look at some of their speculations later in this chapter.

The universe was a somewhat more familiar-sounding place by 0.01 s after the beginning. At that time it consisted of a soup of matter and radiation; the matter included protons and neutrons, leftovers from an even younger and hotter universe. Each particle collided rapidly with other particles. The temperature was no longer high enough to allow colliding photons to produce neutrons and protons, but it was sufficient for the production of electrons and positrons (Figure 28.12a). There was probably also a sea of exotic subatomic particles that would later play a role as dark matter. All the particles jiggled about on their own; it was still much too hot for protons and neutrons to combine to form the nuclei of atoms.

Our picture is of a seething cauldron of a universe, with photons colliding and interchanging energy, sometimes being destroyed to create a pair of particles. The particles in the universe also collide with one another. Frequently, a matter particle and an antimatter particle

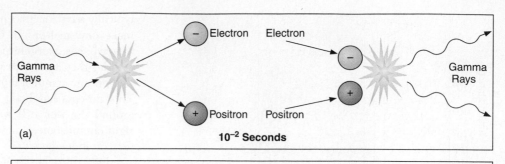

Figure 28.12
A summary of some of the interactions involving particles in the early universe. (a) In the first fractions of a second, when the universe was very hot, energy was converted to particles and antiparticles. The reverse reaction also happened: a particle and antiparticle could collide and produce energy. (b) As the temperature of the universe decreased, the energy of typical photons became too low to create matter. Instead, existing particles fused to create such nuclei as deuterium and helium. (c) Later it became cool enough for electrons to settle down with nuclei and make neutral atoms. Most of the universe was still hydrogen.

meet and turn each other into a burst of gamma-ray radiation.

Among the particles created in the early phases of the universe was the ghostly neutrino (see Chapter 15), which today interacts only very rarely with ordinary matter. In the crowded conditions of the very early universe, however, neutrinos ran into so many electrons and positrons that they experienced frequent interactions despite their "antisocial" natures.

By the time the universe was a little more than 1 s old, the density had dropped to the point where neutrinos no longer interacted with matter, but simply traveled freely through space. In fact, these neutrinos should now be all around us. Since they have been traveling through space unimpeded (and hence unchanged) since the universe was 1 s old, measurement of their properties would offer one of the best tests of the Big Bang model. Unfortunately, the very characteristic that makes them so use-

ful—the fact that they interact so weakly with matter that they have survived unaltered for all but the first second of time—also renders them undetectable, at least with present techniques. Perhaps someday someone will devise a way to capture these elusive messengers from the past.

Atoms Form

When the universe was about 3 min old and its temperature was down to about 900 million K, protons and neutrons could combine without being immediately disrupted by high-energy photons. They began to form the simplest stable nuclei—deuterium (heavy hydrogen), helium, and lithium. At higher temperatures these atomic nuclei had immediately been blasted apart by interactions with photons, and so could not survive. But at the temperatures and densities reached between 3 and 4 min after the beginning, deuterium lasted long enough that collisions could convert some of it to helium, which consists of two protons and two neutrons (Figure 28.12b). In essence, the entire universe was acting the way centers of stars do today—fusing new elements from simpler components.

This burst of cosmic fusion was only a brief interlude, however. The universe was expanding and cooling down. This meant that no elements beyond lithium could form (it got too cool too fast), and even the light elements stopped forming after a few minutes. In the cool universe we know and love today, the fusion of new elements is limited to the centers of stars and the explosions of supernovae.

Still, the fact that the Big Bang model allows the creation of a good deal of helium is the answer to a long-standing mystery in astronomy. Put simply, there is just too much helium in the universe to be explained by what happens inside stars. All the generations of stars that have produced helium in their centers just cannot account for the quantity of helium we observe. Furthermore, even the oldest stars and the most-distant galaxies show significant amounts of helium. These observations find a natural explanation in the synthesis of helium by the Big Bang itself during the first few minutes of time. We estimate that ten times more helium was manufactured in the first 3 min of the universe than in all the generations of stars during the succeeding 10 to 15 billion years.

There is even more we can learn from the way the early universe made atomic nuclei. It turns out that *all* of the deuterium in the universe was formed during the first 3 min. In stars, any region hot enough to fuse two protons to form a deuterium nucleus is also hot enough to change it further—either by destroying it through a collision with an energetic photon, or by converting it to helium through nuclear reactions.

The amount of deuterium produced in the first 3 min of creation depends on the density of the universe at the time deuterium was formed. If the density was high, nearly all the deuterium would have been converted to helium through interactions with protons, just as it is in stars. If the density was low, then the universe expanded and thinned out rapidly enough that some deuterium survived. The amount of deuterium we see today thus gives us a clue to the density of the universe when it was about 3 min old. Theoretical models can relate the density then to the density now; thus measurements of the abundance of deuterium today can give us an estimate of the current density of the universe.

The deuterium measurements indicate that the present-day density is about 10^{-31} g/cm^3, suggesting that the universe will expand forever. There is a possible loophole, however. The deuterium abundance is determined by the density of protons and neutrons, since these are the particles that interact to form it. From the deuterium abundance we know that not enough protons and neutrons are present, by a factor of about 100, to produce a critical-density universe. If, however, there are dark matter particles of some other kind that are not involved in nuclear reactions, then it is still possible that we live in a critical-density universe. This is an additional reason for thinking that dark matter may be made of some exotic, unknown kind of particle, and not combinations of protons and neutrons like the readers of this book. We discuss this possibility in the last section of the chapter.

The Universe Becomes Transparent

Although fusion stopped after a few minutes, the universe continued to resemble the interior of a star for a few hundred thousand years: it remained hot and opaque, with radiation being scattered from one particle to another. It was still too hot for electrons to "settle down" and become associated with a particular nucleus. And electrons are especially effective at scattering photons, thus ensuring that no radiation ever got very far in the early universe without having its path changed. In a way, the universe was like an enormous crowd right after a popular concert; if you get separated from a friend, even if he is wearing a flashing button, it is impossible to see through the dense crowd to spot him. Only after the crowd clears is there a path for the light from his button to reach you.

Not until a few hundred thousand years after the Big Bang, when the temperature had dropped to about 3000 K and the density of atomic nuclei to about 1000 per cubic centimeter, did the electrons and nuclei combine to form stable atoms of hydrogen and helium (Figure 28.12c). With no free electrons to scatter photons, the universe became transparent for the first time in cosmic history. From this point on, matter and radiation interacted much less frequently; we say that they *decoupled* from each other and evolved separately. If we are to detect the light of the early universe, it will come from this decoupling time when radiation was first allowed to move over significant distances.

One billion years after the Big Bang, stars and galaxies had begun to form. Deep in the interiors of stars matter was reheated, nuclear reactions were ignited, and the more gradual synthesis of the heavier elements began. In the meantime, the radiation from the decoupling time continued to cool as space stretched and all radiation became redshifted (thus carrying less and less energy). As billions of years passed, the afterglow of the Big Bang faded away.

We conclude this quick tour of our model of the early universe with a reminder. You must not think of the Big Bang as a *localized* explosion *in space*—like an exploding superstar. There were no boundaries and no site of the explosion. It was an explosion *of space* (and matter and energy) that happened everywhere in the universe. All matter and energy that exist today, including the particles of which you are made, came from the Big Bang. We were, and still are, in the midst of the Big Bang; it is all around us.

Discovery of the Cosmic Background Radiation

In the late 1940s Ralph Alpher and Robert Herman (see Figure 28.10) realized that just before the universe became transparent, it must have been radiating like a blackbody at a temperature of 3000 K (see Chapter 4). If we could have seen that radiation just after neutral atoms formed, it would have resembled radiation from a reddish star. It was as if a giant fireball filled all of the universe.

But that was at least 10 billion years ago, and in the meantime the scale of the universe has increased a thousandfold. This expansion has increased the wavelength of the radiation by a factor of 1000, and, according to Wien's law, has correspondingly lowered the temperature by a factor of 1000 (see Section 4.2). Alpher and Herman predicted that the glow from the fireball should now be at radio wavelengths, and should resemble the radiation from a blackbody at a temperature only a few degrees above absolute zero. Since the fireball was everywhere throughout the universe, the radiation left over from it should also be everywhere. There was no way at the time they published their conclusion to observe such radiation from space, so the prediction was forgotten.

In the mid-1960s, in Holmdel, New Jersey, Arno Penzias and Robert Wilson of AT&T's Bell Laboratories were using a delicate microwave antenna (Figure 28.13) to measure the intensity of radio radiation all around the sky (to check on possible sources of radiation that might interfere with communications satellites). They were plagued with some unexpected background noise, just like static on a radio, that they could not get rid of. The puzzling thing about this radiation was that it seemed to be coming from all directions at once. This is very unusual in astronomy; after all, most radiation has a specific direction where it is strongest—the direction of the Sun, or a supernova remnant, or the disk of the Milky Way, for example.

Penzias and Wilson at first thought that any radiation appearing to come from all directions must be coming from inside their telescope, so they took everything apart to look for the source of the noise. They even found that some pigeons had roosted inside the big horn and had left (as Penzias delicately put it) "a layer of white, sticky, dielectric substance coating the inside of the antenna." However, nothing they did could reduce the background radiation to zero, and they reluctantly came to accept that it must be real, and coming from space.

Penzias and Wilson were not cosmologists, but as they began to discuss their puzzling discovery with other scientists, they were quickly put in touch with a group of astronomers and physicists at Princeton University (a short drive away) who had—as it happened—been redoing the calculations of Gamow's group from the 1940s, and realized that the radiation from the decoupling time should be detectable as a faint afterglow of radio waves. Their work predicted that the temperature corresponding to this **cosmic background radiation (CBR)** should be about 3 K; Penzias and Wilson found the intensity of the radiation they had discovered to match a blackbody with just that temperature.

Many other experiments on Earth and in space soon confirmed the discovery: the radiation was indeed coming from all directions (it was isotropic) and matched the predictions of the Big Bang theory with remarkable precision. Penzias and Wilson had inadvertently observed the glow from the primeval fireball. They received the Nobel Prize for their work in 1978. And just before his death in 1966, Lemaître learned that his "vanished brilliance" had been discovered and confirmed.

Figure 28.13
Robert Wilson (right) and Arno Penzias pose in front of the horn-shaped antenna with which they discovered the cosmic background radiation. The photo was taken in 1978, just after they received the Nobel Prize in physics. (AT&T Bell Laboratories)

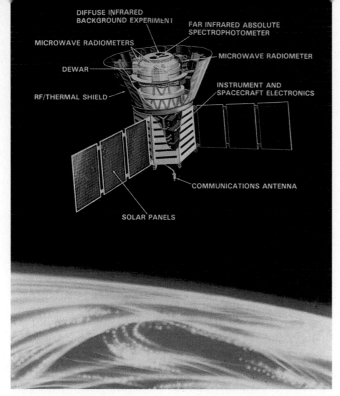

DIFFUSE INFRARED
BACKGROUND EXPERIMENT
FAR INFRARED ABSOLUTE
SPECTROPHOTOMETER
MICROWAVE RADIOMETERS
MICROWAVE RADIOMETER
DEWAR
RF/THERMAL SHIELD
INSTRUMENT AND
SPACECRAFT ELECTRONICS
COMMUNICATIONS ANTENNA
SOLAR PANELS

Figure 28.14
An artist's conception of COBE, the satellite designed to explore the cosmic background radiation at infrared and microwave wavelengths. Various instruments aboard the probe are marked. (NASA)

Properties of the Background Radiation

Accurate measurements of the CBR have now been made with a satellite orbiting the Earth. Named the Cosmic Background Explorer (COBE), it was launched by NASA on November 18, 1989 (Figure 28.14). The data it received quickly showed that the CBR closely matches that expected from a blackbody with a temperature of 2.73 K

(Figure 28.15). This is exactly the result expected if the CBR is indeed redshifted radiation emitted by a hot gas that filled all of space shortly after the universe began.

The first important conclusion from measurements of the CBR, therefore, is that the universe we have today has evolved from a hot, uniform state. This observation provides direct support for the idea that we live in an evolving universe.

A second result (seen with other instruments and now confirmed by COBE) is that the CBR appears to be slightly hotter in one direction than in the exact opposite direction in the sky. This difference comes about because of our own motion through space. If you approach a blackbody, its radiation is slightly Doppler-shifted to shorter wavelengths and therefore resembles radiation from a slightly hotter blackbody. If you move away from a blackbody, its radiation is redshifted and resembles that from a slightly cooler object. Since the CBR fills the universe, we can determine how we are moving by noting the directions in which it is redshifted and blueshifted, and the amount of the shift.

The small temperature differences of the CBR indicate that the Sun, the Milky Way, and the whole Local Group of galaxies are moving at a speed of about 500 km/s in the general direction of the constellation Hydra. Note that this motion is in addition to the motion of these galaxies resulting from the overall expansion of the universe. As we saw in Chapter 27, the extra motion is probably caused by the gravitational attraction of an unusually dense concentration of luminous dark matter (the Great Attractor) that pulls the Local Group toward Hydra.

A third conclusion from the COBE observations is that the early universe was a quiet place. The intensity of the CBR follows a blackbody curve with no excess radiation at any wavelength. If there had been violent events when the universe was very young, those events would

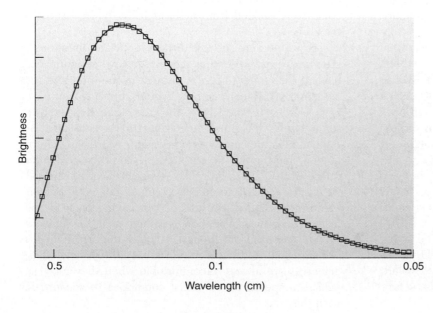

Brightness

0.5   0.1   0.05

Wavelength (cm)

Figure 28.15
The solid line shows how the intensity of radiation should change with wavelength for a blackbody with a temperature of 2.73 K. The boxes show the intensity of the cosmic background radiation as measured at various wavelengths by COBE's instruments. The fit is perfect. When this graph was first shown at a meeting of astronomers, they gave it a standing ovation.

have distorted the smooth curve shown in Figure 28.15. Astronomers had, for example, speculated that there might have been a generation of massive stars that formed and completed their evolution through the supernova phase prior to the formation of galaxies.

It was known even before the launch of COBE that the CBR is extremely *isotropic*. In fact, its uniformity is one of the best confirmations of the cosmological principle. Ground-based measurements showed that if we look at places in the sky that differ in direction by less than a degree, any fluctuations in the intensity of the CBR are less than a few parts in ten thousand.

But according to our theories, the temperature could not have been *perfectly* uniform when the CBR was emitted. After all, the CBR is radiation that was scattering from the particles in the universe at the time of decoupling. If the radiation were completely smooth, then all those particles must have been distributed through space absolutely evenly. Yet it is those particles that have become all the galaxies and stars (and astronomy students) that now inhabit the cosmos. Had they been completely smoothly distributed, they could not have formed all the large-scale structure now present in the universe—the clusters and superclusters of galaxies discussed in the last few chapters.

The early universe must have had tiny density fluctuations from which such structure could evolve. Regions of higher-than-average density would have attracted additional matter and eventually grown into the galaxies and clusters that we see today. These regions would appear to us to have lower-than-average temperatures. (This is because a denser region has more gravity and thus redshifts the radiation, causing it to look a tiny bit cooler.) Therefore, if the seeds of present-day galaxies existed at the time the CBR was emitted, we should see some slight changes in the CBR temperature as we look in different directions in the sky.

Scientists working with the data from the COBE satellite have indeed detected very subtle temperature differences present in the CBR (see the opening figure for this chapter). The temperature variations are typically only 16 millionths of a degree K. The regions of lower-than-average temperature come in a variety of sizes, but even the smallest of the cool areas detected by COBE is far too large to be the precursor of an individual galaxy, or even a supercluster of galaxies. Thus, while the COBE results are a dramatic confirmation that fluctuations did exist in the density of the primeval fireball, astronomers still need to find a way to turn these fluctuations into the structure we presently observe.

New ground-based experiments have found CBR temperature fluctuations on scales of about 20 arcmin, which correspond approximately to the size of a supercluster. Instruments on Earth and in space will be trying to find even smaller variations in years to come. However these observations turn out, it is remarkable that we are now able to see the earliest stages of the birth of structure in the universe.

The Inflationary Universe

Problems with the Standard Big Bang Model

The hot Big Bang model that we have been describing is remarkably successful. It accounts for the expansion of the universe, explains the observations of the CBR, and correctly predicts the abundances of the light elements. As it turns out, this model also predicts that there should be exactly three types of neutrinos in nature, and this prediction has been confirmed by experiments with high-energy accelerators.

The theory is not complete, however. The standard Big Bang model does not explain why there is more matter than antimatter in the universe, nor does it account for the origin of the density fluctuations that ultimately grew into galaxies. It also does not explain the remarkable *uniformity* of the universe. The CBR is the same, no matter in which direction we look, to an accuracy of about 1 part in 100,000. This sameness might be expected if all the parts of the visible universe were in contact at some point, and had time to come to the same temperature. In the same way, if we put some ice into a glass of water and wait a while, the ice will melt and the water will cool down until they are the same temperature.

However, if we accept the standard Big Bang model, all parts of the visible universe were *not* in contact at any time. The fastest that information can go from one point to another is the speed of light. There is a maximum distance that light can have traveled from any point since the time the universe began. This distance is called that point's *horizon distance,* because anything farther away is "below its horizon"—unable to make contact with it. (We saw a similar use of the word *horizon* when we discussed black holes in Chapter 23.) One region of space separated by more than the horizon distance from another has been completely isolated from it through the entire history of the universe.

If we measure the CBR in two opposite directions in the sky, we are observing regions that were significantly beyond each other's horizon distance at the time the CBR was emitted. *We* can see both regions, but *they* can never have seen each other! Why, then, are their temperatures so precisely the same? According to the standard Big Bang model, they have never been able to exchange information, and there is no reason they should have identical temperatures. (It's a little like seeing the clothes students wear at two schools in different parts of the world become identical, without the students ever having been in contact.) The only explanation is simply that the universe somehow started out being absolutely uniform (which is like saying all students were born liking the same clothes). Scientists are always uncomfortable when they must appeal to a special set of initial conditions to account for what they see.

Another problem with the standard Big Bang model is that it does not explain why the density of matter in the universe is so close to the critical density. As we have seen, current observations are unable to tell us whether the expansion of the universe will continue forever or come to a halt and reverse itself. The interesting point, however, is that we find ourselves in a universe that is so nearly balanced between these two possibilities that we cannot yet determine which is correct. The density could have been, after all, so low that it would be obvious that the expansion of the universe will continue forever. Alternatively, there could have been so much matter that the universe was already beginning to contract. Instead, the amount of matter present is within a factor of ten or so of the value that corresponds to precise balance between these two situations. The standard Big Bang model offers no explanation of why this should be the case.

To understand the new ideas that seek to explain these characteristics of the universe, we must first digress and talk about the forces acting on subatomic particles. Then we will return to discussing the grand picture of how the universe might have evolved.

Grand Unified Theories

In physical science, the term *force* (see Chapter 2) is used to describe anything that can change the motion of a particle or body. One of the remarkable discoveries of modern science is that all known physical processes can be described through the action of four forces—gravity, electromagnetism, the strong nuclear force, and the weak nuclear force (Table 28.4).

Although gravity is perhaps the most familiar to you, and certainly appears strong if you jump off a tall building, the force of gravity between two elementary particles—say two protons—is by far the weakest of the four forces. Electromagnetism, which includes both magnetic and electrical forces, holds atoms together and produces the electromagnetic radiation that we use to study the universe. The weak nuclear force is only weak in comparison to its strong "cousin," but is in fact much stronger than gravity.

Both the weak and strong nuclear forces differ from the first two forces in that they act only over very small distances—those comparable to the size of an atomic nucleus or less. The weak force is involved in radioactive decay and in reactions that result in the production of neutrinos. The strong force holds protons and neutrons together in an atomic nucleus (as we saw in Chapter 15).

Physicists have wondered why there are four forces in the universe—why not 300 or, preferably, just one? An important hint comes from the name *electromagnetic* force. For a long time scientists thought that the forces of electricity and magnetism were separate, but James Clerk Maxwell (see Chapter 4) was able to *unify* these forces— to show that they are aspects of the same phenomenon. In the same way, many scientists (including Einstein) have wondered if the four forces we now know could also be unified. Physicists have developed models, called **grand unified theories (GUTs),** to unify three of the four forces.

In these theories, the strong, weak, and electromagnetic forces, are not three independent forces, but instead are different manifestations or aspects of what is, in fact, a single force. The theories predict that at high enough temperatures there would be only one force. At lower temperatures (like the ones in the universe today), however, this single force has changed into three different forces (Figure 28.16). Just as different gases freeze at different temperatures, we can say that the different forces "froze out" of the unified force at different temperatures. Unfortunately, the temperatures at which the three forces were one are so high that they cannot be reached in any terrestrial laboratory. Only the early universe, at times prior to 10^{-35} s, was hot enough to unify these forces. (Many physicists think that gravity is also unified with the other forces at still-higher temperatures.)

The Inflationary Hypothesis

Some forms of the GUTs predict that a remarkable event occurred when the universe was about 10^{-35} s old and the forces were starting to separate. The equations of general relativity, combined with the special state of matter at that time, predict that gravity could briefly have been a repulsive force. In our own time, gravity is an attractive force that slows the expansion of the universe, but for a brief instant near 10^{-35} s after the expansion began, gravity could actually have accelerated the expansion. It is as if the cosmological constant had, for a brief instant, not been equal to zero, and hence there was a tremendous repulsion in the universe.

TABLE 28.4
The Forces of Nature

Force	Relative Strength Today	Range	Important Applications
Gravity	1	Whole universe	Motions of planets, stars, galaxies
Electromagnetism	10^{36}	Whole universe	Atoms, molecules, electricity, magnetic fields
Weak nuclear	10^{33}	10^{-17}m	Radioactive decay
Strong nuclear	10^{38}	10^{-15}m	The existence of atomic nuclei

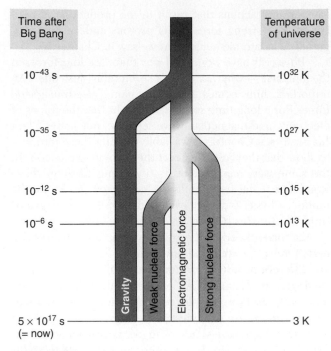

Figure 28.16
The strength of the four forces depends on the temperature of the universe. This diagram shows that at very early times when the temperature of the universe was very high, all four forces resembled one another and were indistinguishable. As the universe cooled, the forces took on separate and distinctive characteristics.

A model universe in which this rapid, early expansion occurs is called an **inflationary universe.** The inflationary universe is identical to the Big Bang universe for all time after the first 10^{-30} s. Prior to that, there was a brief period of extraordinarily rapid expansion or inflation during which the scale of the universe increased by a factor of about 10^{50} times more than predicted by standard Big Bang models (Figure 28.17). As the universe expanded, its temperature dropped below the critical value at which all three forces behave in a symmetrical fashion. In the cooler, asymmetrical universe the nuclear forces dominated the electromagnetic force; they continue to do so in our world today.

Prior to the inflation, all the parts of the universe that we can now see were so small and close to each other that they could exchange information. That is, the horizon distance included all of the universe that we can now observe. Before inflation occurred, there was adequate time for the observable universe to homogenize itself and come to the same temperature. Then inflation expanded those regions tremendously, so that many parts of the universe are now beyond each other's horizon.

Another appeal of the inflationary model is its prediction that the density of the universe should be exactly equal to the critical density. To see why this is so, remember that the density of the universe is intimately connected with the curvature of space. A high-density uni-

verse, for example, would have a closed geometry (like the surface of the balloon we discussed). But the period of inflation was equivalent to blowing up the balloon to a tremendous size. The universe became so big that from our vantage point, no curvature should be visible. (In the same way, the Earth's surface is so big that it looks flat to us in any location.)

If the universe inflated to look completely flat locally, what is the density that goes with a flat geometry? Recall that it is the critical density. Thus a universe that looks flat should have just the right average density to give us the critical value. This prediction is a two-edged sword. It explains why the universe is so close to critical density. But if it turns out that the universe is *not* at critical density, it would be a serious blow to the inflationary model.

So what is the density of the universe? Table 28.5 summarizes our earlier discussion of the amount of matter estimated to be present in various types of astronomical objects. Luminous matter in galaxies contributes less than 1 percent of the mass required to reach critical density. Even if we add the invisible dark matter that is detected through its gravitational influence on luminous objects, we are up to only 10 to 20 percent of the critical density.

Furthermore, the abundance of deuterium indicates that protons and neutrons amount to no more than about 5 percent, and perhaps as little as 1 percent, of the critical density. The only observational evidence that immediately supports the idea that we live in a critical-density universe is the motion of nearby galaxies toward the Great

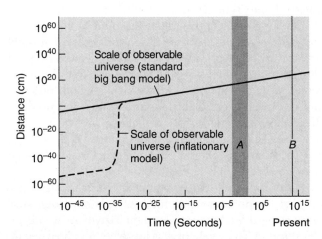

Figure 28.17
How the scale factor of the observable universe changes with time for the standard Big Bang model (solid line) and for the inflationary model (dashed line). (Note that the time scale at the bottom is extremely compressed.) During inflation, regions that were very small and in contact with each other are suddenly blown up to be much larger and outside each other's horizon distance. The two models are the same for all times after 10^{-30} s. Electrons, positrons, and the lightest atomic nuclei are formed during the time interval labeled A. The universe becomes transparent to radiation at the time designated B. "Now" is at the right edge of the figure.

TABLE 28.5
What Different Kinds of Objects Contribute to the Density of the Universe (Dark Matter Is Included)

Object	Density as a Percent of Critical Density
Stars	Less than 1%
Individual galaxies	1–3%
Rich clusters of galaxies	10–20%
Large unseen masses such as the Great Attractor	50–100%

Attractor—implying that some large amount of mass is pulling us in that direction. These observations are so new that some independent confirmation will be required before we can really be confident about the conclusion. But if they are right, the additional mass in the universe must not be made of protons or neutrons.

Dark Matter Possibilities

What we have concluded, then, is that in order for the universe to be at critical density, as inflationary models require, most of its matter must be invisible, *and* cannot be made of protons and neutrons. How should astronomers go about looking for such dark matter? The techniques depend on what we think it might be made of.

One possibility is that the dark matter is made of neutrinos with sufficient mass to produce a critical-density universe. The mass of the neutrino is surely very small, but some tentative experimental evidence suggests that it is not actually zero (see Chapter 15). Attempts are now being made with high-energy accelerators to measure the mass of the neutrino by detecting the change in momentum or energy of another particle that collides with it. Only after the mass is determined will it be possible to calculate the total mass of neutrinos in the universe.

Another experiment involves the study of neutrinos emitted by supernovae. If a supernova were to go off in our own Galaxy, then we could measure the arrival times of the three different types of neutrinos. If they arrive at different times, then at least some of them must have mass. If neutrinos have zero mass, then all of the types will travel at the speed of light and arrive at the same time.

Another possible form dark matter can take is some type of elementary particle that we have not yet detected here on Earth—a particle that has mass and exists in sufficient abundance to reach critical density. GUTs predict the existence of such particles. One class of particles has been given the name WIMPs, which stands for *weakly interacting massive particles.* Since these particles do not participate in nuclear reactions leading to the production of deuterium, the deuterium abundance puts no limits on how many WIMPs might be in the universe. (A number

of other exotic particles have also been suggested as prime constituents of dark matter, by the way.)

If WIMPs do exist, then some of them should be passing through our physics laboratories right now. The trick is to catch them. Since by definition they interact only weakly (infrequently) with other matter, the chances that they will have a measurable effect are small. Nevertheless, physicists are now devising experiments to try to detect them. The basic idea is that a WIMP moving through such a detector might collide with an atomic nucleus and cause it to move. This motion would then be transferred to other particles in the detector, causing a very(!) small change in temperature. Experiments based on this idea will be conducted in the next decade, but are extremely difficult to carry out.

Dark Matter and the Formation of Galaxies

Inflationary models require dark matter for another reason, one related to the formation of galaxies. As we have seen, galaxies must have grown from density fluctuations in the early universe. The observations with COBE give us information on the size of those fluctuations. It turns out that the density variations are too small, at least according to our current theories, to have formed galaxies in the first billion years or so after the Big Bang. Yet observations of quasars suggest that galaxies were indeed formed that early.

The COBE data, however, give us information about density fluctuations only for the type of matter that interacts with radiation. Suppose there is a type of matter that does not interact with light at all—namely dark matter. This matter could have much greater variations in density, which we would not be able to detect because dark matter does not affect the intensity of the radiation measured by COBE. This dark matter might form a kind of gravitational trap that could have begun to attract ordinary matter immediately after the universe became transparent. As ordinary matter became increasingly concentrated, it could have turned into galaxies more quickly thanks to these traps.

"I CAN'T TELL YOU WHAT'S IN THE DARK MATTER SANDWICH. NO ONE KNOWS WHAT'S IN THE DARK MATTER SANDWICH."

For an analogy, imagine a boulevard with traffic lights every half-mile or so. Suppose you are part of a motorcade of cars accompanied by police who lead you past each light, even if it is red. So, too, when the early universe was opaque, radiation carried ordinary matter with it, sweeping past the concentrations of dark matter. Now suppose the police leave the motorcade, and the lights all turn red at the same time. The red lights act as traffic traps; approaching cars now have to stop, and so they bunch up. Likewise, after the early universe became transparent, ordinary matter interacted with radiation only occasionally and so could fall into the dark-matter traps.

The size of the gravitational traps depends on the nature of the dark matter. Suppose it is moving near the speed of light—astronomers call this *hot dark matter*—as neutrinos would. Then small-scale density fluctuations are smoothed out by the rapidly streaming particles as they move from high- to low-density regions. In this case, large-scale structure would form first. If, on the other hand, the dark matter moves slowly—we call this *cold dark matter*—then the particles do not have time to move far enough to smooth out small-scale density fluctuations. In this case, relatively small structures, the size of globular clusters or individual galaxies, are likely to form first.

Neither hot nor cold dark matter is entirely successful in explaining the distribution of galaxies discussed in Chapter 27. Hot dark matter models predict that all galaxies should be found in large sheet-like structures, which is not seen. Cold dark matter cannot produce voids, walls, and long structures such as the Great Wall. Now theories are being developed that contain both hot and cold dark matter. Even though current models are not adequate to explain how galaxies form, it is important to note that galaxies are difficult to form at all unless a substantial amount of dark matter of some kind is present.

To sum up, attempts to understand what the standard Big Bang model does not explain—isotropy, the density of the universe, the formation and distribution of galaxies—have suggested radical new ideas about the universe, which must be tested by observations. If it is really true that most of the matter in the universe is made of some type of particle that we have not yet discovered, then we must accept the challenge of trying to detect it. The search is on—in huge accelerators, in university laboratories around the world, and in deep underground mines, where scientists are trying to trap elusive dark matter particles just as they once succeeded in capturing neutrinos. There are exciting years ahead in cosmology.

Conclusion

Thus the explorations of space end on a note of uncertainty. . . . With increasing distance our knowledge fades . . . and we search among ghostly errors of measurement for landmarks that are scarcely substantial.

—E. Hubble in *The Realm of the Nebulae* (1936)

You may have found this brief discussion of dark matter and cosmology a bit frustrating. We have offered glimpses of theories and observations, but have been unable to provide satisfying answers to some of the problems we have raised. These ideas are at the forefront of modern science, where questions are more numerous than answers, and much more work is needed before we can see clearly. Bear in mind that less than a century has passed since Hubble demonstrated the existence of other galaxies. The quest to understand how the universe of galaxies came to be will keep astronomers busy for a long time to come.

Summary

28.1 Cosmology is the study of the organization and evolution of the universe. The universe is expanding, and this is one of the starting observational points for modern cosmological theories. Before Hubble showed that the universe was expanding, Einstein introduced a **cosmological constant** into his equations to counterbalance gravity and make the universe static. From the expansion rate of the universe we can estimate that all of the matter within it was concentrated in an infinitesimally small volume roughly 9 to 15 billion years ago, a time we call the **Big Bang.** The age of the universe is uncertain because we do not know how much gravity *decelerates* the universe.

28.2 The mass in the universe curves the fabric of space-time. A curved universe has neither center nor edge. In a **closed universe,** which is finite, if you keep going straight ahead you eventually return to where you started. In a **open universe,** which is infinite, not only do you not return to your starting point, but more space than you expect opens up. The Big Bang happened throughout space, everywhere at once. Galaxy redshifts are a result of the stretching of space.

28.3 The factor that controls the ultimate fate of the universe is the density of matter (and energy), which is related to the curvature. The **critical density** is that required to stop the expansion at a time infinitely far in the future. If the density is higher than this, then the rate of expansion will slow and reverse direction so that the galaxies all come together again (a closed universe). Observations, however, suggest that the density is actually so low that the expansion will continue forever (an open universe). A universe with critical density would have zero curvature; we call this model a **flat universe.**

28.4 The universe cools as it expands. The energy of photons is determined by their temperature, and calculations show that in the hot early universe photons had so much energy that when they collided with one another they could produce material particles. As the universe expanded and cooled, protons and neutrons formed first; then came electrons and positrons. Next, fusion reactions produced deuterium, helium, and lithium nuclei. Finally, the universe became cool enough to form neutral hydrogen atoms. At this *decoupling* time the universe became transparent to radiation. Scientists have detected the **cosmic background radiation (CBR)** from the hot early universe. Measurements with the COBE satellite show that the CBR is a blackbody with a temperature of 2.735 K. Tiny fluctuations in the CBR may be showing us the seeds of large-scale structure in the universe.

28.5 The Big Bang model does not explain why the CBR has the same temperature in all directions. Neither does it explain why there was originally more matter than anti-matter, nor why the density of the universe is so close to critical density. New **grand unified theories (GUTs),** which predict a period of very rapid expansion, or **inflation,** when the universe was 10^{-35} s old, are being developed to try to explain these observations. One prediction of these new theories is that the density of the universe should be exactly equal to the critical density. Most observations are inconsistent with such a high density. Determination of the quantity and composition of dark matter is crucial to understanding the early history of the universe.

Review Questions

1. What are the basic observations about the universe that any theory of cosmology must explain?

2. Describe three possible futures for the universe. What property of the universe determines which of these possibilities is the correct one?

3. Which formed first in the early universe—protons and neutrons, or electrons and positrons? Why?

4. Which formed first—hydrogen nuclei or hydrogen atoms? Explain the sequence of events that led to each.

5. Describe at least two characteristics of the universe that are explained by the standard Big Bang model.

6. Describe two properties of the universe that are not explained by the standard Big Bang model (without inflation). How does inflation explain these two properties?

7. Why do astronomers believe there must be dark matter that is not in the form of atoms with protons and neutrons?

Thought Questions

8. What is the most useful probe of the early evolution of the universe—a giant elliptical galaxy, or an irregular galaxy such as the Large Magellanic Cloud? Why?

9. What are the advantages and disadvantages of using quasars to probe the early history of the universe?

10. Suppose someone proposed a model with the Great Attractor as the center of the universe, and with all the galaxies and galaxy clusters falling in toward it. Give some arguments against this idea.

11. Suppose the universe expands forever. Describe what will become of the radiation from the primeval fireball. What will the future evolution of galaxies be like? Could life as we know it survive forever in such a universe? Why?

12. Some theorists argue that the universe is at just the critical density. Do the current observations support this hypothesis?

13. Summarize the evidence for the existence of dark matter in the universe.

14. In this text we have discussed numerous motions of the Earth as it travels through space with the Sun. Describe as many of these as you can.

15. There are a variety of ways of estimating the ages of various objects in the universe. Describe some of these ways, and indicate how well they agree with one another and with the age of the universe itself as estimated by its expansion.

16. Since the time of Copernicus, each revolution in astronomy has moved humans farther from the center of the universe. Now it appears that we may not even be made of the most common form of matter. Trace the changes in scientific thought about the central nature of the Earth, the Sun, and our Galaxy on a cosmic scale. Explain how the notion that most of the universe is made of dark matter continues this "Copernican tradition."

17. Construct a time line for the universe, and indicate when various significant events, from the beginning of the expansion to the formation of the Sun to the appearance of humans on Earth, occurred.

Problems

18. The Andromeda galaxy is approaching the Sun at a velocity of about 300 km/s. Does this indicate that the universe is not expanding? Compare this velocity with that of the Sun in its orbit around the center of the Galaxy. Suppose Andromeda is orbiting the Milky Way with a period of 10 billion years. What velocity would it have?

19. Show that if $H = 25$ km/s per million LY, then the maximum age of the universe is approximately 13 billion years.

20. Suppose $H = 15$ km/s per million LY. What is the maximum age of the universe?

21. It is possible to derive the age of the universe, given the value of the Hubble constant and the distance to a galaxy. Consider a galaxy at a distance of 400 million LY, receding from us at a velocity, v. If the Hubble constant is 25 km/s per million LY, what is its velocity? How long ago was that galaxy right next door to our own Galaxy if it has always been receding at its present rate? Express your answer in years. Since the universe began when all galaxies were very close together, this number is an estimate for the age of the universe.

Suggestions for Further Reading

Boslaugh, J. *Masters of Time: Cosmology at the End of Innocence.* 1992, Addison-Wesley. A journalist reports from the frontiers of cosmological research.

Brush, S. "How Cosmology Became a Science" in *Scientific American,* Aug. 1992, p. 62.

Croswell, K. "A Milestone in Fornax" in *Astronomy,* Oct. 1995, p. 42. On efforts to determine the age of the universe.

Davies, P. *The Last Three Minutes.* 1994, Basic Books. Introduction to the ultimate fate of the universe.

Davies, P. "Everyone's Guide to Cosmology" in *Sky & Telescope,* March 1991, p. 250. Good introductory article.

Ferris, T. *The Red Limit,* 2nd ed. 1983, Morrow. An eloquent history of modern cosmology.

Fienberg, R. "COBE Confronts the Big Bang" in *Sky & Telescope,* July 1992, p. 34. Good summary of the temperature fluctuations discovery.

Gribbin, J. *In Search of the Big Bang.* 1986, Bantam. Excellent beginner's introduction to ideas in cosmology.

Guth, A. and Steinhardt, P. "The Inflationary Universe" in *Scientific American,* May 1984, p. 116.

Halliwell, J. "Quantum Cosmology and the Creation of the Universe" in *Scientific American,* Dec. 1991, p. 76.

Harrison, E. *Cosmology.* 1981, Cambridge U. Press. A fine, erudite textbook, full of good examples.

Overbye, D. *Lonely Hearts of the Cosmos.* 1991, Harper Collins. Wonderful introduction to cosmology today, with a focus on the people involved.

Roth, J. and Primack, J. "Cosmology: All Sewn Up or Coming Apart at the Seams?" in *Sky & Telescope,* Jan. 1996, p. 20.

Smoot, G. and Davidson, K. *Wrinkles in Time.* 1993, Morrow. The full story of the COBE discoveries, by one of the team leaders and a science journalist.

1. Display only spiral galaxies and set the *Limiting Magnitude* to 10. Center the *Display* on the galaxy M106 and set the *Zoom Factor* to 0.9. Print the map. Reset the *Zoom Factor* to 0.925 and print a second map. Imagine the two maps represent different ages of the universe; the zoom factors then correspond to different amounts of expansion by the universe.

Overlay the two maps and notice the "expansion" of the galaxies. (You may need to darken the center of each galaxy marker and hold the sheets toward a light.) Compare the expansion of a galaxy near and one far from M106.

Can you explain the expansions in terms of Hubble's redshift law?

Pick a galaxy other than M106 and line up the two markers for the galaxy.

Does the "expansion" still occur?

Can you use the expansion to locate the center of the universe? Explain your answer.

2. Display only spiral galaxies and set the *Limiting Magnitude* to 10. Center the display on the galaxy M106 and set the *Zoom Factor* to 0.9. Print the map. (You can use the map from Exercise 1.) Reset the *Zoom Factor* to 1.2 and print a second map.

Pick five widely separated galaxies contained in both maps and number them. Select one galaxy to represent the Galaxy; use a millimeter ruler to measure the distances of the other galaxies from the "Galaxy," first on the 0.9-zoom map and then on the other map. Because the distances represent galaxy positions at different ages of the universe, the difference between the distances for each galaxy is a velocity of expansion. To find the rates of expansion, take the difference between the distances for each galaxy and divide it by the distance for the 1.2-zoom map.

How is this related to the Hubble constant?

Choose a different galaxy to represent the Galaxy. Repeat the measurements and find the expansion rate. Did it change?

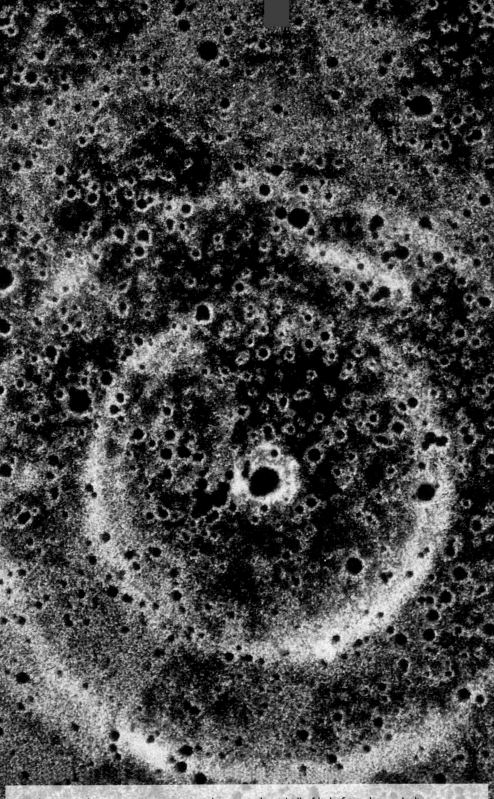

Light echoes from Supernova 1987A. The expanding shell of light from the exploding star reflects from two dusty regions around the supernova, making two circles of light. To bring out the faint light echoes, David Malin subtracted from the echo image a photo of the area taken before the explosion. Since the stars and unrelated tendrils of gas have not changed, they subtract out, leaving black areas, while the yellow light of the supernova echo stands out clearly in contrast. Most of the atoms in our bodies were recycled through the action of such supernovae. (© Anglo-Australian Telescope Board)

Epilogue: Cosmic Evolution and Life Elsewhere

Our voyages have taken us through billions of years of time and billions of light years of space. It is time to sum up where we have been and what we have learned, and to ask one remaining question. Are we the only creatures in the universe thinking about its origin and evolution, or could there possibly be other intelligent life-forms among the stars who also enjoy considering such questions?

We shall not cease from exploration,
And the end of all our exploring,
Will be to arrive where we started,
And know the place for the first time.

T. S. Eliot, *Little Gidding* (from *The Four Quartets* in *The Collected Poems of T. S. Eliot,* 1934, 1936)

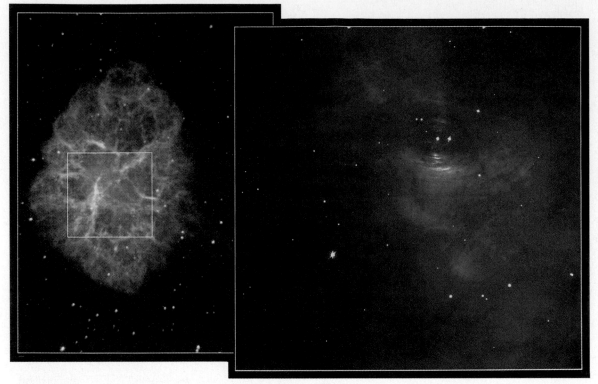

Figure E.1

The Crab Nebula is the remnant of a supernova first seen in July 1054. Now almost 11 LY across, this remnant still glows with tremendous energy in many bands of the electromagnetic spectrum. It is powered inside by a pulsar, a spinning neutron star whose beams sweep over the Earth like a lighthouse 30 times a second. The left-hand image, taken with a ground-based telescope, shows the full nebula. The blue glow in the center is powered by the pulsar. The right-hand image, taken with the Hubble Space Telescope, is a close-up of the Crab's central region. The pulsar itself can be seen as the left member of a pair of stars near the center of the frame. The Hubble was able to observe wisps of material streaming away from the pulsar at half the speed of light! (J. Hester, P. Scowen, and NASA)

What Were the Atoms in Your Body Doing Billions of Years Ago?

Like all good voyages, this one—we hope—has taught you as much about yourself as about the places we have visited. Let's review what we have learned about the history of the cosmos by examining the history of the atoms in your body before they became a part of you.

After the universe cooled sufficiently for atoms to exist, all matter consisted of hydrogen and helium (with a very small amount of lithium). The hydrogen atoms in the water and other hydrogen-rich molecules in your body formed at this early time; they are the oldest atoms that are part of you. But it is not possible to make an organism as complex and interesting as you with only the first three elements. Before something like you could evolve, several generations of stars had to go through their cycles of birth, life, and death.

As we have seen throughout this book, the only place new elements can be synthesized (now that the universe has cooled) is inside stars. We know from observations of distant quasars that stars must have formed within the first billion years or so, because quasar spectra already show the presence of some of the heavier elements. As the years went on, generations of stars produced more and more of the heavier elements. The most-massive stars not only produced the greatest variety of new nuclei, but then had the courtesy to explode (Figure E.1), scattering the newly minted atoms into space.

Over the years, thanks to supernovae, the gas between the stars became increasingly enriched with heavier materials. In the cooler outer layers of old stars, atoms frequently combined into solid particles that we call interstellar dust. The next generations of stars and planets, containing atoms of carbon, nitrogen, silicon, iron, and the rest of the familiar elements, then formed from reservoirs of enriched gas and dust. One of the most remarkable discoveries of modern astronomy is that life on Earth is mostly composed of just those elements that stars find easiest to make.

About 5 billion years ago, possibly prodded by the shock wave of a nearby supernova, a cloud of gas and dust in this cosmic neighborhood began to collapse under its own weight. Out of this cloud formed the Sun and its

Figure E.2
Comet Hyakutake captured by amateur astrophotographer Robert Provin of California State University, Northridge. This 12-min exposure was taken on a clear night in the Mojave Desert with a camera and telephoto lens mounted on a telescope tripod. As the moving comet was held steady in the camera, the stars appeared to streak. (R. Provin)

centers or explosions of stars—in this case assembled rather cleverly into brains, kidneys, fingers, and faces. We might say that through the thoughts of human beings, the matter in the universe can become aware of itself.

The atoms in your body are merely on loan to you from the lending library of atoms that make up our local corner of the universe. Atoms of many kinds circulate through your body, and then leave it with the breath you exhale and the food you eat and excrete. Even the atoms that take up more permanent residence in your tissues will not be part of you much longer than you are alive. Ultimately, you will return your atoms to the vast reservoir of the Earth, where they will be incorporated into other structures and maybe even other living things in the millennia to come.

This picture of *cosmic evolution,* of our descent from the stars, has been obtained through the efforts of scientists in many fields over many decades. Some of its details are, as we saw throughout the book, still tentative and incomplete, but we feel reasonably confident in its broad outlines. While we do not claim to know the full story of how the universe and we ourselves evolved over the course of cosmic history, it is remarkable how much we *have* been able to learn in the short time we have had instruments to probe the physical nature of the stars.

The Copernican Principle

Our study of astronomy has also shown that we have always been wrong in the past whenever we have claimed that the Earth is somehow unique. Copernicus and

planets, together with all the smaller bodies, such as comets, that also orbit the Sun (Figure E.2). The third planet from the Sun, as it cooled, developed an atmosphere that served to moderate temperature extremes and allow the formation of large quantities of liquid water on its surface. The chemicals available on the cooling Earth may have been further enriched by the addition of molecules frozen in the nuclei of comets that eventually collided with our planet.

The chemical variety and moderate conditions on Earth eventually led to the formation of self-reproducing molecules and the beginnings of life. Over the billions of years of Earth history, that life slowly evolved and became more complex. The course of evolution was punctuated by occasional planet-wide changes caused by collisions with those planetesimals (or their fragments) that had not been incorporated into the Sun or one of its accompanying worlds. Mammals may owe their domination of the Earth's surface to just such a collision 65 million years ago.

Through many twisting turns, the course of evolution on Earth produced a creature with self-consciousness, able to ask questions about its own origins and place in the cosmos (Figure E.3). Like most of the Earth, this creature is composed of atoms that were forged in the

Figure E.3
Human beings have the intellect to wonder about their planet and what lies beyond it. Through them, the universe becomes aware of itself. (Photo by A. Fraknoi)

Galileo showed that the Earth was not the center of the solar system, but merely one of a number of bodies orbiting the Sun. Our study of the stars has demonstrated that the Sun itself is a rather undistinguished star, living peacefully through its main-sequence stage like so many billions of others. There seems nothing special about our position in the Milky Way, and nothing surprising about our Galaxy's position in either its own group or its supercluster.

The recent discovery of planets around other stars confirms our ideas that our planetary system is not the only one, and that the formation of planets is probably a natural consequence of the formation of many kinds of stars. Now that our technology is sufficiently advanced to detect planets elsewhere, they seem to be turning up with impressive regularity in the nearby star systems where we can most easily search. While our current techniques of finding planets allow us only to identify Jupiter-mass bodies, there is no reason to believe that other planetary systems could not contain planets like the Earth as well.

Philosophers of science sometimes call this idea— that there is nothing all that special about our place in the universe—the *Copernican principle*. Although it may be tempting to consider ourselves the central focus of all creation, no evidence for such a belief is found in any of the observations discussed in this book.

Most scientists, therefore, would be surprised if the beginning or evolution of life were absolutely limited to the surface of our planet, and had happened nowhere else. There are, after all, billions of stars in our Galaxy with main-sequence lifetimes long enough for life to have developed on a planet around them. Astronomers and biologists have thus long conjectured that a series of events similar to those on the early Earth has probably led to living organisms around other stars. And, where conditions are right, such life may well have evolved to become what we would call intelligent—that is, aware of and interested in its own cosmic history. (In this sense, we must conclude—with tongue firmly in cheek—that taking an astronomy class is the supreme example of intelligent behavior in the universe!)

Such arguments from the Copernican principle— however interesting they may be for philosophers—are nonetheless insufficient for scientists. We would like to find actual evidence for the existence of intelligent life elsewhere. Despite the sensationalistic claims in the tabloid media, no such evidence has yet been found. But because many scientists feel that such a discovery would be a defining moment in the history of the human species, a number of searches for extraterrestrial life have already been carried out, and others are under way.

The Building Blocks of Life

While no unambiguous evidence has yet been found for life beyond Earth, its chemical building blocks have been detected in a wide range of extraterrestrial environments.

Meteorites, such as the one found near Murchison, Australia, have yielded a variety of amino acids (the building blocks of proteins) whose chemical structures clearly denote extraterrestrial origins (see Chapter 13). When we examine the evaporated material around comets such as Comet Halley, we find a number of *organic* molecules— those that on Earth are associated with the chemistry of life.

One of the most surprising results of modern radio astronomy, as we saw in Chapter 19, was the discovery of organic molecules in the giant gas and dust clouds between stars. We find such molecules most readily in regions where the interstellar dust is most abundant—precisely those regions in which star formation (and probably planet formation) happens most easily.

Starting in the early 1950s, scientists have tried to duplicate in their laboratories the chemical pathways that led to life on our planet. In a series of experiments pioneered by Stanley Miller and Harold Urey at the University of Chicago, biochemists have simulated conditions on the early Earth and been able to produce many of the fundamental building blocks of life—including those that go into forming proteins and nucleic acids (Figure E.4). While their accomplishment is still far from making life in the laboratory, it does show that the first steps on the long road to life may not be as difficult as once thought.

Already there are suggestions that these steps once led to life on our neighbor planet Mars. As we saw in Chapter 9, the Viking spacecraft did not find any indication of life existing there today, although its observations suggest that conditions on the red planet were more conducive to the formation of life billions of years ago. In 1996, detailed laboratory analysis of the one ancient martian rock sample available on earth—the 4.5-billion-year-old meteorite called ALH 84001—showed that this bit of martian crust experienced wet conditions about 3.5 billion years ago. At that time, water flowed on Mars (see Chapter 9) and left traces of carbonate minerals and some organic compounds embedded in the meteorite. Some scientists even suspect that tiny structures seen in this rock at high magnification might be the fossilized remains of ancient microbial life. If there are indeed fossils in ancient martian rock, then life would have begun on Mars independently of the Earth and it would be one of the most spectacular discoveries in the history of science. But even if these specific features turn out to be of inorganic origin, the work on ALH 84001 has already verified that wet, apparently Earth-like conditions once existed on Mars. The atmospheres of the jovian planets and of Saturn's satellite Titan do show evidence of organic molecules, but none that have developed into life-forms. It may be that conditions more similar to the early Earth (or early Mars) are required for life to develop, or that we simply do not understand the many difficulties involved in the formation of molecules that can make copies of themselves. In laboratories around the world, biochemists are trying to find answers to these questions.

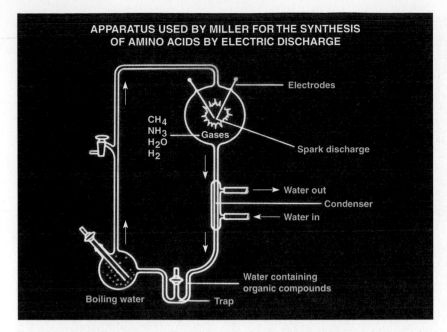

APPARATUS USED BY MILLER FOR THE SYNTHESIS OF AMINO ACIDS BY ELECTRIC DISCHARGE

Electrodes

CH_4
NH_3
H_2O
H_2
Gases

Spark discharge

Water out
Condenser
Water in

Water containing organic compounds

Boiling water
Trap

Figure E.4
The Miller–Urey experiment, performed in 1953, simulated conditions on the early Earth. An "atmosphere" consisting of methane, ammonia, water vapor, and hydrogen was subjected to electrical sparks to simulate lightning. Water at the bottom of the apparatus provided an "ocean" into which materials synthesized in the atmosphere could fall. When its contents were analyzed, they were found to contain a variety of amino acids, the building blocks of proteins. (F. Drake and the Astronomical Society of the Pacific)

Searching for Life Elsewhere

Suppose, for the sake of argument, that we take an optimistic view and assume that life has developed around many other stars, perhaps on planets that resemble the Earth in chemically significant ways. Suppose further that in some of these cases, life has evolved to be intelligent (self-aware and interested in what is happening elsewhere in the universe). How can we find out about, and perhaps even make contact with, such life-forms?

This problem is similar to making contact with people who live in a remote part of the Earth. If students in the United States want to converse with students in Australia, for example, they have two choices. Either one group gets on an airplane and travels to meet the other, or they communicate via some message medium (today, probably by telephone, fax, e-mail, or short-wave radio). Given how expensive airline tickets are, most students would probably select the message route.

In the same way, if we want to get in touch with intelligent life around other stars, we can travel, or we can try to exchange messages. Because of the great distances involved, interstellar space travel is either very slow or very expensive. The fastest spacecraft the human species has built so far would take almost 80,000 years to get to the nearest star. While we could certainly design a faster craft, the more quickly we require it to travel, the greater the energy cost involved. To reach neighboring stars in less than a human life span, we would have to travel close to the speed of light. In that case, however, the expense would become truly astronomical.

The late Bernard Oliver, vice president of the Hewlett–Packard Corporation and an engineer with an abiding interest in life elsewhere, made a revealing calculation about the costs of rapid space travel. Since we do not know what sort of technology we (or other civilizations) might someday develop, Oliver considered a trip to the nearest star in a spaceship with a "perfect engine" — one that would convert its fuel into energy with 100 percent efficiency. (No future technology can possibly do better than this. In reality, nature is unlikely to yield an efficiency even close to the perfect value; just think how much of the energy released by the fuel in a car is wasted.) Even with a perfect engine, the energy cost of a single round-trip journey at 70 percent the speed of light would be equivalent to about *500,000 years worth of total U.S. electrical energy consumption!* Congress is unlikely to fund such a program in the near future.

In case you are wondering why this figure is so high, you must remember that the voyagers could not depend on finding "gas stations" open at their destination. Therefore they would have to carry the fuel for the return legs of the journey with them, and getting all that fuel up to 70 percent the speed of light would be very expensive. The important thing about Oliver's calculation is that it does not depend on present-day technology (since it assumes a perfect engine), but only on the known laws of science. What it shows is that no matter who does the traveling, it is expensive to go fast enough to get to the stars within the course of a single human life.

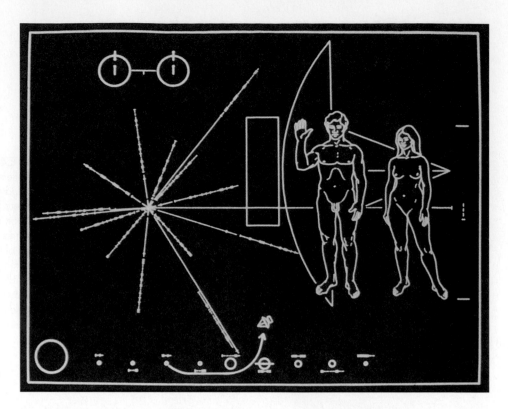

Figure E.5
The image engraved on the plaques aboard the Pioneer 10 and 11 spacecraft. The human figures are drawn in proportion to the spacecraft, which is shown behind them. The Sun and planets in the solar system can be seen at the bottom, with the trajectory that the spacecraft followed. The lines and markings in the left center show the positions and pulse periods for a number of pulsars, to locate the spacecraft's launch in space and time. (Recall that the periods of pulsars lengthen as time goes on, so the exact timing of the pulses might tell a sophisticated group of alien astronomers when the craft began its journey.) (NASA)

This is one reason astronomers are so skeptical about claims that unidentified flying objects (UFOs) are spaceships from extraterrestrial civilizations. Given the distance and expense involved, it seems unlikely that the dozens of UFOs (and, recently, even UFO abductions) reported each year could all be visitors from other stars so fascinated by Earth civilization that they are willing to expend fantastically large amounts of energy or time to reach us.

In fact, a sober evaluation of UFO reports often converts them to IFOs (identified flying objects), or NFOs (not-at-all flying objects). While some are hoaxes, others are natural phenomena such as ball lightning, fireballs, bright planets, or even flocks of birds with reflective bellies. Still others are human craft, such as private planes with some lights missing, or classified military airplanes. In fact, not a single UFO has ever left behind any physical evidence that can be tested in an Earth laboratory and thus shown to be of nonterrestrial origin.[1]

Some visionaries have suggested that we might someday overcome the long-time/large-energy demands of space travel by hollowing out asteroids and sending them to the stars with large colonies of people on board

(equipped with all the necessary provisions for a really long trip). While the original settlers would never see the stars, their far-future descendants might arrive and start a settlement on an Earth-like planet somewhere. This is, for now, only the stuff of science fiction.

Messages on Spacecraft

In the real world we do have four spacecraft—two Pioneers and two Voyagers—which, having finished their program of planetary exploration, are now leaving the solar system. At their coasting speeds, they may well take hundreds of thousands or millions of years to get anywhere close to another star. On the other hand, they were the first products of human technology to leave our home system, and so we wanted to put messages on board to show where they came from.

Each Pioneer carries a plaque with a pictorial message engraved on a gold-anodized aluminum plate (Figure E.5). The Voyagers, launched in 1977, have audio and video records attached (Figure E.6), which allowed the inclusion of over 100 photographs and a selection of music from around the world. (Included among the excerpts from Bach, Beethoven, folk music, tribal chants, etc., is one piece of rock and roll—"Johnny Be Goode" by Chuck Berry.) Given the enormity of the space between stars in our section of the Galaxy, it is very unlikely that these messages will ever be received by anyone. They are more like a note in a bottle thrown into the sea by a shipwrecked sailor, with no realistic expectation of its being found, but a slim hope that perhaps someday, somehow, someone will know of his fate.

[1] If you are interested in pursuing the topic of what UFOs are and aren't, we recommend the following books: Klass, P. *UFO Abductions: A Dangerous Game* (1988, Prometheus Books); Peebles, C. *Watch the Skies: A Chronicle of the Flying Saucer Myth* (1994, Smithsonian Institution Press); and Shaeffer, R. *The UFO Verdict: Examining the Evidence* (1981, Prometheus Books).

The Voyager Message

Communicating with the Stars

If direct visits to stars are unlikely, we must turn to the other alternative for making contact—exchanging messages. Here the news is a lot better. As we have seen throughout this book, we already know (and have learned to use) a messenger—electromagnetic radiation—that moves through space at the fastest speed in the universe. Traveling at the speed of light, radiation reaches the nearest star in only four years, and does so at a fraction of the cost of sending material objects. These advantages are so clear and obvious that we assume they will occur to any other species of intelligent beings who develop technology.

However, we have access to a wide spectrum of electromagnetic radiation, ranging from the longest-wavelength radio waves to the shortest-wavelength gamma rays. Which would be the best for interstellar communication? It would not be smart to select a wavelength that is easily absorbed by interstellar gas and dust, or one that is unlikely to penetrate the atmosphere of a planet like ours. Nor would we want to pick a wave that has lots of compe-

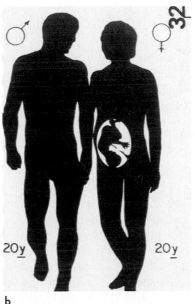

a   b

Figure E.6
(a) Encoded onto a gold-coated copper disk, the Voyager record contains 118 photographs, 90 min of music from around the world, greetings in almost 60 languages, and other audio material. It is a summary of the sights and sounds of Earth. (b) One of the images encoded onto the record. Originally, the team devising the record wanted to send a photograph, from a medical book, of a nude man and a pregnant woman. However, NASA was concerned about offending some people on Earth, and so artist Jon Lomberg drew a silhouette version, which allowed the fetus inside the woman to be shown as well. (NASA)

tition for attention in our neighborhood. For example, it would be difficult (and pretty dumb) for us to put together a signal from our civilization in the visible light region of the spectrum. How could it compete with our extremely strong local source of light, the Sun?

One final criterion makes the selection an easy one: we want the radiation to be inexpensive to produce in large quantities. When we consider all these requirements, radio waves win hands down. Being the lowest-frequency (and lowest-energy) band of the spectrum, they are not very expensive to produce (permitting us to use them extensively for communications on the Earth). They are not significantly absorbed by interstellar dust and gas (which is what allows us to use them to map the structure of our Galaxy). With some exceptions, they easily pass through the Earth's atmosphere, and through the atmospheres of the other planets we are acquainted with. And the Sun does not put out a large quantity of radio waves, which means that a radio message has a realistic chance of being "heard" above the local noise.

The Cosmic Haystack

For these reasons, many astronomers have decided that the radio band is probably the best place in the spectrum for communication among intelligent civilizations. Having made such a decision, however, we still have many questions and a daunting task ahead of us. Shall we *send* a message, or try to *receive* one? Obviously, if every civilization decides to receive only, then no one will be sending, and everyone will be disappointed. On the other hand, it may be appropriate for us to *begin* by listening, since we are likely among the most primitive civilizations in the Galaxy who are interested in exchanging messages.

We do not make this statement to insult the human species (which—with certain exceptions—we are rather fond of). Instead, we base it on the fact that we have had the ability to receive (or send) a radio message across interstellar distances for only a few decades. Compared to the ages of the stars and the Galaxy, this is a mere instant. If there are civilizations out there who are ahead of us in development by even a short time (in the cosmic sense), they are likely to have a head start of many, many years. If there are civilizations behind us, chances are they are sufficiently far behind that they have not yet developed radio communications. In other words, we, who have just started, may well be the "youngest" species in the Galaxy with this capability. Just as the youngest members of a community are often told to be quiet and listen to their elders for a while before they say something foolish, so we may want to begin our exercise in extraterrestrial communication by listening.

Even restricting our activities to listening, however, leaves us with an array of challenging questions. For example, you know from your own experience with radio transmissions that a typical signal only comes in on one channel (which means it is carried by one small frequency

TABLE E.1
The Cosmic Haystack Problem: Some Questions About an Extraterrestrial Message

- From what direction (which star) is the message coming?
- On what channels (or frequencies) is the message being broadcast?
- How wide in frequency is the channel?
- How strong is the signal (can our radio telescopes detect it)?
- Is the signal continuous, or does it shut off at times (as, for example, a lighthouse beam does when it turns away from us)?
- Does the signal drift (change) in frequency because of the changing relative motion of the source and the receiver?
- How is the message encoded in the signal (how do we decipher it)?
- Can we even recognize a message from a completely alien species? Might it take a form we don't at all expect?

band of radio waves). The owners of your favorite station are confident—because there aren't that many channels on the radio dial—that you will find them despite this (although a few actually transmit over two different frequencies, one on the AM band and one on the FM band). Many of us, when we first arrive in a new city, scan up and down the radio band until we find the stations we like. But if we have only an AM radio in our car, and the stations playing our favorite music are all FM, then we are out of luck.

In the same way, it would be very expensive (and perhaps even ill-mannered) for an extraterrestrial civilization to broadcast on a huge number of channels. Most likely, they select one or a few channels for their particular message. But the radio band of the electromagnetic spectrum contains an astronomically large number of possible channels. How can we know in advance which one they have selected, and how they have coded their message into the radio signal?

If your radio has a poor antenna, then it may not pick up the signal from a weak station some distance away. You may not learn about the existence of that station (and others like it) until you buy better equipment. The same will be true for interstellar transmissions. If an extraterrestrial civilization's signal is just too weak for our present-day radio telescopes, they may be broadcasting their little alien hearts out, but we will miss it completely.

Table E.1 summarizes these and other factors that scientists must grapple with when trying to tune into radio messages from distant civilizations. Because their success depends on either guessing right about so many factors, or else searching through all the possibilities for each factor, some scientists have compared their quest to looking for a needle in a haystack. Thus they like to say that the list of factors in Table E.1 defines the *cosmic haystack problem*.

Figure E.7
A 25th anniversary photo of some members of the Project Ozma team standing in front of the 85-ft radio telescope with which the 1960 search for extraterrestrial messages was performed. Frank Drake is in the back row, second from the right. (National Radio Astronomy Observatory)

Radio Searches

Although the cosmic haystack problem seems daunting, many other research problems in astronomy also require a large investment of time, equipment, and patient effort. And, as several astronomers have pointed out, if we don't search, we're sure not to find anything. Thus several groups of radio astronomers have undertaken searches for extraterrestrial messages during the last three decades.

The very first search for such radio signals was conducted by astronomer Frank Drake in 1960, using the 85-ft antenna at the National Radio Astronomy Observatory (Figure E.7). Called Project Ozma, after the queen of the exotic land of Oz in the children's stories of Frank L. Baum, his experiment involved looking at about 7200 channels and two nearby stars over a period of 200 hours. Although he found nothing, he demonstrated the feasibility of such a search, and set the stage for the more sophisticated projects that followed. (It is interesting to note

that what took 200 hours in 1960 could be done with today's automated systems in about a thousandth of a second.)

Since 1960 almost 60 radio searches have been carried out by scientists around the world, each exploring a minuscule region of the cosmic haystack. Although a number of interesting signals have been found, none has met the crucial test of being detected more than once, so it could be checked. Scientists are continuing several of the searches, always trying to improve their equipment and beat the odds against finding that elusive needle of a message.

In 1992 NASA began the most comprehensive search for radio messages ever undertaken, only to have Congress cut the funding for their project after less than a year. Using private donations, the nonprofit SETI (Search for Extra-Terrestrial Intelligence) Institute has undertaken to continue the search. They now call it Project Phoenix, since the program has risen from the ashes of its funding crisis.

Using modern electronics and computers, the Phoenix system can "listen in" on 28 million channels simultaneously. Its software checks promising signals and alerts the experimenters if an interesting signal persists on any one channel, or moves between channels because of Doppler shift. The intent of the project (which resumed in 1995) is to search 2 billion channels for each of about 1000 nearby stars. To make it onto their list, a star must be roughly similar to the Sun and at least 3 billion years old. That means it could have had enough time for intelligent life to develop on any Earth-like planets around it. The first 200 stars, visible from the Southern Hemisphere, were searched using the 64-m radio antenna near Parkes, Australia (Figure E.8). So far, no signal has been found; the search will continue using other large radio telescopes.

What If We Succeed?

No one can predict when or whether such searches will be successful. It may well be that civilizations technologically far in advance of our own use other forms of communication that we are not yet aware of. After all, 150 years ago we did not have an inkling of the possibilities of radio communication, while today it is difficult to imagine our civilization without it. On the other hand, we would never dream of giving a preschooler a book like the one you are reading. Young children learning to read are given very simple books until they have mastered the basics. We hope that advanced civilizations remember their own youth and send out messages that even youngsters like us can find and interpret.

What will happen if we do find a radio signal that is unambiguously the product of an extraterrestrial intelligence? The existence of the signal itself will be of tremendous philosophical importance, demonstrating that we are

(text cont. on page 595)

a

b

Figure E.8
(a) The 64-m radio telescope at Parkes, Australia, was used in 1995 to search for radio signals from possible extraterrestrial civilizations around 200 stars. (b) Phoenix project scientists Jill Tarter and Peter Backus are shown at the telescope controls during their observing run. (Photos by Seth Shostak, SETI Institute)

not alone in the cosmos. But unless we are able to interpret the message, there may not be much practical value in the discovery. If we can eventually work out a method of mutual communication, however, an interesting question will arise: who will speak for planet Earth?

Suppose we know that a star 35 LY away has a technological civilization around it, and that they have given us a kind of code by which we can make ourselves understood. (An easy way to begin interacting might be to send pictures.) Who decides what to send? Does the whole planet try to agree on one set of messages, or can any individual or group send a separate communication? There may be many countries, religious groups, cultural organizations, corporations, and individuals who can afford a radio antenna and are interested in getting "their message" out. Among ourselves, we rarely speak with one voice; should we try to do so in addressing the universe? Confronting such questions may be a good test of whether there is intelligent life on Earth.

Conclusion

Whether or not we ultimately turn out to be the only intelligent species in our part of the Galaxy, our exploration of the cosmos will surely continue. A humble acknowledgment of how much we have left to learn is one of the fundamental hallmarks of science. This should not, however, prevent us from feeling exhilarated about how much we have already managed to discover, and curious about what else we might find out.

Our progress report on the ideas of astronomy ends here, but we hope that your interest in the universe does not. We hope you will keep up with developments in astronomy through the media, or by going to an occasional public lecture by a local scientist. Who, after all, can even guess all the amazing things that future research projects will reveal about both the universe and our connection with it?

Appendix 1
Astronomy on the World Wide Web

by David Bruning (*Astronomy* Magazine)
and Andrew Fraknoi (Foothill College)

With its almost unlimited capability of storing information and images, the World Wide Web has captured the imaginations of professional and amateur astronomers. New Web sites are springing up faster than any mere mortal can keep track of, and you can spend many hours surfing the Web without ever returning to Earth (or completing all the homework you have due in your other courses!).

In this Appendix we have listed some of our favorite Web sites, with particular attention to those that will be most useful for students in introductory astronomy classes. Our brief list is a tiny fraction of all the astronomy sites out there. Luckily, one of the best aspects of the Web is the ability to connect sites that are related and to switch between them with the click of a button. Many of these sites will point you to others, which will point you to yet others, which will . . . well, you get the general idea.

If you are not familiar with the Web, check with the library or computer center at your college or university. Most schools now offer access and training for the Web, and several on-line services such as *CompuServe* and *America Online* offer access to the Web. But because surfing the Web can be so hypnotic, be sure you understand in advance how much you are being charged for each hour of connection.

One small warning: unlike a textbook, which is checked for accuracy and reviewed by experts in the field, the Web is unedited. Anyone can post anything they wish, and "search engines" (see below) can turn up postings that could be written by an expert, a beginner, or an out-and-out crackpot. As you gather information from a search on the Web, keep track of the *source* of each piece of information. Notice the institution or person it comes from; if something sounds strange, check with your instructor before accepting everything at face value. Despite what some Web sites claim, for example, there is no face built by aliens on Mars, no record of UFO landings hidden by the U.S. military, and no evidence that ancient astronauts built the pyramids because humans were too stupid to do it themselves!

You can access the Web sites listed below by typing the Uniform Resource Locator (URL) in the Open URL or Open Location window of your Web browser. Each code begins with the name of the protocol by which information is transferred; in most cases, this is the "hyper-text transfer protocol" or http. (In some browsers, you do not even need to type "http://" but only what comes afterwards.) A much longer list of astronomy sites with easy links is kept by one of us (DB) at:
http://www.kalmbach.com/astro/HotLinks/HotLinks.html

Table of Contents

1. Some Astronomical Institutions of Special Interest
2. Telescopes and Observatories
3. The Solar System in General
4. Specific Worlds in the Solar System
5. Comets, Asteroids, Meteorites
6. The Sun
7. Stars and Stellar Evolution
8. Nebulae (Clouds of Raw Material, Planetary Nebulae, Supernova Remnants, etc.)
9. Galaxies
10. Cosmology
11. The Search for Life Elsewhere
12. Observing Sky Events (Eclipses, Conjunctions, Meteor Showers, etc.)
13. Miscellaneous Sites Related to Astronomy
14. Sites with Good Listings of Other Astronomy Sites
15. Search Engines (How to Find More Information)

1. Some Astronomical Institutions of Special Interest

- Anglo-Australian Observatory (includes wonderful color astronomical photographs of nebulae and galaxies taken by David Malin, some of which appear in this book):
  http://www.aao.gov.au
- Astronomical Society of the Pacific (an organization of professional and amateur astronomers dedicated to public education; has interesting links to other sites):
  http://www.physics.sfsu.edu/asp/asp.html
- *Astronomy* magazine (a magazine for amateur astronomers and students; its site has good information on astronomy as a hobby, astronomy clubs, and other astronomy sites):
  http://www.kalmbach.com/astro/astronomy.html
- Explorations in Education (Space Telescope Science Institute) (features electronic picture books on astronomy and space topics, organized by subject):
  http://www.stsci.edu/exined-html/exined-home.html
- Jet Propulsion Laboratory (a NASA center run by the California Institute of Technology; has good information on current projects in planetary exploration):
  http://www.jpl.nasa.gov/
- The Planetary Society (active space interest group founded by Carl Sagan; encourages the exploration of the planets and the search for life elsewhere):
  http://wea.mankato.mn.us/tps/
- *Sky & Telescope* magazine (monthly publication for amateur

astronomers; a site with up-to-date information on
sky events):

    http://www.skypub.com/

- Star*s Family of Astronomy Resources (includes a mammoth
database of astronomical organizations and people which can
be searched):

    http://cdsweb.u-strasbg.fr/starsfamily.html

- Students for Exploration and Development of Space (SEDS)
(an active site with many images and lots of information):

    http://www.seds.org

2. Selected Telescopes and Observatories

- Compton Gamma-Ray Observatory:

    http://cossc.gsfc.nasa.gov/cossc/descriptions/cgro.html

- Dominion Astrophysical Observatory (largest Canadian observatory):

    http://www.dao.nrc.ca/DAO/DAO-homepage.html

- European Southern Observatory (operates a number of large
telescopes):

    http://http.hq.eso.org/eso-homepage.html

- Global Oscillation Network Group (for helioseismology):

    http://helios.tuc.noao.edu/gonghome.html

- Harvard College Observatory:

    http://cfa-www.harvard.edu/hco-home.html

- Infrared Space Observatory (ISO):

    http://isowww.estec.esa.nl

- Keck Observatory (10-m telescopes):

    http://astro.caltech.edu/keck.html

- Mauna Kea Tour (site of many large telescopes in Hawaii):

    http://www.ifa.hawaii.edu/images/aerial-tour/

- Mount Wilson Observatory:

    http://www.mtwilson.edu

- National Optical Astronomy Observatories (includes Kitt
Peak):

    http://www.noao.edu/noao.html

- National Radio Astronomy Observatory:

    http://info.aoc.nrao.edu/

- Royal Greenwich Observatory (England):

    http://cast0.ast.cam.ac.uk/RGO/RGO.html

- Space Telescope Science Institute (where to find all the wonderful Hubble Space Telescope images, including those so recent they didn't get into this book):

    http://www.stsci.edu/public.html

- X-Ray Timing Explorer (XTE):

    http://heasarc.gsfc.nasa.gov/docs/xte/xte_1st.html

- Guide to ground-based solar and astrophysical observatories:

    http://ranier.oact.hq.nasa.gov/Sensors_page/
      Ground Observ.html

3. The Solar System in General

- Tours of the solar system and general information:

    http://seds.lpl.arizona.edu/nineplanets/nineplanets/
      nineplanets.html
    http://www.jpl.nasa.gov/tours
    http://www.c3.lanl.gov/~cjhamil/SolarSystem/
      homepage.html
    http://www.fourmilab.ch/solar/solar.html
    http://www.nosc.mil/planet-earth/planets.html

- Images of the planets:

    http://www-pdsimage.jpl.nasa.gov/PIA/

    http://astrosun.tn.cornell.edu/
    http://cdwings.jpl.nasa.gov/pds/

4. Specific Worlds in the Solar System

- See section 3, above, for sites that also include information on
each world; for example:

    http://www.c3.lanl.gov/~cjhamil/SolarSystem/Venus.html

- **Venus**

    Surface of the planet:

      http://stoner.eps.mcgill.ca/bud/first.html

    Browse the Magellan images:

      http://delcano.mit.edu/cgi-bin/midr-query

- **Earth**

    Create your own views:

      http://www.fourmilab.ch/earthview/vplanet.html

    Images from the Spaceborne Imaging Radar project:

      http://www.jpl.nasa.gov/sircxsar/

    Aurora predictions:

      http://www.pfrr.alaska.edu/~pfrr/AURORA/PREDICT/
        CURRENT.HTML

- **Moon**

    Phases:

      http://dragon.aoc.nrao.edu/casey-cgi/moon.cgi/today

    Clementine mission:

      http://www.nrl.navy.gov/Clementine

- **Mars**

    Interactive Mars map:

      http://www.c3.lanl.gov/~cjhamil/Browse/mars.html

    Mars atlas:

      http://ic-www.arc.nasa.gov/ic/projects/bayes-group/
        Atlas/Mars

    Center for Mars Exploration:

      http://cmex-www.arc.nasa.gov/

- **Jupiter**

    Galileo mission:

      http://www.noao.edu/galileo
      http://www.jpl.nasa.gov:80/galileo/index.html
      http://ccf.arc.nasa.gov/galileo_probe/

- **Saturn**

    Appearance of rings:

      http://ringside.arc.nasa.gov/www/rpx/viewer/
        rpx_viewer.html

    Cassini mission:

      htpp://www.jpl.nasa.gov/cassini/

5. Comets, Asteroids, Meteorites

- **Comets**

    A wealth of information about comets (has comet news, lists
    which comets are currently visible, images of many comets,
    elements of comet orbits, etc.):

      http://encke.jpl.nasa.gov/

    Shoemaker-Levy 9 impact at Jupiter:

      http://newproducts.jpl.nasa.gov/s19/

    Comet Hale-Bopp:

      http://www.halebopp.com/index.html
      http://encke.jpl.nasa.gov/hale_bopp_info.html

    Comets of the past:

      http://medicine.wustl.edu/~kronkg/past_comets.html

    Rosetta comet lander mission:

      http://champwww.jpl.nasa.gov/champollion/index.html

Comet observing tips (British Astronomical Association):
http://www.ast.cam.ac.uk/~jds/
Comets orbiting beyond Neptune (Kuiper belt objects):
http://cfa-www.harvard.edu/cfa/ps/lists/TNOs.html

- **Asteroids**
Minor Planet Center:
http://cfa-www.harvard.edu/cfa/ps/mpc.html
Near-Earth Asteroid Rendezvous Mission:
http://nssdc.gsfc.nasa.gov/planetary/near.html
http://hurlbut.jhuapl.edu/NEAR/
Centaurs:
http://cfa-www.harvard.edu/cfa/ps/lists/
Centaurs.html
- **Meteorites**
General guide:
http://www.c3.lanl.gov/~cjhamil/
SolarSystem/meteorite.html
Meteorites in Antarctica:
http://exploration.jsc.nasa.gov/curator/antmet/
antmet.html

6. The Sun

- Tour of sun (National Solar Observatory):
http://blazing.sunspot.noao.edu/Exhibit/Exhibit.html
- Daily Sun images and space weather:
http://www.sel.noaa.gov/
http://www.sel.bldrdoc.gov/today.html
- Movie of Sun's surface:
http://www.erim.org/algs/PD/pd_home.html
- Satellite missions and tutorials on the Sun:
http://wwwssl.msfc.nasa.gov/ssl/pad/solar
- Solar and Heliospheric Observatory (SOHO) satellite:
http://sohowww.nascom.nasa.gov/
- Ulysses satellite:
http://ulysses.jpl.nasa.gov/

7. Stars and Stellar Evolution

- Nearest stars (list):
http://proxima.astro.virginia.edu/~pai/
Recons/nearest25.html
- Planets around other stars:
http://cannon.sfsu.edu/~gmarcy/
- Stellar properties:
http://www.astro.washington.edu/strobel/
star-props/star-props.html
- Tutorials:
http://altair.syr.edu:2024/SETI/TUTORIAL/
- Variable star observing:
http://www.aavso.org/

8. Nebulae (Clouds of Raw Material, Planetary Nebulae, Supernova Remnants, etc.)

- Images of star-forming regions:
http://donald.phast.umass.edu/gs/wizimlib.html
- Images of nebulae:
Anglo-Australian Observatory (David Malin):

http://www.aao.gov.au/images.html
Bill Arnett's Nebula Page:
http://www.seds.org/billa/twn/
Planetary Nebula Sampler:
http://www.noao.edu/jacoby/pn_gallery.html

9. Galaxies

- Images of Galaxies:
Anglo-Australian Observatory (David Malin):
http://www.aao.gov.au/galaxies.html
Messier catalog (University of Arizona):
http://www.seds.org/messier/index.html
http://zebu.uoregon.edu/messier.html
William Keel (University of Alabama):
http://www.astr.ua.edu/choosepic.html
- Interacting galaxies: http://crux.astr.ua.edu/

10. Cosmology

- Cosmic Background Explorer satellite:
http://www.gsfc.nasa.gov/astro/cobe/cobe_home.html
- Gamma-ray bursts, debate about:
http://antwrp.gsfc.nasa.gov/diamond_jubilee/debate.html
- Images and animations:
http://www.ncsa.uiuc.edu/General/NCSAExhibits.html
- Lab exercise on the Hubble Law:
http://www.gettysburg.edu/project/physics/clea/
CLEAhome.html
- MACHO project (search for dark matter):
http://wwwmacho.anu.edu.au/
- OGLE project (dark matter search):
http://www.astrow.edu.pl
- On-line text on cosmology:
http://uu-gna.mit.edu:8001/uu-gna/text/astro/universe/
index.html
- Simulation of early universe:
http://zeus.ncsa.uiuc.edu:8080/GC3_Home_Page.html
- Tutorial on cosmology:
http://altair.syr.edu:2024/SETI/TUTORIAL/bigbang.html

11. The Search for Life Elsewhere

- Planets around other stars:
http://cannon.sfsu.edu/~gmarcy/
- SETI Institute (rich source of material on radio searches):
http://www.seti-inst.edu/
- Tutorial on SETI (Syracuse University):
http://altair.syr.edu:2024/SETI/seti.html

12. Observing Sky Events (Eclipses, Conjunctions, Meteor Showers, etc.)

- General:
Abrams Planetarium Sky Calendar:
http://www.pa.msu.edu/abrams/
Mt. Wilson On-line Stargazer Map:
http://www.mtwilson.edu/services/starmap.html
Links to backyard astronomy:
http://www.kalmbach.com/astro/astronomy.html

Another monthly calendar of sky events:
   http://www.nscee.edu/~drdale/onOrbit_05_95/
      SkyCalendar.html
Sky almanac from *Astronomy* Magazine:
   http://www.kalmbach.com/astro/SkyEvents/SkyEvents.html
- Eclipses:
  Animation:
     http://ageninfo.tamu.edu/eclipse/
  Guide to eclipses:
     http://www.c3.lanl/gov/~cvjhamil/SolarSystem/education/
        eclipses.html
  Solar eclipse bulletins:
     http://umbra.nascom.nasa.gov/sdac.html
- Meteor showers:
   http://medicine.wustl.edu/~kronkg/meteor_shower.html
- Radio astronomy for amateurs:
   http://www.rmplc.co.uk/eduweb/sites/trao/index.html

13. Miscellaneous Sites Related to Astronomy

- Ancient astronomy:
   http://kira.pomona.claremont.edu/
- Astronomical League (umbrella organization of amateur astronomy clubs in the U.S.):
   http://www.mcs.net/~bstevens/al
- Astronomy software reviews (by John Mosley for *Sky & Telescope*):
   http://www.skypub.com/software/mosley.html
- Black holes and general relativity:
   The National Center for Supercomputing Applications (has images relating to black holes and current research in general relativity theory):
      http://jean-luc.ncsa.uiuc.edu/
   Visual Trip to a black hole:
      http://cossc.gsfc.nasa.gov/htmltest/rjn_bht.html
- Elements, table of:
   http://www.cchem.berkeley.edu/Table/index.html
- History of astronomy links:
   http://aibn55.astro.uni-bonn.de:8000/
      ~pbrosche/astoria.html
- Image processing (obtain the superb free program *NIH Image* from):
   http://rsb.info.nih.gov/nih-image
- Lab exercises (CLEA Project):
   http://www.gettysburg.edu/project/physics.clea/
      CLEAhome.html
- Light pollution (and what we can do about it):
   http://www.ida.org/
- NASA information on:
   NASA itself:
      http://www.nasa.gov
   Space Shuttle missions:
      http://shuttle.nasa.gov
   Overview of NASA information on Web:
      http://www.gsfc.nasa.gov/NASA_homepage.html

Astronauts and human spaceflight:
   http://images.jsc.nasa.gov
History of spaceflight:
   http://www.ksc.nasa.gov/history/history.html
Space movie archive:
   http://www.univ-rennes1.fr/ASTRO/anim-e.html
- Planetaria around the world:
   http://www.lochness.com
   http://www.kalmbach.com/astro/SpacePlaces/
      SpacePlaces.html
- Skyview (show maps of any part of the sky at different wavelengths):
   http://skyview.gsfc.nasa.gov/skyview.html
- This month in space and astronomy history (calendar of significant events):
   http://newproducts.jpl.nasa.gov/calendar/history/html

14. Sites with Good Listings of Other Astronomy Sites

- Astro Web (sorted by category):
   http://fits.cv.nrao.edu/www/astronomy.html
- Astronomical information on the internet:
   http://ecf.hq.eso.org/astro-resources.html
- *Astronomy* Magazine:
   http://www.kalmbach.com/astro//HotLinks.html
- Guide to on-line data in astronomy:
   http://www.hq.eso.org/online-resources-paper/rrn.html
- Kennedy Observatory list of astronomy WWW resources:
   http://www.cs.dal.ca/~andromed/
- Sonoma State University:
   http://yorty.sonoma.edu/people/faculty/tenn/
      Astronomy Links.html
- Space mission acronyms and links:
   http://ranier.oact.hq.nasa.gov/Sensors_page/
      MissionLinks.html
- U.S. Geological Survey:
   http://info.er.usgs.gov/network/science/astronomy/
- WebStars (astrophysics sites):
   http://guinan.gsfc.nasa.gov/WebStars.html

15. Search Engines (How to Find More Information)

- SavvySearch (searches via 24 search engines for you):
   http://savvy.cs.colostate.edu:2000/
- Metacrawler (searches via 8 engines):
   http://metacrawler.cs.washington.edu:8080/home.html
- Alta Vista (powerful search engine):
   http://www.altavista.digital.com/
- Lycos:
   http://lycos.cs.cmu.edu
- Yahoo:
   http://www.yahoo.com

Appendix 2

Sources of Astronomical Information

1. Popular-level Astronomy Magazines

Astronomy Magazine (Kalmbach Publishing, P. O. Box 1612, Waukesha, WI 53187) The largest circulation astronomy publication in the world; it is very colorful and basic.

Griffith Observer (Griffith Observatory, 2800 E. Observatory Rd., Los Angeles, CA 90027) A small magazine specializing in historical topics.

Mercury Magazine (Astronomical Society of the Pacific, 390 Ashton Ave., San Francisco, CA 94112) The magazine of the largest public education society in astronomy with many interesting features.

Planetary Report (The Planetary Society, 65 N. Catalina Ave., Pasadena, CA 91106) The magazine of a large membership organization that promotes the exploration of the solar system.

Stardate Magazine (McDonald Observatory, RLM 15.308, University of Texas, Austin, TX 78712) Accompanies the popular public radio program on astronomy.

Sky & Telescope (P.O. Box 9111, Belmont, MA 02178) Found in most libraries, this is the popular astronomy magazine "of record" with many excellent articles on astronomy and amateur astronomy.

The following magazines (found in most libraries) contain frequent reports on astronomical developments: *Discover, National Geographic, Science News,* and *Scientific American.*

2. Organizations for Astronomy Enthusiasts

American Association of Variable Star Observers, 25 Birch St., Cambridge, MA 02138 (617-354-0484). An organization of amateur astronomers devoted to making serious observations of stars whose brightness changes.

Astronomical League, c/o Berton Stevens, 2112 Kingfisher Lane E., Rolling Meadows, IL 60008 (708-398-0562). Umbrella organization of amateur astronomy clubs in the U.S. Write to them to find out where the club closest to you is located. (First check with your instructor or the September issue of *Sky & Telescope* each year.) If you write to this all-volunteer organization, be sure to enclose a stamped self-addressed envelope.

Astronomical Society of the Pacific, 390 Ashton Ave., San Francisco, CA 94112 (415-337-1100). This international organization brings together scientists, teachers, and people with a general interest in astronomy; its name is merely a reminder of its origins on the West Coast of the U.S. in 1889. Write for their catalog of interesting astronomy materials or information about their national meetings and astronomy expos.

Committee for the Scientific Investigation of Claims of the Paranormal (CSICOP), P. O. Box 703, Buffalo, NY 14226 (716-636-1425). An organization of scientists, educators, magicians, and other skeptics that seeks to inform the public about the rational perspective on such pseudosciences as astrology, UFOs, psychic power, etc. Publishes *The Skeptical Inquirer* magazine, full of great debunking articles, and holds meetings and workshops around the world.

International Dark Sky Association, c/o David Crawford, 3545 N. Stewart, Tucson, AZ 85716. Small non-profit organization devoted to fighting light pollution and educating politicians, lighting engineers, and the public about the importance of not spilling light where it will interfere with astronomical observations.

The Planetary Society, 65 N. Catalina Ave., Pasadena, CA 91106 (818-793-1675). Large national membership organization founded by Carl Sagan and others; lobbies for more planetary exploration and SETI; publishes a colorful magazine and has a catalog of reasonably priced slides, videos, and gift items.

The Royal Astronomical Society of Canada, 136 Dupont St., Toronto, Ontario M5R 1V2 (416-924-7973). The main organization of amateur astronomers in Canada, with local centers in each province. Write for information on contacting the center near you.

3. Selected Institutions with Services and Materials for the Public

Jet Propulsion Laboratory, Public Information Office, 4800 Oak Grove Dr., Pasadena, CA 91109. This NASA center will usually respond to public inquiries about planetary missions with information pamphlets and lithographs.

NASA Headquarters, Public Information Branch, Washington, DC 20546. You can obtain information about NASA missions and projects. Leave plenty of time to get a response.

Space Telescope Science Institute, Public Information Office, 3700 San Martin Dr., Baltimore, MD 21218. Will often respond to intelligently worded requests for information or pictures from the Hubble Space Telescope.

Appendix 3
Glossary

NOTE: The number in parentheses is the chapter where the term is first defined or explained (P means the prologue; E means the epilogue).

absolute brightness (magnitude) *See* luminosity.

absolute zero A temperature of −273°C (or 0 K), where all molecular motion stops. (4)

absorption spectrum Dark lines superimposed on a continuous spectrum. (4)

accelerate To change velocity; to speed up, slow down, or change direction. (1)

accretion Gradual accumulation of mass, as by a planet forming by the building up of colliding particles in the solar nebula or gas falling into a black hole. (13)

accretion disk A disk of matter spiraling in toward a massive object; the disk shape is the result of conservation of angular momentum. (23)

active galactic nucleus A galaxy is said to have an active nucleus if unusually violent events are taking place in its center, emitting large quantities of electromagnetic radiation. Seyfert galaxies and quasars are examples of galaxies with active nuclei. (26)

active region Areas on the Sun where magnetic fields are concentrated; sunspots, prominences, and flares all tend to occur in active regions. (14)

albedo The fraction of incident sunlight that a planet or minor planet reflects.

alpha particle The nucleus of a helium atom, consisting of two protons and two neutrons.

amplitude The range in variability, as in the light from a variable star.

angular diameter Angle subtended by the diameter of an object.

angular momentum A measure of the momentum associated with motion about an axis or fixed point. (2)

antapex (solar) Direction away from which the Sun is moving with respect to the local standard of rest.

Antarctic Circle Parallel of latitude 66°30′S; at this latitude the noon altitude of the Sun is 0° on the date of the summer solstice.

antimatter Matter consisting of antiparticles: antiprotons (protons with negative rather than positive charge), positrons (positively charged electrons), and antineutrons. (15)

aperture The diameter of an opening, or of the primary lens or mirror of a telescope. (5)

apex (solar) The direction toward which the Sun is moving with respect to the local standard of rest.

aphelion Point in its orbit where a planet (or other body) is farthest from the Sun. (2)

apogee Point in its orbit where an Earth satellite is farthest from the Earth. (2)

apparent brightness A measure of the observed light received from a star or other object at the Earth; i.e., how bright an object appears in the sky, as contrasted with its luminosity. (4)

Arctic Circle Parallel of latitude 66°30′N; at this latitude the noon altitude of the Sun is 0° on the date of the winter solstice.

array (interferometer) A group of several telescopes that is used to make observations at high angular resolution. (5)

association A loose group of young stars whose spectral types, motions, or positions in the sky indicate that they have probably had a common origin. (21)

asterism An especially noticeable star pattern in the sky, such as the Big Dipper. (1)

asteroid An object orbiting the Sun that is smaller than a major planet, but that shows no evidence of an atmosphere or of other types of activity associated with comets. Also called a minor planet. (6)

asteroid belt The region of the solar system between the orbits of Mars and Jupiter in which most asteroids are located. The main belt, where the orbits are generally the most stable, extends from 2.2 to 3.3 AU from the Sun. (11)

astrology The pseudoscience that deals with the supposed influences on human destiny of the configurations and locations in the sky of the Sun, Moon, and planets; a primitive belief system that had its origin in ancient Babylonia. (1)

astronomical unit (AU) Originally meant to be the semimajor axis of the orbit of the Earth; now defined as the semimajor axis of the orbit of a hypothetical body with the mass and period that Gauss assumed for the Earth. The semimajor axis of the orbit of the Earth is 1.000000230 AU. (2)

atom The smallest particle of an element that retains the properties that characterize that element.

atomic mass unit *Chemical:* one-sixteenth of the mean mass of an oxygen atom. *Physical:* one-twelfth of the mass of an atom of the most common isotope of carbon. The atomic mass unit is approximately the mass of a hydrogen atom, 1.67×10^{-27} kg.

atomic number The number of protons in each atom of a particular element.

atomic weight The mean mass of an atom of a particular element in atomic mass units.

aurora Light radiated by atoms and ions in the ionosphere, mostly in the magnetic polar regions. (7)

autumnal equinox The intersection of the ecliptic and celestial equator where the Sun crosses the equator from north to south. A time when every place on Earth has 12 hours of daylight and 12 hours of night. (3)

axis An imaginary line about which a body rotates.

azimuth The angle along the celestial horizon, measured eastward from the north point to the intersection of the horizon with the vertical circle passing through an object.

Balmer lines Emission or absorption lines in the spectrum of hydrogen that arise from transitions between the second (or first excited) and higher energy states of the hydrogen atom. (4)

bands (in spectra) Emission or absorption lines, usually in the spectra of chemical compounds, so numerous and closely spaced that they coalesce into broad emission or absorption bands.

bar A force of 100,000 newtons acting on a surface area of 1 square meter is equal to 1 bar. The average pressure of the Earth's atmosphere at sea level is equal to 1.013 bars. (7)

barred spiral galaxy Spiral galaxy in which the spiral arms begin from the ends of a "bar" running through the nucleus rather than from the nucleus itself. (25)

barycenter The center of mass of two mutually revolving bodies.

basalt Igneous rock, composed primarily of silicon, oxygen, iron, aluminum, and magnesium produced by the cooling of lava. Basalts make up most of Earth's oceanic crust and are also found on other planets that have experienced extensive volcanic activity. (7)

Big Bang theory A theory of cosmology in which the expansion of the universe is presumed to have begun with a primeval explosion. (28)

binary star Two stars revolving about each other. (17)

binding energy The energy required to separate completely the constituent parts of an atomic nucleus. (15)

blackbody A hypothetical perfect radiator, which absorbs and re-emits all radiation incident upon it. (4)

black dwarf A presumed final state of evolution for a low-mass star, in which all of its energy sources are exhausted and it no longer emits significant radiation. (22)

black hole A collapsed massive star (or other collapsed body) whose velocity of escape is equal to or greater than the speed of light; thus no radiation can escape from it. (23)

Bohr atom A particular model of an atom, invented by Niels Bohr, in which the electrons are described as revolving about the nucleus in circular orbits. (4)

brown dwarf An object intermediate in size between a planet and a star. The approximate mass range is from about 1/100 of the mass of the Sun up to the lower mass limit for self-sustaining nuclear reactions, which is 0.08 solar mass. (17)

carbonaceous meteorite A primitive meteorite made primarily of silicates but often including chemically bound water, free carbon, and complex organic compounds. Also called carbonaceous chondrites. (13)

carbon-nitrogen-oxygen (CNO) cycle A series of nuclear reactions in the interiors of stars involving carbon as a catalyst, by which hydrogen is transformed to helium. (15)

Cassegrain focus An optical arrangement in a reflecting telescope in which light is reflected by a second mirror to a point behind the primary mirror.

CBR *See* cosmic background radiation.

CCD *See* charge-coupled device.

cD galaxy A supergiant elliptical galaxy frequently found at the center of a cluster of galaxies.

celestial equator A great circle on the celestial sphere 90° from the celestial poles; where the celestial sphere intersects the plane of the Earth's equator. (1)

celestial meridian An imaginary line on the celestial sphere passing through the north and south points on the horizon and through the zenith.

celestial poles Points about which the celestial sphere appears to rotate; intersections of the celestial sphere with the Earth's polar axis. (1)

celestial sphere Apparent sphere of the sky; a sphere of large radius centered on the observer. Directions of objects in the sky can be denoted by their position on the celestial sphere.

center of gravity Center of mass.

center of mass The average position of the various mass elements of a body or system, weighted according to their distances from that center of mass; that point in an isolated system that moves with constant velocity, according to Newton's first law of motion. (17)

cepheid variable A star that belongs to a class of yellow supergiant pulsating stars. These stars vary periodically in brightness, and the relationship between their periods and luminosities is useful in deriving distances to them. (18)

Chandrasekhar limit The upper limit to the mass of a white dwarf (equals 1.4 times the mass of the Sun). (22)

charge-coupled device (CCD) An array of electronic detectors of electromagnetic radiation, used at the focus of a telescope (or camera lens). A CCD acts like a photographic plate of very high sensitivity. (5)

chemical condensation sequence The calculated chemical compounds and minerals that would form at different temperatures in a cooling gas of cosmic composition; used to infer the composition of grains that formed in the solar nebula at different distances from the protosun. (13)

chromosphere That part of the solar atmosphere that lies immediately above the photospheric layers. (14)

circular satellite velocity The critical speed that a revolving body must have in order to follow a circular orbit. (2)

circumpolar zone Those portions of the celestial sphere near the celestial poles that are either always above or always below the horizon.

closed universe A model of the universe in which the curvature of space is such that straight lines eventually curve back upon themselves; in this model, the universe expands from a big bang, stops, and then contracts to a big crunch. (28)

cluster of galaxies A system of galaxies containing several to thousands of member galaxies. (27)

color index Difference between the magnitudes of a star or other object measured in light of two different spectral regions, for example, blue minus visual ($B - V$) magnitudes. (16)

coma (of comet) The diffuse gaseous component of the head of a comet; i.e, the cloud of evaporated gas around a comet nucleus. (12)

comet A small body of icy and dusty matter that revolves about the Sun. When a comet comes near the Sun, some of its material vaporizes, forming a large head of tenuous gas, and often a tail. (6)

compound A substance composed of two or more chemical elements.

conduction The transfer of energy by the direct passing of energy or electrons from atom to atom.

conservation of angular momentum The law that the total amount of angular momentum in a system remains the same (in the absence of any force not directed toward or away from the point or axis about which the angular momentum is referred).

constellation One of 88 sectors into which astronomers divide the celestial sphere; many constellations are named after a prominent group of stars within them that represents a person, animal, or legendary creature from ancient mythology. (P)

continental drift A gradual movement of the continents over the surface of the Earth due to plate tectonics. (7)

continuous spectrum A spectrum of light composed of radiation of a continuous range of wavelengths or colors rather than only certain discrete wavelengths. (4)

convection The transfer of energy by moving currents in a fluid. (7)

core (of a planet) The central part of a planet, consisting of higher density material. (7)

corona (of Galaxy) A region lying above and below the plane of the Galaxy out to much greater distances than the material that gives off electromagnetic radiation. (24)

corona (of Sun) Outer atmosphere of the Sun. (14)

coronal hole A region in the Sun's outer atmosphere where visible coronal radiation is absent. (14)

cosmic background radiation (CBR) The microwave radiation coming from all directions that is believed to be the redshifted glow of the Big Bang. (28)

cosmic rays Atomic nuclei (mostly protons) that are observed to strike the Earth's atmosphere with exceedingly high energies. (19)

cosmological constant A term in the equations of general relativity that represents a repulsive force in the universe. The cosmological constant is usually assumed to be zero. (28)

cosmological principle The assumption that, on the large scale, the universe at any given time is the same everywhere—isotropic and homogeneous. (27)

cosmology The study of the organization and evolution of the universe. (28)

crater A circular depression (from the Greek word for cup), generally of impact origin.

crescent moon One of the phases of the Moon when it appears less than half full.

critical density In cosmology, the density that provides enough gravity to bring the expansion of the universe just to a stop after infinite time. (28)

crust The outer layer of a terrestrial planet. (6)

dark matter Nonluminous mass, whose presence can be inferred only because of its gravitational influence on luminous matter. Dark matter may constitute as much as 99 percent of all the mass in the universe. The composition of the dark matter is not known. (24)

dark nebula A cloud of interstellar dust that obscures the light of more distant stars and appears as an opaque curtain.

declination Angular distance north or south of the celestial equator. (3)

degenerate gas A gas in which the allowable states for the electrons have been filled; it behaves according to different laws from those that apply to "perfect" gases and resists further compression. (22)

density The ratio of the mass of an object to its volume. (2)

deuterium A "heavy" form of hydrogen, in which the nucleus of each atom consists of one proton and one neutron. (15)

differential galactic rotation The rotation of the Galaxy, not as a solid wheel, but so that parts adjacent to one another do not always stay close together. (24)

differentiation (geological) A separation or segregation of different kinds of material in different layers in the interior of a planet. (6)

disk (of Galaxy) The central plane or "wheel" of our Galaxy, where most of the luminous mass is concentrated. (24)

dispersion Separation, from white light, of different wavelengths being refracted by different amounts. (4)

Doppler effect Apparent change in wavelength of the radiation from a source due to its relative motion away from or towards the observer. (4)

Earth-approaching asteroid An asteroid with an orbit that crosses the Earth's orbit or that will at some time cross the Earth's orbit as it evolves under the influence of the planets' gravity. *See also* near-Earth object.

eccentricity (of ellipse) Ratio of the distance between the foci to the major axis. (2)

eclipse The cutting off of all or part of the light of one body by another; in planetary science, the passing of one body into the shadow of another. (3)

eclipsing binary star A binary star in which the plane of revolution of the two stars is nearly edge-on to our line of sight, so that the light of one star is periodically diminished by the other passing in front of it. (17)

ecliptic The apparent annual path of the Sun on the celestial sphere. (1)

effective temperature *See* temperature (effective).

ejecta Material excavated from an impact crater, such as the blanket of material surrounding lunar craters and crater rays.

electromagnetic force One of the four fundamental forces or interactions of nature; the force that acts between charges and binds atoms and molecules together.

electromagnetic radiation Radiation consisting of waves propagated through the building up and breaking down of electric and magnetic fields; these include radio, infrared, light, ultraviolet, x rays, and gamma rays. (4)

electromagnetic spectrum The whole array or family of electromagnetic waves, from radio to gamma rays. (4)

electron A negatively charged subatomic particle that normally moves about the nucleus of an atom.

element A substance that cannot be decomposed, by chemical means, into simpler substances.

elementary particle One of the basic particles of matter. The most familiar of the elementary particles are the proton, neutron, and electron. (15)

ellipse A curve for which the sum of the distances from any point on the ellipse to two points inside (called the foci) is always the same. (2)

elliptical galaxy A galaxy whose appearance resembles a solid made of a series of ellipses and that contains no conspicuous interstellar material.

ellipticity The ratio (in an ellipse) of the major axis minus the minor axis to the major axis.

emission line A discrete bright line in the spectrum. (4)

emission nebula A gaseous nebula that derives its visible light from the fluorescence of ultraviolet light from a star in or near the nebula.

emission spectrum A spectrum consisting of emission lines. (4)

energy level A particular level, or amount, of energy possessed by an atom or ion above the energy it possesses in its least energetic state; also used to refer to the states of energy an electron can have in an atom. (4)

epicycle A circular orbit of a body in the Ptolemaic system, the center of which revolves about another circle (the deferent). (1)

equator A great circle on the Earth, 90° from (or equidistant from) each pole. (1)

equinox One of the intersections of the ecliptic and celestial equator; one of the two times during the year when the length of the day and night are the same. (3)

equivalence principle Principle that a gravitational force and a suitable acceleration are indistinguishable within a sufficiently local environment. (23)

escape velocity The velocity a body must achieve to break away from the gravity of another body and never return to it. (2)

eucrite meteorite One of a class of basaltic meteorites believed to have originated on the asteroid Vesta.

event horizon The surface through which a collapsing star passes when its velocity of escape is equal to the speed of light; that is, when the star becomes a black hole. (23)

excitation The process of imparting to an atom or an ion an amount of energy greater than it has in its normal or least-energy state. (4)

exclusion principle *See* Pauli exclusion principle.

extinction Reduction of the light from a celestial body produced by the Earth's atmosphere, or by interstellar absorption.

extragalactic Beyond our own Milky Way Galaxy.

eyepiece A magnifying lens used to view the image produced by the objective of a telescope.

fall (of meteorites) Meteorites seen in the sky and recovered on the ground. (13)

fault In geology, a crack or break in the crust of a planet along which slippage or movement can take place, accompanied by seismic activity. (7)

field A mathematical description of the effect of forces, such as gravity, that act on distant objects. For example, a given mass produces a gravitational field in the space surrounding it, which produces a gravitational force on objects within that space.

find (of meteorites) A meteorite that has been recovered but was not seen to fall. (13)

fireball A spectacular meteor, seen for more than an instant in the sky. (13)

fission The breakup of a heavy atomic nucleus into two or more lighter ones. (15)

flare A sudden and temporary outburst of light from an extended region of the Sun's surface. (14)

fluorescence The absorption of light of one wavelength and re-emission of it at another wavelength; especially the conversion of ultraviolet into visible light.

flux The rate at which energy or matter crosses a unit area of a surface.

focal length The distance from a lens or mirror to the point where light converged by it comes to a focus. (5)

focus (of ellipse) One of two fixed points inside an ellipse from which the sum of the distances to any point on the ellipse is a constant. (2)

focus (of telescope) Point where the rays of light converged by a mirror or lens meet.

forbidden lines Spectral lines that are not usually observed under laboratory conditions because they result from atomic transitions that are highly improbable.

force That which can change the momentum of a body; numerically, the rate at which the body's momentum changes. (2)

Fraunhofer line An absorption line in the spectrum of the Sun or of a star.

Fraunhofer spectrum The array of absorption lines in the spectrum of the Sun or a star.

frequency Number of vibrations per unit time; number of waves that cross a given point per unit time (in radiation). (4)

fusion The building up of heavier atomic nuclei from lighter ones. (15)

galactic cannibalism The process by which a larger galaxy strips material from a smaller one. (27)

galactic cluster An "open" cluster of stars located in the spiral arms or disk of the Galaxy. (21)

galaxy A large assemblage of stars; a typical galaxy contains millions to hundreds of billions of stars.

Galaxy The galaxy to which the Sun and our neighboring stars belong; the Milky Way is light from remote stars in the disk of the Galaxy. (24)

gamma rays Photons (of electromagnetic radiation) of energy higher than those of x rays; the most energetic form of electromagnetic radiation. (4)

general relativity theory Einstein's theory relating acceleration, gravity, and the structure (geometry) of space and time. (23)

geocentric Centered on the Earth.

giant (star) A star of large luminosity and radius. (16)

giant molecular cloud Large, cold interstellar clouds, with diameters of dozens of light years and typical masses of 10^5 solar masses; found in the spiral arms of galaxies, these clouds are where massive stars form. (20)

globular cluster One of about 120 large spherical star clusters that form a system of clusters centered on the center of the Galaxy. (21)

grand unified theories (GUTs) Physical theories that attempt to describe the four interactions (forces) of nature as different manifestations of a single force. (28)

granite The type of igneous silicate rock that makes up most of the continental crust of the Earth. (7)

granulation The rice-grain-like structure of the solar photosphere; granulation is produced by upwelling currents of gas that are slightly hotter, and therefore brighter, than the surrounding regions, which are flowing downward into the Sun. (14)

gravity The mutual attraction of material bodies or particles.

gravitational energy Energy that can be released by the gravitational collapse, or partial collapse, of a system; i.e., by particles that fall in toward the center of gravity.

gravitational lens A configuration of celestial objects, one of which provides one or more images of the other by gravitationally deflecting its light. (27)

gravitational redshift The redshift of electromagnetic radiation caused by a gravitational field. The slowing of clocks in a gravitational field. (23)

great circle Circle on the surface of a sphere that is the curve of intersection of the sphere with a plane passing through its center. (3)

greenhouse effect The blanketing (absorption) of infrared radiation near the surface of a planet by, for example, carbon dioxide in its atmosphere. (7)

ground state The lowest energy state of an atom.

H or H$_0$ *See* Hubble constant.

H I region Region of neutral hydrogen in interstellar space. (19)

H II region Region of ionized hydrogen in interstellar space. (19)

half-life The time required for half of the radioactive atoms in a sample to disintegrate. (6)

halo (of galaxy) The outermost extent of our Galaxy or another, containing a sparse distribution of stars and globular clusters in a more or less spherical distribution. (21)

heavy elements In astronomy, usually those elements of greater atomic number than helium.

helio- Prefix referring to the Sun.

heliocentric Centered on the Sun. (1)

helium flash The nearly explosive ignition of helium in the triple-alpha process in the dense core of a red giant star. (21)

Herbig-Haro (HH) object Luminous knots of gas in an area of star formation, which are set to glow by jets of material from a protostar. (20)

hertz A unit of frequency: one cycle per second. Named for Heinrich Hertz, who first produced radio radiation.

Hertzsprung–Russell (H–R) diagram A plot of luminosity against surface temperature (or spectral type) for a group of stars. (17)

highlands (lunar) The older, heavily cratered crust of the Moon, covering 83 percent of its surface and composed in large part of anorthositic breccias. (8)

homogeneous Having a consistent and even distribution of matter that is the same everywhere. (27)

horizon (astronomical) A great circle on the celestial sphere 90° from the zenith; more popularly, the circle around us where the dome of the sky meets the Earth. (1)

horoscope A chart used by astrologers, showing the positions along the zodiac and in the sky of the Sun, Moon, and planets at some given instant and as seen from a particular place on Earth—usually corresponding to the time and place of a person's birth. (1)

Hubble constant Constant of proportionality between the velocities of remote galaxies and their distances. The Hubble constant is thought to lie in the range of 15 to 30 km/s per million LY. (25)

Hubble law (or the law of the redshifts) The radial velocities of remote galaxies are proportional to their distances from us. (25)

hydrostatic equilibrium A balance between the weights of various layers, as in a star or the Earth's atmosphere, and the pressures that support them. (15)

hypothesis A tentative theory or supposition, advanced to explain certain facts or phenomena, which is subject to further tests and verification. (P)

igneous rock Any rock produced by cooling from a molten state. (7)

inclination (of an orbit) The angle between the orbital plane of a revolving body and some fundamental plane—usually the plane of the celestial equator or of the ecliptic.

inertia The property of matter that requires a force to act on it to change its state of motion; the tendency of objects to continue doing what they are doing in the absence of outside forces. (2)

inertial system A system of coordinates that is not itself accelerated, but that either is at rest or is moving with constant velocity.

inflationary universe A theory of cosmology in which the universe is assumed to have undergone a phase of very rapid expansion during the first 10^{-30} s. After this period of rapid expansion, the Big Bang and inflationary models are identical. (28)

infrared cirrus Patches of interstellar dust, which emit infrared radiation and look like cirrus clouds on the images of the sky produced by the Infrared Astronomy Satellite. (19)

infrared radiation Electromagnetic radiation of wavelength longer than the longest (red) wavelengths that can be perceived by the eye, but shorter than radio wavelengths. (4)

interference A phenomenon of waves that mix together such that their crests and troughs can alternately reinforce and cancel one another. (5)

international date line An arbitrary line on the surface of the Earth near longitude 180° across which the date changes by one day. (3)

interstellar dust Tiny solid grains in interstellar space, thought to consist of a core of rock-like material (silicates) or graphite surrounded by a mantle of ices. Water, methane, and ammonia are probably the most abundant ices. (19)

interstellar extinction The attenuation or absorption of light by dust in the interstellar medium. (19)

interstellar medium *or* interstellar matter Interstellar gas and dust. (19)

inverse-square law (for light) The amount of energy (light) flowing through a given area in a given time (flux) decreases in proportion to the square of the distance from the source of energy or light. (4)

ion An atom that has become electrically charged by the addition or loss of one or more electrons. (4)

ionization The process by which an atom gains or loses electrons.

ionosphere The upper region of the Earth's atmosphere in which many of the atoms are ionized.

ion tail (of comet) *See* plasma.

irregular galaxy A galaxy without rotational symmetry; neither a spiral nor an elliptical galaxy. (25)

irregular satellite A planetary satellite with an orbit that is retrograde, or of high inclination or eccentricity.

isotope Any of two or more forms of the same element, whose atoms all have the same number of protons but different numbers of neutrons. (4)

isotropic The same in all directions. (27)

joule The metric unit of energy; the work done by a force of 1 newton (N) acting through a distance of 1 m.

jovian planet *or* giant planet Any of the planets Jupiter, Saturn, Uranus, and Neptune in our solar system, or planets of roughly that mass and composition in other planetary systems. (6)

kinetic energy Energy associated with motion.

Kuiper belt A reservoir of cometary material just beyond the orbit of Pluto. (12)

laser An acronym for *l*ight *a*mplification by *s*timulated *e*mission of *r*adiation; a device for amplifying a light signal at a particular wavelength into a coherent beam.

latitude A north-south coordinate on the surface of the Earth; the angular distance north or south of the equator measured along a meridian passing through a place. (3)

law of areas Kepler's second law: the radius vector from the Sun to any planet sweeps out equal areas in the planet's orbital plane in equal intervals of time. (2)

law of the redshifts *See* Hubble law.

leap year A calendar year with 366 days, inserted approximately every 4 years to make the average length of the calendar year as nearly equal as possible to the tropical year. (3)

light *or* **visible light** Electromagnetic radiation that is visible to the eye.

light curve A graph that displays the time variation of the light from a variable or eclipsing binary star. (17)

light year The distance light travels in a vacuum in one year; 1 LY = 9.46×10^{12} km, or about 6×10^{12} mi. (P)

line broadening The phenomenon by which spectral lines are not precisely sharp but have finite widths.

line profile A plot of the intensity of light versus wavelength across a spectral line.

Local Group The cluster of galaxies to which our Galaxy belongs. (28)

local standard of rest A coordinate system that shares the average motion of the Sun and its neighboring stars about the galactic center.

Local Supercluster The supercluster of galaxies to which the Local Group belongs. (28)

longitude An east-west coordinate on the Earth's surface; the angular distance, measured east or west along the equator, from the Greenwich meridian to the meridian passing through a place. (3)

luminosity The rate at which a star or other object emits electromagnetic energy into space. (16)

luminosity class A classification of a star according to its luminosity within a given spectral class. Our Sun, a G2V star, has luminosity class V. (18)

luminosity function The relative numbers of stars (or other objects) of various luminosities. (17)

lunar eclipse An eclipse of the Moon. (3)

Lyman lines A series of absorption or emission lines in the spectrum of hydrogen that arise from transitions to and from the lowest energy states of the hydrogen atoms.

Magellanic Clouds Two neighboring galaxies visible to the naked eye from southern latitudes.

magma Mobile, high-temperature molten state of rock, usually of silicate mineral composition and with dissolved gases and other volatiles. (7)

magnetic field The region of space near a magnetized body within which magnetic forces can be detected.

magnetic pole One of two points on a magnet (or the Earth) at which the greatest density of lines of force emerge. A compass needle aligns itself along the local lines of force on the Earth and points more or less toward the magnetic poles of the Earth.

magnetosphere The region around a planet in which its intrinsic magnetic field dominates the interplanetary field carried by the solar wind; hence, the region within which charged particles can be trapped by the planetary magnetic field. (7)

magnitude A measure of the amount of light flux received from a star or other luminous object. (1)

main sequence A sequence of stars on the Hertzsprung–Russell diagram, containing the majority of stars, that runs diagonally from the upper left to the lower right. (17)

major axis (of ellipse) The maximum diameter of an ellipse. (2)

mantle (of Earth) The greatest part of the Earth's interior, lying between the crust and the core. (7)

mare (pl. maria) Latin for "sea"; name applied to the dark, relatively smooth features that cover 17 percent of the Moon. (8)

mass A measure of the total amount of material in a body; defined either by the inertial properties of the body or by its gravitational influence on other bodies.

mass extinction The sudden disappearance in the fossil record of a large number of species of life, to be replaced by new species in subsequent layers. Mass extinctions are indications of catastrophic changes in the environment, such as might be produced by a large impact on the Earth. (7)

mass-light ratio The ratio of the total mass of a galaxy to its total luminosity, usually expressed in units of solar mass and solar luminosity. The mass-light ratio gives a rough indication of the types of stars contained within a galaxy and whether or not substantial quantities of dark matter are present. (26)

mass-luminosity relation An empirical relation between the masses and luminosities of many (principally main-sequence) stars. (17)

Maunder Minimum The interval from 1645 to 1715 when solar activity was very low. (14)

mean solar day Average length of the apparent solar day.

merger (of galaxies) When galaxies (of roughly comparable size) collide and form one combined structure. (27)

meridian (celestial) The great circle on the celestial sphere that passes through an observer's zenith and the north (or south) celestial pole. (3)

meridian (terrestrial) The great circle on the surface of the Earth that passes through a particular place and the North and South Poles of the Earth.

Messier catalog A catalog of nonstellar objects compiled by Charles Messier in 1787 (includes nebulae, star clusters, and galaxies).

metals In general, any element or compound whose electron structure makes it a good conductor of electricity. In astronomy, all elements beyond hydrogen and helium. (16)

metamorphic rock Any rock produced by the physical and chemical alteration (without melting) of another rock that has been subjected to high temperature and pressure. (7)

metastable level An energy level in an atom from which there is a low probability of an atomic transition accompanied by the radiation of a photon.

meteor The luminous phenomenon observed when a small piece of solid matter enters the Earth's atmosphere and burns up; popularly called a "shooting star." (13)

meteorite A portion of a meteoroid that survives passage through the atmosphere and strikes the ground. (6)

meteoroid A particle or chunk of typically rocky or metallic material in space before any encounter with the Earth.

meteor shower Many meteors appearing to radiate from a common point in the sky caused by the collision of the Earth with a swarm of solid particles, typically from a comet. (13)

micron Old term for micrometer (10^{-6} meter).

microwave Shortwave radio wavelengths.

Milky Way The band of light encircling the sky, which is due to the many stars and diffuse nebulae lying near the plane of the Galaxy. (24)

minerals The solid compounds (often primarily silicon and oxygen) that form rocks.

minor planet *See* asteroid.

model atmosphere *or* **photosphere** The result of a theoretical calculation of the run of temperature, pressure, density, and so on, through the outer layers of the Sun or a star.

molecule A combination of two or more atoms bound together; the smallest particle of a chemical compound or substance that exhibits the chemical properties of that substance.

momentum A measure of the inertia or state of motion of a body; the momentum of a body is the product of its mass and velocity. In the absence of a force, momentum is conserved. (2)

near-Earth object (NEO) A comet or asteroid whose path intersects the orbit of the Earth. (11)

nebula Cloud of interstellar gas or dust. (19)

neutrino A fundamental particle that has little or no rest mass and no charge but that does have spin and energy. Neutrinos rarely interact with ordinary matter. (15)

neutron A subatomic particle with no charge and with mass approximately equal to that of the proton.

neutron star A star of extremely high density composed almost entirely of neutrons. (12)

nonthermal radiation *See* synchrotron radiation.

nova A star that experiences a sudden outburst of radiant energy, temporarily increasing its luminosity by hundreds to thousands of times. (22)

nuclear Referring to the nucleus of the atom.

nuclear bulge Central part of our Galaxy.

nuclear transformation The change of one atomic nucleus into another, as in nuclear fusion. (15)

nucleosynthesis The building up of heavy elements from lighter ones by nuclear fusion. (21)

nucleus (of atom) The heavy part of an atom, composed mostly of protons and neutrons, and about which the electrons revolve. (4)

nucleus (of comet) The solid chunk of ice and dust in the head of a comet. (11)

nucleus (of galaxy) Central concentration of matter at the center of a galaxy. (24)

occultation The passage of an object of large angular size in front of a smaller object, such as the Moon in front of a distant star or the rings of Saturn in front of the Voyager spacecraft. (11)

Oort comet cloud The large spherical region around the Sun from which most "new" comets come; a reservoir of objects with aphelia at about 50,000 AU, or extending about a third of the way to the nearest other stars. (11)

opacity Absorbing power; capacity to impede the passage of light. (15)

open cluster A comparatively loose or "open" cluster of stars, containing from a few dozen to a few thousand members, located in the spiral arms or disk of the Galaxy; sometimes referred to as a galactic cluster. (21)

open universe A model of the universe in which gravity is not strong enough to bring the universe to a halt; it expands forever. In this model the geometry of spacetime is such that if you go in a straight line, you not only can never return to where you started, but even more space opens up than you would expect from Euclidean geometry. (28)

optical In astronomy: relating to the visible-light band of the electromagnetic spectrum. Optical observations are those made with visible light.

optical double star Two stars at different distances that are seen nearly lined up in projection so that they appear close together, but that are not really gravitationally associated.

orbit The path of a body that is in revolution about another body or point.

oscillation A periodic motion; in the case of the Sun, a periodic or quasi-periodic expansion and contraction of the whole Sun or some portion of it. (14)

ozone A heavy molecule of oxygen that contains three atoms rather than the more normal two. Designated O_3.

parabola A conic section of eccentricity 1.0; the curve of the intersection between a circular cone and a plane parallel to a straight line in the surface of the cone.

parallax An apparent displacement of a nearby star that results from the motion of the Earth around the Sun; numerically, the angle subtended by 1 AU at the distance of a particular star. (18)

parsec A unit of distance in astronomy, equal to 3.26 light years. At a distance of 1 parsec, a star has a parallax of one arcsecond.

Pauli exclusion principle Quantum mechanical principle by which no two particles of the same kind can have the same position and momentum.

peculiar velocity The velocity of a star with respect to the local standard of rest; that is, its space motion, corrected for the motion of the Sun with respect to our neighboring stars.

penumbra The outer, not completely dark part of a shadow; the region from which the source of light is not completely hidden. (3)

perfect radiator or blackbody A body that absorbs and subsequently re-emits all radiation incident upon it.

periastron The place in the orbit of a star in a binary-star system where it is closest to its companion star.

perigee The place in the orbit of an Earth satellite where it is closest to the center of the Earth. (2)

perihelion The place in the orbit of an object revolving about the Sun where it is closest to the Sun's center. (2)

period-luminosity relation An empirical relation between the periods and luminosities of certain variable stars. (18)

perturbation The disturbing effect, when small, on the motion of a body as predicted by a simple theory, produced by a third body or other external agent. (2)

photochemistry Chemical changes caused by electromagnetic radiation. (10)

photometry The measurement of light intensities.

photon A discrete unit of electromagnetic energy. (4)

photosphere The region of the solar (or a stellar) atmosphere from which continuous radiation escapes into space. (14)

pixel An individual picture element in a detector; for example, a particular silicon diode in a CCD.

plage A bright region of the solar surface observed in the monochromatic light of some spectral line. (14)

Planck's constant The constant of proportionality relating the energy of a photon to its frequency.

planet Any of the nine largest bodies revolving about the Sun, or any similar bodies that may orbit other stars. Unlike stars, planets do not (for the most part) give off their own light, but only reflect the light of their parent star. (1)

planetarium An optical device for projecting on a screen or domed ceiling the stars and planets and their apparent motions in the sky.

planetary nebula A shell of gas ejected from, and enlarging about, a certain kind of extremely hot star that is nearing the end of its life. (22)

planetesimals The hypothetical objects, from tens to hundreds of kilometers in diameter, that formed in the solar nebula as an intermediate step between tiny grains and the larger planetary objects we see today. The comets and some asteroids may be leftover planetesimals. (6)

plasma A hot ionized gas.

plate tectonics The motion of segments or plates of the outer layer of the Earth over the underlying mantle. (7)

polar axis The axis of rotation of the Earth; also, an axis in the mounting of a telescope that is parallel to the Earth's axis.

Population I and II Two classes of stars (and systems of stars), classified according to their spectral characteristics, chemical compositions, radial velocities, ages, and locations in the Galaxy. (24)

positron An electron with a positive rather than negative charge; an antielectron. (15)

potential energy Stored energy that can be converted into other forms; especially gravitational energy.

precession (of Earth) A slow, conical motion of the Earth's axis of rotation, caused principally by the gravitational pull of the Moon and Sun on the Earth's equatorial bulge. (1)

precession of the equinoxes Slow westward motion of the equinoxes along the ecliptic that results from precession.

pressure Force per unit area; expressed in units of atmospheres or pascals.

prime focus The point in a telescope where the objective focuses the light.

prime meridian The terrestrial meridian passing through the site of the old Royal Greenwich Observatory; longitude 0°. (3)

primitive In planetary science and meteoritics, an object or rock that is little changed, chemically, since its formation, and hence representative of the conditions in the solar nebula at the time of formation of the solar system. Also used to refer to the chemical composition of an atmosphere that has not undergone extensive chemical evolution. (7)

primitive meteorite A meteorite that has not been greatly altered chemically since its condensation from the solar nebula; called in meteoritics a chondrite (either ordinary chondrite or carbonaceous chondrite). (13)

primitive rock Any rock that has not experienced great heat or pressure and therefore remains representative of the original condensates from the solar nebula—never found on any object large enough to have undergone melting and differentiation. (7)

principle of equivalence Principle that a gravitational force and a suitable acceleration are indistinguishable within a sufficiently local environment. (23)

prism A wedge-shaped piece of glass that is used to disperse white light into a spectrum.

prominence A phenomenon in the solar corona that commonly appears like a flame above the limb of the Sun. (14)

proper motion The angular change per year in the direction of a star as seen from the Sun. (16)

proton A heavy subatomic particle that carries a positive charge; one of the two principal constituents of the atomic nucleus.

proton–proton cycle A series of thermonuclear reactions by which nuclei of hydrogen are built up into nuclei of helium. (15)

protoplanet *or* -star *or* -galaxy The original material from which a planet (or a star or galaxy) condensed. (20)

pulsar A variable radio source of small angular size that emits very rapid radio pulses in very regular periods that range from fractions of a second to several seconds. (22)

pulsating variable A variable star that pulsates in size and luminosity. (18)

quantum efficiency The ratio of the number of photons incident on a detector to the number actually detected.

quantum mechanics The branch of physics that deals with the structure of atoms and their interactions with one another and with radiation.

quasar An object of very high redshift that looks like a star, presumed to be extragalactic and highly luminous; an active galactic nucleus. (26)

radar The technique of transmitting radio waves to an object and then detecting the radiation that the object reflects back to the transmitter; used to measure the distance to, and motion of, a target object. (5)

radial velocity The component of relative velocity that lies in the line of sight; motion toward or away from the observer. (4)

radial velocity curve A plot of the variation of radial velocity with time for a binary or variable star.

radiant (of meteor shower) The point in the sky from which the meteors belonging to a shower seem to radiate. (13)

radiation A mode of energy transport whereby energy is transmitted through a vacuum; also the transmitted energy itself. (4, 15)

radiation pressure The transfer of momentum carried by electromagnetic radiation to a body that the radiation impinges upon.

radioactive dating The technique of determining the ages of rocks or other specimens by the amount of radioactive decay of certain radioactive elements contained therein; something you do when you are desperate on Saturday night. (6)

radioactivity (radioactive decay) The process by which certain kinds of atomic nuclei naturally decompose, with the spontaneous emission of subatomic particles and gamma rays. (6)

radio galaxy A galaxy that emits greater amounts of radio radiation than average. (26)

radio telescope A telescope designed to make observations in radio wavelengths.

reddening (interstellar) The reddening of starlight passing through interstellar dust, caused because dust scatters blue light more effectively than red. (19)

red giant A large, cool star of high luminosity; a star occupying the upper right portion of the Hertzsprung–Russell diagram.

redshift A shift to longer wavelengths of light, typically a Doppler shift caused by the motion of the source away from the observer.

reducing In chemistry, referring to conditions in which hydrogen dominates over oxygen, so that most other elements form compounds with hydrogen. In very reducing conditions free hydrogen (H_2) is present and free oxygen (O_2) cannot exist.

reflecting telescope A telescope in which the principal optical component (objective) is a concave mirror. (5)

reflection nebula A relatively dense dust cloud in interstellar space that is illuminated by reflected starlight. (19)

refracting telescope A telescope in which the principal optical component (objective) is a lens or system of lenses. (5)

refraction The bending of light rays passing from one transparent medium (or a vacuum) to another.

relativistic particle (or electron) A particle (electron) moving at nearly the speed of light.

relativity A theory formulated by Einstein that describes the relations between measurements of physical phenomena by two different observers who are in relative motion at constant velocity (the special theory of relativity) or that describes how a gravitational field can be replaced by a curvature of space-time (the general theory of relativity).

resolution The degree to which fine details in an image are separated, or the smallest detail that can be discerned in an image. (5)

resonance An orbital condition in which one object is subject to periodic gravitational perturbations by another, most commonly arising when two objects orbiting a third have periods of revolution that are simple multiples or fractions of each other. (11)

retrograde (rotation or revolution) Backward with respect to the common direction of motion in the solar system; counterclockwise as viewed from the north, and going from east to west rather than from west to east. (1)

retrograde motion An apparent westward motion of a planet on the celestial sphere or with respect to the stars.

revolution The motion of one body around another.

rift zone In geology, a place where the crust is being torn apart by internal forces, generally associated with the injection of new material from the mantle and with the slow separation of tectonic plates. (7)

right ascension A coordinate for measuring the east-west positions of celestial bodies; the angle measured eastward along the celestial equator from the vernal equinox to the hour circle passing through a body. (3)

rotation Turning of a body about an axis running through it.

RR Lyrae variable One of a class of giant pulsating stars with periods less than one day. (18)

runaway greenhouse effect A process whereby the heating of a planet leads to an increase in its atmospheric greenhouse effect and thus to further heating, thereby quickly altering the composition of its atmosphere and the temperature of its surface. (9)

satellite A body that revolves about a planet.

Schwarzschild radius *See* event horizon.

scientific method The procedure scientists follow to understand the natural world: (1) the observation of phenomena or the results of experiments; (2) the formulation of hypotheses that describe these phenomena and that are consistent with the body of knowledge available; (3) the testing of these hypotheses by noting whether or not they adequately predict and describe new phenomena or the results of new experiments; (4) the modification or rejection of hypotheses that are not confirmed by observations or experiment. (P)

sedimentary rock Any rock formed by the deposition and cementing of fine grains of material. (7)

seeing The unsteadiness of the Earth's atmosphere, which blurs telescopic images. Good seeing means the atmosphere is steady. (5)

seismic waves Vibrations traveling through the Earth's interior that result from earthquakes. (7)

seismology (solar) The study of small changes in the radial velocity of the Sun as a whole or of small regions on the surface of the Sun. Analyses of these velocity changes can be used to infer the internal structure of the Sun. (15)

seismology (terrestrial) The study of earthquakes, the conditions that produce them, and the internal structure of the Earth as deduced from analyses of seismic waves. (7)

semimajor axis Half the major axis of a conic section, such as an ellipse.

SETI The search for extraterrestrial intelligence, usually applied to searches for radio signals from other civilizations. (E)

Seyfert galaxy A galaxy belonging to the class of those with active galactic nuclei; one whose nucleus shows bright emission lines; one of a class of galaxies first described by C. Seyfert. (26)

shepherd satellite Informal term for a satellite that is thought to maintain the structure of a planetary ring through its close gravitational influence. (11)

sidereal period The period of revolution of one body about another measured with respect to the stars.

sidereal time Time on Earth measured with respect to the stars, rather than the Sun; the local hour angle of the vernal equinox. (3)

sidereal year Period of the Earth's revolution about the Sun with respect to the stars. (3)

sign (of zodiac) Astrological term for any of 12 equal sections along the ecliptic, each of length 30°. Because of precession, these signs today are no longer lined up with the constellations from which they received their names. (1)

singularity A theoretical point of zero volume and infinite density to which any object that becomes a black hole must collapse, according to the general theory of relativity. (23)

SNC meteorite One of a class of basaltic meteorites now believed by many planetary scientists to be impact-ejected fragments from Mars.

solar activity Phenomena of the solar atmosphere: sunspots, plages, and related phenomena. (14)

solar antapex Direction away from which the Sun is moving with respect to the local standard of rest.

solar apex The direction toward which the Sun is moving with respect to the local standard of rest.

solar eclipse An eclipse of the Sun by the Moon, caused by the passage of the Moon in front of the Sun. Solar eclipses can occur only at the time of new moon. (3)

solar motion Motion of the Sun, or the velocity of the Sun, with respect to the local standard of rest.

solar nebula The cloud of gas and dust from which the solar system formed. (6)

solar seismology The study of pulsations or oscillations of the Sun in order to determine the characteristics of the solar interior. (14)

solar system The system of the Sun and the planets, their satellites, the minor planets, comets, meteoroids, and other objects revolving around the Sun. (6)

solar time A time based on the Sun; usually the hour angle of the Sun plus 12 h. (3)

solar wind A flow of hot charged particles leaving the Sun. (14)

solstice Either of two points on the celestial sphere where the Sun reaches its maximum distances north and south of the celestial equator; time of the year when the daylight is the longest or the shortest. (3)

spacetime A system of one time and three spatial coordinates, with respect to which the time and place of an event can be specified. (23)

space velocity *or* space motion The velocity of a star with respect to the Sun.

spectral class (or type) The classification of stars according to their temperatures using the characteristics of their spectra; the types are O B A F G K M. (16)

spectral line Radiation at a particular wavelength of light produced by the emission or absorption of energy by an atom.

spectral sequence The sequence of spectral classes of stars arranged in order of decreasing temperatures of stars of those classes. (16)

spectrometer An instrument for obtaining a spectrum; in astronomy, usually attached to a telescope to record the spectrum of a star, galaxy, or other astronomical object. (4)

spectroscopic binary star A binary star in which the components are not resolved optically, but whose binary nature is indicated by periodic variations in radial velocity, indicating orbital motion. (17)

spectroscopic parallax A parallax (or distance) of a star that is derived by comparing the apparent magnitude of the star with its absolute magnitude as deduced from its spectral characteristics.

spectroscopy The study of spectra.

spectrum The array of colors or wavelengths obtained when light (or other radiation) from a source is dispersed, as in passing it through a prism or grating. (4)

speed The rate at which an object moves without regard to its direction of motion; the numerical or absolute value of velocity.

spicule A jet of rising material in the solar chromosphere. (14)

spiral arms Arms (or long denser regions) of interstellar material and young stars that wind out in a plane from the central nucleus of a spiral galaxy. (24)

spiral density wave A mechanism for the generation of spiral structure in galaxies; a density wave interacts with interstellar matter and triggers the formation of stars. Spiral density waves are also seen in the rings of Saturn.

spiral galaxy A flattened, rotating galaxy with pinwheel-like arms of interstellar material and young stars winding out from its nucleus. (25)

spring tide The highest tidal range of the month, produced when the Moon is near either the full or the new phase. (3)

standard bulb An astronomical object of known luminosity; such an object can be used to determine distances.

star A sphere of gas shining under its own power.

star cluster An assemblage of stars held together by their mutual gravity.

Stefan-Boltzmann law A formula from which the rate at which a blackbody radiates energy can be computed; the total rate of energy emission from a unit area of a blackbody is proportional to the fourth power of its absolute temperature.

stellar evolution The changes that take place in the characteristics of stars as they age.

stellar model The result of a theoretical calculation of the physical conditions in the different layers of a star's interior.

stellar parallax *See* parallax.

stellar wind The outflow of gas, sometimes at speeds as high as hundreds of kilometers per second, from a star.

stony-iron meteorite A type of meteorite that is a blend of nickel-iron and silicate materials. (13)

stony meteorite A meteorite composed mostly of stony material. (13)

stratosphere The layer of the Earth's atmosphere above the troposphere (where most weather takes place) and below the ionosphere. (7)

strong nuclear force *or* strong interaction The force that binds together the parts of the atomic nucleus. (15)

subduction zone In terrestrial geology, a region where one crustal plate is forced under another, generally associated with earthquakes, volcanic activity, and the formation of deep ocean trenches. (7)

summer solstice The point on the celestial sphere where the Sun reaches its greatest distance north of the celestial equator; the day with the longest amount of daylight. (3)

Sun The star about which the Earth and other planets revolve.

sunspot A temporary cool region in the solar photosphere that appears dark by contrast against the surrounding hotter photosphere. (14)

sunspot cycle The semiregular 11-year period with which the frequency of sunspots fluctuates. (14)

supercluster A large region of space (more than 100 million LY across) where groups and clusters of galaxies are more concentrated; a cluster of clusters of galaxies. (27)

supergiant A star of very high luminosity and relatively low temperature. (17)

supernova An explosion that marks the final stage of evolution of a star. A Type I supernova occurs when a white dwarf accretes enough matter to exceed the Chandrasekhar limit, collapses, and explodes. A Type II supernova marks the final collapse of a massive star. (22)

surface gravity The weight of a unit mass at the surface of a body.

synchrotron radiation The radiation emitted by charged particles being accelerated in magnetic fields and moving at speeds near that of light.

tail (of a comet) *See* dust tail of a comet *and* plasma.

tangential (transverse) velocity The component of a star's space velocity that lies in the plane of the sky.

tectonic Activity and motion that result from expansion and contraction of the crust of a planet.

temperature A measure of how fast the particles in a body are moving or vibrating in place; a measure of the average heat energy in a body.

temperature (Celsius; formerly centigrade) Temperature measured on scale where water freezes at 0° and boils at 100°.

temperature (color) The temperature of a star as estimated from the intensity of the stellar radiation at two or more colors or wavelengths.

temperature (effective) The temperature of a blackbody that would radiate the same total amount of energy that a particular object, such as a star, does.

temperature (excitation) The temperature of a star as estimated from the relative strengths of lines in its spectrum that originate from atoms in different stages of excitation.

temperature (Fahrenheit) Temperature measured on a scale where water freezes at 32° and boils at 212°.

temperature (ionization) The temperature of a star as estimated from the relative strengths of lines in its spectrum that originate from atoms in different stages of ionization.

temperature (Kelvin) Absolute temperature measured in Celsius degrees, with the zero point at absolute zero.

temperature (radiation) The temperature of a blackbody that radiates the same amount of energy in a given spectral region as does a particular body.

terrestrial planet Any of the planets Mercury, Venus, Earth, or Mars; sometimes the Moon or Pluto are included in the list. (6)

theory A set of hypotheses and laws that have been well demonstrated to apply to a wide range of phenomena associated with a particular subject.

thermal energy Energy associated with the motions of the molecules or atoms in a substance.

thermal equilibrium A balance between the input and outflow of heat in a system.

thermal radiation The radiation emitted by any body or gas that is not at absolute zero.

thermonuclear energy Energy associated with thermonuclear reactions or that can be released through thermonuclear reactions.

thermonuclear reaction A nuclear reaction or transformation that results from encounters between particles that are given high velocities (by heating them). (15)

tidal force A differential gravitational force that tends to deform a body. (3)

tidal stability limit The distance—approximately 2.5 planetary radii from the center—within which differential gravitational forces (or tides) are stronger than the mutual gravitational attraction between two adjacent orbiting objects. Within this limit, fragments are not likely to accrete or assemble themselves into a larger object. Also called the Roche limit.

tide Deformation of a body by the differential gravitational force exerted on it by another body; in the Earth, the deformation of the ocean surface by the differential gravitational forces exerted by the Moon and Sun. (3)

transition region The region in the Sun's atmosphere where the temperature rises very rapidly from the relatively low temperatures that characterize the chromosphere to the high temperatures of the corona. (14)

triple-alpha process A series of two nuclear reactions by which three helium nuclei are built up into one carbon nucleus. (21)

tropical year Period of revolution of the Earth about the Sun with respect to the vernal equinox.

Tropic of Cancer The parallel (circle) of latitude 23.5° N.

Tropic of Capricorn The parallel (circle) of latitude 23.5° S.

troposphere Lowest level of the Earth's atmosphere, where most weather takes place. (7)

turbulence Random motions of gas masses, as in the atmosphere of a star.

21-cm line A line in the spectrum of neutral hydrogen at the radio wavelength of 21 cm. (19)

ultraviolet radiation Electromagnetic radiation of wavelengths shorter than the shortest visible wavelengths; radiation of wavelengths in the approximate range 10 to 400 nm. (4)

umbra The central, completely dark part of a shadow. (3)

uncertainty principle Heisenberg uncertainty principle. It is fundamentally impossible to make simultaneous measurements of a particle's position and velocity with infinite accuracy.

universe The totality of all matter, radiation and space; everything accessible to our observations.

variable star A star that varies in luminosity. (18)

velocity The speed and the direction a body is moving; e.g., 44 km/s toward the north galactic pole.

velocity of escape The speed with which an object must move in order to enter a parabolic orbit about another body (such as the Earth), and hence move permanently away from the vicinity of that body. (2)

vernal equinox The point on the celestial sphere where the Sun crosses the celestial equator passing from south to north; a time in the course of the year when the day and night are roughly equal. (3)

very-long-baseline interferometry (VLBI) A technique of radio astronomy whereby signals from telescopes thousands of kilometers apart combined to obtain very high resolution by letting waves from different sites interfere with each other. (5)

visual binary star A binary star in which the two components are telescopically resolved. (17)

void A region between clusters and superclusters of galaxies that appears relatively empty of galaxies. (27)

volatile materials Materials that are gaseous at fairly low temperatures. This is a relative term, usually applied to the gases in planetary atmospheres and to common ices (H_2O, CO_2, and so on), but it is also sometimes used for elements such as cadmium, zinc, lead, and rubidium that form gases at temperatures up to 1000 K. (These are called volatile elements, as opposed to refractory elements.)

volume A measure of the total space occupied by a body. (2)

watt A unit of power (energy per unit time).

wavelength The spacing of the crests or troughs in a wave. (4)

weak nuclear force *or* **weak interaction** The nuclear force involved in radioactive decay. The weak force is characterized by the slow rate of certain nuclear reactions—such as the decay of the neutron, which occurs with a half-life of 11 min.

weight A measure of the force due to gravitational attraction.

white dwarf A star that has exhausted most or all of its nuclear fuel and has collapsed to a very small size; such a star is near its final stage of life. (17)

Wien's law Formula that relates the temperature of a blackbody to the wavelength at which it emits the greatest intensity of radiation. (4)

winter solstice Point on the celestial sphere where the Sun reaches its greatest distance south of the celestial equator; the time of the year with the shortest amount of daylight. (3)

Wolf-Rayet star One of a class of very hot stars that eject shells of gas at very high velocity.

x rays Photons of wavelengths intermediate between those of ultraviolet radiation and gamma rays.

x-ray stars Stars (other than the Sun) that emit observable amounts of radiation at x-ray frequencies.

year The period of revolution of the Earth around the Sun. (1)

Zeeman effect A splitting or broadening of spectral lines due to magnetic fields. (14)

zenith The point on the celestial sphere opposite to the direction of gravity; or the direction opposite to that indicated by a plumb bob; the point directly above the observer. (1)

zero-age main sequence Main sequence on the H–R diagram for a system of stars that have completed their contraction from interstellar matter and are now deriving all their energy from nuclear reactions, but whose chemical composition has not yet been altered by nuclear reactions. (21)

zodiac A belt around the sky 18° wide centered on the ecliptic. (1)

zone of avoidance A region near the Milky Way where obscuration by interstellar dust is so heavy that few or no exterior galaxies can be seen.

Appendix 4
Powers-of-Ten Notation

In astronomy (and other sciences), it is often necessary to deal with very large or very small numbers. In fact, when numbers get truly large in everyday life, such as the national debt in the U.S., we call them astronomical. Among the ideas astronomers must routinely deal with is that the Earth is 150,000,000,000 m from the Sun, and the mass of the hydrogen atom is 0.00000000000000000000000000167 kg. No one in his or her right mind would want to continue writing so many zeros!

Instead, scientists have agreed on a kind of shorthand notation, which is not only easier to write, but (as we shall see) makes multiplication and division of large and small numbers much easier. If you have never used this *powers-of-ten notation* or *scientific notation,* it may take a bit of time to get used to it, but you will soon find it much easier than keeping all those zeros.

Writing Large Numbers

The convention in this notation is that we generally have *only one number to the left of the decimal point.* If a number is not in this format, it must be changed. The number 6 is already in the right format, because for integers, we understand there to be a decimal point to the right of them. So 6 is really 6., and there is indeed only one number to the left of the decimal point. But the number 165 (which is 165.) has three numbers to the left of the decimal point, and is thus ripe for conversion.

To change 165 to proper form, we must make it 1.65 and then keep track of the change we have made. (Think of the number as a weekly salary, and suddenly it makes a lot of difference whether we have $165 or $1.65.) We keep track of the number of places we moved the decimal point by expressing it as a power of ten. So 165 becomes 1.65×10^2 or 1.65 multiplied by ten to the second power. The small raised 2 is called an *exponent,* and it tells us how many times we moved the decimal point to the left.

Note that 10^2 also designates 10 squared or 10×10, which equals 100. And 1.65×100 is just 165, the number we started with. Another way to look at scientific notation is that we separate out the messy numbers out front, and leave the smooth units of ten for the exponent to denote. So a number like 1,372,568 becomes 1.372568 times a million (10^6) or 1.372568 times 10 multiplied by itself 6 times. We had to move the decimal point six places to the left (from its place after the 8) to get the number into the form where there is only one digit to the left of the decimal point.

The reason we call this powers-of-ten notation is that our counting system is based on increases of ten; each place in our numbering system is ten times greater than the place to the

right of it. As you have probably learned, this got started because human beings have ten fingers and we started counting with them. It is interesting to speculate that if we ever meet intelligent life-forms with only eight fingers, their counting system would probably be a powers-of-eight notation!

So, in the example we started with, the number of meters from the Earth to the Sun is 1.5×10^{11}. Elsewhere in the book, we mention that a string a light year long would fit around the Earth's equator 236 million or 236,000,000 times. In scientific notation, this would become 2.36×10^8. Now if you like expressing things in millions, as the annual reports of successful companies do, you might like to write this number as 236×10^6. However, the usual convention is to have only one number to the left of the decimal point.

Writing Small Numbers

Now take a number like 0.00347, which is also not in the standard (agreed-to) form for scientific notation. To get it into that format, we must make the first part of it 3.47 by moving the decimal point three places to the right. Note that this motion to the right is the opposite of the motion to the left that we discussed above. To keep track, we call this change negative and put a minus sign in the exponent. Thus 0.00347 becomes 3.47×10^{-3}.

In the example we gave at the beginning, the mass of the hydrogen atom would then be written as 1.67×10^{-27} kg. In this system, one is written as 10^0, a tenth as 10^{-1}, a hundredth as 10^{-2}, and so forth. Note that any number, no matter how large or how small, can be expressed in scientific notation.

Multiplication and Division

The powers-of-ten notation is not only compact and convenient, it also simplifies arithmetic. To multiply two numbers expressed as powers of ten, you need only multiply the numbers out front and then add the exponents. If there are no numbers out front, as in $100 \times 100,000$, then you just add the exponents ($10^2 \times 10^5 = 10^7$). When there are numbers out of front, you do have to multiply them, but they are much easier to deal with than numbers with many zeros in them.

Some examples:

$$3 \times 10^5 \times 2 \times 10^9 = 6 \times 10^{14}$$

$$0.04 \times 6,000,000 = 4 \times 10^{-2} \times 6 \times 10^6$$
$$= 24 \times 10^4 = 2.4 \times 10^5$$

Note in the second example that when we added the exponents, we treated negative exponents as we do in regular arithmetic (−2 plus 6 equals 4). Also, notice that the first result we got had a 24 in it, which was not in the usual form, having two places to the left of the decimal point, and we therefore changed it.

To divide, you divide the numbers out front and subtract the exponents. Here are several examples:

$$1,000,000 \div 1000 = 10^6 \div 10^3 = 10^{6-3} = 10^3$$

$$9 \times 10^{12} \div 2 \times 10^3 = 4.5 \times 10^9$$

$$2.8 \times 10^2 \div 6.2 \times 10^5 = 0.452 \times 10^{-3} = 4.52 \times 10^{-4}$$

If this is the first time that you have met scientific notation, we urge you to practice using it (you might start by solving the exercises below). Like any new language, the notation looks complicated at first, but gets easier as you practice it.

Exercises

1. On April 8, 1996, the Galileo spacecraft was 775 million kilometers from Earth. Convert this number to scientific notation. How many astronomical units is this? (An astronomical unit is the distance from the Earth to the Moon; see above, but remember to keep your units consistent!)

2. During the first six years of its operation, the Hubble Space Telescope circled the Earth 37,000 times, for a total of 1,280,000 km. Use scientific notation to find the number of kilometers in one orbit.

3. In a college cafeteria, a soybean-vegetable burger is offered as an alternative to regular hamburgers. If 489,875 burgers were eaten during the course of a school year, and 997 of them were veggie-burgers, what fraction of the burgers does this represent?

4. In a June 1990 Gallup poll, 27 percent of adult Americans thought that alien beings have actually landed on Earth. The number of adults in the U.S. in 1990 was (according to the census) about 186,000,000. Use scientific notation to determine how many adults believe aliens have visited the Earth.

5. In 1995, 1.7 million degrees were awarded by colleges and universities in the U.S. Among these were 41,000 PhD degrees. What fraction of the degrees were PhD's? Express this number as a percent.

Appendix 5
Units Used in Science

In the American system of measurement (originally developed in England), the fundamental units of length, weight, and time are the foot, pound, and second, respectively. There are also larger and smaller units, which include the ton (2240 lb), the mile (5280 ft), the rod (16½ ft), the yard (3 ft), the inch (1/12 ft), the ounce (1/16 lb), and so on. Such units, whose origins in decisions by British royalty have been forgotten by most people, are quite inconvenient for conversion and arithmetic computation.

In science, therefore, it is more usual to use the metric system, which has been adopted in virtually all countries except the United States. Its great advantage is that every unit increases by a factor of ten, instead of the strange factors in the American system. The fundamental units of the metric system are

length: 1 meter (m)

mass: 1 kilogram (kg)

time: 1 second (s)

A meter was originally intended to be 1 ten-millionth of the distance from the equator to the North Pole along the surface of the Earth. It is about 1.1 yd. A kilogram is the mass that on Earth results in a weight of about 2.2 lb. The second is the same in metric and American units.

The most commonly used quantities of length and mass of the metric system are the following:

Length

1 kilometer (km) = 1000 meters = 0.6214 mile
1 meter (m) = 0.001 km = 1.094 yards = 39.37 inches
1 centimeter (cm) = 0.01 meter = 0.3937 inch
1 millimeter (mm) = 0.001 meter = 0.1 cm
1 micrometer (μm) = 0.000001 meter = 0.0001 cm
1 nanometer (nm) = 10^{-9} meter = 10^{-7} cm

To convert from the American system, here are a few helpful factors:

1 mile = 1.6093 km
1 inch = 2.5400 cm

Mass

Although we don't make the distinction very carefully in everyday life on Earth, strictly speaking the kilogram is a unit of mass (measuring how many atoms a body has) and the pound is a unit of weight (measuring how strongly the Earth's gravity pulls on a body).

1 metric ton = 10^6 grams = 1000 kg (and it produces a weight of 2.2046×10^3 lbs on Earth)
1 kg = 1000 grams (and it produces a weight of 2.2046 lbs on Earth)
1 gram (g) = 0.0353 oz (and the equivalent weight is 0.0022046 lb)
1 milligram (mg) = 0.001 g

And a weight of 1 lb is equivalent on Earth to a mass of 0.4536 kg, while a weight of 1 oz is produced by a mass of 28.3495 g.

Temperature

Three temperature scales are in general use:

1. Fahrenheit (F); water freezes at 32°F and boils at 212°F.
2. Celsius or centigrade° (C); water freezes at 0°C and boils at 100°C.
3. Kelvin or absolute (K); water freezes at 273 K and boils at 373 K.

All molecular motion ceases at −459°F = −273°C = 0 K, a temperature called *absolute zero*. Kelvin temperature is measured from this lowest possible temperature. It is the temperature scale most often used in astronomy. Kelvins are degrees that have the same value as centigrade or Celsius degrees, since the difference between the freezing and boiling points of water is 100 degrees in each.

On the Fahrenheit scale, the difference between the freezing and boiling points of water is 180 degrees. Thus, to convert Celsius degrees or Kelvins to Fahrenheit, it is necessary to multiply by 180/100 = 9/5. To convert from Fahrenheit to Celsius degrees or Kelvins, it is necessary to multiply by 100/180 = 5/9.

The full conversion formulas are:

K = °C + 273
°C = 0.555 × (°F − 32)
°F = (1.8 × °C) + 32

°Celsius is now the name used for centigrade temperature; it has a more modern standardization but differs from the old centigrade scale by less than 0.1°.

Appendix 6
Some Useful Constants for Astronomy

Physical Constants

speed of light (c) = 2.9979×10^8 m/s

gravitational constant (G) = 6.672×10^{11} N m²/kg²

Planck's constant (h) = 6.626×10^{-34} joule·s

mass of a hydrogen atom (m_H) = 1.673×10^{-27} kg

mass of an electron (m_e) = 9.109×10^{-31} kg

Rydberg constant (R) = 1.0974×10^7 per m

Stefan-Boltzmann constant (σ) =
    5.670×10^{-8} joule/(s·m²·deg⁴)

constant in Wien's law ($\lambda_{max}T$) = 2.898×10^{-3} m·deg

electron volt (energy) (eV) = 1.602×10^{-19} joules

energy equivalent of 1 ton TNT = 4.3×10^9 joules

Astronomical Constants

astronomical unit (AU) = 1.496×10^{11} m

light year (LY) = 9.461×10^{15} m

parsec (pc) = 3.086×10^{16} m = 3.262 LY

sidereal year (yr) = 3.158×10^7 s

mass of Earth (M_E) = 5.977×10^{24} kg

equatorial radius of Earth (R_E) = 6.378×10^6 m

obliquity of ecliptic (ϵ) = 23° 27′

surface gravity of Earth (g) = 9.807 m/s²

escape velocity of Earth (v_E) = 1.119×10^4 m/s

mass of Sun (M_{Sun}) = 1.989×10^{30} kg

equatorial radius of Sun (R_{Sun}) = 6.960×10^8 m

luminosity of Sun (L_{Sun}) = 3.83×10^{26} watts

solar constant (flux of energy received at Earth) S =
    1.37×10^3 watts/m²

Hubble constant (H_0) =
    approximately 25 km/sec per million LY

Appendix 7
Physical Data for the Planets

Planet	Diameter (km)	Diameter (Earth = 1)	Mass (Earth = 1)	Mean Density (g/cm³)	Rotation Period (days)	Inclination of Equator to Orbit (°)	Surface Gravity (Earth = 1)	Velocity of Escape (km/s)
Mercury	4,878	0.38	0.055	5.43	58.6	0.0	0.38	4.3
Venus	12,104	0.95	0.82	5.24	−243.0	177.4	0.91	10.4
Earth	12,756	1.00	1.00	5.52	0.997	23.4	1.00	11.2
Mars	6,794	0.53	0.107	3.9	1.026	25.2	0.38	5.0
Jupiter	142,800	11.2	317.8	1.3	0.41	3.1	2.53	60
Saturn	120,540	9.41	94.3	0.7	0.43	26.7	1.07	36
Uranus	51,200	4.01	14.6	1.2	−0.72	97.9	0.92	21
Neptune	49,500	3.88	17.2	1.6	0.67	29	1.18	24
Pluto	2,200	0.17	0.0025	2.0	−6.387	118	0.09	1

Orbital Data for the Planets

Planet	Semimajor Axis AU	Semimajor Axis 10⁶ km	Sidereal Period Tropical Years	Sidereal Period Days	Mean Orbital Speed (km/s)	Orbital Eccentricity	Inclination of Orbit to Ecliptic (°)
Mercury	0.3871	57.9	0.24085	87.97	47.9	0.206	7.004
Venus	0.7233	108.2	0.61521	224.70	35.0	0.007	3.394
Earth	1.0000	149.6	1.000039	365.26	29.8	0.017	0.0
Mars	1.5237	227.9	1.88089	686.98	24.1	0.093	1.850
(Ceres)	2.7671	414	4.603		17.9	0.077	10.6
Jupiter	5.2028	778	11.86		13.1	0.048	1.308
Saturn	9.538	1427	29.46		9.6	0.056	2.488
Uranus	19.191	2871	84.07		6.8	0.046	0.774
Neptune	30.061	4497	164.82		5.4	0.010	1.774
Pluto	39.529	5913	248.6		4.7	0.248	17.15

Adapted from *The Astronomical Almanac* (U.S. Naval Observatory), 1981.

Appendix 8
Satellites of the Planets

Planet	Satellite Name	Discovery	Semimajor Axis (km × 1000)	Period (days)	Diameter (km)	Mass (10^{20} kg)	Density (g/cm³)
Earth	Moon	—	384	27.32	3476	735	3.3
Mars	Phobos	Hall (1877)	9.4	0.32	23	1×10^{-4}	2.0
	Deimos	Hall (1877)	23.5	1.26	13	2×10^{-5}	1.7
Jupiter	Metis	Voyager (1979)	128	0.29	20	—	—
	Adrastea	Voyager (1979)	129	0.30	40	—	—
	Amalthea	Barnard (1892)	181	0.50	200	—	—
	Thebe	Voyager (1979)	222	0.67	90	—	—
	Io	Galileo (1610)	422	1.77	3630	894	3.6
	Europa	Galileo (1610)	671	3.55	3138	480	3.0
	Ganymede	Galileo (1610)	1,070	7.16	5262	1482	1.9
	Callisto	Galileo (1610)	1,883	16.69	4800	1077	1.9
	Leda	Kowal (1974)	11,090	239	15	—	—
	Himalia	Perrine (1904)	11,480	251	180	—	—
	Lysithea	Nicholson (1938)	11,720	259	40	—	—
	Elara	Perrine (1905)	11,740	260	80	—	—
	Ananke	Nicholson (1951)	21,200	631 (R)	30	—	—
	Carme	Nicholson (1938)	22,600	692 (R)	40	—	—
	Pasiphae	Melotte (1908)	23,500	735 (R)	40	—	—
	Sinope	Nicholson (1914)	23,700	758 (R)	40	—	—
Saturn	Unnamed	Voyager (1985)	118.2	0.48	15?	3×10^{-5}	—
	Pan	Voyager (1985)	133.6	0.58	20	3×10^{-5}	
	Atlas	Voyager (1980)	137.7	0.60	40	—	—
	Prometheus	Voyager (1980)	139.4	0.61	80	—	—
	Pandora	Voyager (1980)	141.7	0.63	100	—	—
	Janus	Dollfus (1966)	151.4	0.69	190	—	—
	Epimetheus	Fountain, Larson (1980)	151.4	0.69	120	—	—
	Mimas	Herschel (1789)	186	0.94	394	0.4	1.2
	Enceladus	Herschel (1789)	238	1.37	502	0.8	1.2
	Tethys	Cassini (1684)	295	1.89	1048	7.5	1.3
	Telesto	Reitsema et al. (1980)	295	1.89	25	—	—
	Calypso	Pascu et al. (1980)	295	1.89	25	—	—
	Dione	Cassini (1684)	377	2.74	1120	11	1.4
	Helene	Lecacheux, Laques (1980)	377	2.74	30	—	—
	Rhea	Cassini (1672)	527	4.52	1530	25	1.3
	Titan	Huygens (1655)	1,222	15.95	5150	1346	1.9
	Hyperion	Bond, Lassell (1848)	1,481	21.3	270	—	—
	Iapetus	Cassini (1671)	3,561	79.3	1435	19	1.2
	Phoebe	Pickering (1898)	12,950	550 (R)	220	—	—
Uranus	Cordelia	Voyager (1986)	49.8	0.34	40?	—	—
	Ophelia	Voyager (1986)	53.8	0.38	50?	—	—
	Bianca	Voyager (1986)	59.2	0.44	50?	—	—
	Cressida	Voyager (1986)	61.8	0.46	60?	—	—
	Desdemona	Voyager (1986)	62.7	0.48	60?	—	—
	Juliet	Voyager (1986)	64.4	0.50	80?	—	—
	Portia	Voyager (1986)	66.1	0.51	80?	—	—
	Rosalind	Voyager (1986)	69.9	0.56	60?	—	—

(Table continued on page A-24)

Satellites of the Planets (Continued)

Planet	Satellite Name	Discovery	Semimajor Axis (km × 1000)	Period (days)	Diameter (km)	Mass (10²⁰ kg)	Density (g/cm³)
	Belinda	Voyager (1986)	75.3	0.63	60?	—	—
	Puck	Voyager (1985)	86.0	0.76	170	—	—
	Miranda	Kuiper (1948)	130	1.41	485	0.8	1.3
	Ariel	Lassell (1851)	191	2.52	1160	13	1.6
	Umbriel	Lassell (1851)	266	4.14	1190	13	1.4
	Titania	Herschel (1787)	436	8.71	1610	35	1.6
	Oberon	Herschel (1787)	583	13.5	1550	29	1.5
Neptune	Naiad	Voyager (1989)	48	0.30	50	—	—
	Thalassa	Voyager (1989)	50	0.31	90	—	—
	Despina	Voyager (1989)	53	0.33	150	—	—
	Galatea	Voyager (1989)	62	0.40	150	—	—
	Larissa	Voyager (1989)	74	0.55	200	—	—
	Proteus	Voyager (1989)	118	1.12	400	—	—
	Triton	Lassell (1846)	355	5.88 (R)	2720	220	2.1
	Nereid	Kuiper (1949)	5,511	360	340	—	—
Pluto	Charon	Christy (1978)	19.7	6.39	1200	—	—

Appendix 9
Upcoming (Total) Eclipses

1. Total Eclipses of the Sun

Date	Duration of Totality (min)	Where Visible
1997 March 9	2.8	Siberia, Arctic
1998 Feb. 26	4.4	Central America
1999 Aug. 11	2.6	Central Europe, Central Asia
2001 June 21	4.9	Southern Africa
2002 Dec. 4	2.1	South Africa, Australia
2003 Nov. 23	2.0	Antarctica
2005 April 8	0.7	South Pacific Ocean
2006 March 29	4.1	Africa, Asia Minor, U.S.S.R.
2008 Aug. 1	2.4	Arctic Ocean, Siberia, China
2009 July 22	6.6	India, China, South Pacific
2010 July 11	5.3	South Pacific Ocean
2012 Nov. 13	4.0	Northern Australia, South Pacific
2013 Nov. 3	1.7	Atlantic Ocean, Central Africa
2015 March 20	4.1	North Atlantic, Arctic Ocean
2016 March 9	4.5	Indonesia, Pacific Ocean

Date	Duration of Totality (min)	Where Visible
2017 Aug. 21	2.7	Pacific Ocean, U.S.A., Atlantic Ocean
2019 July 2	4.5	South Pacific, South America
2020 Dec. 14	2.2	South Pacific, South America, South Atlantic Ocean
2021 Dec. 4	1.9	Antarctica
2023 April 20	1.3	Indian Ocean, Indonesia
2024 April 8	4.5	South Pacific, Mexico, East U.S.A.
2026 Aug. 12	2.3	Arctic, Greenland, North Atlantic, Spain
2027 Aug. 2	6.4	North Africa, Arabia, Indian Ocean
2028 July 22	5.1	Indian Ocean, Australia, New Zealand
2030 Nov. 25	3.7	South Africa, Indian Ocean, Australia

2. Some Upcoming Total Lunar Eclipses

1997 Sep. 16	2003 May 16	2007 March 3
2000 Jan. 21	2003 Nov. 9	2007 Aug. 28
2000 July 16	2004 May 4	2008 Feb. 21
2001 Jan. 9	2004 Oct. 28	

Appendix 10
The Nearest Stars

Name (Catalog number)	Distance (LY)	Spectral Type	Location[1] RA	Location[1] Dec	Luminosity (Sun = 1)
Sun	—	G2V	—	—	1.0
Proxima Centauri	4.2	M5V	14 30	−62 41	6×10^{-6}
Alpha Centauri A	4.3	G2V	14 33	−60 50	1.5
Alpha Centauri B	4.3	K0V	14 33	−60 50	0.5
Barnard's Star	6.0	M4V	17 57	+04 33	4×10^{-4}
Wolf 359 (Gliese 406)	7.8	M6V	10 56	+07 03	2×10^{-5}
Lalande 21185 (HD 95735)	8.2	M2V	11 04	+36 02	5×10^{-3}
Luyten 726-8 A	8.6	M5V	01 38	−17 58	6×10^{-5}
Luyten 726-8 B (UV Ceti)	8.6	M6V	01 38	−17 58	4×10^{-5}
Sirius A	8.6	A1V	06 45	−16 43	24
Sirius B	8.6	w.d.[2]	06 45	−16 43	3×10^{-3}
Ross 154 (Gliese 729)	9.6	M4V	18 50	−23 49	5×10^{-4}
Ross 248 (Gliese 905)	10.3	M6V	23 42	+44 12	1×10^{-4}
Epsilon Eridani	10.7	K2V	03 33	−09 27	0.3
Ross 128 (Gliese 447)	10.8	M4V	11 48	+00 49	3×10^{-4}
Luyten 789-6 A	11.1	M5V[3]	22 39	−15 20	1×10^{-4}
Luyten 789-6 B	11.1	—	22 39	−15 20	—
Luyten 789-6 C	11.1	—	22 39	−15 20	—
BD +43°44 A (Gliese 15 A)	11.3	M1V	00 18	+44 61	6×10^{-3}
BD +43°44 B (Gliese 15 B)	11.3	M3V	00 18	+44 61	4×10^{-4}
Epsilon Indi	11.3	K5V	22 03	−56 47	0.14
61 Cygni A	11.3	K5V	21 07	+38 45	8×10^{-2}
61 Cygni B	11.3	K7V	21 07	+38 45	4×10^{-2}
BD +59°1915 A (Gliese 725 A)	11.4	M3V	18 43	+59 37	3×10^{-3}
BD +59°1915 B (Gliese 725 B)	11.4	M4V	18 43	+59 37	2×10^{-3}
Tau Ceti	11.4	G8V	01 44	−15 56	0.45
Procyon A	11.4	F5IV	07 39	+05 13	7.7
Procyon B	11.4	w.d.[2]	07 39	+05 13	6×10^{-4}
CD −36°15693 (Lacaille 9352)	11.5	M2V	23 06	−35 52	1×10^{-2}
GJ 1111 (G51-15)	11.8	M7V	08 29	+26 47	1×10^{-5}
GJ 1061	12.0	M5V	03 36	−44 30	8×10^{-5}
Luyten 725-32 (YZ Ceti)	12.2	M5V	01 12	−18 04	3×10^{-4}
BD +5°1668 (Gliese 273)	12.3	M4V	07 28	+05 17	1×10^{-3}
CD −39°14192 (Gliese 825)	12.6	M0V	21 17	−38 52	3×10^{-2}
Kapteyn's Star	12.6	M0V	05 11	−44 56	4×10^{-3}

[1] Location (right ascension and declination) given for Epoch 2000.0
[2] White dwarf star
[3] The stars in this system are so close to each other, it is not possible to measure their spectral types and luminosities separately.

With many thanks to Dr. Todd Henry, Space Telescope Science Institute, and the RECONS team for updated information. For the latest data on nearby star systems, see their web page at http://proxima.astro.virginia.edu/~pai/Recons

Appendix 11

The Brightest Stars

NOTE: These are the stars that *appear* the brightest visually, as seen from our vantage point on Earth. They are not necessarily the stars that are intrinsically the brightest.

Name[1]	Luminosity (Sun = 1)	Distance[2] (LY)	Spectral Type	ProperMotion (arcsec/yr)	Right Ascension (Epoch 2000.0) (h)	(m)	Declination (Epoch 2000.0) (deg)	(min)
Sirius (α CMa)	40	9	A1V	1.33	06	45.1	−16	43
Canopus (α Car)	1500	98	F01	0.02	06	24.0	−52	42
Alpha Centauri	2	4	G2V	3.68	14	39.6	−60	50
Arcturus (α Boo)	100	36	K2III	2.28	14	15.7	+19	11
Vega (α Lyr)	50	26	A0V	0.34	18	36.9	+38	47
Capella (α Aur)	200	46	G5III	0.44	05	16.7	+46	00
Rigel (β Ori)	8×10^4	815	B8Ia	0.00	05	12.1	−08	12
Procyon (α Cmi)	9	11	F5IV-V	1.25	07	39.3	+05	13
Betelgeuse (α Ori)	10^5	500	M2Iab	0.03	05	55.2	+07	24
Achernar (α Eri)	500	65	B3V	0.10	01	37.7	−57	14
Beta Centauri	9300	300	B1III	0.04	14	03.8	−60	22
Altair (α Aql)	10	17	A7IV-V	0.66	19	50.8	+08	52
Aldebaran (α Tau)	200	20	K5III	0.20	04	35.9	+16	31
Spica (α Vir)	6000	260	B1V	0.05	13	25.2	−11	10
Antares (α Sco)	10^4	390	M1Ib	0.03	16	29.4	−26	26
Pollux (β Gem)	60	39	K0III	0.62	07	45.3	+28	02
Fomalhaut (α PsA)	50	23	A3V	0.37	22	57.6	−29	37
Deneb (α Cyg)	8×10^4	1400	A2Ia	0.00	20	41.4	+45	17
Beta Crucis	10^4	490	B0.5IV	0.05	12	47.7	−59	41
Regulus (α Leo)	150	85	B7V	0.25	10	08.3	+11	58

[1] The brightest stars typically have names from antiquity (although most fainter stars are not given names, but merely catalog designations). Next to each star's ancient name, we have put its name in the system originated by Bayer (see the "Astronomy Basics" box in Chapter 18.) The abbreviations of the constellations are given in Appendix 14.

[2] The distances of the more remote stars are estimated from their spectral types and apparent brightnesses and are only approximate. The luminosities for those stars are approximate to the same degree.

Appendix 12

The Brightest Members of the Local Group

Galaxy	Type[1]	Right Ascension[2] (h)	(m)	Declination (Degrees)	Distance[3] (1000 LY)	Absolute Magnitude	Apparent Magnitude	Diameter (1000 LY)
Milky Way	Sbc	17	46	−29		−20.6		130
Andromeda;								
M31; NGC224	Sb	00	43	+41	2200	−21.6	4.4	200
M33; NGC598	Sc	01	34	+31	2500	−19.1	6.3	45
Large								
Magellanic Cloud	Irr	05	24	−70	160	−18.4	0.6	20
Small								
Magellanic Cloud	Irr	00	53	−73	300	−17.0	2.8	15
IC10	Irr	00	20	+59	4000	−16.2	11.7	6
NGC205	E5pec	00	40	+42	2200	−15.7	8.6	10
M32; NGC221	E2	00	43	+41	2200	−15.5	9.0	5
NGC6822	Irr	19	45	−15	1700	−15.1	9.3	8
WLM	Irr	00	02	−15	2000	−15.0	11.3	7
IC5152	Sd	22	03	−51	2000	−14.6	11.7	5
NGC185	E3pec	00	39	+48	2200	−14.6	10.1	6
IC1613	Irr	01	05	+02	2500	−14.5	10.0	12
NGC147	E5	00	33	+48	2200	−14.4	10.4	10
Leo A	Irr	09	59	+31	5000	−13.5	12.7	7
Pegasus	Irr	23	29	+15	5000	−13.4	12.4	8
Fornax	E3	02	40	−34	500	−12.9	8.5	3
GR8	Irr	12	59	+14	4000	−11.0	14.6	0.2
DDO210	Irr	20	47	−13	3000	−11.0	15.3	4
Sagittarius Dwarf[3]	Dwarf E	18	50	−32	80	?	?	25
Sagittarius	Irr	19	30	−18	4000	−10.6	15.6	5
Sculptor	E3	00	60	−34	300	−10.6	9.1	1
Andromeda I	E3	00	46	+38	2200	−10.6	14.0	2
Andromeda III	E5	00	35	+36	2200	−10.6	14.0	3
Andromeda II	E2	01	16	+33	2200	−10.6	14.0	2.3
Pisces; LGS3	Irr	01	03	+22	3000	−9.7	15.5	0.5
Leo I	E3	10	09	+12	600	−9.6	11.8	1
Leo II	E0	11	14	+22	600	−9.2	12.3	0.5
Ursa Minor	E5	15	09	+67	300	−8.2	11.6	1
Draco	E3	17	20	+58	300	−8.0	12.0	0.5
Carina	E4	06	42	−51	300	>−5.5	>13.0	0.5
Sgr(Anon)	Epec				50			

[1] S mean spiral, Sb means barred spiral, E means elliptical, Irr means irregular; the numbers represent subgroups into which Hubble and others divided these broad categories.

[2] Coordinates are given for Epoch 2000.0

[3] Many of the distances (and therefore the diameters calculated from them) are only approximate.

[4] This close neighbor galaxy is so extended on the sky that giving a magnitude would not make sense.

Data courtesy of Paul Hodge, University of Washington.

Appendix 13
The Chemical Elements

Element	Symbol	Atomic Number	Atomic Weight* (Chemical Scale)	Number of Atoms per 10^{12} Hydrogen Atoms
Hydrogen	H	1	1.0080	1×10^{12}
Helium	He	2	4.003	8×10^{10}
Lithium	Li	3	6.940	2×10^{3}
Beryllium	Be	4	9.013	3×10^{1}
Boron	B	5	10.82	9×10^{2}
Carbon	C	6	12.011	4.5×10^{8}
Nitrogen	N	7	14.008	9.2×10^{7}
Oxygen	O	8	16.00	7.4×10^{8}
Fluorine	F	9	19.00	3.1×10^{4}
Neon	Ne	10	20.183	1.3×10^{8}
Sodium	Na	11	22.991	2.1×10^{6}
Magnesium	Mg	12	24.32	4.0×10^{7}
Aluminum	Al	13	26.98	3.1×10^{6}
Silicon	Si	14	28.09	3.7×10^{7}
Phosphorus	P	15	30.975	3.8×10^{5}
Sulfur	S	16	32.066	1.9×10^{7}
Chlorine	Cl	17	35.457	1.9×10^{5}
Argon	Ar(A)	18	39.944	3.8×10^{6}
Potassium	K	19	39.100	1.4×10^{5}
Calcium	Ca	20	40.08	2.2×10^{6}
Scandium	Sc	21	44.96	1.3×10^{3}
Titanium	Ti	22	47.90	8.9×10^{4}
Vanadium	V	23	50.95	1.0×10^{4}
Chromium	Cr	24	52.01	5.1×10^{5}
Manganese	Mn	25	54.94	3.5×10^{5}
Iron	Fe	26	55.85	3.2×10^{7}
Cobalt	Co	27	58.94	8.3×10^{4}
Nickel	Ni	28	58.71	1.9×10^{6}
Copper	Cu	29	63.54	1.9×10^{4}
Zinc	Zn	30	65.38	4.7×10^{4}
Gallium	Ga	31	69.72	1.4×10^{3}
Germanium	Ge	32	72.60	4.4×10^{3}
Arsenic	As	33	74.91	2.5×10^{2}
Selenium	Se	34	78.96	2.3×10^{3}
Bromine	Br	35	79.916	4.4×10^{2}
Krypton	Kr	36	83.80	1.7×10^{3}
Rubidium	Rb	37	85.48	2.6×10^{2}
Strontium	Sr	38	87.63	8.8×10^{2}
Yttrium	Y	39	88.92	2.5×10^{2}
Zirconium	Zr	40	91.22	4.0×10^{2}
Niobium (Columbium)	Nb(Cb)	41	92.91	2.6×10^{1}
Molybdenum	Mo	42	95.95	9.3×10^{1}
Technetium	Tc(Ma)	43	(99)	—
Ruthenium	Ru	44	101.1	68
Rhodium	Rh	45	102.91	13
Palladium	Pd	46	106.4	51
Silver	Ag	47	107.880	20
Cadmium	Cd	48	112.41	63
Indium	In	49	114.82	7

* Where mean atomic weights have not been well determined, the atomic mass numbers of the most stable isotopes are given in parentheses.

(Table continued on page A-30)

The Chemical Elements (continued)

Element	Symbol	Atomic Number	Atomic Weight* (Chemical Scale)	Number of Atoms per 10^{12} Hydrogen Atoms
Tin	Sn	50	118.70	1.4×10^2
Antimony	Sb	51	121.76	13
Tellurium	Te	52	127.61	1.8×10^2
Iodine	I(J)	53	126.91	33
Xenon	Xe(X)	54	131.30	1.6×10^2
Cesium	Cs	55	132.91	14
Barium	Ba	56	137.36	1.6×10^2
Lanthanum	La	57	138.92	17
Cerium	Ce	58	140.13	43
Praseodymium	Pr	59	140.92	6
Neodymium	Nd	60	144.27	31
Promethium	Pm	61	(147)	—
Samarium	Sm(Sa)	62	150.35	10
Europium	Eu	63	152.00	4
Gadolinium	Gd	64	157.26	13
Terbium	Tb	65	158.93	2
Dysprosium	Dy(Ds)	66	162.51	15
Holmium	Ho	67	164.94	3
Erbium	Er	68	167.27	9
Thulium	Tm(Tu)	69	168.94	2
Ytterbium	Yb	70	173.04	8
Lutecium	Lu(Cp)	71	174.99	2
Hafnium	Hf	72	178.50	6
Tantalum	Ta	73	180.95	1
Tungsten	W	74	183.86	5
Rhenium	Re	75	186.22	2
Osmium	Os	76	190.2	27
Iridium	Ir	77	192.2	24
Platinum	Pt	78	195.09	56
Gold	Au	79	197.00	6
Mercury	Hg	80	200.61	19
Thallium	Tl	81	204.39	8
Lead	Pb	82	207.21	1.2×10^2
Bismuth	Bi	83	209.00	5
Polonium	Po	84	(209)	—
Astatine	At	85	(210)	—
Radon	Rn	86	(222)	—
Francium	Fr(Fa)	87	(223)	—
Radium	Ra	88	226.05	—
Actinium	Ac	89	(227)	—
Thorium	Th	90	232.12	1
Protactinium	Pa	91	(231)	—
Uranium	U(Ur)	92	238.07	1
Neptunium	Np	93	(237)	—
Plutonium	Pu	94	(244)	—
Americium	Am	95	(243)	—
Curium	Cm	96	(248)	—
Berkelium	Bk	97	(247)	—
Californium	Cf	98	(251)	—
Einsteinium	E	99	(254)	—
Fermium	Fm	100	(253)	—
Mendeleevium	Mv	101	(256)	—
Nobelium	No	102	(253)	—
Lawrencium	Lr	103	(262)	—
Rutherfordium	Rf	104	(261)	—
Hahnium	Ha	105	(262)	—
Seaborgium	Sg	106	(263)	—
Nielsbohrium	Ns	107	(262)	—
Hassium	Hs	108	(264)	—
Meitnerium	Mt	109	(266)	—
Ununnilium	Uun	110	(269)	—
Unununium	Uuu	111	(272)	—

* Where mean atomic weights have not been well determined, the atomic mass numbers of the most stable isotopes are given in parentheses.

Appendix 14
The Constellations

Constellation (Latin Name)	Genitive Case Ending	English Name or Description	Abbre-viation	Approximate Position α h	δ °
Andromeda	Andromedae	Princess of Ethiopia	And	1	+40
Antila	Antilae	Air pump	Ant	10	−35
Apus	Apodis	Bird of Paradise	Aps	16	−75
Aquarius	Aquarii	Water bearer	Aqr	23	−15
Aquila	Aquilae	Eagle	Aql	20	+5
Ara	Arae	Altar	Ara	17	−55
Aries	Arietis	Ram	Ari	3	+20
Auriga	Aurigae	Charioteer	Aur	6	+40
Boötes	Boötis	Herdsman	Boo	15	+30
Caelum	Caeli	Graving tool	Cae	5	−40
Camelopardus	Camelopardis	Giraffe	Cam	6	+70
Cancer	Cancri	Crab	Cnc	9	+20
Canes Venatici	Canum Venaticorum	Hunting dogs	CVn	13	+40
Canis Major	Canis Majoris	Big dog	CMa	7	−20
Canis Minor	Canis Minoris	Little dog	CMi	8	+5
Capricornus	Capricorni	Sea goat	Cap	21	−20
Carina*	Carinae	Keel of Argonauts' ship	Car	9	−60
Cassiopeia	Cassiopeiae	Queen of Ethiopia	Cas	1	+60
Centaurus	Centauri	Centaur	Cen	13	−50
Cepheus	Cephei	King of Ethiopia	Cep	22	+70
Cetus	Ceti	Sea monster (whale)	Cet	2	−10
Chamaeleon	Chamaeleontis	Chameleon	Cha	11	−80
Circinus	Circini	Compasses	Cir	15	−60
Columba	Columbae	Dove	Col	6	−35
Coma Berenices	Comae Berenices	Berenice's hair	Com	13	+20
Corona Australis	Coronae Australis	Southern crown	CrA	19	−40
Corona Borealis	Coronae Borealis	Northern crown	CrB	16	+30
Corvus	Corvi	Crow	Crv	12	−20
Crater	Crateris	Cup	Crt	11	−15
Crux	Crucis	Cross (southern)	Cru	12	−60
Cygnus	Cygni	Swan	Cyg	21	+40
Delphinus	Delphini	Porpoise	Del	21	+10
Dorado	Doradus	Swordfish	Dor	5	−65
Draco	Draconis	Dragon	Dra	17	+65
Equuleus	Equulei	Little horse	Equ	21	+10
Eridanus	Eridani	River	Eri	3	−20
Fornax	Fornacis	Furnace	For	3	−30
Gemini	Geminorum	Twins	Gem	7	+20
Grus	Gruis	Crane	Gru	22	−45
Hercules	Herculis	Hercules, son of Zeus	Her	17	+30
Horologium	Horologii	Clock	Hor	3	−60
Hydra	Hydrae	Sea serpent	Hya	10	−20
Hydrus	Hydri	Water snake	Hyi	2	−75
Indus	Indi	Indian	Ind	21	−55
Lacerta	Lacertae	Lizard	Lac	22	+45

(Table continued on A–32)

The Constellations (Continued)

Constellation (Latin Name)	Genitive Case Ending	English Name or Description	Abbreviation	α (h)	δ (°)
Leo	Leonis	Lion	Leo	11	+15
Leo Minor	Leonis Minoris	Little lion	LMi	10	+35
Lepus	Leporis	Hare	Lep	6	−20
Libra	Librae	Balance	Lib	15	−15
Lupus	Lupi	Wolf	Lup	15	−45
Lynx	Lyncis	Lynx	Lyn	8	+45
Lyra	Lyrae	Lyre or harp	Lyr	19	+40
Mensa	Mensae	Table Mountain	Men	5	−80
Microscopium	Microscopii	Microscope	Mic	21	−35
Monoceros	Monocerotis	Unicorn	Mon	7	−5
Musca	Muscae	Fly	Mus	12	−70
Norma	Normae	Carpenter's level	Nor	16	−50
Octans	Octantis	Octant	Oct	22	−85
Ophiuchus	Ophiuchi	Holder of serpent	Oph	17	0
Orion	Orionis	Orion, the hunter	Ori	5	+5
Pavo	Pavonis	Peacock	Pav	20	−65
Pegasus	Pegasi	Pegasus, the winged horse	Peg	22	+20
Perseus	Persei	Perseus, hero who saved Andromeda	Per	3	+45
Phoenix	Phoenicis	Phoenix	Phe	1	−50
Pictor	Pictoris	Easel	Pic	6	−55
Pisces	Piscium	Fishes	Psc	1	+15
Piscis Austrinus	Piscis Austrini	Southern fish	PsA	22	−30
Puppis*	Puppis	Stern of the Argonauts' ship	Pup	8	−40
Pyxis* (= Malus)	Pyxidus	Compass of the Argonauts' ship	Pyx	9	−30
Reticulum	Reticuli	Net	Ret	4	−60
Sagitta	Sagittae	Arrow	Sge	20	+10
Sagittarius	Sagittarii	Archer	Sgr	19	−25
Scorpius	Scorpii	Scorpion	Sco	17	−40
Sculptor	Sculptoris	Sculptor's tools	Scl	0	−30
Scutum	Scuti	Shield	Sct	19	−10
Serpens	Serpentis	Serpent	Ser	17	0
Sextans	Sextantis	Sextant	Sex	10	0
Taurus	Tauri	Bull	Tau	4	+15
Telescopium	Telescopii	Telescope	Tel	19	−50
Triangulum	Trianguli	Triangle	Tri	2	+30
Triangulum Australe	Trianguli Australis	Southern triangle	TrA	16	−65
Tucana	Tucanae	Toucan	Tuc	0	−65
Ursa Major	Ursae Majoris	Big bear	UMa	11	+50
Ursa Minor	Ursae Minoris	Little bear	VMi	15	+70
Vela*	Velorum	Sail of the Argonauts' ship	Vel	9	−50
Virgo	Virginis	Virgin	Vir	13	0
Volans	Volantis	Flying fish	Vol	8	−70
Vulpecula	Vulpeculae	Fox	Vul	20	+25

* The four constellations Carina, Puppis, Pyxis, and Vela originally formed the single constellation, Argo Navis.

Appendix 15
The Messier Catalog of Nebulae and Star Clusters

M	NGC or (IC)	Right Ascension (1980) h	m	Declination (1980) °	'	Apparent Visual Magnitude	Description
1	1952	5	33.3	+22	01	8.4	"Crab" nebula in Taurus; remains of SN 1054
2	7089	21	32.4	−0	54	6.4	Globular cluster in Aquarius
3	5272	13	41.2	+28	29	6.3	Globular cluster in Canes Venatici
4	6121	16	22.4	−26	28	6.5	Globular cluster in Scorpius
5	5904	15	17.5	+2	10	6.1	Globular cluster in Serpens
6	6405	17	38.8	−32	11	5.5	Open cluster in Scorpius
7	6475	17	52.7	−34	48	3.3	Open cluster in Scorpius
8	6523	18	02.4	−24	23	5.1	"Lagoon" nebula in Sagittarius
9	6333	17	18.1	−18	30	8.0	Globular cluster in Ophiuchus
10	6254	16	56.1	−4	05	6.7	Globular cluster in Ophiuchus
11	6705	18	50.0	−6	18	6.8	Open cluster in Scutum Sobieskii
12	6218	16	46.3	−1	55	6.6	Globular cluster in Ophiuchus
13	6205	16	41.0	+36	30	5.9	Globular cluster in Hercules
14	6402	17	36.6	−3	14	8.0	Globular cluster in Ophiuchus
15	7078	21	28.9	+12	05	6.4	Globular cluster in Pegasus
16	6611	18	17.8	−13	47	6.6	Open cluster with nebulosity in Serpens
17	6618	18	19.6	−16	11	7.5	"Swan" or "Omega" nebula in Sagittarius
18	6613	18	18.7	−17	08	7.2	Open cluster in Sagittarius
19	6273	17	01.4	−26	14	6.9	Globular cluster in Ophiuchus
20	6514	18	01.2	−23	02	8.5	"Trifid" nebula in Sagittarius
21	6531	18	03.4	−22	30	6.5	Open cluster in Sagittarius
22	6656	18	35.2	−23	56	5.6	Globular cluster in Sagittarius
23	6494	17	55.8	−19	00	5.9	Open cluster in Sagittarius
24	6603	18	17.3	−18	26	4.6	Open cluster in Sagittarius
25	(4725)	18	30.5	−19	16	6.2	Open cluster in Sagittarius
26	6694	18	44.1	−9	25	9.3	Open cluster in Scutum Sobieskii
27	6853	19	58.8	+22	40	8.2	"Dumbbell" planetary nebula in Vulpecula
28	6626	18	23.2	−24	52	7.6	Globular cluster in Sagittarius
29	6913	20	23.3	+38	27	8.0	Open cluster in Cygnus
30	7099	21	39.2	−23	16	7.7	Globular cluster in Capricornus
31	224	0	41.6	+41	10	3.5	Andromeda galaxy
32	221	0	41.6	+40	46	8.2	Elliptical galaxy; companion to M31
33	598	1	32.7	+30	33	5.8	Spiral galaxy in Triangulum
34	1039	2	40.7	+42	43	5.8	Open cluster in Perseus
35	2168	6	07.5	+24	21	5.6	Open cluster in Gemini
36	1960	5	35.0	+34	05	6.5	Open cluster in Auriga
37	2099	5	51.1	+32	33	6.2	Open cluster in Auriga
38	1912	5	27.3	+35	48	7.0	Open cluster in Auriga
39	7092	21	31.5	+48	21	5.3	Open cluster in Cygnus
40	—	12	21	+59	—	—	Close double star in Ursa Major
41	2287	6	46.2	−20	43	5.0	Loose open cluster in Canis Major
42	1976	5	34.4	−5	24	4	Orion nebula
43	1982	5	34.6	−5	18	9	Northeast portion of Orion nebula

(Table continued on A-34)

The Messier Catalog of Nebulae and Star Clusters (Continued)

M	NGC or (IC)	Right Ascension (1980) h	m	Declination (1980) °	'	Apparent Visual Magnitude	Description
44	2632	8	39	+20	04	3.9	Praesepe; open cluster in Cancer
45	—	3	46.3	+24	03	1.6	The Pleiades; open cluster in Taurus
46	2437	7	40.9	−14	46	6.6	Open cluster in Puppis
47	2422	7	35.7	−14	26	5.0	Loose group of stars in Puppis
48	2548	8	12.8	−5	44	6.0	"Cluster of very small stars"
49	4472	12	28.8	+8	06	8.5	Elliptical galaxy in Virgo
50	2323	7	02.0	−8	19	6.3	Loose open cluster in Monoceros
51	5194	13	29.1	+47	18	8.4	"Whirlpool" spiral galaxy in Canes Venatici
52	7654	23	23.3	+61	30	8.2	Loose open cluster in Cassiopeia
53	5024	13	12.0	+18	16	7.8	Globular cluster in Coma Berenices
54	6715	18	53.8	−30	30	7.8	Globular cluster in Sagittarius
55	6809	19	38.7	−30	59	6.2	Globular cluster in Sagittarius
56	6779	19	15.8	+30	08	8.7	Globular cluster in Lyra
57	6720	18	52.8	+33	00	9.0	"Ring" nebula; planetary nebula in Lyra
58	4579	12	36.7	+11	55	9.9	Spiral galaxy in Virgo
59	4621	12	41.0	+11	46	10.0	Spiral galaxy in Virgo
60	4649	12	42.6	+11	40	9.0	Elliptical galaxy in Virgo
61	4303	12	20.8	+4	35	9.6	Spiral galaxy in Virgo
62	6266	16	59.9	−30	05	6.6	Globular cluster in Scorpius
63	5055	13	14.8	+42	07	8.9	Spiral galaxy in Canes Venatici
64	4826	12	55.7	+21	39	8.5	Spiral galaxy in Coma Berenices
65	3623	11	17.9	+13	12	9.4	Spiral galaxy in Leo
66	3627	11	19.2	+13	06	9.0	Spiral galaxy in Leo; companion to M65
67	2682	8	50.0	+11	53	6.1	Open cluster in Cancer
68	4590	12	38.4	−26	39	8.2	Globular cluster in Hydra
69	6637	18	30.1	−32	23	8.0	Globular cluster in Sagittarius
70	6681	18	42.0	−32	18	8.1	Globular cluster in Sagittarius
71	6838	19	52.8	+18	44	7.6	Globular cluster in Sagittarius
72	6981	20	52.3	−12	38	9.3	Globular cluster in Aquarius
73	6994	20	57.8	−12	43	9.1	Open cluster in Aquarius
74	628	1	35.6	+15	41	9.3	Spiral galaxy in Pisces
75	6864	20	04.9	−21	59	8.6	Globular cluster in Sagittarius
76	650	1	41.0	+51	28	11.4	Planetary nebula in Perseus
77	1068	2	41.6	−0	04	8.9	Spiral galaxy in Cetus
78	2068	5	45.7	0	03	8.3	Small emission nebula in Orion
79	1904	5	23.3	−24	32	7.5	Globular cluster in Lepus
80	6093	16	15.8	−22	56	7.5	Globular cluster in Scorpius
81	3031	9	54.2	+69	09	7.0	Spiral galaxy in Ursa Major
82	3034	9	54.4	+69	47	8.4	Irregular galaxy in Ursa Major
83	5236	13	35.4	−29	31	7.6	Spiral galaxy in Hydra
84	4374	12	24.1	+13	00	9.4	Elliptical galaxy in Virgo
85	4382	12	24.3	+18	18	9.3	Elliptical galaxy in Coma Berenices
86	4406	12	25.1	+13	03	9.2	Elliptical galaxy in Virgo
87	4486	12	29.7	+12	30	8.7	Elliptical galaxy in Virgo
88	4501	12	30.9	+14	32	9.5	Spiral galaxy in Coma Berenices
89	4552	12	34.6	+12	40	10.3	Elliptical galaxy in Virgo
90	4569	12	35.8	+13	16	9.6	Spiral galaxy in Virgo

The Messier Catalog of Nebulae and Star Clusters (Continued)

M	NGC or (IC)	Right Ascension (1980) h	m	Decli- nation (1980) °	'	Apparent Visual Magnitude	Description
91	omitted	—	—	—	—	—	
92	6341	17	16.5	+43	10	6.4	Globular cluster in Hercules
93	2447	7	43.7	−23	49	6.5	Open cluster in Puppis
94	4736	12	50.0	+41	14	8.3	Spiral galaxy in Canes Venatici
95	3351	10	42.9	+11	49	9.8	Barred spiral galaxy in Leo
96	3368	10	45.7	+11	56	9.3	Spiral galaxy in Leo
97	3587	11	13.7	+55	07	11.1	"Owl" nebula; planetary nebula in Ursa Major
98	4192	12	12.7	+15	01	10.2	Spiral galaxy in Coma Berenices
99	4254	12	17.8	+14	32	9.9	Spiral galaxy in Coma Berenices
100	4321	12	21.9	+15	56	9.4	Spiral galaxy in Coma Berenices
101	5457	14	02.5	+54	27	7.9	Spiral galaxy in Ursa Major
102	5866(?)	15	05.9	+55	50	10.5	Spiral galaxy (identification in doubt)
103	581	1	31.9	+60	35	6.9	Open cluster in Cassiopeia
104*	4594	12	39.0	−11	31	8.3	Spiral galaxy in Virgo
105*	3379	10	46.8	+12	51	9.7	Elliptical galaxy in Leo
106*	4258	12	18.0	+47	25	8.4	Spiral galaxy in Canes Venatici
107*	6171	16	31.4	−13	01	9.2	Globular cluster in Ophiuchus
108*	3556	11	10.5	+55	47	10.5	Spiral galaxy in Ursa Major
109*	3992	11	56.6	+53	29	10.0	Spiral galaxy in Ursa Major
110*	205	0	39.2	+41	35	9.4	Elliptical galaxy; companion to M31

* Not in Messier's original (1781) list; added later by others.

Index

Note: Figures, tables, and footnotes are indicated by *italics*, "t", and "n" respectively.

A

A ring, Saturn, 242, *242*, 242t, 244
Absolute temperature scale, 91
Absorption line spectrum, 288
Absorption lines
    elements, 332
    star, *332*, 334, *334*
Accelerating environment
    gravity, 461
Acceleration, 45–46
    Galileo, 34, 37
Accretion, 279, 283
Accretion disk, 473, *474*, 493
Acetaldehyde, 383
Adams, John Couch, 54, *54*, 56
Advanced X-Ray Astrophysics Facility, 127
Agamemnon, 254
Airy, George, 54
Alcor, 345
Aldrin, Buzz, *166*
Algol, 350–351, 369
Allende meteorite, *274*, 276
Alpha Centauri, 343, 344, 364, 365, 366, 373
Alpha Orionis, *see* Betelgeuse
Alpher, Ralph, 570, *570*, 574
Alps, 153, *153*
ALSEPs, 169
Amalthea, Jupiter, 230
Amino acids
    early Earth, *589*
    meteorites, 276, 588
Ammonia
    Jupiter and Saturn, 216, 217
Andromeda galaxy, 12, *13*, 484, 487, *488*, 502, 507–508, 528, 547
Angles
    measurements, 21
Angular momentum
    conservation, 47
    definition, 47, 55
Anorthosites, 171
Antarctic
    meteorites recovered, 273–274, *274*, 588
Antarctic Circle, 63
Antares, 376, *436*
Antenna
    radio, *119*
Antimatter, 312, 322
Aphelion, 49, 56
Aphrodite Terra, Venus, *190*, 190–191
Apogee, 49, 56

Apollo 17
    Earth image from, *148*
Apollo flights, 168–169, 169t
Apollo rockets, 169, *170*
Apparent brightness, 328, 337
Arclets, 533, *533*
Arcsec, 128
    definition, 110
Arctic Circle, 63
Arecibo Observatory, 122, *122*
Aristarchus of Samos, 24
Aristotle, 24, 32, 364
Armstrong, Neil, 167, 169, 172, 196
Associations
    stars, 422, 431
Asterism
    definition, 23
Asteroid belt, 50, 56, 250, 265
Asteroid(s), 135, *135*, 144, 280
    and comets, 249–267
    C (carbonaceous), 251, 265
    classification, 251–252, *252*, 265
    close up, 252–254
    comets and planets
        orbits, 50, *50*
    composition, 138, 251
    definition, 249–250
    discovery, 250
    Earth approaching, 255–256
    families, 251
    Flora family, 253
    Gaspra, 253, *253*
    Ida, *135*, 253, *253*, 277
    impact, 160–161, *161*
    largest, 251t
    M (metal), 251–252, 265
    motions, *250*, 250–251
    names, 250
    outer solar system, 255
    positions, 250, *251*
    primitive, 251
    S (stony composition), 251, 265
    Toutatis, 256, *256*, 265
    Trojan, 254, *255*, 265
    World Wide Web, A3
Astrology, 17–18
    beginnings, 28–29
    definition, 28, 36–37
    natal, 29, 30, 37
    personality characteristics, 30
    sun-sign, 29, 30
    tests, 30
    today, 29

Astronaut(s)
    on Moon, *166*
    "weightless," 48–49
Astronomical institutions
    World Wide Web, A1–A2, A5
Astronomical instruments, 107–129
Astronomical unit, 8, 44, 55, 362–363
Astronomy
    ancient, 23–28
        world, 23
    basics, 21–23, 138–139
    birth, 17
    definition, 2
    modern
        birth, 30–35
    nature, 2–3
    numbers, 5
    poets, 55
    radio
        origins, 118–119
    spectroscopy, 92–94
    study, 2
    World Wide Web, A1–A4
    x-ray, 125
Astronomy magazines, A5
Astronomy organizations, A1, A5
Astrophysical Journal, The
    Hubble/Humason paper, 513
Atmosphere(s)
    comets, *260*, 260, *261*
    Earth, 154–157, *155*, 194t
        observations outside, 122–127
    jovian planets, 216–222, 226, 282
    Jupiter, 216–217, 218
    Mars, 194t, 200, 205
    Neptune, 218–219, *219*
    planets, 282
    Saturn, 216–217, 218, 220
    Sun, 288–290, *289*, 291, *292*
    terrestrial planets, 282
    Titan, 235–236, *236*, 282
    Uranus, 218
    Venus, 194t, 194–196, *195*
Atomic clock
    experiment, 467
Atomic nucleus(i), 95, 103, 313–314
Atom(s), 15
    excited state, 98, 103, 380
    formation
        early universe, 572–573
    ground state, 95–96, *96*, 98, 103
    human body, 587
    ionized, 100, 103

Atom(s) (*Continued*)
  probing, 95, *95*
  structure, 94–97
Aurora(ae), 149, 293, *294*
Autumnal equinox, 65
Average density
  universe, 566
Axis
  Earth, 19, 26, *26*

B

B ring, Saturn, 242, *242*, 242t
Baade, Walter, 487, *488*, 509–510
Baade's window, *484*
Babylonians
  astrology, 28
Backus, Peter, *594*
Balmer series, hydrogen, 99
Bar, 154, 163
Barnard 86, *385*, 385–386, *418*
Barnard, E.E., 385, 386
Barnard's star, 335, *335*, 386
Barred spiral galaxies, 504, *505*
Basalts, 148, 163, 171, 190, 193, 200
  Vesta, 252
Baseline
  triangulation, 363
Basins
  Mars, 197
  Mercury, 179–180
  Moon, 171
Bayer, Johann, 366
Bell, Jocelyn, 450, *450*
Bellerophon, 410–412
Bessel, Friedrich, 364, *365*
Beta Lyrae, 369
Beta Orionis, *342*, 366
Beta Pictoris, 409
Beta Pictoris disk
  Hubble Space Telescope, *408*, 409
Betelgeuse, *342*, 366, *400*, *420*, 429
Big Bang, 13, *14*, 15, 559–583
  standard model, 570, 578, *578*
    problems, 576–577
Big Dipper, 335, *335*, *500*
Binary star systems
  black hole candidates, 473, 474t, 475
  evolution, 452–454
Binary stars, 345–347, 357
  eclipsing, *349*, 349–350, 357
    diameters, 351, *351*, 351, 358
    light curve, 351, *351*
  mass calculation from orbit,
    346–347
  revolution, 345, *345*
  spectroscopic, 345, 357
  visual, 345, 357
Binding energy
  nucleus, 314
Binoculars, 109, *110*
Biot, Jean Baptiste, 272
Black dwarfs, 441
Black hole(s), 13, 443, 459, 468–472, 475
  accretion disks around, *529*, 530
  candidates, 473, 474t
  classical collapse, 468–469
  collapse with relativity, 469–470

Black hole(s) (*Continued*)
  energy and, 527
  energy production around, 528
  event horizons, 470, 528–529
  evidence, 472–475
  fuel
    galaxies collisions, 530
  growth, 475
  M87, 527
  Milky Way Galaxy, 443–494
  myth, 470
  NGC 3115, 528
  NGC 4261, 528, *528*
  production, 472
  requirements, 473
  supermassive, 475
  trip into, 470–472
Blackbody(ies), 90–92, *91*, 103
  radiation, 575, *575*
Blink comparator, 239
Blueshift, 101
Bohr, Niels, 96–97
Bohr atom, 96–97
Bonner Durchmusterung, 366
Brahe, Tycho, 42, *42*, 55, 364, 444
Brightness
  apparent, 87, 103, 328, 337
  image, 110
  stars
    variations, 510
  units, 330
British Isles
  ancient astronomy, 23
Brown dwarfs, 347–348, 358, 406
Burbidge, Geoffrey, *430*
Burbidge, Margaret, *430*
Bursters, 127, 454
Butler, Paul, 411–412

C

3C 219, *525*
3C 273, 520, *521*
3C 275.1, *523*
4C 41.17, *566*
C ring, Saturn, 242, *242*, 242t
Calculus, 48
Calendar(s)
  early, 69–70
  functions, 68
  Gregorian, 70
  Julian, 69, 70
  units, 68
Callisto, Jupiter, 230, *230*, 231–232, *232*,
    245
Caloris basin, Mercury, 178, *180*
Canals
  Mars, 186, *186*, 187, 204
Cannon, Annie, 332–333, *333*
Caracol, Mayan observatory, 69
Carbon-14
  solar activity, 303
Carbon atoms, fusion of, 426
Carbon dioxide
  Earth, 155, 156, 158, 282
    martian polar caps, 201, 205
  terrestrial planets, 282
  Venus, 194, 195, 205

Carbon monoxide, *490*
Carbon-nitrogen-oxygen cycle, 316
Carswell, Robert, 531
Carter, Jimmy, 591
Cartwheel galaxy, *549*
Cassini, Gian Domenico, 242
Cassini Division, Saturn, 242, 242t, 244
Cassini spacecraft mission, 212–213, 245
Castalia, 256, 265
Castelli, Benedetto, 305
Castor, 345
Cat's Eye Nebula, *440*
CCD's, 112, *112*, 128
Celestial equator, 20
Celestial poles, 19–20, *20*, 36
"Celestial police," 250
Celestial sphere, 61
  around Earth, 18–19, *19*, 36
Centaurus, 366
Center of mass
  stars, 345–346, *346*
Cepheid variable stars, 368–369, 370,
  371, 373, 482, 502
  distance measurements, 509–510
Ceres, 49t, 50, 250, 251, 265
CFCs, 155
Challis, James, 54
Chandrasekhar, Subrahmanyan, 437, 438,
  438, 533
Chandrasekhar limit, 437, 455
Charge-coupled devices, 112, *112*, 128
Charon, 238–239, *240*, 245
Chemical condensation sequence
  solar nebula, 279, *279*
Chemical elements. *See* Elements.
Chicxulub crater, 160–161, *162*
Chinese
  ancient astronomy, 23
Chiron, 255
Chlorine, 321, 322
Christy, James, 238
Chromosphere, Sun's, *290*, 290–291,
  306
  emission lines, 291
  spicules, 291
Chryse Planitia, Mars, 196
Circular satellite velocity, 56
  definition, 50
Circumpolar zone, 21
Circumstellar disks, 142
Clark, David, 444
Clementine spacecraft, 170, *170*
Climate
  definition, 157
  Earth, 156–157
    Sun variability, 303–304, *304*
  Mars
    change, 203
Clouds
  dust, 385–387, *387*, *388*
  interstellar. *See* Interstellar clouds.
  jovian planets, 217–219, *218*
  Jupiter, 217, *217*, 226
  Magellanic, 369–370, *370*
  Mars, *197*, 200–201
  molecular, 383
    giant, 398, 413
    Orion, 399, 399–401, *400*

Clouds (*Continued*)
    stellar nurseries, *398*, 398–399, *401*
    neutral hydrogen, 380
    protogalactic, 494–496
    Saturn, 217, *217*
    Titan, 236, *236*
    Venus, 188, *188*, 190–191
Clough, Arthur Hugh, *The New Sinai*, 55
Clusters. *See* Galaxy clusters *or* star clusters
Colombo, Giuseppe, 179
Color index(indices), 330, 337
Coma cluster, *1*, 539–540, *541*, 546, *546*
Comet(s), 135–136, 144
    and asteroids, 249–267
    appearance, 256–257
    asteroids and planets
        orbits, 50, *50*
    atmosphere, *260*, 260, *261*
    composition, 138
    definition, 250
    "dirty snowball" model, 258
    dust, 258-259, *259*
    head, 260, *260*
    hunting as hobby, 264
    naming, 136
    nucleus, 258–259, 266
    orbits, 257
    origin and evolution, 261–264
    parts, *258*
    tail, 256, 260, *261*, 266
    well-known, 258t
    World Wide Web, A2–A3
Comet Halley, 248, 257, 257–258, 259, *259*, 260, *260*, 261, 264, 266, 588
Comet Mrkos, *261*
Comet Shoemaker-Levy 9, 161–162, *162*
    and Jupiter, 139, *139*, 263, *263*, *265*, 266
Comet Swift-Tuttle, 272
Comet West, 263
Communication
    interstellar, 591–592
Compton, Arthur Holly, 127
Compton Gamma Ray Observatory, 127, *127*, 128
Comte, Auguste, 336
Conduction, 318
Conic sections, 43, *43*
Constants, A21
    gravitation, 48
Constellation(s), 22–23 and A31–A32
    definition, 12, 22, 23, 36
    naming, 23
Convection, 151, 163, 318, 324
    stellar, 318, *318*
Coordinates system, 59
Copernican principle, 588
Copernicus, Nicolaus, *31*, 31–32, 55, 362, 587–588
    De Revolutionibus Orbium Coelestium, 31, *31*, 32, 37
Cordelia, Uranus, 245
Corona(ae)
    Milky Way Galaxy, 490, 497
    Sun, 77, *78*, 80, 291–293, *292*, *293*, *294*, 306

Corona(ae) (*Continued*)
    Venus, 193, *193*
Coronal holes, 293, 306
Coronal mass ejections, 299
Coronium, 292
Cosmic Background Explorer Satellite (COBE), 124, 484, 558, 575, *575*, 579, 581
Cosmic background radiation, *558*, 574–576, 579
Cosmic calendar, *14*, 15
Cosmic dust, 136, 144, *259*, 269, 272
Cosmic evolution, 585
Cosmic haystack problem, 592, 592t
Cosmic influences
    evolution of Earth, 159–163, 445–446
Cosmic rays, 391–393
    composition, 392
    genetic mutations, 445
    origin, 392–393, 445
Cosmological constant, 560, 581
Cosmological principle, 538–539, 556
Cosmology, 559–581
    early Greek and Roman, 23–24
    World Wide Web, A3
Crab Nebula, *434*, 444, 450, *451*, 452, *586*
    pulsar, 450, 451, *451*, 452, 454
Craters
    Callisto, 232, *232*
    Earth, 159–160
    Ganymede, 232, *232*
    Mercury, 179, 180, *180*
    Moon. *See* Moon, craters.
    Venus, 191–192
Critical density, 568, 581
    universe, 566, *567*, 569, 578
Cyanoacetylene, 383
61 Cygni, 364, 365, 366
Cygnus, 479
Cygnus arm, 486
Cygnus Rift, 486
Cygnus X-1, 473

D

Dactyl, 253
Dark matter, 12, 508, 509, 543–544
    cold, 580
    composition, 555, 556, 579
    galactic formation, 579–580
    galaxy clusters, 554
    hot, 580
    Milky Way Galaxy, 490, 497, 553, 555
    research, 555
    search
        gravitational lenses, 533, *533*, 555
    spiral galaxies, 553–554, *554*
    superclusters, 554–555
    universe, 553–555, 556
Darwin, Charles, 311
Darwin, George, 75, *75*, 176
Davis, Raymond, Jr., 321
Day
    defining, 178
    length, 66
    solar, 66
    sidereal, 66

Dean, Geoffrey, 30
Declination, 59, 61, 79
Degenerate stars, 437, 443
Deimos, Mars, 254, *254*
Delta Cephei, 368, *368*, 369
Density(ies)
    average
        universe, 566
    critical, 568, 581
        universe, 566, *567*, 569, 578
    definition, 46–47, 56
    materials, 47t
    specific, 47n
    weight, 47n
Detectors
    optical, 111–114
Deuterium
    abundance, 578, 579
    formation, 315, 322, 572–573
    present-day measurements, 573
Differential galactic rotation, 486
Differentiation, 138, 144, 276
Diffraction grating, 114, 495
Dimensions, 563–564
Dinosaurs
    extinction, 160–162
Diomedes, 254
Dispersion, 92, 103
Distance(s), 361–373, 511–513
    fundamental units, 362
    galaxies
        estimation methods, 511t
    Hubble law, 513
    units, 5–6
    within solar system, 362–363
Distance methods
    parallax, 363–367
    spectroscopic parallax, 371–372
    Type I supernovae, 510, *510*
    variable stars, 368–371, 509–510
Distance scale
    extragalactic, 509–511
Dog Star, 357, *357*
Donne, John, *Anatomy of the World*, 55
Doppler, Christian, 100
Doppler effect, 100–101, *101*, 103, 178–179, 512, 514, 520, 561
30 Doradus, *13*, 348, 506, *507*
Drake, Frank, 593, *593*
Draper, Henry, 336
Dreyer, John, *New General Catalog (NGC) of Nebulae and Clusters*, 379
Dumbbell Nebula, *439*
Dust
    cometary, 258–259, *259*, 271–272
    Galaxy, 485
    solar system, 136, 144, *259*, 269, 272, 278
Dust clouds, 385–387, *387*, 388
Dwarf galaxy(ies), 539, *540*
    elliptical, 506, *506*
Dwingeloo 1, 539, *541*

E

Earth
    approaching asteroids, 255–256
    as planet, 147–165

Earth (*Continued*)
    atmosphere, 154–157, *155*, 194t
        blue, 388, *390*
        evolution, 158
        observations outside, 122–127
    axis, 19
        precession, 26, *26*
    basic properties, 148, *148*, 148t, 188t
    carbon dioxide, 155, 156, 158, 282
    celestial sphere around, 18–19, *19*, 36
    climate, 156–157
        Sun variability, 303–304, *304*
    composition, 137
    craters, 159–160
    crust, 146, 148, *149*, 150–154
    evolution, 587
        cosmic influences, 159–163
        Sun influence, 428
    fault zones, *152*, 152–153
    geological activity, 139, 280–281
    gravity, 281–282
    greenhouse effect, *158*, 158–159, 163, 282
    impacts
        life on Earth, 159–163
        Moon origin, 176–177, *177*, 182
    interior, 148–149, *149*
    life and chemical evolution, 157–159
    locating places, 59
    magnetic field, 149
    magnetosphere, 149, *150*
    measurement
        Eratosthenes, 24–25, *25*, 36
    motion(s), 59, 61, 66
    mountain building, 153, *153*
    oxygen, 155, 158
    plate tectonics, 151, 280–281
    rift and subduction zones, 151–152, *152*
    rotation period, 66, 68
    seasons, 59, 61–66
    size, relative, 7, 8
    slowing, 75
    tidal forces, 472
    tides, 73–75
    volcanoes, *153*, 153–154
    water, 282
    weather, *156*, 156–157
    World Wide Web, A2
Earth satellite(s)
    launching, 50–51, *51*
    motion, 50–52
    orbits
        and space flight, 50–52
        friction, 51–52
    Skylab, 293
    unclassified
        and satellite debris, *51*
Eclipse(s)
    Moon, 24, *76*, 78, *78*, 80
    Sun, 24, *76*, 76–77, 79, 80, 291, *292*
        total, 466
    upcoming, A25
Ecliptic, 21, *22*
Eddington, Arthur, 357, 438, 466
Egyptians
    ancient astronomy, 23

Einstein, Albert, 312, *312*, 313, *313*, 459, 460, 462, 463, 465, 472, 514, 560
Einstein cross, *532*
Einstein Observatory, 125
Einstein ring, *532*
Electric charge, 84
Electric field, 84
Electrical repulsion
    nuclear attraction versus, 314–315
Electromagnetic force, 577
Electromagnetic radiation, 85, *86*, 87–92, 89t, 103, 318
    interstellar communication, 591–592
Electromagnetic spectrum, 87–92, *88*, 90t, 103
Electromagnetism, 15, 577
    Maxwell's theory, 84–85, 96
Electron(s), 84, 95, 312
    degenerate, 455
        massive star, 442–443
    early universe, 571–572, 573
    neutrino, 322
Elementary particles, 324
    matter, 312–313
    properties, 313t
Elements, A29–A30
    absorption lines, 332
    abundances, 334
    cosmically abundant, 15
    globular clusters, 430
    heavy
        globular star clusters, 430
        nuclear fusion, 431
        open star clusters, 429–430
        Sun, 429
    light
        cosmic rays, 392
    making new, 314, 315–316, 419–420, 426
        massive stars, 429, 431, 445
    open clusters, 429–430
    outer solar system, 210, 210t
    Sun, 288, *289*, 334
Ellipse, 43, 55
    drawing, 43, *43*
    eccentricity, 43, 55
    foci, 43
    major axis, 43
    semimajor axis, 43, 55
    shape, 43
Elliptical galaxies. See Galaxy(ies), elliptical
Enceladus, Saturn, *231*
Energy
    and mass, 312
    black holes, 527, 528
    conservation
        law, 310
    fusion, 323
    fusion reaction, 442
    gravitational, 310–312
    into matter, 571
    levels, 98–99, *99*
    quasars, 522
    reactions in Sun, 316
    thermal, 310–312
Engineering
    spacecraft, 212

Epicycle(s), 36
    definition, 28, *28*
Epsilon ring, Uranus, 243, 245
Equal areas
    law, 44, *44*
Equant point
    definition, 28
Equivalence principle, 460–461, 475
Equivalent mass, 566n
Eratosthenes
    earth measurement, 24–25, *25*, 36
40 Eridani B, 357
Eros, 254
Escape velocity, 468
    definition, 52, 56
Eta Carinae, 429, *429*
Ethyl alcohol
    clouds, 382
Europa, Jupiter, 230, *230*, *233*, 233–234, 245
European Southern Observatory, 116
    Very Large Telescope, *116*, 117t, 118, *560*
Event horizon, 469, 470, 475
Evolution
    cosmic, 585
    stars, 418–420
    universe
        early, 571–572, 576
Ewen, Harold, 381, *381*
Excited state, atoms, 98, 103
Explosions
    supernova. *See* Supernova(ae).
    white dwarf
        mild, 453–454
        violent, 454
Exponent, 5, A19
    negative, 5, A19
Extragalactic distance scale, 509–511
Extraterrestrial life, *see* Life, elsewhere
Extraterrestrial message(s)
    questions, 592, 592t
    searches, 593–595

F

F ring, Saturn, 242t, 243, 244, 245, *245*
Faults, *152*, 152–153, 163
Fireball, 270, *270*
Fisher, Richard, 511
Fission, nuclear, 314, *314*
Flamsteed, John, 366
Flares
    solar, 299, *299*, 306
Fluorescence, 379–380
Focal length, 109
Focus(i)
    ellipse, 43, 55
    lens, 109
Force(s), 45, 577
    electromagnetic, 577
    nature, 46, 577t, *578*
    nuclear
        strong, 314
Fornax, *561*
Foucault, Jean, 61
Foucault pendulum, 61
Fowler, William, 429, *430*

Fraunhofer, Joseph, 93, 330
Frequency, 86
Friction
    and satellite orbit, 51–52
Frozen stars. *See* Black hole(s).
Fusion
    cold, 323
    energy, 323
    nuclear, 314, 315–316, 419–420,
        426, 442, 445
    on Earth, 323

G

Galactic rotation
    differential, 486
Galaxies, 10, *10*, 11, 12, 14, 479, 519
    active, 524–527, 534
        elliptical, 525–526
    age, 546
    blue, *538*, 551
    cannibalism, 550–551, *551*, 556
    clusters. *See* Galaxy clusters.
    collisions, 547–551, *549*, *550*
        black hole fuel, 530
        galactic evolution, 548
    colors, 545, *545*
    composition, 546
    distances, 509–513, *561*
        estimation methods, 511t
    distribution, 538–545, 580
    dwarf, 506, *506*, 539, *540*
    elliptical, *505*, 505–506, *506*, 515,
        546, 556
        active, 525–526
        dwarf, 506, *506*
        gas and dust, 546
        giant, *505*, 506
            formation, 553
        mass
            measurement, 508
        surface brightness
            distance measurement,
                511, 511t
    evolution, 506–507, 538, 545–551,
        556
        bottom-up theories, 551, *552*,
            552–553
        top-down theories, 551–552,
            552
    formation, 573
        dark matter, 579–580
    host
        quasar, 523, *523*
    Hubble Space Telescope images,
        500, *523*, 524
    irregular, 506, 516
    Local Group, 539, *540*, 556, A35
    mass-to-light ratio, 508, 516, 555
    masses, 507–508
    merger, 550, 556
    Milky Way. *See* Milky Way Galaxy.
    nucleus
        active, 526–527
    properties, 507–509
    radio, 525
    recession, 512–514
    Seyfert, *524*, 524–525

Galaxies (*Continued*)
    shapes, 545
    spectrum(a), 508, 512, 545
    spiral, 485, *485*, *503*, 503–505, *505*,
        515, 547
        barred, 504
        dark matter, 553–554, *554*
        formation, 553
        mass-to-light ratios, 511
        speeds, 512
        Tully-Fisher distance measure-
            ment method, 511, 511t, 516
    types, 503–505, 508t, 515
    velocity-distance relation, 512–513,
        513
    World Wide Web, A3
Galaxy clusters, 539, 547, 556
    dark matter, 554
Galilean satellites, 230, *230*, 231–235
Galileo, Galilei, 33, *33*, 37, 55, 59, 108,
    188, 230, 295, 305, 480, 587–588
    astronomical observations, 34–35
    beginning modern science, 33–34
Galileo Probe, 212, *213*, 218
Galileo spacecraft, 52, 211–212, 230, 253
    images of asteroid, *135*, *253*, *277*
Galle, Johann, 54
Gallium, 322
Gamma rays, 87, 103
Gamma Sagittarii, 484
Gamow, George, 570, *570*
Ganymede, Jupiter, *138*, 230, *230*, 232,
    233, 245
Gas(es)
    between stars, 586
    degenerate, 437
    elliptical galaxies, 546
    hot
        Milky Way Galaxy, 492, *492*
        spiral galaxies, 546–547
    interstellar, *379*, 379–383, 383t
    interstellar clouds, *384*, 384–385
    Sun, made of, 288
    supernovae, 382
Gaspra, 253, *253*
Gaugelin, Michel, 30
Geller, Margaret, 543, 544, *544*
General relativity theory. *See* Relativity.
Geocentric view, 18, 36
Giant impact theory, Moon, 176–177, *177*
Giant molecular clouds, 398, 413
Giant planets. *See* Planet(s), jovian.
Giant stars, 333, 337, 351–352, 358
    development, 419
    evolution, 426–427
    mass loss, 427
    supergiant and, 351–352, 356, 358
Gilbert, Grove K., *173*, 173–174, 182
Giotto spacecraft, 259
Gliese 229, 348
Global Oscillations Network Group
    (GONG), 320, *321*
Global warming, 159, *159*
Globular star clusters, *360*, 421, *421*, *422*,
    425, *425*,431, 482, 489, 505, *505*, 562, *562*
    heavy elements, 430
    Milky Way Galaxy, 481–483, *483*,
        494, 496

Glossary, A6–A17
Goodricke, John, 351, 368, 369, *369*
Grand unified theories, 577, 581
    predictions, 577–579
Granites, 148, 163
Granulation, 294, *295*, 306
    convection currents of gases,
        294–295
Gravitation (*see also* Gravity)
    constant, 48
    more than two bodies, 52–54, 56
Gravitational energy, 310–312
Gravitational lenses, 518, 531–533, 555
    multiple images, 531, *532*
    search for dark matter, 533, *533*, 555
    two images, 531, *532*
Gravitational redshift, 468
Gravity, 15, 460–465, 563, 577
    accelerating environment, 461
    and mass, 48
    general relativity, 460–465
    law, 47–49, 56
    light, 469
    Milky Way Galaxy, 489–490, 494,
        494
    spacetime, 463–465
    sphere of influence, star's, 261
    terrestrial planets, 281–282
    time machines and, 471
    universal, 47–49
Great Attractor, *554*, 555
Great circle, 59
Great Dark Spot
    Neptune, 221, *221*, 222, *222*, 226
Great Red Spot
    Jupiter, 220–221, *221*, 226
Great Wall, 543, *543*
Greeks
    astrology, 28–29
    early cosmology, 23–24
    earth measurement, 24
    stellar parallax, 24
Greenhouse effect
    Earth, *158*, 158–159, 163, 282
    runaway
        definition, 195
    Venus, 195–196, 205, 282
Gregory, James, 109
Ground state
    atom, 95–96, *96*, 98, 103

H

Hale, George Ellery, 115, *115*, 336, 482
Hale-Bopp, Comet, 258
Half-life
    radioactive
        definition, 140, 144
Halley, Edmund, 257, *257*, 266
Halley's Comet. *See* Comet Halley.
Halo
    Milky Way Galaxy, 483, 496
Halo stars
    Milky Way Galaxy, *494*
Haute-Provence Observatory, France,
    410–411
Hawaii
    hot spots under, 153

Hawaii (*Continued*)
    Mauna Kea observatory, 114–116, 117–118, *118*, 336
HD 114762, 412
Heat, defined, 90
Hektor, 254
Heliocentric model
    solar system, *31*, 31, *32*, 32–33
Helium, 94, 137, 210, 291
    core
        red giants, 426
    following helium flash, 426–427, *427*
    formation, 572–573
    hydrogen fusion, 315–316, *316*
    Jupiter and Saturn, 215, *216*
    stars, 429, 430, 436
        main-sequence, 418
Helium flash, 426, 431
Helix Nebula, *439*
Hellas, Mars, 197
Henderson, Thomas, 364
Herbig-Haro objects, 403, 413
Herman, Robert, *570*, 574
Herschel, John, 54, 379, 422
Herschel, William, 53, 88-89, 136, 345, 379, 480, *481*
Hertz, Heinrich, 86
Hertzsprung, Ejnar, 352, *353*, 370
Hess, Victor, 391, *392*
Hewish, Antony, 450, *450*
Highlands
    lunar, 171, *171*
    martian, 197, *197*
Hipparchus, 25–26, 27, 36, 329
Hipparcos, 367, *367*
Hirayama, Kiyotsuga, 251
Horizon
    definition, 18, 36
Horizon distance, 576
Horoscope
    definition, 29, 37
    interpretation, 29, 30
    zodiac signs, 29
Hourglass Nebula, *440*
Hoyle, Fred, *430*
HR 3522, 412
H-R diagram(s), 352–357, *354*, *355*, 358, 371–373, *372*, *373*, 418
    evolution star like Sun, 426, *427*
    extremes of stellar luminosities, diameters, and densities, 356–357
    main sequence, 354, 355, 358
    older clusters, 424–425, *425*, 431
    star final-stages life, *441*
    stellar evolution, 405–407
        main sequence through red giant, 419–420, *420*
    stellar total mass and composition, 356
    young clusters, 423–424, *424*, 431
H II regions, 379–380, 393
Hubble, Edwin, 124, 502, 503, *503*, 508, 510, 512–513, 538–539, 561
Hubble constant, 513, 516, 561, 562
    value, 513, 562
Hubble law, 512–513, 516, 564, 569
Hubble Space Telescope, *7*, 58, 110, *124*, 124t, 124–125, *125*, 128, 348, 367, 370, 371, *420*, 429

Hubble Space Telescope (*Continued*)
    and Jupiter, *210*
    asteroid Vesta, 252, *253*
    Beta Pictoris disk, *408*, 409
    black holes in galaxies, 527, 528
    Comet Shoemaker-Levy 9, *263*
    deepest image, *500*
    Eta Carinae, *429*
    galactic distances, *561*
    galaxies, *500*, *523*, *524*
    Herbig-Haro objects, 403, *404*
    images distant galaxies, 545
    long-exposure image, *500*, *538*
    Orion Nebula, *396*, 399, *399*, *400*, 407
    planetary nebulae, *440*
    Shuttle mission repair, *560*
    spiral galaxy M100, *509*
    Supernova 1987A, 447, *447*
Hubble time, 560–562
Huchra, John, 543
Huggins, Sir William, 330, *331*, 334–335
Humason, Milton, *512*, 512–513
Huygens, Christian, 235
Hyades, 422
Hyakutake comet, 258, *587*
Hydrogen
    cold clouds, 380–382
    energy-level diagram, 98–99, *99*
    helium fusion, 315–316, *316*
    interstellar gas, 379, 380
    Jupiter and Saturn, 137, 210, 215, *216*, 280
    Milky Way Galaxy, 485, *485*
    neutral clouds, 380
    spectrum, 97–98, *98*
    stars, 331, 334, 429, 430
        main-sequence, 418
    Sun, 288
Hydrogen maser, 468
Hydrostatic equilibrium, 317–318, 324
Hypotheses, 4

I

Iben, Icko, 419
Ice Age
    Little, 303, *304*
Ice ages, 157
Ida, *135*, 253, *253*, 277
Igneous rock, 150, *150*, 163
Impact basins
    lunar, 171, *172*
    martian, 197
    mercurian, 179–180
Impacts
    Earth, 160, 275
        evolution of life, 162–163
    Jupiter, 263, *265*
Inertia
    law, 45
Infrared Astronomy Satellite (IRAS), *123*, 123–124, 386–387, *400*
Infrared cirrus, 387, *387*, 387, 393
Infrared observations, 112–113, 390
Infrared radiation, 88–89, 103, *478*
Infrared Space Observatory, 124
Infrared telescopes
    airborne and space, 122–124

Ingersoll, Andrew, 218
Institutions
    services and materials for public, A5
Instruments
    astronomical, 107–129
Integration time
    eye, 112
Interferometer array, 121, 128
Interferometer(s), 121
    very long baseline, 121–122, 128
Interferometry, 466–467
    radio, 120–121, 128
International Astronomical Union, 136
International date line, 67–68, *68*, 79
International Geophysical Year (IGY), 224
International Ultraviolet Explorer, 125
Interplanetary debris
    collision with Earth, 160
Interstellar clouds, 378
    evolution, 385
    gas pressures and temperature, *384*, 384–385
    structure and distribution, 384–385
Interstellar dust, 385–387, *387*, *388*
    components, 390–391, *391*
Interstellar extinction, 387, 393
Interstellar gas, 379, 379–383, 383t
    model, 383–385
Interstellar grains, 378, *378*, 390–391, *391*, 393
Interstellar matter, 377–378, 393
    around Sun, 385
    Milky Way Galaxy, 484–485
Interstellar medium, 378–379
Interstellar molecules, 382–383, 588
Interstellar reddening, 388–390, 393
Inverse-square law, 87, 103
Io, Jupiter, 224–225, 230, *230*, 233, *234*, 234–235, *235*, 245
    heat source, 280
    Pele volcano, *244*
Ionization, 100
Iron, 314, 442, 445
Irregular galaxies, 506, 516
Islam
    astronomy, 30–31
Isotopes, 96, *96*, 103

J

Jansky, Karl G., 118
Jeffers, Robinson, *Star Swirls*, 55
Jewel Box, *422*
Joule, 310
Jovian planets. *See* Planet(s), jovian.
Jupiter, 8, *9*, 132
    appearance, 36
    atmosphere, 216–217, 218
    basic properties, 213t, 213–214, 226
    clouds, 214, *214*, 217, 217, 219, 226
    colors, 214, *214*
    Comet Shoemaker-Levy 9, 139, *139*, 263, *263*, *265*, 266
    composition, 137, *137*, 215, *216*, 280
    finding from another star, 413
    internal energy source, 215–216
    internal structure, 215, *216*
    magnetic field, 222–225, *223*

Jupiter (*Continued*)
mass, 210, *210*, 348
moons, 230
ring, *241*, 241t
satellites, *217*, 230, *230*, 231–235,
232, *233*, 263
found by Galileo, 34–35
spacecraft exploration, 210–211
storms, 220–221, *221*
winds, 219–220, *220*
World Wide Web, A2

K

Ka'aba, 272
Kant, Immanuel, 502
Keck, Howard, 336
Keck Telescope(s), Mauna Kea, Hawaii,
117–118, *118*, 336
Kelvin temperature scale, 91
Kepler, Johannes, *42*, 42–43, 44, 55, 291,
362, 444
Kepler's law(s), 178, 473
first, 44
second, 44, *44*, 47
third, 44, 49, 56, 346, 486, 489, 492,
507
Kilometers
light years, 5–6
King Crater, Moon, *175*
Kippenhahn, R., 423
Kirchhoff, Gustav, 94
Kitt Peak National Observatory, Arizona,
114, 116, 451
Kopal, Zdenek, 369
Kruger 60, 345, *345*
Kuiper Airborne Observatory, 123, *123*
Kuiper belt, 262, *262*

L

Laplace, Pierre Simon, 468–469
Large Magellanic Cloud, *13*, 369–370,
370, 446, *447*, 506, *507*
Latitude, 60, *60*, 61, 65
Law(s)
conservation of energy, 310
gravity, 47–49, 56
Hubble, 512–513, 516
distances, 513
inertia, 45
inverse-square, 87, 103
Kepler's. *See* Kepler's law(s).
motion
Newton's, 45, 55
interpretation, 45–46
motion of planets, 44
nature, 4–5
radiation, 90–92
Stefan-Boltzmann, 92
Wien's, 91, 103, 330, 386, 574
Leavitt, Henrietta, 369, 370, 502
Lee, T.D., 438
Lemaitre, Georges, 570, *570*, 574
Lens(es)
focal length, 109, 128
gravitational. *See* Gravitational lenses.
images
formation, *109*, 109–110

Lens(es) (*Continued*)
properties, 110–111
Leo I, 506, *506*, A35
Leverrier, Urbain Jean Joseph, 54, 56,
465–466
Levy, David, 264, *264*
Lick, James, 336, *336*
Lick Observatory, 115, 336, 386, 411
Life
chemical building blocks, 588
elsewhere, 588–589
search for, 588–595, A3
evolution
impacts, 162–163
origin, *157*, 157–158
Light
absorption, 94
interstellar dust, 391
continuous spectrum, 88, *89*
gravity, 469
inverse-square law, 87, 103
nature, 84–87
optical properties, *89*, 92, *92*
photons, 86–87
pollution, 495
propagation, 87
sources for cities
spectra, 495, *495*
visible, 88, 92, 103
wave-light characteristics, 85–86, *86*
Light curve, 368, *368*, 368, 373
Light second, 362
Light travel time
consequences, 6–7
Light year(s), 5–6, 342, 365, 373
definition, 5
Line broadening, 508
Lippershey, Hans, 34
Lithium
formation, 572, 573
Little Ice Age, 303, *304*
Little Maunder Minimum, 301–303
Local Bubble, 385, 393
Local Fluff, 385, 393
Local Group, 12, *13*, 539, *540*, 556
brightest members, A35
movement, 554–555
Local Supercluster, 542, *542*
Longitude, 60, *60*, 61, 67, 68
Lookback time(s), 569, 569t
Lowell, Percival, 186–187, *187*,
237–238, 512
Lowell Observatory, Flagstaff, Arizona,
187, 512
Luminosity(ies), 326, 328–329, 330, 337
quasars, 521–522
stars, 372, *372*, 372, 373
distribution, 343, *344*, 357
supernovae, 445
Luminosity classes, 372
Luminosity function
stars, *343*, 343–344
Lyman series, hydrogen, 99

M

M4, 376, 379
M13, 421, *421*
M20, 388, *389*

M31, 12, *13*, 484, 487, *488*, 502, 503, 505,
507, 510, 512, 548
M32, *488*, 506
M41, 424, *424*
M42. *See* Orion Nebula.
M51, *548*
M74, *503*
M78, *366*
M83, *11*, *502*
M84, *541*
M86, *541*
M87, 458, 475, *505*, 506, *525*, 527, *527*
black hole in center, 527
M100, *371*, *509*
M104, *504*
Machholz, Don, 264
MACHOs, 555
Magellan, Ferdinand, 12, 506
Magellan spacecraft, *189*, 189–194
Magellanic Cloud(s), 12, *13*, 369–370, *370*
Large, 446, *447*, 506, *507*
Small, 489, 506, *507*
Magma, 153, 163
"Magnetic bottle," 323
Magnetic field(s), 149
Earth, 149–150
Galaxy, 392–393
Jupiter, 222–225, *223*
Mercury, 178
Neptune, 225, *225*
neutron star, 451, *451*
Saturn, 225
Sun, 292, 296, *297*, 299
Uranus, 225, *225*
Magnetism, defined, 84
Magnetosphere(s), 149, *150*, 163
definition, 222
jovian planets, 222–225, *223*, 226
Magnification, 126
Magnitude scale, 329, *329*
Magnitudes, 329, 337
stars, 26, 36
Main-sequence stars. *See* Stars, main-
sequence.
Making Connections
astronomy and days of week, 70
astronomy and future tourism,
243–244
astronomy and mythology, 350
astronomy and philanthropy, 336
astronomy and poets, 55
choosing telescope, 126
comet hunting as hobby, 264
defining day, 178
engineering and space science, 212
ethyl alcohol in clouds, 382
fusion on Earth, 323
gravity and time machines, 471
light pollution and Milky Way, 495
parallax and space astronomy, 367
quasars and attitudes of astronomers,
533
rainbow, 102
red giant Sun and fate of Earth, 428
solar system names, 136
striking meteorites, 275
supernovae in history, 444
testing astrology, 30
universe in distant future, 568

Malin, David, 342, 400
Manned Maneuvering Unit, 40
Mantle, Earth's, 148–149, 163
Marcy, Geoffrey, 411–412
Mare(Maria), 171, *172*, 179, 190
Mariner 4 spacecraft, 196
Mariner 9 spacecraft, 196
Mariner 10 spacecraft, 180, 181
Mars, 130
    appearance, 36, *186*, 186–187
    atmosphere, 194t, 200, 205
    basic properties, 188t, 188–189
    brightness, 186
    canals, 186, *186*, 187, 204
    canyonlands, 184
    channels, 186
        floods, *202*, 202–203
    climate change, 203
    clouds, *197*, 200–201
    composition, 137, 280
    cracks and canyons, *198*, 198–199
    eccentricity, 49–50
    geology, *196*, 196–200, 205, 280
    global properties, 196–197, *197*
    gravity, 281–282
    mountains, 281, *281*
    orbit, 43, 44
    polar caps, *201*, 201–202
    rotation, 188, 204
    samples, 200, *200*
    satellites, *254*
    search for life, 203, *204*, 588
    spacecraft exploration, 196
    surface
        view from, *199*, 199–200, *200*
    volcanism, *197*, 197–198, 281, *282*
    World Wide Web, A2
Martin, Jim, *204*
Maser
    hydrogen, 468
Mass, 45, 46
    and energy, 312
    and gravity, 48
    center of
        stars, 345-346, *346*
    definition, 46
    equivalent, 566n
    loss, in stars, 436, 441–442
Mass extinctions
    Earth, 160–161, 163
Mass-luminosity relation
    stars, 348, *349*, 358
Mass-to-light ratio(s), 508–509, 516, 555, 555t
Matter
    decoupling, 573
    density
        universe, 577
    elementary particles, 312–313
    energy into, 571
    in general relativity, 462
Mauna Kea Observatory, 114, 117, 118, *257*
Mauna Kea volcano, 154
Mauna Loa volcano, *153*, 282
Maunder, E.W., 301
Maunder Minimum, 301, 303, 306
    Little, 301–303

Maxwell, James Clerk, 84, *84*, 191, 577
    theory of electromagnetism, 84–85, 96
Maxwell Mountains, 191, 193
Mayan culture
    ancient astronomy, 23
Mayan observatory, Caracol, 69
Mayer, Michel, 410–411
McCandless, Bruce, 40
McMath-Pierce Solar Telescope, 286, 308, 326
Megaton, 160n
Mercury
    appearance, 36
    basins, 179–180
    Caloris basin, 178, *180*
    composition and structure, 168t, 177–178, *179*, 182
    craters, 179, 180, *180*
    geological activity, 139, 280
    motion, 465–466
    orbit(s), 49, 177, 465
    origin, 181
    properties, 168t
    rotation, 178–179, *179*
    scarps, 180, *180*
    surface, *134*, 179–181, *180*, 182
Meridians, 59, *59*, 79
Messier, Charles, 366, 378–379
Messier catalogue
    nebulae and star clusters, A34–A35
Metamorphic rocks, 150, 163
Meteor Crater, Arizona, 160, *161*, 274
Meteor shower(s), 270–272, *271*, *272*, 283
    Leonid, 271–272
    major annual, 271, 271t
    observation, 273
    Perseid, 271, 272
Meteorite(s), 136, 144, 269, 272–276, 283
    ages, 274–276
    amino acids, 276, 588
    carbonaceous, 276
    classification, 274, *274*, 274t, 276, 283
    compositions, 276
    definition, 272
    differentiated, 276
    extraterrestrial origin, 272–273
    falls, 273
    finds, 273, *273*–274, *274*
    from Mars, 200, 588
    from Vesta, 252, *252*
    primitive, 276
        Allende and Murchison, *274*, 276
    shower, 272–273
    SNC, 200, *200*
    striking property, 275, *275*
    World Wide Web, A3
Meteors, 136, *270*, 270–272, 283
    definition, 19, 270
    speed and size, 270
Meter, 362, A21
    modern redefinitions, 362
Methane
    Jupiter and Saturn, 216, 217
    Neptune, 218, *221*

Methane (*Continued*)
    Pluto, *244*
    Titan, 236
    Triton, 235, 236
Metric system, 362
    international, 5
"Mice" galaxies, *549*
Michell, John, 468
Microwave antenna, Bell Labs, 574, *574*
Microwaves, 89, 103
Milankovich, Milutin, 157
Milky Way, 34, 479–480, *480*
    light pollution and, 495
Milky Way Galaxy, 10, 11, *11*, *478*, 480, *483*, 483–484, *484*, 496, 539, *540*
    age, 546
    architecture, 480–485, *481*
    black hole at center, 491–494
    central region, 490–494
    corona, 490, 497
    dark matter, 489–490, 497, 553, 555
    disk
        stars, 483, 484
        gas and dust, 484–485
    formation, 494–496, 553
    gas and dust, 378, 485, 547
    globular clusters, 481–483, *483*, 494, 496
    gravity, 489–490, 494, *494*
    halo stars, *494*, 496
    Herschel's diagram, 480–481, *481*
    hydrogen, 485, *485*
    inner part, 484, *484*
    interstellar matter, 484–485
    mass, 489-490
    matter concentrations near, *554*, 554–555
    molecular clouds, 491, *491*
    nucleus, 490–494, *491*
    satellite galaxies
        distances, 548, A35
    spiral arms, *483*, 486, 496
    spiral structure formation, *486*, 486–487
    stellar populations, 487–489
Miller, Stanley, 588
Miller-Urey experiment, *589*
Mimas, Saturn, 244
Mirrors
    image formation by, *109*, 109–110
    largest, 116–118
Mizar, 345
Molecular clouds, 383
    Milky Way Galaxy, 491, *491*
Molecule(s), 15
    interstellar, 382–383
Momentum
    angular
        conservation, 47
        definition, 47, 55
    change, 45
    conservation, 48
    definition, 55–56
    factors influencing, 45, 46
Moon, 7, 21, 167–183
    acceleration
        orbit, 48
    appearance, 171

Moon (*Continued*)
    basic facts, 168, *168*
    composition and structure, 170–171, 182
    craters, 173–176, 182
        counts, 175–176, *176*, 182
        impact, 173–176, *174*
        origin, 173–174, *174*
        process producing, *174*, 174–175, *175*
    Earth and, 7–8, *8*
    eclipse(s), 24, *76*, 78, *78*, 80
    exploration, 168–170
    general properties, 168t, 168–171
    geological activity, 139, 280
    geological features, *171*, 171–172
    highlands, 171, *171*, 182
    impact basins, 171, *172*
    maria, 171, *172*, 182
    motion, 22, 59
    mountains, 172, *173*
    observing, 181
    origin
        theories, 176–177, *177*, 182
    phases, 70–73, *71*
        appearance, 181
    pull on Earth, *73*, 73–74
    revolution and rotation, *72*, 73
    rocks
        ages, 171
    samples, 168–169, *169*
    surface, 171–173, 172–173, *173*
    tides, 73–75
    volcanoes, 171–172, *172*
    World Wide Web, A2
Motion
    Newton's laws, 45, 55
        interpretation, 45–46
    proper, 335
Mount Wilson Observatory, 115, 502, 538
Mountains
    building
        Earth, 153, *153*
        Moon, 172, *173*
    origins, 281, 283
    Venus, 191, 193, *194*
Murchison meteorite, 276
Mutations
    genetic
        cosmic rays, 445
MyCn, *440*

N

Natal astrology, 29, 30, 37
National Aeronautics and Space Adminis-
    tration (NASA)
    airborne observatories, 123
    Hubble Space Telescope, 124–125
    lunar exploration, 168–169
    satellite observatories, 127
    search for radio messages, 593
National Radio Astronomy Observatory,
    121, 121t, 593, *593*
Nature
    forces, 46, 577t
    laws, 4–5

Near-Earth Objects (NEOs), 255–256, 265
NEAR (Near Earth Asteroid Ren-
    dezvous), 254
Nebula(ae), 377, 393, 502
    dark, 385–387
    Messier catalogue, A34–A35
    nomenclature, 378–379
    planetary, 437–439, *439*, 455
    reflection, 387–388, *388*
    spiral, 502. *See* Galaxy(ies).
    World Wide Web, A3
Negative exponent, 5, A19
Neptune, 49, 237, 238, 466, 553
    atmosphere, 218–219, *219*
    basic properties, 213, 226
    composition, 137, 280
    discovery, 53–54, *54*
    internal structure, 215, *216*
    magnetic fields, 225, *225*
    rings, *241*, 241t, 243, *244*, 245
    satellites, 231, 236–237, 240
    storms, 221, *221*, 222, *222*
    weather, 220
Nereid, 240
Neutrino(s), 312–313, 315, *319*, 324, 443,
    576
    dark matter, 579
    early universe, 571–573
    interactions, *321*, 321–322
    mass, 322
    solar, 321
    Supernova 1987A, 449, *449*
    supernovae, 448, 449, 579
    total mass
        universe, 579
Neutrino detector(s), 449, *449*
Neutrino electron, 322
Neutrino "telescopes," 449, *449*
Neutron(s), 312
    ball, 442–443
    degenerate
        massive star, 443
    early universe, 571, 573
Neutron stars, 412, 443, 449–450, 455,
    473
    companions, 454
    discovery, 450–451
    magnetic field, 451, *451*
    properties, 450t
    spinning, 451
Newton, Isaac, 44–45, *45*, 47–48, 55, 88,
    257, 260, 460, 463, 465
    first law, 45
    law of gravity, 47–49, 56
    laws of motion, 45, 55
        interpretation, 45–46
    satellite orbit, 50, *51*
    second law, 45
    third law, 45, 46
Newton's great synthesis, 44–47
NGC 205, *488*, 506
NGC 1068, 524, *524*
NGC 1189, *545*
NGC 1190, *545*
NGC 1199, *545*
NGC 1265, *547*
NGC 1365, *504*

NGC 1566, *524*
NGC 2264, 423, *423*
NGC 2997, 485, *485*
NGC 3115
    black hole, 528
NGC 3293, 423, *424*
NGC 3603, *378*
NGC 4038, *536*, *550*
NGC 4039, *536*, *550*
NGC 4261
    black hole, 528, *528*
NGC 4676A, *549*
NGC 4676B, *549*
NGC 4755, *422*
NGC 5746, *554*
NGC 6166, *551*
NGC 6240, *549*
NGC 6251, *526*
NGC 6520, 385, *418*
NGC 6543, *440*
NGC 7027, *440*
NGC 7252, *549*
NGC 7293, *439*
Nickel
    in Earth's core, 149
    radioactive
        supernovae, 448
1992 QB, 262, *262*
Niven, Larry, *World Out of Time*, 471
North Star, *20*, 24, 26, 369
Nova(ae), 454, 455
    binary star systems, 454
Nuclear attraction
    versus electrical repulsion,
        314–315
Nuclear bulge
    Milky Way Galaxy, 483–484, *484*,
        488, 496
    spiral galaxies, 504, *505*
Nuclear fission, 314, *314*, 315–316,
    419–420, 426, 442, 445
Nuclear force
    strong, 314
Nuclear fusion, 314,
    heavy elements, 442
Nucleosynthesis, 429, 431
Nucleus(i)
    active galactic, 524, 526–527, 534
    atomic, 95–96, 313–316
    comets, 258–259, 266
    Milky Way Galaxy, 490–494, *491*,
        497

O

Observatory(ies)
    high-energy, 125–127
    modern astronomical, *114*,
        114–118
    pretelescopic, 108, *108*
    World Wide Web, A2
Occultation, 246
    definition, 242
Oliver, Bernard, 589
Olympus Mons, Mars, 197, *197*, 198, 205,
    281, *282*
Omega Centauri, 421, *422*
Oort, Jan, 261–262, 266

Oort comet cloud, 261–262, 266, 522
Opacity, 318
Open star clusters, 421–422, *422, 424,* 431
  heavy elements, 429–430
Ophelia, Uranus, 245
Optical detectors, 111–114
Optical double stars, 345
Optical window, 107
Orbits
  motion and mass, 49
  planets
    range data, 49t, 49–50
    solar system, 49–50
      description, 49
Orientale
  lunar impact basin, *172*
Orion, 327, *327, 342, 366,* 399, *399, 400*
  constellation, 22
Orion arm, 486
Orion molecular cloud, *399,* 399–401, *400*
Orion Nebula, *6, 399, 400,* 402, 486, 502
  Hubble Space Telescope, *396, 399, 407*
58 Orionis, *342,* 366
Oscillating theory of universe, 567
Oscillation, 84–85, *85*
Ostro, Steven, 256
Oxidation, 137, 310, *310*
Oxygen
  Earth, 155, 158
Ozone, 155, 163, 304

P
Pallas, 250, 251
Pan, Saturn, 245
Parallax
  definition, 24
  measurements, 367
  space astronomy, 367
  spectroscopic, 372
  stellar, 364–365, *365,* 373
  triangulation, 363
Parsec, 365
Particles
  elementary. *See* Elementary particles.
  weakly interacting massive, 579
Patroclus, 254
Pauli, Wolfgang, 312
Payne, Cecilia, 333
Peebles, James, 544
51 Pegasi, 411
Pegasus, 411
Pele volcano, Io, *244*
Penumbra, 295
Penzias, Arno, 574, *574*
Perigee, 49, 56
Perihelion, 49, 56, 465, 466
Perseus, 350
Perseus arms, 486
Perturbations
  calculation, 53
Phobos, Mars, 254, *254*
Pholus, 255
Photochemistry, 218, 226

Photographic plate, 112, *112*
Photometry, 329
Photon(s), 86, 97, 103
  absorption, 94, 318
  definition, 86
  early universe, 571
  generation, 97, 318, *319*
  light as, 86–87
Photosphere, 288–290, *289*
  activity above, *297,* 297–301
  gases, 288, 289, 289t
  properties, 289t, 306
Photosynthesis, 158
Piazzi, Giovanni, 250
Pickering, Edward C., 333, 336, 345, 369
Pigott, Edward, 369
Pioneer spacecraft, 210–211
  magnetosphere of Jupiter, 223, *223*
  messages aboard, 590, *590*
Plages, 297, *298,* 306
Planck, Max, 97
Planck's constant, 97
Planetary nurseries, 142
Planetary Society, 143
Planetary system, 132n
  overview, 132–137
Planetesimals, 142, *142,* 144, *268,* 279, 280
Planetology
  comparative, 132
Planet(s), 135t, 348, 406
  atmospheres, 282
  beyond Solar system
    search and discovery, 409–413
  imaging, 412–413, 414
  comets
    asteroids
      orbits, 50, *50*
  composition and structure, 137–139
  definition, 21
  differentiation, 138, 144
  discovery, 410–412
    around other stars, 412, 412t
  Earth, 147–165
  elevation differences, *281,* 281–282, *282*
  evolution, 280–282, 283
    divergent, 203–204, 205
  extra-solar, 409–413
  formation, 142–143, 586–587, 588
  formation around other stars, 407–409
  geological activity, 139, 280–281
  giant. *See* Planet(s), jovian.
  jovian, *134,* 134, 137, 144, 222–225, *223, 226*
    appearance and rotation, 214–215
    atmospheres, 216–222, 226
    basic properties, 213t, 213–214, 226
    clouds and atmospheric structure, 217–219, *218*
    composition and structure, 215, *216*
    formation, 279–280
    internal heat sources, 215–216
    spectroscopic observations, 216

Planet(s) *(Continued)*
    storms, 220–222, *221, 222*
    winds and weather, 219–220
  motion, 22
    laws, 42–44
  naming, 136
  observation, 36
  orbital data, A22
  orbiting Sun
    plane, 133, *133*
  orbits, 10, 49t, 49–50
    spacing, 134
  outer
    abundances, 210, 210t
    exploration, 210–213
  physical data, A22
  retrograde motion, 27, *27,* 36
  rings, 214, *214,* 231, 240–245, 241t, 246
  satellites. *See* Satellite(s).
  second generation, 412
  temperatures, 138, 139
  terrestrial, 133, *134,* 137–138, 144
    formation, 279
Plate tectonics
  concept, 151, 154, 163
  Earth, 151
Pleiades, 348, 387, *388,* 421–422, 441
Plutarch, 291
Pluto, 8–9, 49, 134, 522
  discovery, 237–238, *238,* 245
  motion, 238
  nature, 239–240, *240*
  orbit
    speed, 49, 50
  origin, 240
  satellite, 238–239, *240*
Poets
  astronomy, 55
Pohl, Fred, *Gateway,* 471
Polar caps
  Mars, *201,* 201–202
  Triton, 236–237, *237*
Polaris, *20,* 21, 24, 26, 369
Polycyclic aromatic hydrocarbons, 383
Polynesians
  ancient astronomy, 23
Pope, Alexander, *An Essay on Man,* 55
Population I stars, 487, 488, 496–497
Population II stars, 487–488, 496
Positron(s), 312, 392, 393
  early universe, 571–573
Pound, Robert, 467
Powers-of-ten notation, 5 and A18–A19
P-P chain, *315,* 315–316, 322, 324
PPL 15, 348
Precession, 26, 29, 36
Primeval atom, 570
Primitive rocks, 150, 163
Prism, *89, 92, 92*
Project Ozma, *593*
Project Phoenix, 593, *594*
Prometheus, Saturn, 245
Prominences, 297, *298, 299,* 306
Proper motion, 335
Proportionality constants, 513
Proto-planetary disks
  evolution, 408

Protogalactic cloud, 494–496
Proton-proton cycle, *315*, 315–316, 322, 324
Proton(s), 15, 84, 95, 96, 312
    early universe, 571–572
    temperature for fusion, 314–315
Protoplanets, 279, 280
Protostar(s), 402, *403*, 405, 413
    disks around, *407*, 407–408, 413
    evolutionary tracks, 405, *406*
Proxima Centauri, 9, 344, 366–367
PSR 1257+12, 412
Psyche, 252
Ptolemy, Claudius, 18, 27–28, 29, 364
    Almagest, 27, 36
    Tetrabiblos, 29, 37
Pulsar(s), 412, 450, 455
    Crab Nebula, 450, 451, *451*, 452, 454
    evolution, 452, *453*
    spinning lighthouse model, 451–452
Pulsating variable stars, 368, 373
Purcell, Edward, 381, *381*
Pythagoras, 23

Q

QSO 0957+561, 531, *531*
QSO 1007+417, *526*
QSO 1229+204, *523*
Quantum mechanics, 86, 96–97, 313
Quasars, 13, 519, *520*, 520–523, 534
    attitudes of astronomers, 533
    distances, 522, 523
    energy, 522
    evolution, 529–531
    galaxies evolution, 546
    host galaxy, 523, *523*
    light, 531
    luminosity, 521–522
    numbers
        redshifts, 530, *530*
    power behind, 527–531
    redshifts, 520, 521, 522, 523
    sizes, 521
    spectra, 520–521, 529, 531
Quasi-stellar objects. *See* Quasars.
Quasi-stellar radio sources. *See* Quasars.
Queloz, Didier, 410–411

R

R 136, *348*, *416*
Radar, 122, 362, *363*
Radar astronomy, 122, 128, 178–179, 189–190
Radial velocity, 101, 103, 320, 334, 337
Radial-velocity curve
    stars, 346, *347*, 347
Radiant, 271, *271*
Radiation, 318, 324
    and temperature, 90
    blackbody, 575, *575*
    cosmic background
        discovery, 574
        isotropic, 576
        Penzias and Wilson, 574

Radiation (*Continued*)
    properties, *575*, 575–576, 581
    temperature fluctuations, *558*, 576
    uniformity, 576
decoupling, 573
definition, 85
early universe, 571–572
electromagnetic, 85, *86*, 87–92, 89t, 103, 318
    interstellar communication, 591–592
infrared, 88–89, 103, *478*
radio. *See* Radio radiation.
synchrotron
    Jupiter, 222–223, *223*, 226
ultraviolet, 88, 103
Radiation laws, 90–92
Radio antenna(s), 118–122, 466–467
Radio astronomy
    origins, 118–119
Radio galaxies, 525
Radio image, *120*
Radio interferometry, 120–121, 128
Radio jets, 529
Radio lobes, 525
Radio radiation
    detection 119–120
    instruments, 118–122
    Milky Way Galaxy, 493
    Sun, 466
Radio stars, 520
Radio telescope(s), 118–122, *120*, 121t
    Arecibo, Puerto Rico, 122, *122*
    Bonn, Germany, *120*
    New Mexico, *82*, 121
    Parkes, Australia, 593, *594*
Radio window, 107
Radioactive decay, 141, *141*
    dating rocks, 141, 141t
Radioactive rocks, 140–141, 144
Rainbow, 102, *102*
Raman, C.V., 438
Reber, Grote, 119, *119*
Rebka, Glenn, 467
Red dwarfs, 357
Red giant(s), 420, 421, *424*
    development, 419
    evolution
        models, 419–420
        helium core, 426
Redshift(s), 101, 512, 516, 560t, 565
    gravitational, 468
    lookback times, 569t
    quasars, 520, 521, 522, 523
Reducing atmosphere, defined, 137
Reflection nebulae, 387-388, *388*
Refraction, 66
Relativity
    theory, 312
        general, 313, 459, 460–468, 469–470, 514
            predictions, 467
            tests, 465–467
            tests of time, 467
            time, 467–468
        special, 333
Resolution, 110

Resonance(s), 246
    definition, 244
Retrograde motion
    planets, 27, *27*, 36
Rho Ophiuchi, 376
Riccioli, John Baptiste, 345
Ride, Sally, 302
Rift zones, 151–152, *152*, 163
Rigel, *342*, 366
Right Ascension, 59, 61, 79
Ring Nebula, *439*
Rings
    Jupiter, *241*, 241t, 245
    Neptune, 231, *241*, 241t, 243, *244*, 245
    outer solar system, 230, 245
    planetary, 214, *214*, 231, 240–245, 246
        causes, 241–242
        properties, 241t
    -satellite interactions, 243–245
    Saturn, 135, *135*, 231, 241, *241*, 241t, 242, *242*, 242t, 244, 245, *245*, 246
    Uranus, 214, *214*, 231, 241, *241*, 241t, 242–243, *243*, 245, 246
Rockets, 46, *46*
Rock(s)
    age
        measurement, 140
    igneous, 150, *150*, 163
    lunar
        ages, 171
    metamorphic, 150, 163
    primitive, 150, 163
    radioactive, 140–141, 144
    sedimentary, 150, 163
Roentgen, Wilhelm, 127
Roentgensatellit, 127, *127*
Romans
    early cosmology, 23–24
Rosat, 127, *127*
Rosette Nebula, *12*
Rotation
    galaxies, 504, 506
    Galaxy, 489–490
    planets, 133
    stars, 335–336, 337
Royal Greenwich Observatory, 59, *59*
RR Lyrae, 370
RR Lyrae variable stars, 368, 370–371, *371*, 373, 482, 483, 487–488, 489
Russell, Henry Norris, 352, *352*, 353, *353*, 438, 482
Rutherford, Ernest, 95

S

S106, 403, *403*
Sagan, Carl, 15, 143
Sagittarius A°
    Milky Way Galaxy, 493
Sagittarius-Carina arms, 486
Sagittarius dwarf galaxy, 12, A35
San Andreas fault, 152, *152*, 153
Sandage, Allan, 520
Satellite(s), 135 and A23–A24
    artificial Earth. *See* Earth satellite(s).
    Galilean, 230, *230*, 231–235, *232*, *233*, *234*, *235*, 245

Satellite(s) (*Continued*)
  Jupiter, 230, *230*, 231–235, *232, 233*
  Mars, *254*
  Neptune, 231, 240
  outer solar system, 230, 245
  Pluto, 238–239, *240*
  -ring interactions, 243–245
  Saturn, 230–231, *231*, 244, 245
  Uranus, 214, *214*, 231, 245
Saturn
  appearance, 36
  atmosphere, 216–217, 218, 220
  basic properties, 213t, 213–214, 226
  clouds, 217, *217*
  composition, 215, *216*, 280
  internal energy source, 216
  internal structure, 215, *216*
  magnetic fields, 225
  rings, 135, *135*, 231, 241, *241*, 241t,
    242, *242*, 242t, 244, 245, *245*, 246
  satellites, 230–231, *231*, 244, 245
  storms, 222, *222*
  winds, 219–220, *220*
  World Wide Web, A2
Scarps
  Mercury, 180, *180*
Schiaparelli, Giovanni, 186
Schmidt, Maarten, 520, *520*, 533
Schmidt telescope, 562
Schmitt, Jack, 168, 169
Schwabe, Heinrich, 296
Schwarzschild, Karl, 470, *470*
Schwarzschild radius, 470
Science
  nature, 3–4
  units, A20
Scientific laws, 4
Scientific notation, 5, A19–20
Search for Extra-Terrestrial Intelligence
  (SETI), 588–595
Seasons
  different latitudes, 65
  Earth, 59, 61–66
  Mars, 188, 201
  Uranus, 214–215, *215*
Sedimentary rocks, 150, 163
Seeing
  atmospheric, 116, 128
Seeing For Yourself
  Earth, round, 35
  observing Moon, 181
  observing planets, 36
  observing Sun, 305
  showering with stars, 273
  solar eclipse observation, 79
Seismic waves
  definition, 148, 163
Seismology
  solar, 320, 324
SETI Institute, 593
Seyfert, Carl, 524
Seyfert galaxy(ies), 524, 524–525
Shapley, Harlow, 353, 370, 481, 482, *482*,
  501
Sharp, Nigel, 451
Shelton, Ian, 446
Shoemaker-Levy 9 comet. See Comet
  Shoemaker-Levy 9.

"Shooting stars." *See* Meteors.
Sidereal day, 66, *67*, 79
Sigma Herculis, *340*
Silicates, 137
Singularity, 472, 475
Sirius, 23, 343, 357, *357*, 366, 367, 441
Sirius B, 357, *357*
Sky
  dome, 18, *18*
  events
    World Wide Web, A3–A4
  locating places, 59–60
  turning
    observing, *20*, 20–21
Slipher, Vesto M., 512, *512*
Small Magellanic Cloud, 369–370, 489
SNC meteorites, 200, *200*
Soil
  lunar, 172–173, *173*
Solar day, 66, *67*, 79
Solar eclipse. *See* Eclipse(s).
Solar nebula, 142, *142*, 144
  chemical condensation sequence,
    279, *279*
  solar system formation, *278*,
    278–279, 283
Solar seismology, 320, 324
Solar system, 8, 131–145, 132n, A25
  distances within, 362–363
  formation
    age constraints, 278
    chemical constraints, *276*,
      276–277
    motion constraints, 276
  heliocentric model, *31*, 31, *32*,
    32–33
  members, 132–134
    mass, 132t
    smaller, 135–136
  moons
    composition, 138
  names, 136
  origin, 141–143, 269
  outer
    asteroids, 255
    ring and satellite systems, 230,
      245
    spacecraft exploration,
      210–211, 211t, 226
  satellites, 229 and A27–A28
  scale model, 136–137
  World Wide Web, A2
Solar time
  apparent, 66, 79
  mean, 66–67, 79
Solar wind, 149, *261*, 293
Sombrero galaxy, *504*
Space Infrared Telescope Facility
  (SIRTF), 124
Space Shuttle, *46*, 52, 58, 124, *125*, 560
Spacecraft, 589. *See also specific spacecraft.*
  exploration
    Mars, 196
    outer solar system, 210–211,
      211t, 226
    Venus, *189*, 189–190
  interplanetary, 52

Spacetime, 463, 475
  examples, *463*, 463–465, *464*
  geometry, 563
  gravity, 463–465
  two-dimensional, *467*, 467–468
Specific density, 47n
Spectral classes, 331–333, 332t, 337
Spectral lines
  formation, 97–100, *99*
  formation 21-cm line, 380–381, *381*
Spectrometer, 92, 103, *113*, 113–114, *114*
Spectroscopic parallax, 372
Spectroscopy, 92–94, 113–114
Spectrum(a)
  electromagnetic, 87–92, *88*, 90t,
    103
  galaxies, 508, 512, 545
  hydrogen, 97–98, *98*
  quasars, 520–521, 529, 531
  stars. *See* Stars, spectrum(a).
  types, 94, 103
Speed
  and momentum, 45
Spica, 446
Spicules, 291
Spiral arms
  formation, *486*, 486–487
  Milky Way Galaxy, *483*, 486, 496
Spiral density wave model, 487, 496
Spiral galaxy(ies), 485, *485*, *503*,
  503–505, *505*, 515, 542
  barred, 504
Spiral nebulae, 502
Sporer, Gustav, 301
Sputnik, 50
Standard time, 66–67
Star clusters, 12, *12*, 420–422
  characteristics, 421t
  chemical composition
    differences, 429–430
  evolution, 552–553
  globular. *See* Globular star clusters.
  Messier catalogue, A34–A35
  older
    H-R diagrams, 424–425, *425*
  open, 421–422, *422*, *424*, 431
    heavy elements, 429–430
  young
    H-R diagrams, *423*, 423–424
Stars, 2, 417–433
  absorption lines, *332*, 334, *334*
  ages, 488, 562–563
  apparent brightness, 328
  approaching death, 430
  associations, 422, 431
  basic facts, 398t
  binary. *See* Binary stars.
  birth, 401–402, *402*
  brightest, A27
  brightness, 328–329
  center of mass, 345–346, *346*
  characteristics
    measurement, 352t
  classification, 327
  collisions, 547–548
  colors, 330
  core
    contracting, 436–437

Stars (*Continued*)
  massive
      escape velocity, 468, *469*
  critical mass, 406
  death, 435–457, 445t
  degenerate, 437
  density, 344
  diameters, 349–352
    eclipsing binary stars, 351, *351*
    stars blocked by Moon, 349
  discovery of planets around, 412, 412t
  distance(s), 364–365, *365*
    H-R diagram, 371
    measurement difficulties,
      372–373
    measurements, 342–343,
      368–371
    units, 365
  double, 345
  energy source, 318
  evolution, 418–420, 428t
    final stages, 425–426
    H-R diagram, 405–407
  evolutionary time scales, 406–409
  evolutionary tracks, 405–406, *406*
  fixed and wandering, 21–22
  formation, 398–405, *548*, 548–549,
    551, 573
    continuous process, 397
    elliptical galaxies, 546
    gravity and pressure, 399
    spiral galaxies, 546
    through molecular cloud, 401,
      401
  frozen. *See* Black hole(s).
  gas and dust, 378
  giant. *See* Giant stars.
  globular clusters. *See* Globular star
    clusters.
  gravitational sphere of influence,
    261
  heat transfer, 318
  hydrogen, 331, 334
  low-mass
    death, 436–442
  luminosity function, *343*, 343–344
  luminosity(ies), 326, 328–329, 330,
    337, 368
    classes, 372, *372*, 373, *373*
    distribution, 343, *344*, 357
  magnitude scale, 329
  magnitudes, 26, 36
  main-sequence, 405–406, 413,
    418–419, 419t
    characteristics, 355–356, 356t,
      358
  main-sequence to red giant, 419
  mass-luminosity relation, 348, *349*,
    358
  mass(es), 344–348
    evolution, 406
    loss, 436
      evidence, 441–442
    range, 347–348, *348*
  massive
    collapse and explosion, 443
    core, 442, *442*, 443
    elements, making new, 429, 431

Stars (*Continued*)
    evolution, 427–430, 442–449
  Milky Way Galaxy disk, 483, 484
  Milky Way Galaxy nucleus, 492, *493*,
    497
  model, 319
  nearest, A26
  neutron. *See* Neutron stars.
  nomenclature, 365–366
  nuclear bulge, 483–484, *484*, 488,
    496
  optical double, 345
  planetary formation around, 407–409
  Population I, 487, 488, 496–497
  Population II, 487–488, 496
  populations
    Milky Way Galaxy, 487–489
  position
    measurement, 409
  radial-velocity curve, 346, *346*, 347,
    347
  reddened, 388–390, 393
  rotation, 335–336, *337*
  size(s), 327
    effects, 333–334
  source of elements, 427–429
  spectral types (classes), 331–333,
    332t, 337, 371–372
  spectrum(a), 330–336, *331*
    classification, 331–333
    formation, 331
    value, 92–94, *93*, *94*
  stable, 317–318
  supergiant. *See* Supergiants.
  superluminous, 356
  temperature/color, 330, 331, *332*,
    332t
  triangulation, 363–364
  variable, 368–371
    distance measurements,
      509–510
    period-luminosity relationship,
      369–370, *371*, 371, *371*, 371,
      373
  World Wide Web, A3
  zero-age main sequence, 418, 431
Stefan-Boltzmann law, 92
Stellar parallax, 24, 36
Stellar wind(s), *402*, 403–405, 413
Stephenson, Richard, 444
Stickney, *254*
Stockhausen, Karlheinz, *Ylem*, 570
Stonehenge, 23, 65, 69, *69*
Storms
  jovian planets, 220–222, *221*, *222*
  Jupiter, 220–221, *221*
  Neptune, 221, *221*, 222, *222*
  Saturn, 222, *222*
Stratosphere, 155, 163
Stromatolites, 157, *157*
Struve, Friedrich, 364
Subduction zones, 151–152, *152*, 163
Sulfur dioxide
  Io, 234, *234*
  Venus, 194
Summer solstice, 63, *64*
Sun, 8, 11, 12, 132, *132*, 309–325, 410
  active regions, 299–301, *301*, 306

Sun (*Continued*)
  activity
    carbon-14 studies, 303
  and Earth, 8, *9*
    climate, 303–304
  as variable star, 301–304
  atmosphere, 288–290, *289*
    temperatures, 291, *292*
  basic characteristics, 287, 288t
  chromosphere, *290*, 290–291
  composition, 288
  convection currents, 318, *318*
  corona, 77, 78, 80, 291–293, *292*,
    293, *294*, 306
  diameter, 349
  distance, 362
  eclipses, 24, 76, 76–77, 79, 80, 291,
    292
  ecliptic, 21, *22*
  elements, 288, *289*, 334
  energy conservation, 310–311
  energy radiation, 324
  energy sources, 310, 315–316,
    316
  escape velocity, 468
  fate of Earth, 428
  flares, 299, *299*, 306
  formation, 310, 586–587
  fusion reactions, 314–316
  gas, 318
    pressure, 317, *317*
  gravitational contraction as energy
    source, 311–312
  gravitational energy conversion to
    heat, 311
  heat transfer, 318
  heavy elements
    percent mass, 429
  interior
    observations, 320–322
    theoretical model, 319,
      319
    theory, 316–319
  interstellar matter around, 385
  mass conversion to energy, 312
  mass-to-light ratio, 508
  Milky Way Galaxy, 481, *481*
  observing, 305
  orbit and orbital speed, 489
  outer layers, 288–293, 306
  photosphere. *See* Photosphere.
  photospheric granulation, 294–295,
    295
  plages and prominences, 297, *298*,
    299
  planets'
    temperatures, 138
  pulsations, 320
  radio waves, 466
  reactions
    nuclear, 315–316
  reddened, 390, *390*
  rising and setting, 21
  rotation period, 295
  spectrum, *93*
  stability, 317
  stars around, *10*, 10–11
  structure, *290*, 290–291

Sun *(Continued)*
    sunspots, 295, *295*
    transition region, 291, 306
    variability
        Earth's climate, 303–304, *304*
    velocity changes, 320, *320*
    wind, 149, *261*, 293
    World Wide Web, A3
    zodiacal cloud, 413
Sun-sign
    astrology, 29, 30
Sunshine
    seasons, *62*, 62–65, *64–65*
Sunspots, *286*, 295, *295*, 306
    cycle, *296*, 296–297, 306
    magnetic fields, 296–297, *297*
    number
        variations, 301–303, *302*
    photosphere, *289*
Superclusters, 13, 542, 556
    dark matter, 554–555
    evolution, 552
    Local, 542, *542*
Supercomputers
    NASA Ames Reseach Center, 53
Supergiants, 351–352, 356, 358, 429
Supernova 1987A, 413, 446–448, *447,*
    448, 584
    behavior predictions, 448
    brightness, 448, *448*
    neutrinos, 449, *449*
Supernova(ae), 382, 443, 454, 455
    cosmic-ray particles source, 445
    creative contribution, 445, *446*
    dangers in proximity, 445–446
    discovery, 443–445
    fusion of elements, 442, 445
    history, 444
    luminosity, 445
    neutrinos, 448, 449, 579
    pulsars. *See* Pulsar(s).
    Tycho, 444, *444*, 454
    types, 454
    Type I, 454
        distance measurements, 510,
        510
Synchrotron radiation
    Jupiter, 222–223, *223*, 226
Szilard, Leo, 313

T
Tarantula Nebula, *416*, 446. *See* 30
    Doradus.
Tarter, Jill, *594*
Tectonics
    Earth, 151, 154, 163
    Venus, 193, *193, 194*, 204–205
Teide 1, 348
Telescope(s), 108–111, 116–127
    adaptive optics, 111
    aperture, 108
    Canada-France-Hawaii, 106
    choosing own, 126
    complete, 111
    construction, 115
    first astronomical, 108
    functions, 108

Telescope(s) *(Continued)*
    Galileo, 34, *34*
    gamma-ray, 127
    Hale, Palomar Mountain, 111, *111*
    Hubble Space. *See* Hubble Space
      Telescope.
    infrared
        airborne and space, 122–124
    Keck, 117–118, *118*, 336
    Lowell Observatory, 187, 512
    major new, 116–118, *117*, 117t
    McMath-Pierce Solar, 286
    Mount Wilson, California, 502, 538
    neutrino, 449, *449*
    radio, 118–122, *120*
        New Mexico, *82*, 121
        Parkes, Australia, 593, *594*
    reflecting, 109–110, *110*
    refracting, 109, *110*
    Schmidt, 562
    Sun observation, 305
    use, 111
    World Wide Web, A2
    x-ray, 302
Temperature(s)
    and radiation, 90
    fluctuations
        cosmic background radiation, 576
    forces of nature, 577
    interstellar clouds, *384*, 384–385
    planets, 138, 139
    proton fusion, 314–315
    solar atmosphere, 291, *292*
    universe
        early, 571, *571*, 576
Temperature scales, A20
Terraforming, 203–204
Terrestrial planets, 133, *134*, 137–138,
    144
    formation, 279
Tharsis bulge, Mars, 197, 198, 205, 280
Thermal energy, 310–312
Thomson, J.J., 95
Thomson, William(Lord Kelvin), 311,
    311
Thuban, 26
Tidal forces
    Earth, 472
Tides
    bulges, 74, *74*
    formation, 74–75, *75*
    Moon, 73–75
Time
    measurement, 66–68
Time machines
    gravity and, 471
Titan, Saturn, 230, *231*, 235, 235–236,
    245, 282
Titius-Bode rule, 134–135
Tombaugh, Clyde, 238, 239, *239*, 245
Tonry, John, 511
Toutatis, 256, *256*, 265
Trapezium star cluster, *400*, 401
Triangulation, 363–364, *364*
Trifid Nebula, 388, *389*
Triple alpha process, 426, 431
Triton, Neptune, 231, 236–237, *237*, 240,
    245

Trojans, 254, *255*, 265
Tropic of Cancer, 63
Tropic of Capricorn, 64
Troposphere, 155, 163
47 Tucanae, 360, 425, *425, 562*
Tully, Brent, 511
Tully-Fisher distance measurement
    spiral galaxies, 511, 511t, 516
Tunguska explosion, 160, *160*
Tycho's Supernova, 444, *444*, 454

U
Uhuru satellite, 125
Ultraviolet radiation, 88, 103
Umbra, 295
Unidentified flying objects (UFOs), 590,
    590n
United Kingdom Infrared Telescope,
    Mauna Kea
      Orion Nebula, *400*
United Nations Educational, Scientific,
    and Cultural Organization (UNESCO),
    482
Units used in science, A20
Universal gravity, 47–49
Universe
    age, 561
    average density, 566
    balloon analogy, 564, *564*
    beginning, 561, 569–576
    change in scale, 565
    closed, 564, *567*, 567, 568, 581
    critical density, 566, *567*, 569, 578
    dark matter, 553–555, 556
    deceleration, 562, 566
    density
        contributors, 578, 579t
    density of matter, 577
    distant future, 568
    early
        evolution, 571–572, 576
        interacting particles, 571–572,
        572
    expanding, 511–515, 560–563,
      565–566, *566*
        models, *514*, 514–515, *515*
    flat, 567–568, 581
    homogeneous, 539, 556
    inflationary, 577–579, *578*, 581
    isotropic, 539, 556
    large scale, 12–14
    models, 565–569, 569t
    observed characteristics, 560t
    open, 564, *567*, 567–568, 569, 581
    organization, 537–557
    slices, 542–544, *543*
    structure, 544–545
    temperatures, 571, *571*
    tour, 7–12
    uniformity, 576
47 Urae Majoris, 412
Uranus, 5, 237–238, 466
    atmosphere, 218
    basic properties, 213, 226
    composition, 137, 280
    internal structure, 215, *216*
    magnetic fields, 225, *225*

Uranus (*Continued*)
    orbit, 53–54
    orbital deviations, 553
    rings, 214, *214*, 231, 241, *241*, 241t,
        242–243, *243*, 245, 246
    satellites, 214, *214*, 231, 245
    seasons, 214–215, *215*
Urey, Harold, 588
Utopia, Mars, 196

V

Valles Marineris, Mars, *2*, 198, *198*, 205
Van Allen, James, 224, *224*
Van Allen belts, 149, 224
Variable stars, 368–371
    period-luminosity relationship,
        369–370, *371*, 371, 373
Vega, 26, 364, 365, 366
Vega 1, 259
Vega 2, 259
Vela supernova remnant, *3*
Vela satellites, 127
Velocity
    and momentum, 45
    radial, 320, 334, 337
Velocity-distance relation
    galaxies, 512–513, *513*
Venera spacecraft, 189, 193, 205
Venus
    Aphrodite Terra, *190*, 190–191
    appearance, 36
    atmosphere, 194t, 194–196, *195*
    basic properties, 188, 188t
    brightness, 186
    carbon dioxide, 194, 195, 205
    clouds, 188, *188*, 190–191
    composition, 137, 280
    coronae, 193, *193*
    craters
        impact, *191*, 191–192
    geology, 189–194, 281
    gravity, 281–282
    heliocentric theory, *32*, 33
    lava plains, 190, 192
    mountains, 191, 193, *194*, 281, *281*
    rotation, 188
    spacecraft exploration, *189*,
        189–190
    sulfur dioxide, 194
    surface, 191, 193–194, *194*,
        204–205
    tectonics, 193, *193*, *194*, 204–205
    temperature
        surface, 195
    volcanoes, *192*, 192–193
    World Wide Web, A2
Vernal equinox, 61, 65
Very Large Array, *116*, 121
    M87 image, *525*
Very Long Baseline Array, 121–122, *122*
    M87 image, *525*
Vesta, 250, 252, *253*, 265
Viking lander spacecraft, *132*, 196
    landing sites, 199, *199*
    recordings and measurements, *199*,
        199–200, *200*, *202*
    search for life, 203, *204*, 205

Viking lander spacecraft (*Continued*)
    testing general relativity, *467*,
        467–468
Viking orbiters, 196
70 Virginis, 412
Virgo cluster, 13, 539, *541*
Vogel, Hermann, 351
Voids, 542, 543, 556
Volatiles
    definition, 168
Volcanism
    Earth, *153*, 153–154, 281
    Io, 234, *235*
    Mars, *197*, 197–198, 281
    Moon, 171–172, *172*
    terrestrial planets, 281, *281*
    Triton, 236–237, *237*
    Venus, *192*, 192–193, 281
Volume
    definition, 46
von Helmholtz, Hermann, 311, *311*
Voyager spacecraft, 52, *52*, 211, *211*, 212,
    229, 234–235
    magnetosphere of Jupiter, 223,
        223
    messages aboard, 590–591, *591*
Voyagers in Astronomy
    Barnard, E.E., 386
    Cannon, Annie, 333
    Darwin, George, 75
    Einstein, Albert, 313
    Geller, Margaret, 544
    Goodricke, John, 369
    Hale, George Ellery, 115
    Halley, Edmund, 257, *257*
    Hubble, Edwin, 503
    Lowell, Percival, 187
    Russell, Henry Norris, 353
    Sagan, Carl, 143
    Shapley, Harlow, 482
    Tombaugh, Clyde, 239
    Van Allen, James, 224
    Walker, Art, 302
    Wegener, Alfred, 154
Vulcan, 466
Vulpecula, 450

W

Walker, Art, 302, *302*
Walsh, Dennis, 531
Water
    terrestrial planets, 282
Water ice
    martian polar caps, 201, *201*,
        205
Watt, 310
Wave(s)
    creation, 84, 85
    frequency, 86, 103
    radio, 89
    repeating phenomenon, 86
Wavelength(s), 334
    definition, 86, 103
Weakly Interacting Massive Particles
    (WIMPs), 579
Weather
    Earth, *156*, 156–157

Weather (*Continued*)
    jovian planets, 219–220
    Mars, 199, 200–201
    Neptune, 220
Week
    days, 70
Wegener, Alfred, 154
Weight
    density, 47n
Wells, H.G., 187
Weymann, Ray, 531
Wheeler, John, 468, *469*, 567
Whipple, Fred L., 258, *259*, 266
Whirlpool galaxy, *548*
White dwarfs, 354, 357, 358, 437, *437*,
    455
    core, 442–443
    explosions
        mild, 453–454
        violet, 454
    production
        massive stars, 442
    properties, 450t
    ultimate fate, 441, *441*
Wien's law, 91, 103, 330, 386, 574
Wildt, Rupert, 215
Wilson, Robert, 574, *574*
WIMPs, 579
Wind(s)
    jovian planets, 219–220
    Jupiter, 219–220, *220*
    Mars, 200
    Saturn, 219–220, *220*
    solar, 149, *261*, 293
    stellar, *402*, 403–405, 413
Winter solstice, 63–64, *65*
W.M. Keck Observatory, 117, *117*
Wolf, Max, 250
Wollaston, William, 93
World Wide Web
    astronomy, A1–A4

X

X-ray astronomy, 125
X-ray telescopes, 302
    black holes, 473
X rays, 87, 89, 103
    Coma cluster, 546, *546*

Y

Yang, C.N., 438
Year
    definition, 21
Yerkes, Charles, 336
Yerkes Observatory, 115, 386

Z

Zeeman effect, 296, *297*
Zenith
    definition, 18, 36
Zero-age main sequence stars, 418,
    431
Zodiac
    definition, 22, 36
    signs, 29

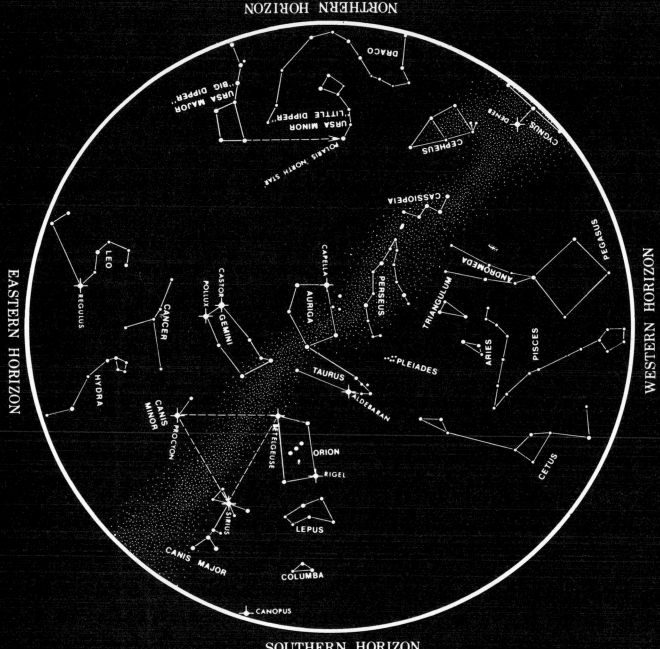

NORTHERN HORIZON

DRACO
URSA MAJOR "BIG DIPPER"
URSA MINOR "LITTLE DIPPER"
CEPHEUS
CYGNUS
DENEB
POLARIS "NORTH STAR"
CASSIOPEIA
EASTERN HORIZON
LEO
REGULUS
CANCER
POLLUX
CASTOR GEMINI
HYDRA
CAPELLA
AURIGA
PERSEUS
ANDROMEDA
PEGASUS
TRIANGULUM
ARIES
PISCES
WESTERN HORIZON
PLEIADES
TAURUS
ALDEBARAN
CANIS MINOR
PROCYON
BETELGEUSE
ORION
RIGEL
CETUS
SIRIUS
LEPUS
CANIS MAJOR
COLUMBA
CANOPUS

SOUTHERN HORIZON

THE NIGHT SKY IN JANUARY

Latitude of chart is 34°N, but it is
practical throughout the continental
United States.

To use: Hold chart vertically and turn
it so the direction you are facing
shows at the bottom.

Chart time (Local Standard):

10 p.m. First of month

9 p.m. Middle of month

8 p.m. Last of month

Star Chart from *GRIFFITH OBSERVER*, Griffith Observatory, Los Angeles

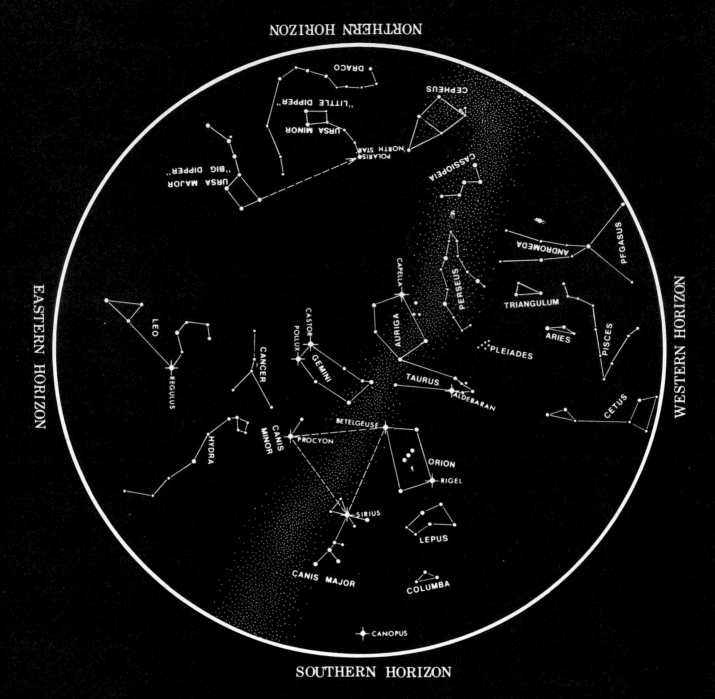

NORTHERN HORIZON

EASTERN HORIZON

WESTERN HORIZON

DRACO
CEPHEUS
"LITTLE DIPPER"
URSA MINOR
CASSIOPEIA
POLARIS "NORTH STAR"
"BIG DIPPER"
URSA MAJOR
PEGASUS
ANDROMEDA
PERSEUS
TRIANGULUM
CAPELLA
ARIES
PISCES
LEO
AURIGA
PLEIADES
CASTOR
POLLUX
GEMINI
CANCER
REGULUS
TAURUS
ALDEBARAN
CETUS
BETELGEUSE
CANIS MINOR
PROCYON
HYDRA
ORION
RIGEL
SIRIUS
LEPUS
CANIS MAJOR
COLUMBA
CANOPUS

SOUTHERN HORIZON

THE NIGHT SKY IN FEBRUARY

Latitude of chart is 34°N, but it is
practical throughout the continental
United States.

To use: Hold chart vertically and turn
it so the direction you are facing
shows at the bottom.

Chart time (Local Standard):

10 p.m. First of month
9 p.m. Middle of month
8 p.m. Last of month

Star Chart from *GRIFFITH OBSERVER*, Griffith Observatory, Los Angeles

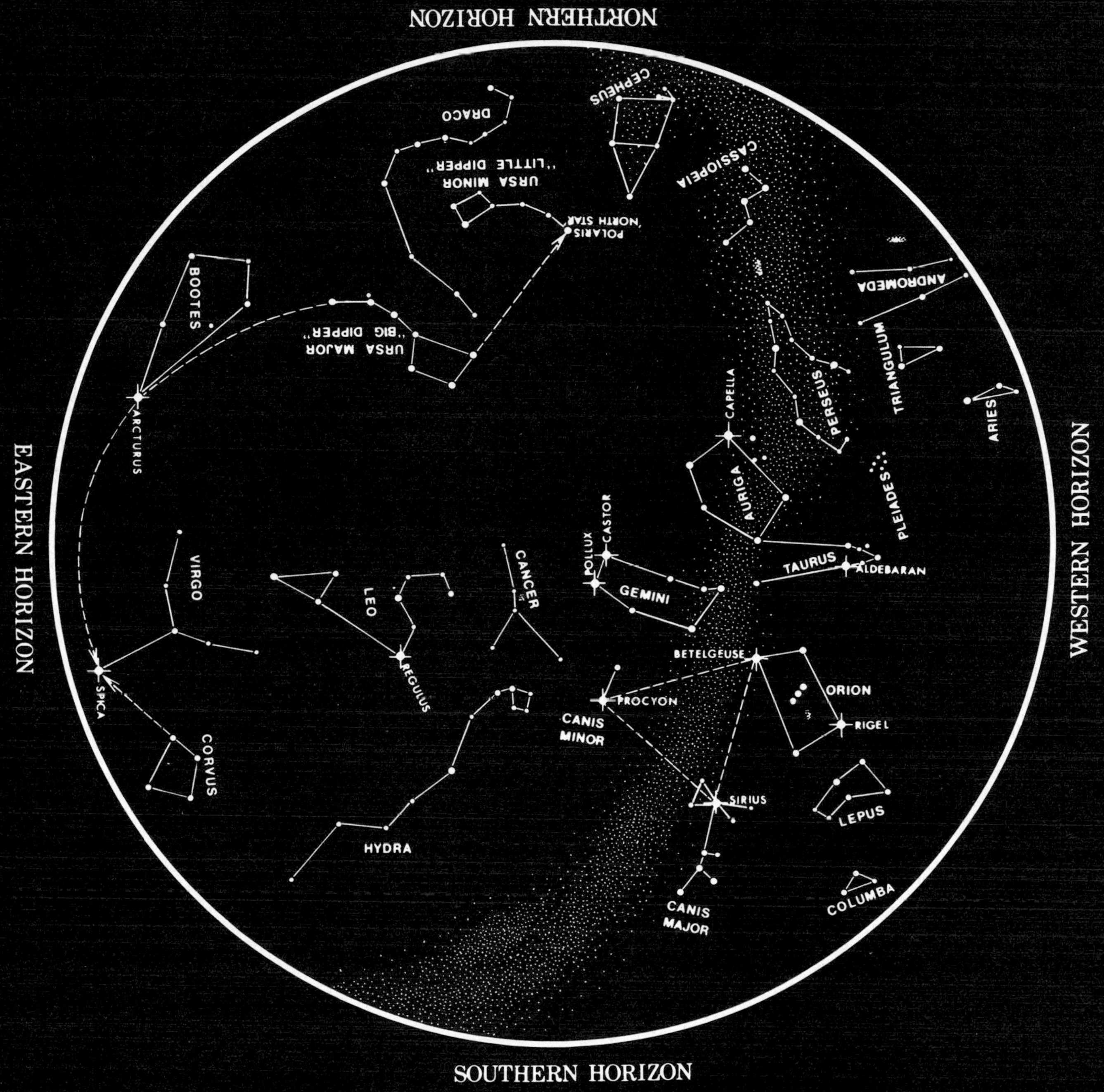

THE NIGHT SKY IN MARCH

Latitude of chart is 34°N, but it is practical throughout the continental United States.

To use: Hold chart vertically and turn it so the direction you are facing shows at the bottom.

Chart time (Local Standard):

10 p.m. First of month
9 p.m. Middle of month
8 p.m. Last of month

Star Chart from *GRIFFITH OBSERVER*, Griffith Observatory, Los Angeles

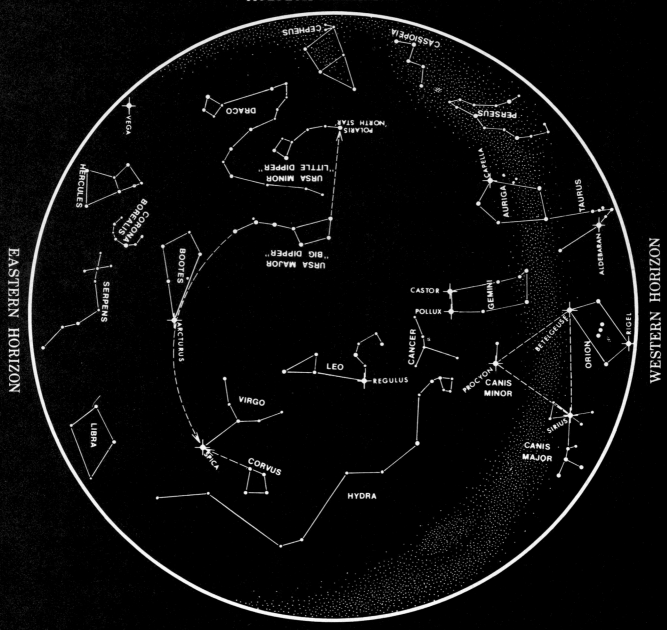

NORTHERN HORIZON

CEPHEUS

CASSIOPEIA

DRACO

VEGA

POLARIS 'NORTH STAR'

PERSEUS

HERCULES

URSA MINOR "LITTLE DIPPER"

CAPELLA

AURIGA

TAURUS

CORONA BOREALIS

BOOTES

URSA MAJOR "BIG DIPPER"

ALDEBARAN

SERPENS

CASTOR

GEMINI

POLLUX

CANCER

BETELGEUSE

RIGEL

ARCTURUS

LEO

ORION

REGULUS

PROCYON

CANIS MINOR

LIBRA

VIRGO

SIRIUS

SPICA

CORVUS

CANIS MAJOR

HYDRA

EASTERN HORIZON

WESTERN HORIZON

SOUTHERN HORIZON

THE NIGHT SKY IN APRIL

Latitude of chart is 34°N, but it is practical throughout the continental United States.

To use: Hold chart vertically and turn it so the direction you are facing shows at the bottom.

Chart time (Local Standard):

10 p.m. First of month

9 p.m. Middle of month

8 p.m. Last of month

Star Chart from *GRIFFITH OBSERVER*, Griffith Observatory, Los Angeles

NORTHERN HORIZON

EASTERN HORIZON

WESTERN HORIZON

SOUTHERN HORIZON

THE NIGHT SKY IN MAY

Latitude of chart is 34°N, but it is practical throughout the continental United States.

To use: Hold chart vertically and turn it so the direction you are facing shows at the bottom.

Chart time (Local Standard):

10 p.m. First of month

9 p.m. Middle of month

8 p.m. Last of month

Star Chart from *GRIFFITH OBSERVER*, Griffith Observatory, Los Angeles

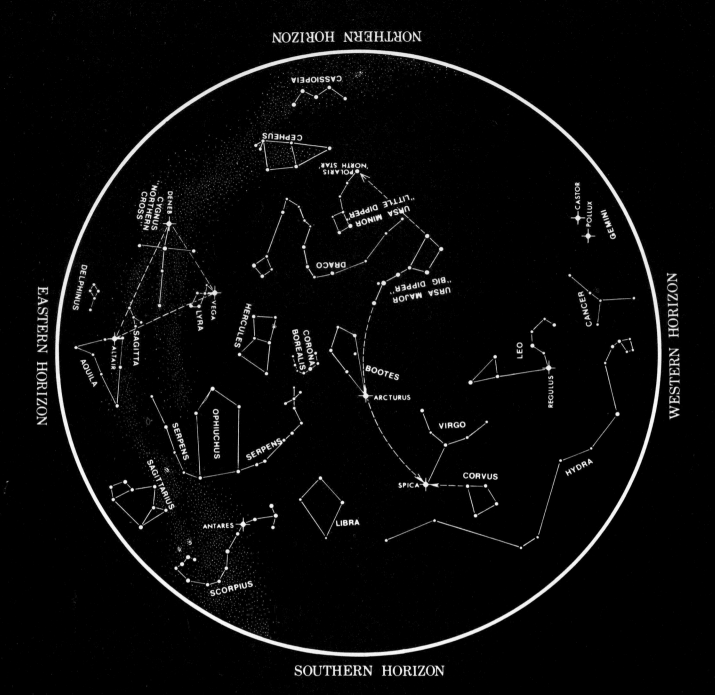

SOUTHERN HORIZON

THE NIGHT SKY IN JUNE

Latitude of chart is 34°N, but it is
practical throughout the continental
United States.

To use: Hold chart vertically and turn
it so the direction you are facing
shows at the bottom.

Chart time (Local Standard):

10 p.m. First of month

9 p.m. Middle of month

8 p.m. Last of month

Star Chart from *GRIFFITH OBSERVER*, Griffith Observatory, Los Angeles

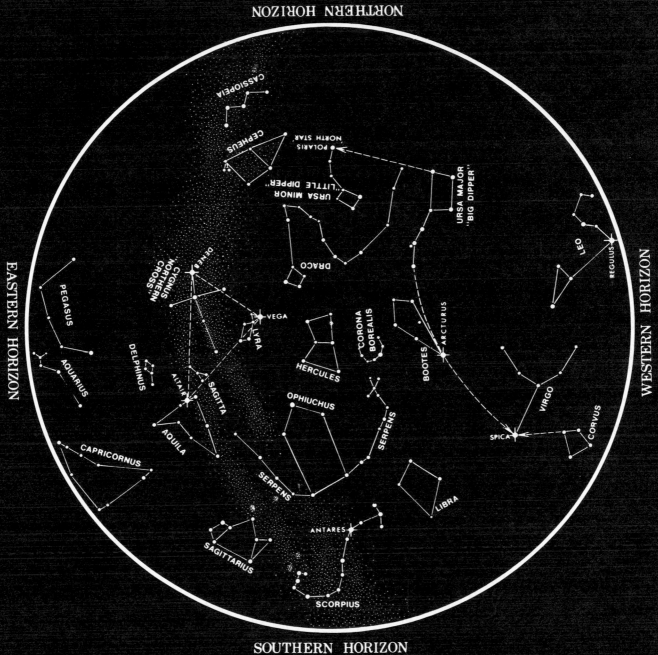

NORTHERN HORIZON

EASTERN HORIZON

WESTERN HORIZON

SOUTHERN HORIZON

THE NIGHT SKY IN JULY

Latitude of chart is 34°N, but it is practical throughout the continental United States.

To use: Hold chart vertically and turn it so the direction you are facing shows at the bottom.

Chart time (Local Standard):

10 p.m. First of month

9 p.m. Middle of month

8 p.m. Last of month

Star Chart from *GRIFFITH OBSERVER*, Griffith Observatory, Los Angeles

THE NIGHT SKY IN AUGUST

Latitude of chart is 34°N, but it is practical throughout the continental United States.

To use: Hold chart vertically and turn it so the direction you are facing shows at the bottom.

Chart time (Local Standard):

10 p.m. First of month

9 p.m. Middle of month

8 p.m. Last of month

Star Chart from *GRIFFITH OBSERVER*, Griffith Observatory, Los Angeles

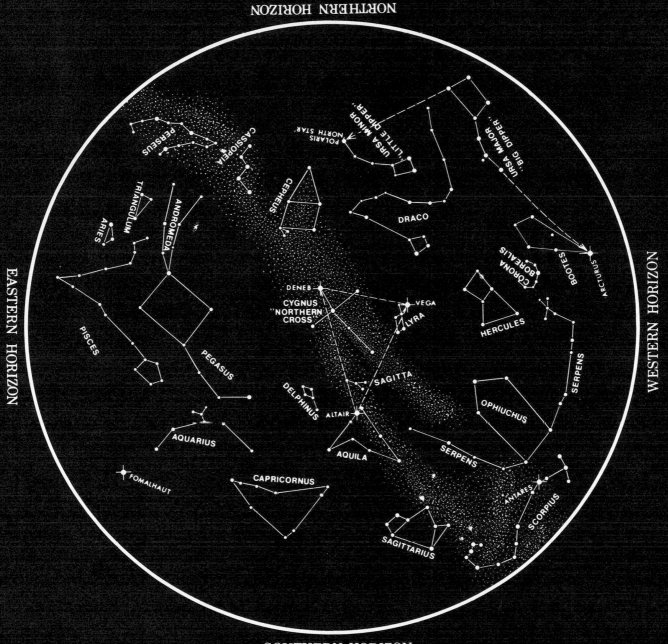

NORTHERN HORIZON

EASTERN HORIZON

WESTERN HORIZON

PERSEUS
CASSIOPEIA
URSA MINOR
POLARIS "NORTH STAR"
LITTLE DIPPER
URSA MAJOR "BIG DIPPER"
TRIANGULUM
ANDROMEDA
ARIES
CEPHEUS
DRACO
BOOTES
ARCTURUS
CORONA BOREALIS
PISCES
DENEB
CYGNUS "NORTHERN CROSS"
VEGA
LYRA
HERCULES
PEGASUS
SERPENS
SAGITTA
DELPHINUS
ALTAIR
OPHIUCHUS
AQUILA
SERPENS
AQUARIUS
FOMALHAUT
CAPRICORNUS
ANTARES
SCORPIUS
SAGITTARIUS

SOUTHERN HORIZON

THE NIGHT SKY IN SEPTEMBER

Latitude of chart is 34°N, but it is practical throughout the continental United States.

To use: Hold chart vertically and turn it so the direction you are facing shows at the bottom.

Chart time (Local Standard):

10 p.m. First of month

9 p.m. Middle of month

8 p.m. Last of month

Star Chart from *GRIFFITH OBSERVER*, Griffith Observatory, Los Angeles

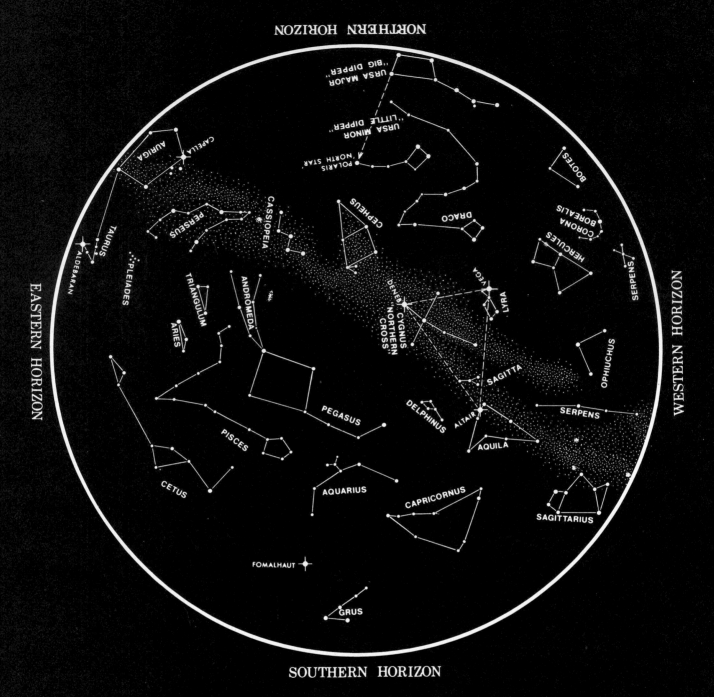

NORTHERN HORIZON

"BIG DIPPER"
URSA MAJOR

"LITTLE DIPPER"
URSA MINOR

POLARIS "NORTH STAR"

BOOTES

CEPHEUS

DRACO

CORONA BOREALIS

CASSIOPEIA

AURIGA   CAPELLA

HERCULES

SERPENS

PERSEUS

TAURUS

VEGA

LYRA

ALDEBARAN

·.PLEIADES

TRIANGULUM

ANDROMEDA

DENEB CYGNUS "NORTHERN CROSS"

OPHIUCHUS

ARIES

SAGITTA

SERPENS

PISCES

PEGASUS

DELPHINUS

ALTAIR

AQUILA

CETUS

AQUARIUS

CAPRICORNUS

SAGITTARIUS

FOMALHAUT

GRUS

EASTERN HORIZON

WESTERN HORIZON

SOUTHERN HORIZON

THE NIGHT SKY IN OCTOBER

Latitude of chart is 34°N, but it is practical throughout the continental United States.

To use: Hold chart vertically and turn it so the direction you are facing shows at the bottom.

Chart time (Local Standard):

10 p.m. First of month

9 p.m. Middle of month

8 p.m. Last of month

Star Chart from *GRIFFITH OBSERVER*, Griffith Observatory, Los Angeles

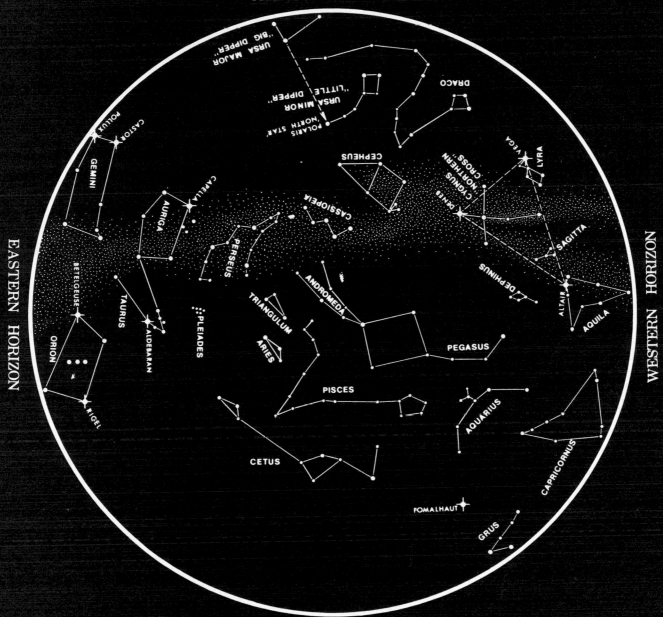

NORTHERN HORIZON

EASTERN HORIZON

WESTERN HORIZON

SOUTHERN HORIZON

THE NIGHT SKY IN NOVEMBER

Latitude of chart is 34° N, but it is
practical throughout the continental
United States.

To use: Hold chart vertically and turn
it so the direction you are facing
shows at the bottom.

Chart time (Local Standard):

10 p.m. First of month

9 p.m. Middle of month

8 p.m. Last of month

Star Chart from *GRIFFITH OBSERVER*, Griffith Observatory, Los Angeles

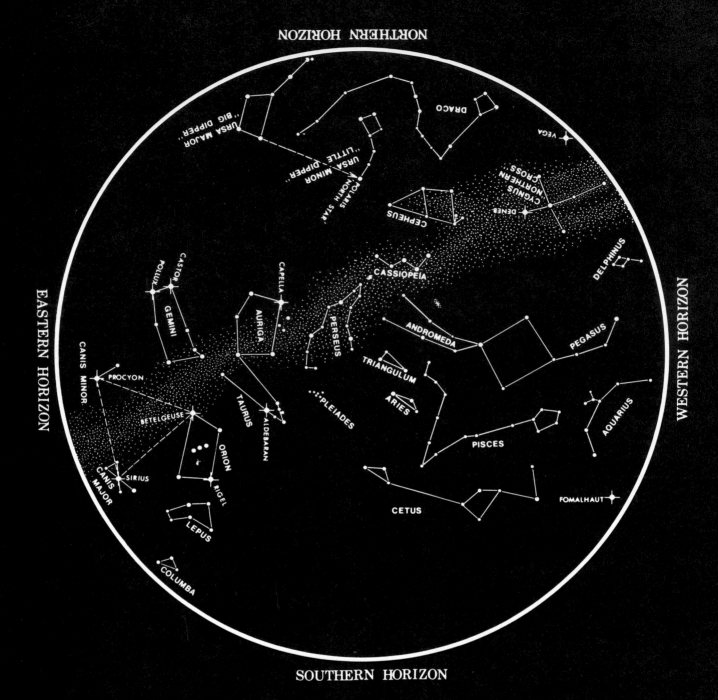

NORTHERN HORIZON

EASTERN HORIZON

WESTERN HORIZON

SOUTHERN HORIZON

THE NIGHT SKY IN DECEMBER

Latitude of chart is 34°N, but it is practical throughout the continental United States.

To use: Hold chart vertically and turn it so the direction you are facing shows at the bottom.

Chart time (Local Standard):

10 p.m. First of month

9 p.m. Middle of month

8 p.m. Last of month

Star Chart from *GRIFFITH OBSERVER*, Griffith Observatory, Los Angeles